# Radar
# Cross Section
# Handbook

## Volume 1

# Radar
# Cross Section
# Handbook

by
**George T. Ruck**
**Donald E. Barrick**
**William D. Stuart**
**Clarence K. Krichbaum**

**Editor:**
**George T. Ruck**

**Volume 1**

Peninsula Publishing • Los Altos, California

# Radar Cross Section Handbook
## Volume 1

Copyright 1970  Kluwer Academic/Plenum Publishers
New York

ISBN 0-932146643

Reprint of the original edition in 2002 published by:
   Peninsula Publishing
   P. O. Box 867
   Los Altos, California 94023 USA
   Telephone: 1-650-948-2511
   Fax: 1-650-948-5004
   Email: sales@peninsulapublishing.com
   Web site: www.peninsulapublishing.com

Printed in the United States of America

# PREFACE

As any author can testify, the writing of a book is a long and often tedious task, one that cannot possibly be accomplished without the aid and encouragement of many persons. Acknowledgment of all the sources of aid is not possible; however, the authors wish to express their thanks to some of the people who were particularly helpful during the preparation of this book.

Among these are Mr. Wayne Rife and Dr. Samuel Globe of Battelle Memorial Institute, for their encouragement and support during the writing of this book; Mr. Darold Conrad, Jr., for the drafting of most of the figures and illustrations; and Mrs. Mary Butler and the girls of the typing section of Battelle's Reports Processing Department, who typed the manuscript.

The manuscript was prepared under the sponsorship of the Advanced Research Projects Agency of the Department of Defense, and the authors wish to thank Mr. Morris Witow, the ARPA program director, for his patient encouragement and support.

It should be noted that the views expressed in this book are solely those of the authors and do not necessarily represent those of the Department of Defense.

# ACKNOWLEDGMENTS

The figures and tables below are printed with the permissions of the following sources.

*Institute of Electrical and Electronics Engineers, Inc.*

| | | |
|---|---|---|
| Table 10-2 | Figs. 6-7 through 6-11 | Figs. 8-16 through 8-18 |
| Fig. 2-35 | Figs. 6-31 through 6-33 | Fig. 8-75 |
| Fig. 5-5 | Fig. 6-35 | Fig. 10-22 |
| Fig. 5-16 | Fig. 6-39 | Fig. 10-27 |
| Fig. 5-20 | Figs. 6-54 through 6-57 | Fig. 11-10 |
| Fig. 6-2 | Figs. 6-64 through 6-67 | Fig. 11-11 |
| Fig. 6-3 | Fig. 7-33 | |

*Institution of Electrical Engineers*

| | |
|---|---|
| Fig. 10-8 | Figs. 10-11 through 10-15 |

*American Institute of Physics*

| | | |
|---|---|---|
| Figs. 4-5, 4-20 | Fig. 6-84 | Fig. 7-16 |
| Fig. 6-4 | Fig. 6-85 | Figs. 10-19 through 10-21 |

*National Research Council of Canada*

| | | |
|---|---|---|
| Fig. 6-17 | Fig. 10-7 | Figs. 10-29 through 10-33 |

*American Geophysical Union, Inc.*

| | |
|---|---|
| Fig. 10-28 | Figs. 10-35 through 10-38 |

*Radio Science*

Figs. 6-12 through 6-14

*RCA Review*

| | | |
|---|---|---|
| Fig. 10-9 | Fig. 10-10 | Fig. 10-16 |

*Pergamon Press*

Fig. 2-5

*Proceedings of the Physical Society of London*

Fig. 7-32

*Robert G. Kouyoumjian*

| | | |
|---|---|---|
| Fig. 6-75 | Fig. 6-76 | Fig. 6-79 |

*U.S. Atomic Energy Commission*

Fig. 10-34

# NOTATION

The symbols used throughout this book were chosen to conform, as closely as possible, to standard usage in the literature. Unfortunately, it was not possible to assign a unique meaning to every symbol; thus, many symbols have meanings which vary from chapter to chapter. These symbols are defined in each chapter or section in which they are used. Other symbols occur with a common meaning throughout the book, and these are given below:

$A$ = area; projected area; arbitrary function or coefficient, defined where used

Ai = Airy function (Section 2.2.2.5.1)

$a$ = radius; half-width; linear dimension; semimajor axis of ellipse or ellipsoid

$a_{ij}$ = scattering matrix element

$\alpha$ = cone or wedge interior half angle; propagation coefficient [Eq. (7.1-33)]

$\alpha_m$ = surface-ray decay coefficient [Eq. (2.2-64)]; Airy function root $[\text{Ai}(-\alpha_m) = 0]$

$\mathbf{B}$ = magnetic flux density

Bi = Airy function (Section 2.2.2.5.1)

$b$ = linear dimension; semiminor axis of ellipse

$\hat{b}$ = unit binormal vector

$\beta$ = angle; phase angle

$C(x)$ = Fresnel cosine integral

$c$ = linear dimension; speed of light in vacuum; semiminor axis of ellipsoid

$\Gamma$ = effective reflection coefficient [Eq. (8.5-3)]; Gamma function (Section 7.2.2.1); arbitrary function, defined where used

$\gamma$ = wedge exterior angle; average radar cross section per unit surface area [Eq. (9.1-1)]; integration variable; logarithmic derivative [Eqs. (7.2-27) and (7.2-28)]; effective reflection coefficient [Eq. (2.3-95)]; Thomson angle [Eq. (10.2-169)]

$D$ = diffraction coefficient; ambipolar diffusion coefficient; arbitrary function or coefficient, defined where used

$d$ = linear dimension

$\delta$ = angle; loss angle ($\tan \delta = \epsilon''/\epsilon_0$)

$\mathbf{E}$ = electric field intensity

$E_x$ = electric field component

$E(a/b)$ = elliptic integral of the second kind

$\hat{\mathbf{e}}$ = unit vector

$e$ = 2.71828; electron charge (Chapter 10)

$\epsilon$ = complex permittivity

$\epsilon'$ = real part of $\epsilon$

$\epsilon''$ = imaginary part of $\epsilon$

$\epsilon_0$ = permittivity of vacuum ($\epsilon_0 = (1/36\pi) \times 10^{-9}$)

$\epsilon_r$ = relative permittivity ($\epsilon_r = \epsilon/\epsilon_0$)

$\epsilon_m$ = Neuman number ($\epsilon_0 = 1$, $\epsilon_{m \neq 0} = 2$)

$F_I$ = impulse response [Eq. (2.1-41)]

$F$ = scattering amplitude [Eq. (2.1-6)]

$F(a/b)$ = elliptic integral of the first kind

$f$ = frequency; arbitrary parameter, defined where used

$f(x)$ = Fock's current function [Eq. (2.2-182)]

$\zeta$ = surface-height coordinate; angle; phase angle; arbitrary parameter, defined where used

$g(x)$ = Fock's current function [Eq. (2.2-165)]

$\eta$ = intrinsic impedance of a medium ($\eta = \sqrt{\mu/\epsilon}$); arbitrary parameter, defined where used

$H_n^{(1)}(x)$, $H_n^{(2)}(x)$ = Hankel functions of the first and second kinds

$\mathbf{H}$ = magnetic field intensity

$H_x$ = magnetic field component

$h$ = linear dimension, root-mean-square height

$h_n^{(1)}(x)$, $h_n^{(2)}(x)$ = spherical Hankel functions

$\Theta$ = angle; phase function; arbitrary function, defined where used

$\theta$, $\theta'$ = spherical polar coordinate, incident and scattered fields

$I_a$, $I_b$, $I_c$ = ellipsoid coefficients [Eq. (5.1-23)]

$i = \sqrt{-1}$

$\mathbf{J}$ = volume current density

$J_n(x)$ = Bessel function

$j_n(x)$ = spherical Bessel function

$\mathbf{K}$ = surface current density

$K_x$ = surface-current component

$k$ = wave number ($k = \omega\sqrt{\mu\epsilon}$)

$k_0$ = free-space wave number ($\equiv \beta_0$ of Chapter 10)

$\hat{\mathbf{k}}$ = unit vector in the direction of propagation of a plane wave

$\lambda$ = wavelength

$\lambda_0$ = free-space wavelength

$l =$ distance, correlation length

$m =$ index of refraction ($m = \sqrt{\mu_r \epsilon_r}$); dimensionless parameter occurring in Fock theory [$m = (ka/2)^{1/3}$]; summation index

$m_e =$ mass of the electron

$\mu =$ complex permeability

$\mu' =$ real part of $\mu$

$\mu'' =$ imaginary part of $\mu$

$\mu_0 =$ permeability of vacuum ($\mu_0 = 4\pi \times 10^{-7}$)

$\mu_r =$ relative permeability ($\mu_r = \mu/\mu_0$)

$N_e =$ electron density

$\hat{\mathbf{n}} =$ unit vector normal to a surface

$n =$ summation index

$\nu_c =$ collision frequency (Chapter 10)

$\nu_m, \bar{\nu}_m =$ creeping wave decay coefficients [Eqs. (2.3-72) and (2.3-73)]

$\xi =$ a parameter occurring in Fock theory ($\xi = \int_0^s (k/2R^2)^{1/3}\, ds$); arbitrary parameter, defined where used

$O(x) =$ of the order of $x$

$P_m{}^n(x) =$ associated Legendre function

$\hat{p}(x) =$ Fock function [Eq. (2.2-203)]

$\pi = 3.14159$

$\rho =$ cylindrical radial coordinate; radius of curvature

$\hat{q}(x) =$ Fock function [Eq. (2.2-204)]

$R =$ reflection coefficient; spherical radial distance; radius of curvature

$r =$ spherical radial distance

$S(x) =$ Fresnel sine integral

$s =$ arc length; arbitrary parameter, defined where used

$\sigma =$ radar cross section; conductivity

$\sigma_{\text{p.c.}} =$ radar cross section of a perfect conductor

$T =$ cylindrical scattering function [Eq. (4.1-8)]

$t =$ time

$U, u =$ scalar field amplitude

$\mathbf{V} =$ charged-particle velocity

$\varphi, \varphi' =$ spherical azimuthal coordinates, incident and scattered fields

$\psi =$ angle, phase angle

$\psi_0 =$ free space Green's function [Eq. (2.2-21)]

$x, y, z =$ Cartesian coordinates

$Y =$ effective input admittance (Section 7.1.2); modal surface admittance (Section 3.4)

$Z =$ effective input impedance (Section 7.1.2); modal surface impedance (Section 3.4)

$z_i$ = complex material impedance [Eqs. (7.1-6) and (7.1-7)]

$z_i'$ = complex material impedance [Eqs. (7.1-15) and (7.1-16)]

$y(x)$ = effective intrinsic admittance of an inhomogeneous medium [Eq. (7.2-24)]

$x(x)$ = effective intrinsic impedance of an inhomogeneous medium [Eq. (7.2-26)]

$\omega$ = angular radar frequency ($\omega = 2\pi f$)

$\omega_P$ = angular plasma frequency [Eq. (10.2-14)]

$\omega_H$ = angular gyromagnetic resonance frequency [Eq. (10.2-48)]

Some of the operational symbols which are used throughout the book are

$\langle \ \rangle$ = average, either time or ensemble

Re $A$ = real part of $A$

Im $A$ = imaginary part of $A$

$|A|$ = magnitude of $A$

$\perp$ = perpendicular

$\parallel$ = parallel

$[A]$ = matrix $A$

lim = "limit of"

exp $a$ = exponential $a$, equivalent to $e^a$

ln = logarithm to the base $e$

$A^*$ = complex conjugate of $A$

# CONTENTS

## Volume 1

## CHAPTER 4

## Cylinders

## CHAPTER 5
## Ellipsoids and Ogives

## CHAPTER 6
## Cones, Rings, and Wedges

Contents                                                                                      xvii

# Volume 2

## CHAPTER 7

## Planar Surfaces

# CHAPTER 8

# Complex Bodies

## CHAPTER 9

## Rough Surfaces

CHAPTER 10

# Ionized regions

## CHAPTER 11
## Radar Cross-Section Measurements

# Radar
# Cross Section
# Handbook

### Volume 1

# Chapter 1
# INTRODUCTION
G. T. Ruck

## 1.1. HISTORY

According to Skolnik,[1] the history of radar might be said to date back to the experiments of Heinrich Hertz in the late 1800's, although the serious development of specific radar equipment began in the middle 1930's, both in the United States and Great Britain. The history of scattering and diffraction investigations also dates back to the same period,[2] when a number of investigators were examining the response of specific objects to an incident electromagnetic field. Of course, the motivation for these investigations was optical problems or the development of mathematical theories; however, the work is in many cases also applicable to determining the radar cross section of specifically shaped targets. During this period, analytically exact solutions were obtained for the scattered fields from a sphere,[3] an infinite circular cylinder,[4] an infinite elliptic cylinder,[5] for normal incidence on an infinite parabolic cylinder,[6] a perfectly conducting half-plane,[7] and a perfectly conducting infinite wedge.[8] Numerous aspects of the scattered or diffracted fields produced by these and other shapes were investigated by many workers during the early 1900's.

Around 1940, a tremendous proliferation occurred in investigations of scattering and diffraction phenomena, motivated primarily by the development and use of radar equipment. The results of many of these investigations were not published until after the war, and a specific emphasis on the determination of the radar cross section or radar response of a body can be noted in these works. From the end of World War II until the present time, the investigation of the radar cross section or radar response of specific targets has continued to be an area of considerable interest to many researchers, and a great deal of published work exists in this field.

## 1.2. OBJECTIVES

The published work pertinent to the determination of the radar cross section of various objects dates back to the early part of the twentieth century, and even earlier. The total number of journal papers and reports pertaining to some aspect of the radar cross section problem is extremely large. (A recent bibliography covering the period from 1957 through 1964

1

contains 1420 references to journal articles and books alone.[9]) There is, at
the present time, no single book available which covers in a relatively comprehensive manner the subject of radar cross section determination. In
light of this, the authors have felt for several years that a need existed for
a handbook which presented the state-of-the-art in radar cross section
determination, summarized the available analytical techniques for estimating
radar cross sections, and presented experimental data typical of the types
of radar targets encountered. This book is an attempt to fill this need.

The intention of the authors has been to write this handbook so that it
can be used easily by persons who are not specialists in the radar field. In
some respects, this goal has not been achieved completely. This is largely
because, over certain frequency ranges for many targets, there are no simple
analytical expressions available for estimating the radar cross sections which
have the correct dependence on the various parameters involved. In addition,
in many cases there are no valid sets of experimental data available which
would allow empirical expressions to be formulated. Thus, in these areas,
it is not an easy task for a specialist to obtain valid cross section estimates,
and would be even more difficult for a nonspecialist.

Sufficient information is presented, however, so that a nonspecialist can,
with perseverance, use the handbook to obtain engineering estimates of the
radar cross section of a particular body when such estimates are possible.

The book has been organized so that the basic definitions underlying
the subject matter and some of the theoretical tools available for radar
cross section analysis are discussed in the following chapter. After that, five
chapters (3 through 7) are devoted to discussions of the scattering from a
number of simple geometric shapes. The remainder of the book discusses
the radar cross sections of complex bodies and scattering from rough surfaces.

## 1.3. UNITS AND TERMINOLOGY

The rationalized MKS system of units is used for all equations in this
book.[10] Thus, the permittivity and permeability of free space have the
values $\epsilon_0 = (1/36\pi) \times 10^{-9}$, and $\mu_0 = 4\pi \times 10^{-7}$. The so-called intrinsic
impedance of free space, designated by $\eta$, has the value $\eta = \sqrt{\mu_0/\epsilon_0} = 120\pi$.

Wherever possible in the figures, the cross section has been normalized
so as to be dimensionless, or is given in square meters with either logarithmic
(decibel) or linear scales being used. The normalization most frequently used
is $\lambda_0^2$, the square of the radar wavelength.

When equations for the scattered fields are presented, these have usually
been normalized with respect to the magnitude of either the incident electric
or magnetic field strength. Complex notation is used throughout the book
with $e^{-i\omega t}$ time dependence assumed, and then suppressed in the equations.

The term "scattered field" is to be interpreted in this book as meaning: at any point in space, the difference between the total field at that point and the assumed incident field at the same point.

Corresponding to conventional usage, the term "diffracted field" has several different meanings depending upon where it is used. In some cases it is synonymous with the scattered field, as in Chapter 3, where it refers to the forward and near-forward scattered field. In other places, it refers to the total field, as in Section 7.3, where it stands for the total field behind an aperture. In still another very common usage, which occurs throughout the book, the diffracted field means that portion of the scattered field which cannot be attributed to specular points on the scatterer.

The authors have attempted to assign unique symbols to the various parameters and functions occurring throughout the book. Certain symbols, however, do have different meanings in different chapters. In order to eliminate any confusion, those symbols which have a common meaning throughout the book are listed and defined in the glossary.

## 1.4. GENERAL BIBLIOGRAPHY

References to the sources of the material used in each chapter are given in the last section of that chapter. In addition to these, there are a number of sources which contain material applicable to the book as a whole or in part. Some of the latter will be presented in the following paragraphs.

A very useful reference paper is a recent and comprehensive bibliography on Radar Reflectivity prepared by Corriher and Pyron

Corriher, H. A., and Pyron, B. O., A Bibliography of Articles on Radar Reflectivity and Related Subjects: 1957-1964, *Proc. IEEE*, **53**: 1025 (1965).

This is an extensive bibliography of books, journal articles, and published symposia papers related to the subject of radar cross section. The items included were published during the period of 1957 through 1964.

A number of good books discuss the theoretical foundations of electromagnetic theory and include discussions of various theoretical aspects of scattering and diffraction theory. Several of the best of these are

Jones, D. S., *The Theory of Electromagnetism*, MacMillan, New York (1964).

Van Bladel, J., *Electromagnetic Fields*, McGraw-Hill, New York (1964).

Harrington, R. F., *Time-Harmonic Electromagnetic Fields*, McGraw-Hill, New York (1961).

Several monographs have been written which discuss theoretical and experimental scattering problems. The oldest of these is a book by Mentzner, in which theoretical techniques for obtaining exact and approximate solutions for the scattered fields from several simple geometric shapes are discussed.

Mentzner, J. R., *Scattering and Diffraction of Radio Waves*, Pergamon, New York (1955).

A more recent book by King and Wu, reviews the theoretical and experimental status of selected scattering and diffraction problems. In addition, a number of laboratory techniques for measuring scattered fields are discussed in this book.

King, R. W. P., and Wu, T. T., *The Scattering and Diffraction of Waves*, Harvard University Press, Cambridge, Mass. (1959).

A series of generally excellent reports pertaining to various topics of this handbook has been prepared by the University of Michigan. One of the reports, by Crispin *et al*, discusses the radar cross sections of various simple and more complex shapes and, although somewhat difficult to use due to the organization of the report, contains a wealth of useful information. The remaining reports are comprehensive reviews of the status of theoretical and experimental information on scattering by a particular geometrical shape.

Crispin, J. W., Jr., Goodrich, R. F., and Siegel, K. M., A Theoretical Method for the Calculation of the Radar Cross Sections of Aircraft and Missiles, University of Michigan, Report 2591–1–M, July 1959, AF 19(604)–1949.

Goodrich, R. F., Harrison, B. A., Kleinman, R. E., and Senior, T. B. A., Studies in Radar Cross Sections XLVII–Diffraction and Scattering by Regular Bodies–I: The Sphere, University of Michigan, Report 3648-I-T, 1961.

Einarsson, O., Kleinman, R. E., Laurin, P., and Uslenghi, P. L. E., Studies in Radar Cross Sections L–Diffraction and Scattering by Regular Bodies IV: The Circular Cylinder, University of Michingan, Report 7133–3–T, February 1966.

Sleator, F. B., Studies in Radar Cross Sections XLIX–Diffraction and Scattering by Regular Bodies III: The Prolate Spheroid, University of Michigan, Report 3648-6-T, February 1964.

Kleinman, R. E., and Senior, T. B. A., Studies in Radar Cross Sections XLVIII–Diffraction and Scattering by Regular Bodies–II: The Cone, University of Michigan, Report 3648-2-T, January 1963.

A book by Van de Hulst is an outstanding source for information concerning the theoretical and experimental results on scattering from certain dielectric bodies, particularly spheres. This book gives very good discussions of Rayleigh, Rayleigh–Gans, and resonance region scattering by dielectric and perfectly conducting spheres, as well as other shapes.

Van de Hulst, H. C., *Light Scattering by Small Particles*, Wiley, New York (1957).

A very useful source of material on the various problems associated with radar cross-section measurements is the proceedings of the RADC Radar Reflectivity Measurements Symposium.

"Radar Reflectivity Measurements Symposium," Rome Air Development Center, RADC-TDR-64-25, Vols. I and II, April 1964, AD-601364 and AD-601365.

## 1.5.  REFERENCES

1. Skolnik, M. I., *Introduction to Radar Systems*, McGraw-Hill, New York (1962).
2. Logan, N. A., Survey of Some Early Studies of the Scattering of Plane Waves by a Sphere, *Proc. IEEE* **53**:769 (1965).
3. Mie, G., Beiträge zur Optik trüber Medien, speziell kolloidaler Metallösungen, *Ann. Phys.* **25**:377 (1908).
4. Lord Rayleigh, On the Electromagnetic Theory of Light, *Phil Mag.* **12**:81 (1881).
5. Sieger, B., *Ann. Phys.* **27**: 626 (1908).
6. Lamb, H., *Proc. Lond. Math. Soc.* **4**:190 (1906).
7. Sommerfeld, A., *Math. Ann.* **47**:317 (1896).
8. Macdonald, H. M., *Electric Waves*, Cambridge University Press, Cambridge, England (1902).
9. Corriher, H. A., and Pyron, B. O., A Bibliography of Articles on Radar Reflectivity and Related Subjects: 1957–1964, *Proc. IEEE* **53**:1025 (1965).
10. Stratton, J. A., *Electromagnetic Theory*, McGraw-Hill, New York (1941).

*Chapter 2*

# THEORY

G. T. Ruck, D. E. Barrick, and W. D. Stuart

## 2.1. BASIC SCATTERING THEORY

### 2.1.1. Radar Cross Section

The power received, $P_r$, at a radar is a function of several parameters. It is useful to attempt to separate the variables so that $P_r$ may be expressed as a function of the transmitter system, the propagation path from the transmitter system to the target, the propagation path from the target to the receiving system, and the receiving system. The functional relationship between these parameters and the received power is discussed in several texts,[1-5] and can be summarized in the radar equation:

$$P_r = \underbrace{\frac{P_t G_t}{L_t}}_{\substack{\text{Transmitting}\\\text{system}}} \underbrace{\frac{1}{4\pi r_t^2 L_{mt}}}_{\substack{\text{Propagating}\\\text{medium}}} \underbrace{\sigma}_{\substack{\text{Target}}} \underbrace{\frac{1}{4\pi r^2 L_{mr}}}_{\substack{\text{Propagating}\\\text{medium}}} \underbrace{\frac{G_r \lambda_0^2}{4\pi L_r}}_{\substack{\text{Receiving}\\\text{system}}} \underbrace{\frac{1}{L_p}}_{\substack{\text{Polarization}\\\text{effects}}} \tag{2.1-1}$$

The parameters of the above equation are defined as

$\qquad P_t$ = transmitter power in watts
$\qquad G_t$ = gain of the transmitting antenna in the direction of the target
$\qquad L_t$ = numerical factor to account for losses in the transmitting system
$\qquad L_r$ = a similar factor for the receiving system
$\qquad r_t$ = range between the transmitting antenna and the target
$\qquad \sigma$ = radar cross section
$L_{mt}, L_{mr}$ = numerical factors which allow the propagating medium to have loss
$\qquad r$ = range between the target and receiving antenna
$\qquad G_r$ = gain of the receiving antenna in the direction of the target
$\qquad \lambda_0$ = radar wavelength
$\qquad L_p$ = numerical factor to account for polarization losses.

Polarization losses occur because the transmitting and receiving antennas are not properly polarized, because the propagation medium changes the original polarization (as for example, in Faraday rotation), or because the

7

target depolarizes the signal. Polarization is discussed in more detail in Sections 2.1.3 and 2.1.4.

This handbook is a discussion of the target characteristic, $\sigma$, which is called the radar cross section. By analyzing the dimensions of the various terms in Eq. (2.1-1), it is seen that $\sigma$ has the dimensions of area, accounting for the equivalent name of radar echoing area. Rearranging Eq. (2.1-1) to equate $\sigma$ to the other variables, one obtains

$$\sigma = \frac{P_r L_r (4\pi)}{G_r \lambda_0^2} \frac{L_t}{P_t G_t} L_{mr} L_{mt} (4\pi)^2 r_t^2 r^2 L_p \qquad (2.1\text{-}2)$$

Examining Eq. (2.1-2), one question arising is whether the variables for the target, receiving system, propagation path, and transmitting system are independent variables; that is, is $\sigma$ a parameter defined only by the target, or is it a function of the means by which it is measured? It is known that if a radar of very large power sounds the ionosphere, the ionospheric parameters will change; thus, the reflectivity will change. This is known as the Luxembourg effect. However, for the radar targets discussed in this handbook the constitutive parameters are considered to be independent of the signal strength. The only parameters of the receiver–transmitter system which influence the observed radar cross section are the radar (or carrier) frequency (or frequencies), the transmitted waveform, the character of the wavefront of the incident field if the target is close to the transmitting system, and the polarization of the radar signal. For all of these parameters the cross section is still a function of only the target body, since any wavefront can be decomposed into a linear combination of plane waves, any waveform can be similarly decomposed into its frequency components, and the cross section is unique for a particular frequency and polarization. Therefore, the cross section may be determined by superposition from a linear combination of frequencies, polarizations, and plane wave responses. The effects of varying these parameters are discussed in the following sections.

Upon reexamining Eq. (2.1-2), it is seen that $\sigma$ can be expressed as $4\pi$ times the power delivered per unit solid angle in the direction of the receiver divided by the power per unit area incident at the target. The factor $4\pi$ enters from the definition of solid angle. Assuming for the moment that the propagation path between the target and the receiving system is lossless, then this power ratio may be expressed as

$$4\pi r^2 \frac{(\mathbf{E}^s \cdot \mathbf{E}^{s*})}{(\mathbf{E}^i \cdot \mathbf{E}^{i*})} = 4\pi r^2 \frac{(\mathbf{H}^s \cdot \mathbf{H}^{s*})}{(\mathbf{H}^i \cdot \mathbf{H}^{i*})} \qquad (2.1\text{-}3)$$

where $\mathbf{E}^s$ and $\mathbf{H}^s$ are the scattered electric and magnetic fields, respectively, and $\mathbf{E}^i$ and $\mathbf{H}^i$ are the incident fields. The scattered field is defined to be the difference between the total field (with the target present) and the incident field $\mathbf{E}^i$ (the total field that would exist if there were no target present). This is summarized as

$$\mathbf{E}^s = \mathbf{E}^T - \mathbf{E}^i \tag{2.1-4}$$

Finding $\sigma$ now becomes a problem in electromagnetic field theory. In order that $\sigma$ be independent of $r$, it is desirable to let $r$ in Eq. (2.1-3) become arbitrarily large so[†]

$$\sigma = 4\pi \lim_{r \to \infty} r^2 \frac{(\mathbf{E}^s \cdot \mathbf{E}^{s*})}{(\mathbf{E}^i \cdot \mathbf{E}^{i*})} = 4\pi \lim_{r \to \infty} r^2 \frac{(\mathbf{H}^s \cdot \mathbf{H}^{s*})}{(\mathbf{H}^i \cdot \mathbf{H}^{i*})} \tag{2.1-5}$$

Some of the problems encountered when the value of $\sigma$ is desired with the receiver close to the target are discussed in Section 2.1.5. Unless otherwise indicated, when $\sigma$ is used throughout the rest of the book, the limiting process indicated in Eq. (2.1-5) will be assumed.

To compute the scattered field one sometimes computes the current induced on the target and then treats the target current distribution in terms of an equivalent aperture distribution. Antennas are often compared to an isotropic antenna, that is, an antenna which radiates uniformly in all directions. This philosophy leads to another, often-quoted version of the definition of radar cross section: given the target echo at the receiving system, $\sigma$ is the area which would intercept sufficient power out of the transmitted field to produce the given echo by isotropic reradiation. This definition is essentially equivalent to the other definitions given.

### 2.1.2. Monostatic–Bistatic Cross Section

For any object other than a sphere one would intuitively expect the radar cross section to be a function of the angle of incidence of the radar signal. For all bodies including the sphere the cross section is also a function of the angle to the receiver. In this book the angles $\theta$ and $\phi$ define the direction of propagation of the incident field, and the angles $\theta'$ and $\phi'$ define the direction of propagation of the scattered field of interest. This coordinate system is illustrated in Figure 2-1.

The case where $\theta = \theta'$ and $\phi = \phi'$ is called monostatic scattering or backscattering. This is the most common situation in radar and, therefore, the emphasis in the literature and in this handbook is on obtaining the backscatter cross section. If the receiving antenna and the transmitting antenna are not the same antenna but are located adjacent to each other the situation is called quasi-monostatic. When the receiver and transmitter are clearly separated the bistatic cross section is observed. Despite the fact that most radars are monostatic or quasi-monostatic, bistatic radar has seen considerable use, for example, in satellite "fences" where there are one

---

[†] Note that according to this definition $\sigma$ is a scalar function. A more general definition, to be discussed in a later section, defines $\sigma$ to be a tensor function.

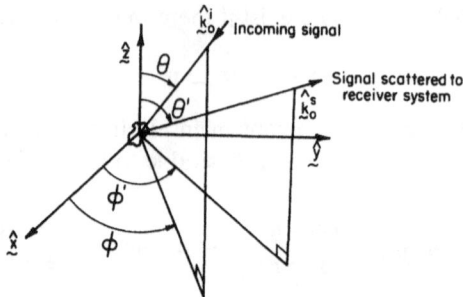

Fig. 2-1.   Coordinate system used thoroughout book.

transmitter and several widely-spaced receivers watching for earth-satellite passages. The use of several receivers is called multistatic radar. However, each radar-receiving system sees a bistatic cross section of the target under observation.

The special case of $\theta' = \pi - \theta$, $\phi' = \phi + \pi$ is called forward scatter. In general, the term forward scatter is taken both in this book and in the literature to imply both the special case of exact forward scatter and the more general case of quasi-forward scatter. Similarly, backscatter includes both the monostatic case and the quasi-monostatic case.

In computing the radar cross section it is sometimes convenient to express the cross section as

$$\sigma(\theta, \phi; \theta'\phi') = \frac{\lambda_0^2}{\pi} \mid F(\theta, \phi; \theta', \phi')\mid^2 = \frac{4\pi}{k_0^2} \mid F(\theta, \phi; \theta', \phi')\mid^2 \quad (2.1\text{-}6)$$

where $F(\theta, \phi; \theta', \phi')$ is called the scattering amplitude function and $k_0$ is the free space wave number, or $k_0 = 2\pi/\lambda_0$. In some cases it is desirable to factor out $(\pi f)^2$, where $f$ is the frequency, from the amplitude function. Some physics and optical texts designate $\sigma$ as the "differential scattering cross section" because the total scattered power per unit incident field intensity is obtained by integrating $\sigma$ over all solid angles as shown in Eq. (2.1-7). The result of this integration is designated the total cross section $\sigma_T$, where

$$\sigma_T = \frac{1}{4\pi} \int \sigma(\theta', \phi')\, d\Omega \qquad (2.1\text{-}7)$$

For a perfectly conducting or lossless dielectric target, $\sigma_T$ is a constant times the imaginary part of the scattering amplitude function evaluated for forward scatter, or[†]

$$\sigma_T = \frac{4\pi}{k_0^2} \operatorname{Im} F(\theta, \phi; \pi - \theta, \phi + \pi) \qquad (2.1\text{-}8)$$

[†] This result is sometimes referred to as the "forward scattering theorem" or the extinction formula, and its derivation can be found in Reference 6.

The total cross section and Eq. (2.1-8) arise frequently in multiple scattering problems.

The differential cross section is written without the modifier "differential" in the remainder of this book. The cross section in the forward direction ($\theta' = \pi - \theta$, $\phi' = \phi + \pi$) can be many times the backscatter cross section. If one builds a convex absorbing body with dimensions identical to a conducting body, the backscatter cross section can be several orders of magnitude below that of the conducting body. However, their high-frequency forward scatter cross sections will be the same.[7] For large convex-shaped conducting bodies, the high-frequency forward scatter cross section increases with frequency while the high-frequency backscatter cross section remains comparatively constant with frequency.[8] The backscatter and forward scatter cross sections are equal for normal incidence on a perfectly conducting flat plate.

Siegel[8] has used the physical optics approximation (discussed in Section 2.2.2.3) to show that if the wavelength ($\lambda_0$) of the radar signal is small compared to the target dimensions, then a target with a projected area $A$ will have a forward scatter cross section of

$$\sigma(\theta' = \pi - \theta; \phi' = \phi + \pi) = \frac{4\pi A^2}{\lambda_0^2} \qquad (2.1\text{-}9)$$

which is the same as the backscatter cross section for a perfectly conducting plate of area $A$ at normal incidence.

Crispin, Goodrich, and Siegel[9] have considered bistatic scattering at angles less than $\pi$ radians and have developed the "monostatic–bistatic equivalence theorem" which gives a method of estimating the bistatic cross section from known monostatic cross sections. Such a method is desirable since most methods of calculating cross section and most published measurements are for the monostatic case. The theorem states: "For perfectly conducting bodies which are sufficiently smooth, in the limit of vanishing wavelength, the bistatic cross section is equal to the monostatic cross section at the bisector of the bistatic angle between the direction to the transmitter and receiver." A corollary of this theorem is that the cross section is unchanged if the positions of the transmitter and receiver are interchanged. Crispin et al. show with the physical optics approximation that the theorem is approximately true if the bistatic angle is considerably less than $\pi$ radians. The theorem is obviously untrue as the bistatic angle approaches $\pi$ radians since, from Eq. (2.1-9), all convex bodies with the same projected area $A$ will have the same forward scatter cross section, while the monostatic cross section at $\pi/2$ radians can be any value, depending upon the body shape chosen. One consequence of the theorem is that, although the actual value of the cross section is probably different for any particular situation, it is noted that the range of cross sections for the bistatic case is approximately the same as the range of cross sections for the monostatic case.

Kell[10] has extended the theorem to more general cases including both metallic- and dielectric-coated targets. As the bistatic angle approaches zero, the bistatic radar cross section approaches the monostatic radar cross section viewed on the bisector of the bistatic angle plus a small correction term. The correction term is a function of the bistatic angle and the target. It is assumed that a target may be approximated by a collection of $M$ discrete scatterers. The scatterers directly illuminated are called "simple centers." Those illuminated by multiple reflection are termed "reflex centers." The cross section of the target is given by

$$\sigma = \left| \sum_{m=1}^{M} \sqrt{\sigma_m}\, e^{i\psi_m} \right|^2 \tag{2.1-10}$$

where $\sigma_m$ is the radar cross section of the $m$th scattering center and $\psi_m$ is the phase difference between the field scattered by the $m$th center and that scattered by the first center.

Assume for the moment that the coordinates are rotated so that $\phi' = \phi$. The bistatic angle is then $(\theta' - \theta)$. The phase quantity $\psi_m$ is now related to the bistatic angle by the following expression:

$$2k_0\, d_m \cos\left(\frac{\theta' - \theta}{2}\right) + \zeta_m \tag{2.1-11}$$

where $k_0$ is the free space wave number $(2\pi/\lambda_0)$, $d_m$ is the distance between the $m$th and the first simple center projected on the bistatic bisector axis, and $\zeta_m$ is the residual phase contribution of the $m$th center (including path-length phase terms if the center is a reflex center).

From Eqs. (2.1-10) and (2.1-11) it is seen that the cross section changes because of changes in the contributions from the individual scattering centers and changes in relative phase between the contributions. Sometimes the amplitude changes so drastically that there is a change in the existence of a center—either the appearance of a new center or the disappearance of one previously present. This occurs more commonly at bistatic angles greater than $\pi/2$ radians. It is in general difficult to include the effects of these centers in Eqs. (2.1-10) and (2.1-11).

The change in phase by the factor $\cos[(\theta' - \theta)/2]$ in Eq. (2.1-11) can also be considered as a change in wavelength. From this fact the results of the monostatic–bistatic equivalence theorem may be improved first by measuring the monostatic cross section as a function of all desired aspect angles at a frequency higher by the factor $\sec[(\theta' - \theta)/2]$ than the frequency at which the cross section is desired. The measured data is then translated by one-half the desired bistatic angle and the frequency is reduced by the factor $\cos[(\theta' - \theta)/2]$.

Kell reports good agreement between results obtained by use of the equivalence theorem and approximate theoretical results for scattering from a metallic cylinder $24\lambda$ long and $4.8\lambda$ in diameter for bistatic angles up to $10°$ at all aspect angles. The generalized monostatic–bistatic equivalence theorem is not applicable to bodies in which multiple scattering is important, nor to convex bodies in which the contributions from creeping waves are relatively important. (Creeping waves are discussed in detail in Section 2.3.2.) In determining the radar cross section of a sphere, creeping waves are very important (see Chapter 3), and Kell reports that the equivalence theorem fails to predict scattering of creeping waves beyond bistatic angles of one degree for spheres with values of $k_0$ times the radius up to 39.62.

### 2.1.3. Polarization Effects

For a plane electromagnetic wave, the electric field vector must always point in a direction perpendicular to the direction of propagation. If this direction is constant, the electric field lies entirely in one plane and the electromagnetic wave is said to be linearly polarized. If the direction rotates with time at a constant rate the wave is said to be elliptically polarized. A circular polarized wave is a special case of elliptical polarization. If the E field vector appears to be rotating clockwise to an observer looking in the direction of propagation the polarization is said to be right circular or right elliptical.[†] Counterclockwise rotation looking in the direction of propagation is, of course, designated left circular or left elliptical, and is so used throughout this book. The electric fields for right and left circular polarization are denoted $E_R$ and $E_L$. They are orthogonal fields, that is, $E_R \cdot E_L = 0$. Elliptical polarization may then be represented as a linear combination of two circularly polarized waves. Circular and linear polarized fields are illustrated in Figure 2-2.

Linear polarization can also be divided into two orthogonal polarizations. In some texts these are called horizontal and vertical polarizations with the planes containing the electric field lying parallel and perpendicular to the earth's surface, respectively. Since the earth's surface is not always a convenient reference, this sometimes leads to ambiguities. In this handbook linear polarization is generally decomposed into perpendicular and parallel polarizations, with the notation for the electric fields of $E_\perp$ and $E_\parallel$, respectively. These fields are orthogonal, $E_\perp \cdot E_\parallel = 0$. In Figure 2-1, consider a plane containing the $z$-axis and the unit vector $\hat{k}_0{}^i$ pointing in the direction of propagation of the incident field. If the $E$ field lies in this plane, the polarization is said to be parallel. When the $E$ field lies perpendicular to this plane the polarization is said to be perpendicular. The target lies at the

[†] This is the IEEE standard definition.[11] The reader should be warned that many authors use the opposite definition.

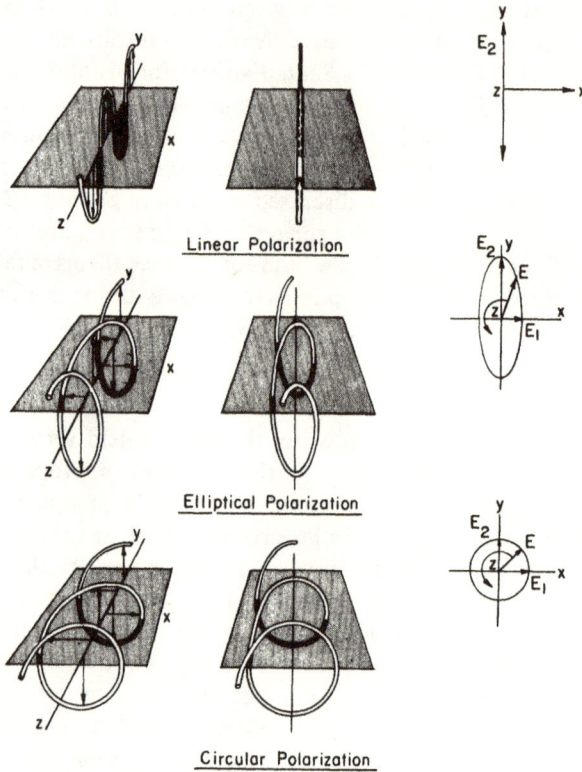

Fig. 2-2. Illustration of elliptical, circular, and linearly polarized fields.

center of the coordinate system with either a principal axis or other unique axis oriented along the $z$ direction. If the target has no principal or other unique axis as, for example, a spherical target, the $z$-axis is chosen in whatever manner is convenient. If a field $\mathbf{E}$ is arbitrarily polarized it may be written as

$$\mathbf{E} = \mathbf{E}_\perp + \mathbf{E}_\parallel = E_\perp \hat{\perp} + E_\parallel \hat{\parallel} e^{i\delta} \qquad (2.1\text{-}12)$$

where $\hat{\perp}$ and $\hat{\parallel}$ are unit vectors defined by $\hat{\perp} \times \hat{\parallel} = \hat{\mathbf{k}}_0$, with $\hat{\mathbf{k}}_0$ a unit vector in the direction of propagation, and $\delta$ the time phase angle by which $\mathbf{E}_\parallel$ leads $\mathbf{E}_\perp$. If $\sin \delta = 0$, the wave is linearly polarized. For other values of $\delta$ the wave is elliptically polarized; and if $E_\perp = E_\parallel$, along with $\cos \delta = 0$, the wave is circularly polarized. If $\delta$ and/or the ratio $E_\parallel/E_\perp$ are not deterministic functions of time, the wave is not "completely polarized." Any wave which is not completely polarized is said to be "partially polarized," and may be represented as the sum of a completely polarized and a "randomly polarized" wave. Randomly polarized signals are of special importance in radio astronomy and at optical frequencies. For the remainder of this handbook, unless otherwise specified, completely polarized fields are assumed.

The arbitrarily polarized field $E$ may also be written as the sum of two circularly polarized fields, as follows:

$$\mathbf{E} = \mathbf{E}_R + \mathbf{E}_L$$
$$= (E_R\hat{\mathbf{R}} + E_L e^{i\psi}\hat{\mathbf{L}}) \tag{2.1-13}$$

where $\psi$ is the time phase angle by which $\mathbf{E}_L$ leads $\mathbf{E}_R$. For a right circularly polarized wave, $\mathbf{E}_R$ may be related to $\mathbf{E}_\perp$ and $\mathbf{E}_\parallel$ as follows:

$$\mathbf{E}_R = \mathbf{E}_\perp + i\mathbf{E}_\parallel = E(\hat{\perp} + i\hat{\parallel})/\sqrt{2} \tag{2.1-14}$$

where $E = E_\perp = E_\parallel$. A left circularly polarized wave $\mathbf{E}_L$ may be similarly related to $\mathbf{E}_\perp$ and $\mathbf{E}_\parallel$, or

$$\mathbf{E}_L = \mathbf{E}_\perp - i\mathbf{E}_\parallel = E(\hat{\perp} - i\hat{\parallel})/\sqrt{2} \tag{2.1-15}$$

where $E = E_\perp = E_\parallel$. From Eqs. (2.1-12) through (2.1-15) it is seen that the definitions of $\hat{\mathbf{R}}$ and $\hat{\mathbf{L}}$ are

$$\hat{\mathbf{R}} = \frac{\hat{\perp} + i\hat{\parallel}}{\sqrt{2}}$$
$$\hat{\mathbf{L}} = \frac{\hat{\perp} - i\hat{\parallel}}{\sqrt{2}} \tag{2.1-16}$$

The components of the electric field in circular representations are obviously

$$E_R = \frac{E_\perp - iE_\parallel}{\sqrt{2}}$$
$$E_L = \frac{E_\perp + iE_\parallel}{\sqrt{2}} \tag{2.1-17}$$

Matrix notation provides a convenient representation of the transformation between an elliptically polarized wave represented in the circularly polarized system and the same wave represented in the linearly polarized system. These transformations are

$$\begin{bmatrix} E_R \\ E_L \end{bmatrix} = \frac{1}{\sqrt{2}} \begin{bmatrix} 1 & -i \\ 1 & +i \end{bmatrix} \begin{bmatrix} E_\perp \\ E_\parallel \end{bmatrix} = [T] \begin{bmatrix} E_\perp \\ E_\parallel \end{bmatrix}$$
$$\begin{bmatrix} E_\perp \\ E_\parallel \end{bmatrix} = \frac{1}{\sqrt{2}} \begin{bmatrix} 1 & 1 \\ +i & -i \end{bmatrix} \begin{bmatrix} E_R \\ E_L \end{bmatrix} = [T]^{-1} \begin{bmatrix} E_R \\ E_L \end{bmatrix} \tag{2.1-18}$$

where $[T]^{-1}$ is the inverse of $[T]$.

The preceding discussion has been primarily in terms of the electric fields. However, associated with each electric field there is a magnetic field which may be determined in general from Maxwell's equations. For a plane wave in a linear, passive, reciprocal, isotropic, and locally homogeneous medium, the magnitude of the magnetic field is related to the magnitude of the electric field by the local impedance of the medium, or $|H| = |E|/\sqrt{\mu/\epsilon} = |E|/\eta$. The direction of the magnetic field is perpendicular to both the electric field and the direction of propagation. As a manner of notation in this book, a magnetic field $H_\parallel$ is one which is perpendicular to an electric field $E_\parallel$. In other words, the subscript refers to the polarization of the field which is defined in terms of the electric field vector. Various other notations are current in the literature.

The linear and the circular representations of an elliptically polarized wave are only two of many methods using orthogonal unit vectors for representing the same wave; however, they will suffice for the purposes of this book. For discussions of numerous other specialized representations of the polarization state of an electromagnetic wave, the reader is referred to the literature.[12-16]

Most targets do not reflect differently polarized incident waves in the same fashion at any particular aspect. Also, for many bodies, the scattered field is polarized differently than the incident field. This phenomenon is known as "depolarization" or "cross polarization." In order to consider the polarization of the scattered field in more detail, it is convenient to examine a perfect reflector. Consider a plane wave incident at an arbitrary angle $\theta$ upon a semi-infinite slab of perfect conductor. The plane wave may have either of the two linear polarizations as shown in Figure 2-3. Let the incident field in case (a) be represented as $\mathbf{E}_\perp{}^i = E_\perp{}^i \hat{\mathbf{x}} \exp(-ik_0 r)$. The scattered field may be represented as $\mathbf{E}_\perp{}^s = E_\perp{}^s \hat{\mathbf{x}} \exp(ik_0 r)$. From the boundary conditions, $E_\perp{}^s = -E_\perp{}^i$. Similarly, for the parallel case, $H_\parallel{}^s = H_\parallel{}^i$. Therefore, a perfect reflector reflects linear polarization in such a manner that an electric field which is parallel polarized remains parallel polarized, and an electric field which is perpendicularly polarized remains polarized in the perpendicular direction but is phase shifted 180°. From the relationship between linear and circular polarization [Eq. (2.1-17)] it is seen that a wave which is right circularly

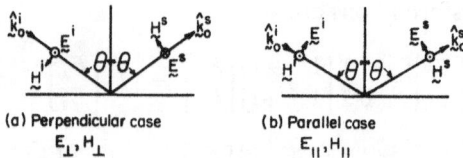

(a) Perpendicular case
$E_\perp, H_\perp$

(b) Parallel case
$E_\parallel, H_\parallel$

Fig. 2-3. Coordinates for a plane wave incident at an angle $\theta$ on a semi-infinite perfectly conducting slab.

polarized becomes left circularly polarized upon reflection, and one which is originally left circularly polarized is right circularly polarized after reflection. Thus, for normal incidence on a perfect reflector, using circularly polarized antennas of the same sense, no signal will be received.

The definition of cross section given assumes the radar to be capable of transmitting and receiving both orthogonal polarizations, either simultaneously or independently in any order, and that the radar can measure both the amplitude and plane of the received signal. The radar cross section observed when a parallel polarized field is transmitted and a parallel polarized field is received is denoted $\sigma_{\parallel \parallel}$ or, for simplicity, just $\sigma_{\parallel}$. The symbol $\sigma_{\perp \parallel}$ denotes the cross-polarized cross section observed when a perpendicular field is received and a parallel field is transmitted. The left-most subscript in $\sigma_{rec\ trans}$ refers to the polarization which is received, while the right-most subscript refers to the incident field component under consideration. Similar definitions apply to $\sigma_{\parallel \perp}$, $\sigma_{\perp \perp}$ (shortened for convenience to $\sigma_{\perp}$), $\sigma_{RR}$, $\sigma_{LL}$, $\sigma_{RL}$, and $\sigma_{LR}$. In discussing radar cross sections for circularly polarized cases, it should be remembered that the cross-polarized terms are the result of receiving with the same polarization as was transmitted. That is, $\sigma_{RR}$ and $\sigma_{LL}$ are cross-polarized cross sections.

The linear cross sections and the circular cross sections are not simply related. A knowledge of $\sigma_{\perp}$, $\sigma_{\parallel \parallel}$, $\sigma_{\parallel \perp}$, and $\sigma_{\parallel}$ alone will not enable a calculation of $\sigma_{RR}$, $\sigma_{RL}$, $\sigma_{LR}$, and $\sigma_{LL}$. More knowledge is needed. This is discussed in more detail in the following section.

### 2.1.4. The Scattering Matrix

In the previous section it was pointed out that the observed radar cross section is a function of the polarization of the field incident upon the target. In Section 2.1.1 it was assumed that the receiving system and the transmitting system were so oriented that little polarization loss occurs. However, a radar is not necessarily restricted to transmitting and receiving a single polarization. The definition of radar cross section from Eq. (2.1-5) implies that the cross section is to be determined by taking the ratio of the magnitudes of the scattered electric field to the incident electric field. Based on this definition $\sigma$ will be a scalar. To remove the dependence of the cross section on the radar polarization it is necessary to define $\sigma$ as a tensor operator which relates the incident field to the scattered field. It is shown in textbooks[17,18] on electromagnetic theory that the power density carried by an electromagnetic field is one-half the real part of the complex Poynting vector. The power density scattered may be determined by

$$\tfrac{1}{2} \text{Re}(\mathbf{E}^s \times \mathbf{H}^{s*}) = \bar{\sigma} \cdot \tfrac{1}{2} \text{Re}(\mathbf{E}^i \times \mathbf{H}^{i*}) \qquad (2.1\text{-}19)$$

where Re signifies the real part, $\mathbf{H}^{s*}$ and $\mathbf{H}^{i*}$ are the complex conjugates of the magnetic field vectors, and $\bar{\sigma}$ is a dyad or second-order tensor. The tensor notation for the radar cross section is convenient when working with anisotropic media. For other cases the radar cross section is more commonly represented by a matrix which is denoted as a scattering matrix by analogy with quantum mechanics. Some authors prefer to use a power scattering matrix. It is, in general, more convenient to use a voltage scattering matrix $S$. If an arbitrarily polarized plane wave is incident upon a target, the scattered field far from the object is given by the following relationships:

$$E_1^s = a_{11}E_1^i + a_{12}E_2^i$$
$$E_2^s = a_{21}E_1^i + a_{22}E_2^i \tag{2.1-20}$$

The $a_{ij}$ are, in general, complex quantities, and the subscripts 1 and 2 represent any two general orthogonal polarization components. It is convenient to write Eq. (2.1-20) in matrix notation

$$\begin{bmatrix} E_1^s \\ E_2^s \end{bmatrix} = [S] \begin{bmatrix} E_1^i \\ E_2^i \end{bmatrix} \tag{2.1-21}$$

where

$$[S] = \begin{bmatrix} a_{11} & a_{12} \\ a_{21} & a_{22} \end{bmatrix}$$

Each of the $a_{ij}$ elements has both an amplitude and phase so that

$$[S] = \begin{bmatrix} |a_{11}| e^{i\phi_{11}} & |a_{12}| e^{i\phi_{12}} \\ |a_{21}| e^{i\phi_{21}} & |a_{22}| e^{i\phi_{22}} \end{bmatrix}$$

The components of the matrix may be related to the cross section as follows:

$$\sigma_{11} = 4\pi r^2 |a_{11}|^2, \qquad \sigma_{12} = 4\pi r^2 |a_{12}|^2$$
$$\sigma_{21} = 4\pi r^2 |a_{21}|^2, \qquad \sigma_{22} = 4\pi r^2 |a_{22}|^2 \tag{2.1-22}$$

The scattering matrix contains, in general, all the available information on the scattering properties of a target. For a particular frequency, orientation of the target and orientation of the transmitting–receiving systems, if $S$ is completely specified, the cross section of the target can be computed for every possible combination of transmitting and receiving system polarizations. Conversely, if the cross section is known for several polarizations, $S$ can be determined uniquely. This fact makes possible the specification of $S$ by several different techniques.

One method of specifying $S$ is to measure the amplitude and the absolute phase for each component. For a monostatic radar $a_{12} = a_{21}$. This relationship is not true for bistatic radar. In the monostatic case there are three amplitudes and three phases to evaluate, while for the bistatic case, four amplitudes and four phases must be evaluated. All amplitudes are always nonnegative, but the phases may be either positive or negative. In the special case of the monostatic return from a rotationally symmetric target, $a_{12} = a_{21} = 0$. (It is assumed that the plane of incidence contains the radar line of sight and the target symmetry axis.) For this special case only two amplitudes and two phases are needed to specify $S$ completely. The absolute phase is a function of the distance from the radar to the target. For the body having one plane of symmetry the absolute phase is also a function of the rotation around this plane. It is almost impossible to measure absolute phase in the laboratory, so this technique is not practical for use with presently available radars in the field.

A second technique involves relative phase measurement. One phase angle, say $\phi_{11}$, is factored out of the matrix and not evaluated:

$$[S] = e^{i\phi_{11}} \begin{bmatrix} \sqrt{\sigma_{11}} & \sqrt{\sigma_{12}}\, e^{i(\phi_{12}-\phi_{11})} \\ \sqrt{\sigma_{21}}\, e^{i(\phi_{21}-\phi_{11})} & \sqrt{\sigma_{22}}\, e^{i(\phi_{22}-\phi_{11})} \end{bmatrix} \frac{1}{\sqrt{4\pi r^2}} \qquad (2.1\text{-}23)$$

With this factorization, seven quantities are left to be evaluated in the general case, five in the monostatic case, and three for monostatic scattering from either metallic or dielectric symmetrical bodies.[19,20] Relative phases are easier to measure than absolute phases since they are not functions of target distance, but it is still advantageous at times to measure only amplitudes. However, this usually requires more amplitude measurements than there are quantities to evaluate because of ambiguities in the signs of the phase terms.

Once the scattering matrix $S$ is determined, the polarization which must be transmitted to maximize the reflected signal power can be found by Mikulski's theorem.[21] His theorem states that if $S$ is conjugated to give $S^*$ and the product $S^*S$ is found, then the polarization desired is determined by the eigenvector of $S^*S$ corresponding to the maximum eigenvalue of $S^*S$.

A scattering matrix specified with respect to a particular coordinate system may be transformed to another form referenced to a different coordinate system by a unitary transformation. The same technique also applies if $S$ is known in one set of orthogonal polarizations and an $S$ is desired for another set of orthogonal polarizations. For example, $S$ is usually expressed in terms of either linear or circular polarizations. Rewriting Eq. (2.1-21) for these two cases gives

$$\begin{bmatrix} E_\perp{}^s \\ E_\|{}^s \end{bmatrix} = \begin{bmatrix} a_\perp & a_{\perp\|} \\ a_{\|\perp} & a_\| \end{bmatrix} \begin{bmatrix} E_\perp{}^i \\ E_\|{}^i \end{bmatrix}$$

and

$$\begin{bmatrix} E_R{}^s \\ E_L{}^s \end{bmatrix} = \begin{bmatrix} a_{RR} & a_{RL} \\ a_{LR} & a_{LL} \end{bmatrix} \begin{bmatrix} E_R{}^i \\ E_L{}^i \end{bmatrix}$$

In Eq. (2.1-18) linear and circular fields were transformed into each other by use of the transformation $[T]$. By simple matrix manipulation the elements of the circular scattering matrix can be determined from the linear scattering matrix elements by[†]

$$\begin{bmatrix} a_{RR} & a_{RL} \\ a_{LR} & a_{LL} \end{bmatrix} = [T] \begin{bmatrix} a_\perp & a_{\perp\parallel} \\ a_{\parallel\perp} & a_\parallel \end{bmatrix} \begin{bmatrix} 1 & 0 \\ 0 & -1 \end{bmatrix} [T]^{-1} \qquad (2.1\text{-}24)$$

The individual elements of the ciruclar scattering matrix are

$$a_{RL} = \frac{+a_\parallel + a_\perp + i(a_{\perp\parallel} - a_{\parallel\perp})}{2}, \qquad a_{RR} = \frac{-a_\parallel + a_\perp - i(a_{\parallel\perp} + a_{\perp\parallel})}{2}$$

$$(2.1\text{-}25)$$

$$a_{LL} = \frac{-a_\parallel + a_\perp + i(a_{\parallel\perp} + a_{\perp\parallel})}{2}, \qquad a_{LR} = \frac{+a_\parallel + a_\perp - i(a_{\parallel\perp} - a_{\perp\parallel})}{2}$$

Similarly, the inverse relationship, giving the linear scattering elements in terms of the circular scattering elements, can be found from

$$\begin{bmatrix} a_\perp & a_{\perp\parallel} \\ a_{\parallel\perp} & a_\parallel \end{bmatrix} = [T]^{-1} \begin{bmatrix} a_{RR} & a_{RL} \\ a_{LR} & a_{LL} \end{bmatrix} \begin{bmatrix} 1 & 0 \\ 0 & -1 \end{bmatrix} [T] \qquad (2.1\text{-}26)$$

with

$$a_\parallel = \frac{-a_{RR} + a_{RL} + a_{LR} - a_{LL}}{2}, \qquad a_{\parallel\perp} = \frac{+i(a_{RR} - a_{LR} + a_{RL} - a_{LL})}{2};$$

$$(2.1\text{-}27)$$

$$a_{\perp\parallel} = \frac{-i(a_{RR} + a_{LR} - a_{RL} - a_{LL})}{2}, \qquad a_\perp = \frac{a_{RR} + a_{LL} + a_{RL} + a_{LR}}{2}.$$

From Eqs. (2.1-22) and (2.1-25) the circular scattering cross sections can be determined in terms of the linear-scattering matrix elements, while from Eqs. (2.1-22) and (2.1-27) the linear cross sections can be determined in terms of the circular matrix elements.

There is one case in which the use of a scalar to represent the radar cross section is always correct. If the target has more than two planes of symmetry intersecting along the radar line-of-sight, the scattering matrix $S$ for orthogonal linear polarization reduces to a scalar times the unit matrix.

---

[†] When describing polarization states in terms of linear unit vectors such as $\hat{\imath}$ and $\hat{\parallel}$, one must account for the fact that the coordinate system changes "handedness" upon reflection from a target. For instance, in a spherical coordinate system, $\hat{\theta}$, $\hat{\varphi}$, and $\hat{k}_0{}^i$, in that order, describe a left-handed system for a wave propagating toward a target at the origin. However, $\hat{\theta}$, $\hat{\varphi}$, $\hat{k}_0{}^s$ describe a right-handed system for a scattered wave propagating away from this target. In order for the matrix operations to be valid, this change upon scatter must be circumvented. Since right-handed systems are used in this book, an additional matrix is introduced in Eq. (2.1-24) immediately preceding the transformation $[T]^{-1}$ operating on the incident wave.

The term scattering matrix in the literature does not always mean a power matrix or a voltage matrix $S$ as used here. For example, Tsu[22] considers an infinite matrix which contains all of the scattering properties of the target. His matrix permits the determination of scattering at any angle for any frequency. The elements of the matrix are determined by a multipole expansion of the expression for scattering from the target. The matrix may be truncated for a small target of the order of a wavelength in size since the scattering is well approximated by the dipole or quadrupole approximation. Even in these cases, the problem is quite complicated if the target is not axially symmetric. For larger bodies this method does not appear to be practical. In the remainder of the book only the voltage matrix $S$ is discussed unless otherwise specified.

### 2.1.5. Near-Field Cross Section

In Section 2.2.1.2 it is shown that the scattered magnetic field from a perfectly conducting object can be obtained by integrating the surface current distribution $K$ over the surface of the scatterer, that is,

$$\mathbf{H}^s(\mathbf{r}) = \frac{1}{4\pi} \oint [\mathbf{K}(\mathbf{r}') \times \nabla'\psi_0] \, ds' \tag{2.1-28}$$

where $s$ is the surface of the scatterer, $\mathbf{r}'$ is a vector from an arbitrary origin to a point on the surface, $\mathbf{r}$ is a vector from the origin to the observation point, and

$$\psi_0 = \frac{e^{ik_0|\mathbf{r}-\mathbf{r}'|}}{|\mathbf{r}-\mathbf{r}'|} \tag{2.1-29}$$

Expressing the source and observation points in terms of the spherical coordinates, $(r, \theta, \phi)$ and $(r', \theta', \phi')$, the function $\psi_0$ can be expanded for $r > r'$ as

$$\psi_0 = ik_0 \sum_{m=0}^{\infty} \sum_{n=0}^{\infty} \epsilon_m(2n+1) \frac{(n-m)!}{(n+m)!} P_n^m(\cos\theta) P_n^m(\cos\theta')$$
$$\times \cos m(\phi-\phi') j_n(k_0 r') h_n^{(1)}(k_0 r) \tag{2.1-30}$$

where $\epsilon_m = 1$ for $m = 0$, $\epsilon_m = 2$ for $m \neq 0$, and $P_n^m(x)$ are the associated Legendre functions discussed in Reference 18, page 192. Examining $h_n^{(1)}(k_0 r)$ as a function of $r$, then

$$h_n^{(1)}(k_0 r) = i^{-n-1} \frac{e^{ik_0 r}}{k_0 r} \sum_{v=0}^{n} \frac{1}{v!} \frac{\Gamma(n+1+v)}{\Gamma(n+1-v)} (-2ik_0 r)^{-v} \tag{2.1-31}$$

Substituting this and Eq. (2.1-30) for $\psi_0$ back into the integral, Eq. (2.1-28), then

$$\mathbf{H}^s(r, \theta, \phi)$$

$$= \frac{1}{4\pi} \frac{e^{ik_0 r}}{r} \sum_{m=0}^{\infty} \sum_{n=0}^{\infty} \sum_{\nu=0}^{n} i^{-n} \epsilon_m$$

$$\times \frac{(2n+1)}{\nu!} \frac{(n-m)!}{(n+m)!} \frac{\Gamma(n+1+\nu)}{\Gamma(n+1-\nu)} P_n^m(\cos\theta)(-2ik_0 r)^{-\nu}$$

$$\times \left\{ \oint_s \mathbf{K}(r', \theta', \phi') \times \nabla'[P_n^m(\cos\theta')\cos m(\phi - \phi') j_n(k_0 r')] \, ds \right\}$$

$$(2.1\text{-}32)$$

The quantity inside the curly brackets { } in this equation is observed to be independent of $r$; thus the $r$ dependence of the scattered magnetic field is an inverse power series in $r$. As $r$ becomes large it is apparent that only the $\nu = 0$ term contributes, so the field varies as $1/r$. This is known as the far zone or "Fraunhofer" region. As $r$ becomes smaller, the $(r)^{-2}$ and higher-order terms become increasingly important. The range of distances, such that these terms cannot be neglected, is called the "Fresnel" or "near-zone" region. Many authors make a distinction between the Fresnel and near-zone regions, however, in this work no such distinction will be made.

The distance at which the $(kr)^{-2}$, $(kr)^{-3}$, and higher-order terms can be neglected is, of course, arbitrary. It is customarily chosen[23] to be at a distance equal to two times the maximum antenna or scatterer dimension squared divided by the wavelength $(2d^2/\lambda_0)$. This distance was chosen because it has been found for targets not essentially flat that if the maximum variation in phase does not exceed $\pi/8$ radians and the maximum amplitude variation does not exceed 1.0 db over the target aperture, the error in $\sigma$ caused by this variation will be less than 1 db at the point of maximum monostatic reflection. The error does change the heights of the minor lobes, whose values are 20 db or more below the maximum return, and fills in the nulls of the target angular-reflectivity pattern.

In the far field there is no radial component of either the electric or magnetic field, and the scattered fields are transverse to the direction of propagation.

Considering the definition of the scattering matrix from Eq. (2.1-21) it is seen that, in the near field, $S$ must be a three by three matrix. Also, the electric and magnetic field components are not related simply by $120\pi$ in the near field as they are in the far field. Perhaps even more significant is the fact that the power density is not a function which decreases monotonically with distance in the near zone. The primary practical effect of this is to increase the difficulty in the laboratory measurement of radar cross section.

The error introduced in the laboratory determination of $\sigma$ by the phase or amplitude variation across the target cannot always be determined analytically. One practical solution is to keep moving the target away from the transmitting system until $\sigma$ of a fixed aspect is constant with distance variation. Fortunately, measurement techniques exist utilizing focused fields, or for small models, utilizing regions near the transmitting antenna where the fields are sufficiently uniform to permit accurate measurements. Kouyoumjian and Peters[24] have given a quantitative discussion of the range requirement and its relation to antenna illumination, relative sizes of the antennas and model, and other factors. Hansen[25] has recently summarized considerable information about the near field of antennas and the effects of various antenna aperture distributions on the power density.

Many theoretical problems involve infinite or semi-infinite bodies. By definition any receiving antenna would be in the near field of such objects if the entire object is illuminated. Since this type of body is not physically possible, and the entire body obviously cannot be illuminated, the near-field effects are usually ignored. Many three-dimensional problems, such as a plane wave normally incident upon an infinite cylinder, reduce to a two-dimensional problem. For this class of problem it is convenient to define an "echoing width" $\sigma^c$, analogous to Eq. (2.1-5),

$$\sigma^c = 2\pi \lim_{r \to \infty} r \, \frac{(\mathbf{E}^s \cdot \mathbf{E}^{s*})}{(\mathbf{E}^i \cdot \mathbf{E}^{i*})} \qquad (2.1\text{-}33)$$

where $\sigma^c$ has the units of length.

An alternate formulation of the cross section of semi-infinite bodies is given by Brysk,[26] who replaces the limit as $r$ goes to infinity of Eq. (2.1-5) with a finite $r$ which is large enough so that when a series expansion in terms of powers of $r$ is made, only the leading term is retained. For finite bodies this term is independent of $r$; for a semi-infinite or infinite body the leading term is linear in $r$. The cross section of an infinite circular cylinder of radius $a$ at high frequencies by this definition is $\pi r a$. By the definition of Eq. (2.1-33), $\sigma^c$ is $\pi a$. In general, the cross section in this book is independent of $r$.

### 2.1.6. Frequency Effects

In the preceding sections of this book, it has been stated that the radar cross section varies with frequency. The variation of the cross section of a fixed body as the frequency is increased can be characterized by three regions —a Rayleigh region, a resonance region, and an optical or high-frequency region. In Section 2.2.2, approximate techniques for determining the cross section in these regions are discussed. The cross section that is the subject of most of the discussion in this book and in the literature is the continuous wave (cw) cross section. This is the radar cross section produced by a monochromatic transmitted signal. Even if an operational radar radiated such a signal, the scattered signal still may not be monochromatic since most radar targets are moving with respect to the radar. The result is that the signal would be Doppler-shifted. The amount of Doppler shift ($\Delta f$) in cycles per second from a body moving with velocity ($\mathbf{v}$) is given by

$$\Delta f = f \frac{\bar{\mathbf{v}} \cdot (\hat{\mathbf{k}}_0{}^s - \hat{\mathbf{k}}_0{}^i)}{c} \tag{2.1-34}$$

where $f$ is the incident frequency, $c$ is the speed of light, and $\Delta f$ may be either positive or negative. In defining radar cross section, it is tacitly assumed that the receiving system is capable of receiving the entire range of possible $f + \Delta f$'s. The cross section is obtained by determining the power in the scattered field at frequency $f + \Delta f$. If different portions of the radar target are moving with different velocities (as could happen if the radar is observing, for example, a nuclear explosion or a rotating body), the scattered signal will, exist at a variety of frequencies, and the radar cross section observed will not, in general, be the same as the cross section which would be observed if the target were stationary.

A similar phenomenon occurs in the reflection of a pulse or pulse train from a target. One approach to computing the scattered field is to expand the transmitted pulse(s) in a Fourier series. At each frequency of the series the scattered field is computed. The complete scattered field is then obtained by summing the series of computed scattered fields. This method gives the steady-state scattered field but neglects some transient effects.

The transient and steady-state effects can be obtained if one calculates the response of a radar target to an incident impulse function. A scattering amplitude function was defined in Eq. (2.1-6). After that equation it was remarked that it is sometimes convenient to factor out $(\pi f)^2$. If the factoring is done, Eq. (2.1-6) becomes

$$\sigma = \pi c^2 \mid F(i\omega)\mid^2 = \pi c^2 \left| \frac{F(\theta, \phi; \theta', \phi')}{\pi f} \right|^2 \tag{2.1-35}$$

where $F(i\omega)$ is the Fourier transform of the impulse response $F_I(t)$, or

$$F(i\omega) = \int_{-\infty}^{\infty} F_I(t)\, e^{i\omega t}\, dt \tag{2.1-36}$$

with

$$F_I(t) = \int_{-\infty}^{\infty} F(i\omega)\, e^{-i\omega t}\, df \tag{2.1-37}$$

and $c$ is the velocity of light.

Therefore, in theory, if one knows the impulse response of an object, one could find the radar cross section at any frequency or combination of frequencies and for any transmitted waveform. In fact, if one knew the impulse response of an object, the optimum waveform for the transmitted signal to maximize the backscattered signal would be the time inverse of the impulse

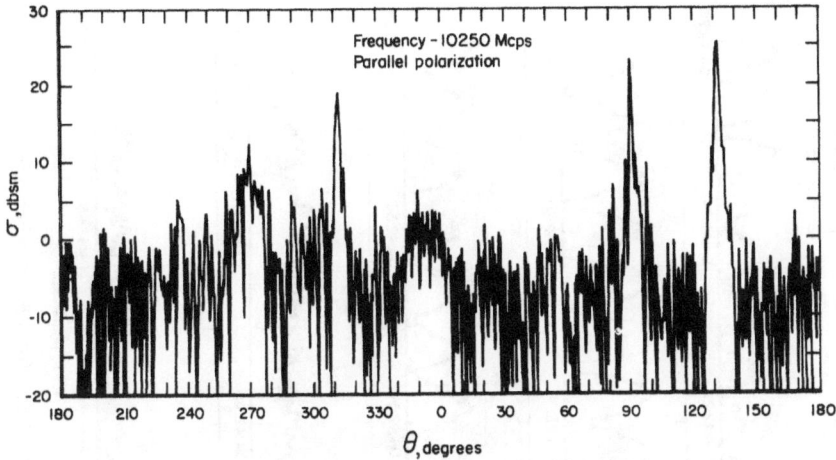

Fig. 2-4.   Measured backscatter cross section of the Telstar satellite.

response. The target would then be a matched filter to the incident field. The scattered signal is the autocorrelation function of the input signal in this case. In general, the scattered signal is the convolution of the input signal with the impulse response. The impulse response is not known exactly at present for any radar target. An approximate technique to determine this response is discussed in Section 2.2.2.4. The impulse response is, of course, a function of aspect angle. In general, the aspect angle of the target is not known *a priori*. Therefore, it is unlikely that the matched filter case would occur.

The effect of motion and transient phenomena are, in general, marked by two other factors. One is the fact that the cw radar cross section is not usually known to a high degree of accuracy. Most of this book is devoted to the discussion of how to predict or to measure the cw cross section, and to a discussion of the accuracies of these methods.

The second factor is that the motion of the body is likely to introduce aspect angle changes which can cause large changes in cross section. For example, Figure 2-4 gives the measured cross section of the Telstar satellite at a frequency of 10,250 Mcps.[27] It is seen that a change of aspect of less than one degree can introduce a change of three orders of magnitude in cross section. The implications of such a rapid change are discussed in the following section.

## 2.1.7.   Dynamic Cross Sections

For most practical situations there is relative motion between the observing radar and the observed target. The radar "sees" a dynamic cross section. However, for most of the calculations and measurements discussed in this handbook the target is assumed to be stationary, that is, the cross

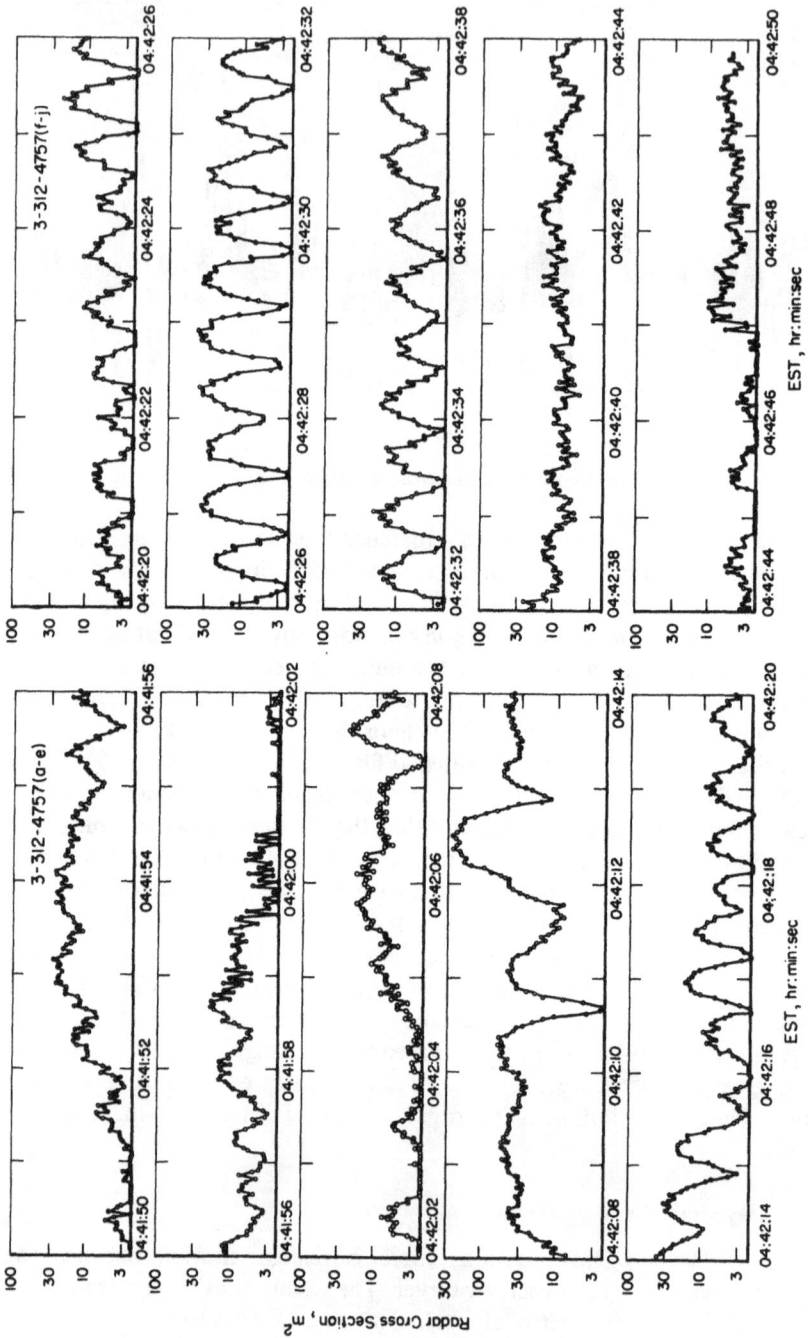

Fig. 2-5. Measured pulse-by-pulse backscatter cross section of the satellite 1957 Beta [from Pettengill et al.[28]].

section under discussion is a static cross section. It is, therefore, pertinent to discuss the differences between the static and dynamic cross sections and to show how one may be obtained from the other.

The relative motion between the radar and the target can cause several effects, one obvious effect being to Doppler shift the radar signals. It is shown in the previous sections that this Doppler shift does not affect the definition or the determination of radar cross section, but does affect its observation, use, and, especially, the way the received signal must be processed. The radar cross section is independent of the signal processing, so that it is assumed that the observing radar is capable of analyzing any Doppler shift. Similar assumptions apply to effects caused by the radar signal not being a plane wave at the target.

A more important effect of the motion is the change of relative aspect angle of the target with respect to the radar. In the previous section, the variation of the measured static radar cross section of Telstar satellite at a frequency of 10,250 Mcps was presented in Figure 2-4. From this figure it is seen that only a small change in aspect angle can cause a very large change in cross section. However, not only is the aspect angle changing, but also the angle of polarization. As a result, the time variation of the observed radar cross section of a complex moving target can be quite difficult to interpret. An example of a dynamic cross section is given in Figure 2-5, where the pulse-by-pulse radar cross section of the satellite 57 beta as measured by MIT's Millstone Hill radar at 440 Mcps is given over a one-minute time interval.[28] Figure 2-5 shows that for this particular target, variations in the observed cross section of up to 20 db occurred over a one-minute interval.

Considerable effort has been expended in investigating the statistics of dynamic cross sections, particularly in the case of aircraft. It can be shown that for a target consisting essentially of a large number of point scatterers in random relative motion, the radar cross section will have an exponential probability density.[†] For a relatively complex target, such as an aircraft at high frequencies, experimental results agree well with this conclusion. Figure 2-6 shows measured cumulative probability distribution curves for an F-86 aircraft at three frequencies, along with the theoretical curves for a Rayleigh distribution.[29] It is apparent that some deviation occurs; however, in general, the agreement is good. At lower frequencies or for more simple targets, the Rayleigh distribution may be a much poorer model, and Swerling[30] assumes a probability density of the form

$$P(\sigma) = \frac{4\sigma}{\langle \sigma \rangle} \, e^{-2\sigma/\langle \sigma \rangle} \qquad (2.1\text{-}38)$$

for targets of this nature.

---

† An exponential density is often referred to as a Rayleigh-power density; see Reference 2, page 51.

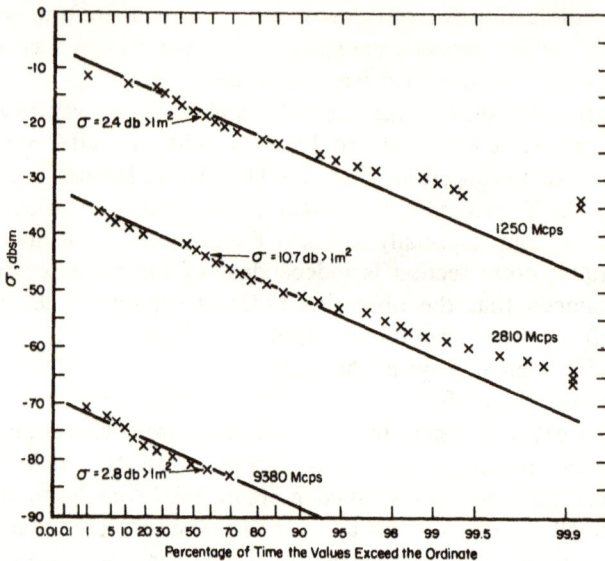

Fig. 2-6.   Measured in-flight cumulative probability distribution
for an F-86 aircraft at three frequencies.

Further discussion of dynamic cross sections is beyond the realm of this
book and the literature should be consulted.[31,32,33]

## 2.2.   ANALYTICAL TECHNIQUES FOR CROSS-SECTION COMPUTATION

There exists a wide variety of analytical techniques for computing radar
cross sections. Some of these lead, in principle, to exact solutions, although
approximations are usually required to obtain numerical results even from
what was originally an exact formulation. A typical example of this type of
technique is the formulation of a scattering problem as a boundary-value
problem with the resultant formally exact solution appearing as an infinite
series. To obtain numerical results, the sum of this series must usually be
approximated in some fashion.

Other techniques involve various simplifying assumptions to arrive at the
basic formulation so that even if the mathematical analysis can be carried
through exactly, only an approximate solution is obtained. A typical example
of this type is the physical optics technique for which it is generally true that
even if the required integration can be performed exactly, only an approximate solution is obtained.

## 2.2.1. Exact Solutions

In this section several techniques will be discussed which, in principle, are capable of providing exact analytical solutions to scattering problems, valid for any size of scatterer or incident field wavelength. Unfortunately, in most cases these solutions involve infinite series, or infinite sets of linear equations, and must be approximated in some fashion in order to obtain numerical results. Nevertheless, exact analytical solutions of this nature are of considerable importance, not only for the numerical results obtainable, but for the insight into the physical phenomena they provide and as a starting point for the generation of approximation techniques.

### 2.2.1.1. *Boundary-Value Problems*

2.2.1.1.1. *The Vector Helmholtz Equation from Maxwell's Equations.* A basic property of electromagnetic fields is their vector nature. At every point in space a given electric field must be assigned both a magnitude and a direction. A propagating field or wave has not only a magnitude and direction for its electric (and magnetic) field at a given instant of time, but also a direction of propagation. This vector nature of the field distinguishes electromagnetic theory from acoustics, for example, where the field is a scalar quantity.

As in any discipline of science, electromagnetic theory is formulated from certain fundamental equations, the well-known Maxwell equations. These equations in their vector form are

$$\nabla \times \mathbf{E} = -\frac{\partial \mathbf{B}}{\partial t} \qquad (2.2\text{-}1)$$

and

$$\nabla \times \mathbf{H} = \frac{\partial \mathbf{D}}{\partial t} + \mathbf{J} \qquad (2.2\text{-}2)$$

These, along with the equation of continuity,

$$\nabla \cdot \mathbf{J} = -\frac{\partial \rho}{\partial t} \qquad (2.2\text{-}3)$$

form the basis of electromagnetics. The vector quantities $\mathbf{E}$ and $\mathbf{H}$ are the electric and magnetic field intensities; $\mathbf{D}$ is known as the electric displacement; and $\mathbf{B}$ is the magnetic induction. $\mathbf{J}$ is the current density, or amount of charge per unit time flowing across a surface normal to $\mathbf{J}$ divided by the area of the surface; while $\rho$ is the charge density or charge per unit volume. Conservation of energy and charge are inherent in electromagnetic theory.

For the scattering problems of interest in this book, an object of finite size, or scatterer, is located in a homogeneous isotropic charge-free medium which is taken to be free space. The incident wave originates far from the scatterer, so that in the vicinity of the scatterer the incident or exciting field may be treated as a plane wave. In all but a few cases, the incident and scattered fields are monochromatic (sinusoidal in time).

With the limitations of the preceding paragraph, Eqs. (2.2-1)–(2.2-3) can be simplified. $\mathbf{D}$ and $\mathbf{B}$ in free space are related to $\mathbf{E}$ and $\mathbf{H}$ by

$$\mathbf{D} = \epsilon_0 \mathbf{E}, \qquad \mathbf{B} = \mu_0 \mathbf{H} \tag{2.2-4}$$

where $\epsilon_0$ and $\mu_0$ are the permittivity and permeability of free space. $\mathbf{J} = 0$ in free space. Differentiation with respect to time can be represented as $\partial/\partial t = -i\omega$. Using these simplifications and elementary vector algebra, one can arrive at a vector partial differential equation, called the vector Helmholtz equation, for $\mathbf{E}$ and $\mathbf{H}$ of the form

$$\nabla^2 \mathbf{E} + k_0{}^2 \mathbf{E} = 0$$
$$\nabla^2 \mathbf{H} + k_0{}^2 \mathbf{H} = 0 \tag{2.2-5}$$

where $k_0{}^2 = \omega^2 \mu_0 \epsilon_0$, and the time factor $e^{-i\omega t}$ has been suppressed. In the preceeding equation, vectors $\mathbf{E}$ and $\mathbf{H}$ can represent the incident fields, the scattered fields, or any linear combination of them outside the scatterer. Obviously, $\mathbf{E}$ and $\mathbf{H}$ are not independent of each other as is seen from Eqs. (2.2-1), (2.2-2), and (2.2-4); consequently, it is generally necessary to solve only one of the preceding equations for the scattered field. The operator $\nabla^2$ in Eq. (2.2-5) refers to the vector Laplacian, and can be written as

$$\nabla^2 \mathbf{F} = \nabla(\nabla \cdot \mathbf{F}) - \nabla \times \nabla \times \mathbf{F} \tag{2.2-6}$$

Only when $\mathbf{F}$ is resolved into components in a rectangular coordinate system does the vector Helmholtz equation separate simply into three scalar partial differential equations of the form

$$\nabla^2 \psi + k_0{}^2 \psi = 0 \qquad \text{or} \qquad \nabla \cdot \nabla \psi + k_0{}^2 \psi = 0 \tag{2.2-7}$$

where now $\psi$ is a scalar quantity and can represent either the $x$, $y$, or $z$ component of the fields of Eq. (2.2-5).

2.2.1.1.2. *Requirements on the Scattered Field Solution.* The scattered electric field, $\mathbf{E}^s$, must satisfy Eq. (2.2-5) but it must also satisfy several other constraints or requirements. These are listed below and are an integral part of the scattering problem.

(a) The divergence of the electric fields (and magnetic fields) in free space must be zero, i.e.,

$$\nabla \cdot \mathbf{E} = 0, \qquad \nabla \cdot \mathbf{H} = 0 \tag{2.2-8}$$

Such must be also true inside the scatterer if its material is homogeneous, isotropic, and contains no space charges. These requirements are derived directly from Maxwell's equations [(2.2-1) and (2.2-2)].

(b) At very large distances from a finite-sized scatterer, the scattered field must diverge spherically; this means that the scattered field at large distances must contain a factor $e^{ik_0r}/r$, where $r$ is the radial distance from a convenient point on or inside the scatterer to the distant point of observation. This requirement arises from conservation of energy, and is often called the radiation condition at infinity.

(c) The incident and scattered fields must satisfy the appropriate boundary conditions on the surface of the scatterer. These boundary conditions result from integration of Maxwell's equations over an infinitesimally small closed area containing a medium whose electrical properties $\mu$, $\epsilon$, and $\sigma$ suffer a discontinuous jump. The boundary conditions for two types of scatterer materials are given here.

## I. The Scatterer Surface is Perfectly Conducting

For a perfectly conducting scatterer the boundary conditions at the surface are

$$\hat{n} \cdot \mathbf{H}^T = 0, \qquad \hat{n} \cdot \mathbf{E}^T = q_s/\epsilon_0$$
$$\hat{n} \times \mathbf{H}^T = \mathbf{K}, \qquad \hat{n} \times \mathbf{E}^T = 0 \tag{2.2-9}$$

where $\mathbf{H}^T = \mathbf{H}^i + \mathbf{H}^s$ and $\mathbf{E}^T = \mathbf{E}^i + \mathbf{E}^s$ represent the total vector fields at the surface of the object, $q_s$ is the surface charge (charge per unit area on the surface), and $\mathbf{K}$ is the surface current (charge per time on the surface moving past a line normal to the flow direction per unit length of the line). $\hat{n}$ is a unit vector, normal to the scatterer surface, and is directed outward.

## II. The Scatterer is Nonperfectly Conducting and Isotropic

In this case, let the properties of the scatterer be $\epsilon_1$ and $\mu_1$, where $\epsilon_1 = \epsilon_1' + i\sigma_1/\omega = \epsilon_1' + i\epsilon_1''$ accounts both for the permittivity, $\epsilon_1'$, of the material and $\sigma_1$ its conductivity. A similar relationship exists for $\mu_1 = \mu_1' + i\mu_1''$, where the imaginary part arises from magnetic losses. Let $\mathbf{E}^T$, $\mathbf{H}^T$ be the total field quantities at the scatterer surface but outside the scatterer, and $\mathbf{E}_1^T$, $\mathbf{H}_1^T$ be the total field quantities just inside the scatterer surface, then the boundary conditions become

$$\hat{n} \cdot (\mu_0 \mathbf{H}^T - \mu_1 \mathbf{H}_1^T) = 0, \qquad \hat{n} \cdot (\epsilon_0 \mathbf{E}^T - \epsilon_1 \mathbf{E}_1^T) = 0$$
$$\hat{n} \times (\mathbf{E}^T - \mathbf{E}_1^T) = 0 \qquad \hat{n} \times (\mathbf{H}^T - \mathbf{H}_1^T) = 0 \tag{2.2-10}$$

No surface charges or currents can exist on the surface of the scatterer in this case. The total fields inside the scatterer must satisfy vector Helmholtz equations of the form

$$\nabla^2 \mathbf{E}_1^T + k_1^2 \mathbf{E}^T = 0$$
$$\nabla^2 \mathbf{H}_1^T + k_1^2 \mathbf{H}^T = 0 \tag{2.2-11}$$

with

$$k_1{}^2 = \omega^2 \epsilon_1 \mu_1 = k_0{}^2 m_1{}^2 \tag{2.2-12}$$

where $m_1$ is called the index of refraction of the scatterer medium.

2.2.1.1.3. *Application of the Boundary-Value Approach.* The so-called boundary value approach to the solution of scattering problems consists of selecting the proper solution for the scattered field, $E^s$, from the many possible solutions which satisfy the vector Helmholtz equation (2.2-5). This "selecting" is accomplished by forcing the proper solution to conform to the three constraints of the preceding section.

The vector Helmholtz equation and the constraints are completely general and are applicable in any coordinate system. Any general orthogonal coordinate system is represented by three sets of coordinate surfaces; each set of surfaces is a function of one of three variables, $\xi_1$, $\xi_2$, or $\xi_3$. A vector at a point in such a coordinate system can be represented by components along three unit vectors, $\hat{\xi}_1$, $\hat{\xi}_2$, and $\hat{\xi}_3$. The direction of $\xi_1$, for example, at a point formed by the intersection of three surfaces with parameters $\xi_1$, $\xi_2$, $\xi_3$, is found by holding $\xi_2$ and $\xi_3$ constant and varying $\xi_1$. With few exceptions, success in solving electromagnetic scattering problems by this technique results from choosing a coordinate system in which the surface of the scatterer is one of the natural coordinate surfaces of the system, say a surface of constant $\xi_1$. Then the normal to the scatterer surface, $\hat{n}$, is equal to $\hat{\xi}_1$ on the surface. Thus the boundary conditions, Eqs. (2.2-9) and (2.2-10), become tractable. The second constraint, the radiation condition at infinity, is satisfied in most solvable boundary-value problems only if the same $\xi_1$ surfaces become spherical of large radius, as $\xi_1$ approaches one of its limits.

Even though the conditions discussed in the preceding paragraph are met in the selection of a proper coordinate system, solution of the vector Helmholtz equation by boundary-value techniques is limited to only a very few problems. This vector equation has been solved only by using solutions $\psi$ of the scalar Helmholtz equation (2.2-7). Then three separate solutions to the vector Helmholtz equation (2.2-5), can be formed as follows:[34]

$$\mathbf{L} = \nabla \psi, \qquad \mathbf{M} = \nabla \times (\mathbf{a}\psi), \qquad \mathbf{N} = \frac{1}{k_0} \nabla \times \mathbf{M} \tag{2.2-13}$$

where $\psi$ is a solution to Eq. (2.2-7) and $\mathbf{a}$ is either a constant vector or the radial vector. Then $\mathbf{L}$, $\mathbf{M}$, and $\mathbf{N}$ are all solutions of the vector Helmholtz equation.

Satisfaction of the first constraint (the zero divergence of the fields) eliminates $\mathbf{L}$, since its divergence is not zero. Thus the scattered $\mathbf{E}$ and $\mathbf{H}$ fields can be written in terms of linear combinations of $\mathbf{M}$ and $\mathbf{N}$. However, the solutions are vectors whose components are resolved along directions determined by **a**.

Solution of the scalar Helmholtz partial differential equation for $\psi$ is attacked in the boundary-value approach by separation of variables. This means that a solution of the form

$$\psi = \psi_1(\xi_1)\,\psi_2(\xi_2)\,\psi_3(\xi_3) \qquad (2.2\text{-}14)$$

is proposed, and an attempt is made to find separate differential equations for $\psi_1(\xi_1)$, $\psi_2(\xi_2)$, and $\psi_3(\xi_3)$. Then each of these three differential equations is solved, and the product of their solutions is formed according to Eq. (2.2-14). However, such separation is possible only in eleven known coordinate systems.[35] The surfaces in these separable orthogonal coordinate systems are all confocal quadric surfaces or their degenerate forms. The most general of these is the ellipsoidal system (in which the surfaces are ellipsoids, hyperboloids of one sheet, and hyperboloids of two sheets). The other systems are rectangular coordinates, circular cylinder coordinates, elliptic cylinder coordinates, parabolic cylinder coordinates, spherical coordinates, conical coordinates, oblate spheroidal coordinates, prolate spheroidal coordinates, paraboloidal coordinates, and parabolic coordinates.

In the separation of $\psi$ in the above systems and solution of the three differential equations in $\psi_1$, $\psi_2$, and $\psi_3$, two arbitrary constants (call them $m$ and $n$) must be introduced. Since these constants are arbitrary and since there are an infinite number of possible solutions involving an infinite variety of $m$ and $n$, it is common to choose values for $m$ and $n$ which will give a discrete or continuous set of solutions for $\psi_1$, for example, those which are orthogonal and complete. If this is done, then any arbitrary solution can be written as a linear combination of members of this orthogonal and complete set of functions. The values of $m$ and $n$ which give orthogonal and complete sets of functions, $\psi_1$, $\psi_2$, and $\psi_3$, are called eigenvalues of the problem; and the corresponding functions, $\psi_{1mn}$, $\psi_{2mn}$, and $\psi_{3mn}$, are called the eigenfunctions. (For example, the sine and cosine functions form an orthogonal complete set over any interval of $2\pi$ radians when the eigenvalues are $m = 0, \pm1, \pm2, \pm3, \ldots$. A representation of an arbitrary function over the interval $2\pi$ by these eigenfunctions is called a Fourier series.)

In rectangular coordinates where $\xi_1 = x$, $\xi_2 = y$, $\xi_3 = z$, for example, the solutions inside a closed rectangular region can be of the form

$$\psi_{p,q} = \exp(i[pk_0 x + qk_0 y + k_0\sqrt{1 - p^2 - q^2}\,z]) \qquad (2.2\text{-}15)$$

where the eigenvalues $p$ and $q$ are determined by the boundary constraints.

Similarly, in rectangular coordinates the solutions over an open rectangular region have the form

$$\psi(k_x, k_y, k_z) = \exp[i(k_x x + k_y y + k_z z)] \qquad (2.2\text{-}16)$$

where

$$k_x{}^2 + k_y{}^2 + k_z{}^2 = k_0{}^2 \qquad (2.2\text{-}17)$$

The eigenvalues $k_x$, $k_y$, $k_z$, are continuous and range from $-\infty$ to $\infty$.

In spherical coordinates where $\xi_1 = r$, $\xi_2 = \theta$, $\xi_3 = \phi$, the solutions can be of the form

$$\psi = j_n(k_0 r) \, P_n{}^m(\cos \theta) \cos m\phi \qquad (2.2\text{-}18)$$

where $j_n(x)$ is a spherical Bessel function of order $n$ and argument $x$, $P_n{}^m(\cos \theta)$ is an associated Legendre function of order $m$, degree $n$, and argument $\cos \theta$. The eigenvalues here take on the values $m = 0, \pm 1, \pm 2, ..., \infty$ and $n = 0, 1, 2, ..., \infty$. In the remaining nine systems, similar eigenfunction solutions exist; most of them are not as familiar as those in the preceding examples.

Other difficulties restrict further the usefulness of most of these eleven systems. Only in the first six systems (i.e., rectangular, circular cylinder, elliptic cylinder, parabolic cylinder, spherical, and conical coordinates) can the vector **a** of Eq. (2.2-13) represent one of the natural unit vectors of the system. In the first, it may be any of the three coordinate unit vectors; in the next three cylindrical systems, it can be the constant axial unit vector; and in the last two systems, it may be the radial unit vector. Thus, in the other systems (for example, prolate spheroidal), **a** cannot be chosen along $\xi_1$, $\xi_2$, or $\xi_3$. Thus, for example, in scattering from a prolate spheroid, the normal to the spheroid surface, $\xi_1$, must be resolved into components along directions determined by the choice of **a**, and satisfaction of the boundary conditions becomes more difficult; nonetheless, results have been obtained for scattering from a prolate (and oblate) spheroid in this manner.

Another difficulty is the failure of the eigenfunction solutions to satisfy easily the radiation condition at infinity as they must if the scatterer is of finite size. For instance, the rectangular system [see Eq. (2.2-15)], along with the three cylindrical systems, has solutions which fail to satisfy this constraint.

A third difficulty is the inability to specify most scattering shapes as a single coordinate surface in one of the eleven systems. For example, a finite circular cylinder with flat ends is made up of two coordinate surfaces in the circular cylindrical system. A finite cone with a spherical cap is made up of two coordinate surfaces in the spherical system. However, the latter scattering problem has been solved[36] despite this added difficulty in describing the scatterer surface.

### 2.2.1.2. *Integral Equations*

Vector solutions to Maxwell's equations can also be formulated as integral equations. In this form the boundary conditions and the radiation condition are automatically satisfied by the solutions to the integral equation. The starting point for most integral equation formulations of scattering

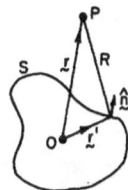

Fig. 2-7.  Coordinates of the field point $P$, and points on the scatterer surface.

problems is a vector analog of Green's theorem which relates volume integrals of vector functions to surface integrals of related functions over the surface enclosing the volume.[37] It is shown in most texts on electromagnetic theory,[38] that the radiated or scattered fields in a source-free region can be expressed in terms of the total fields on the surface of a volume enclosing all the sources of the field. Thus, considering the closed surface of a finite scatterer designated $S$, the scattered fields at some observation point $P$ not on the surface $S$ are given by

$$\mathbf{E}^s(\mathbf{r}) = \frac{1}{4\pi} \int_s [(\hat{\mathbf{n}}' \times \mathbf{E}^T) \times \nabla'\psi_0 + i\omega\mu(\hat{\mathbf{n}}' \times \mathbf{H}^T)\,\psi_0 + (\hat{\mathbf{n}}' \cdot \mathbf{E}^T)\,\nabla'\psi_0]\,ds' \tag{2.2-19}$$

and

$$\mathbf{H}^s(\mathbf{r}) = \frac{1}{4\pi} \int_s [(\hat{\mathbf{n}}' \times \mathbf{H}^T) \times \nabla'\psi_0 - i\omega\epsilon(\hat{\mathbf{n}}' \times \mathbf{E}^T)\,\psi_0 + (\hat{\mathbf{n}}' \cdot \mathbf{H}^T)\,\nabla'\psi_0]\,ds' \tag{2.2-20}$$

where the coordinates are as indicated in Figure 2-7.[†] The integrals in Eqs. (2.2-19) and (2.2-20) are performed over the closed surface $S$ for which $\hat{\mathbf{n}}'$ is a unit vector normal to the surface at the point $\mathbf{r}'$, and the total fields $\mathbf{E}^T$, $\mathbf{H}^T$ are evaluated at the surface point $\mathbf{r}'$. The function $\psi_0$ is called the "free space Green's function," and is given by

$$\psi_0 = \frac{e^{ik_0|\mathbf{r}-\mathbf{r}'|}}{|\mathbf{r} - \mathbf{r}'|} \tag{2.2-21}$$

Equations equivalent to Eqs. (2.2-19) and (2.2-20), but valid over any surface $S$ either closed or open, are

$$\mathbf{E}^s(\mathbf{r}) = \frac{1}{4\pi} \nabla \times \int_s (\hat{\mathbf{n}}' \times \mathbf{E}^T)\,\psi_0\,ds'$$

$$- \frac{1}{4\pi i\omega\epsilon} [\nabla(\nabla\cdot) + k_0^2] \int_s (\hat{\mathbf{n}}' \times \mathbf{H}^T)\,\psi_0\,ds' \tag{2.2-22}$$

---

[†] Equations (2.2-19) and (2.2-20) for the electromagnetic fields are sometimes known as the Stratton–Chu integrals. They differ in sign from those derived by Stratton in Section 8.14 of Reference 37 because Stratton defines his $\hat{\mathbf{n}}$ opposite that used here; the direction of the normal vector chosen here follows accepted convention for scattering problems.

and

$$\mathbf{H}^s(\mathbf{r}) = \frac{1}{4\pi} \nabla \times \int_s (\hat{\mathbf{n}}' \times \mathbf{H}^T) \, \psi_0 \, ds'$$

$$+ \frac{1}{4\pi i \omega \mu} [\nabla(\nabla \cdot) + k_0^2] \int_s (\hat{\mathbf{n}}' \times \mathbf{E}^T) \, \psi_0 \, ds' \qquad (2.2\text{-}23)$$

In most scattering problems, both the scattered fields and the fields on the surface are unknown so that Eqs. (2.2-19) and (2.2-20) cannot be easily used. They can be converted into integral equations in one unknown vector field through the use of Maxwell's equations, or for $r > r'$,

$$\mathbf{E}^T(\mathbf{r}) = \mathbf{E}^i(\mathbf{r}) + \frac{1}{4\pi} \int_s [(\hat{\mathbf{n}}' \times \mathbf{E}^T) \times \nabla'\psi_0$$

$$+ (\hat{\mathbf{n}}' \times \nabla' \times \mathbf{E}^T) \, \psi_0 + (\hat{\mathbf{n}}' \cdot \mathbf{E}^T) \, \nabla'\psi_0] \, ds' \qquad (2.2\text{-}24)$$

and

$$\mathbf{H}^T(\mathbf{r}) = \mathbf{H}^i(\mathbf{r}) + \frac{1}{4\pi} \int_s [(\hat{\mathbf{n}}' \times \mathbf{H}^T) \times \nabla'\psi_0$$

$$+ (\hat{\mathbf{n}}' \times \nabla' \times \mathbf{H}^T) \, \psi_0 + (\hat{\mathbf{n}}' \cdot \mathbf{H}^T) \, \nabla'\psi_0] \, ds' \qquad (2.2\text{-}25)$$

Each of these vector equations represents three coupled integro-differential equations for the three components of the vector field.

In the case of a perfectly conducting scatterer, the preceding equations simplify considerably since $\hat{\mathbf{n}}' \times \mathbf{E}^T = \hat{\mathbf{n}}' \cdot \mathbf{H}^T = 0$, thus

$$\mathbf{E}^T(\mathbf{r}) = \mathbf{E}^i(\mathbf{r}) + \frac{1}{4\pi} \int_s [(\hat{\mathbf{n}}' \times \nabla' \times \mathbf{E}^T) \, \psi_0 + (\hat{\mathbf{n}}' \cdot \mathbf{E}^T) \, \nabla'\psi_0] \, ds' \qquad (2.2\text{-}26)$$

and

$$\mathbf{H}^T(\mathbf{r}) = \mathbf{H}^i(\mathbf{r}) + \frac{1}{4\pi} \int_s [(\hat{\mathbf{n}}' \times \mathbf{H}^T) \times \nabla'\psi_0] \, ds' \qquad (2.2\text{-}27)$$

The scattered field then becomes

$$\mathbf{E}^s(\mathbf{r}) = \frac{1}{4\pi} \int_s [i\omega\mu(\hat{\mathbf{n}}' \times \mathbf{H}^T) \, \psi_0 + (\hat{\mathbf{n}}' \cdot \mathbf{E}^T) \, \nabla'\psi_0] \, ds' \qquad (2.2\text{-}28)$$

and

$$\mathbf{H}^s(\mathbf{r}) = \frac{1}{4\pi} \int_s [(\hat{\mathbf{n}}' \times \mathbf{H}^T) \times \nabla'\psi_0] \, ds' \qquad (2.2\text{-}29)$$

The above equations are useful for computing the scattered fields when the surface fields are known; however, the requirement that $r > r'$ complicates matters when attempts are made to utilize these as integral equations for computing the surface fields. In the event that $r = r'$, the Green's function becomes singular and this must be taken into account. By carefully accounting for this singularity, integral equations have been obtained for the fields on the surface of the scatterer. For a perfectly conducting scatterer, the surface current, $\mathbf{K} = \hat{n} \times \mathbf{H}^T$, can be obtained from the vector form of Maue's integral equation,[39] or

$$\mathbf{K}(\mathbf{r}) = 2[\hat{n} \times \mathbf{H}^i(\mathbf{r})] + \frac{1}{2\pi}\left[\hat{n} \times \int_s (\mathbf{K}(\mathbf{r}') \times \nabla'\psi_0)\, ds'\right] \quad (2.2\text{-}30)$$

In terms of the magnetic field at the surface Eq. (2.2-30) becomes

$$\mathbf{H}^T(\mathbf{r}) = 2\mathbf{H}^i(\mathbf{r}) + \frac{1}{2\pi}\int_s [(\hat{n}' \times \mathbf{H}^T) \times \nabla'\psi_0]\, ds' \quad (2.2\text{-}31)$$

Another integral equation for the surface current can be obtained by starting with Eq. (2.2-24), giving

$$\mathbf{E}^s(\mathbf{r}) - \frac{1}{4\pi}\int_s [(\hat{n}' \times \nabla' \times \mathbf{E}^T)\, \psi_0 + (\hat{n}' \cdot \mathbf{E}^T)\, \nabla'\psi_0]\, ds' = 0 \quad (2.2\text{-}32)$$

or

$$\hat{n} \times \mathbf{E}^i(\mathbf{r}) = \frac{-i}{4\pi}\left(\hat{n} \times \int_s \left[\omega\mu_0\mathbf{K}(\mathbf{r}')\,\psi_0 + \frac{\nabla' \cdot \mathbf{K}(\mathbf{r}')}{\omega\epsilon_0}(\nabla'\psi_0)\right] ds'\right) \quad (2.2\text{-}33)$$

For a dielectric scatterer, integral equations can be written for the equivalent electric and magnetic surface currents defined by†

$$\mathbf{K}_e = \hat{n} \times \mathbf{H}^T, \qquad \mathbf{K}_m = -\hat{n} \times \mathbf{E}^T \quad (2.2\text{-}34)$$

Letting the scatterer be homogeneous of material $\epsilon$, $\mu$ where these may be complex, then on the scatterer surface,[40]

$$\mathbf{K}_e(\mathbf{r}) = \left[\frac{2\mu_0}{\mu + \mu_0}\right][\hat{n} \times \mathbf{H}^i(\mathbf{r})] - \frac{1}{2\pi(\mu + \mu_0)}\hat{n} \times \int_s [\mathbf{K}_e(\mathbf{r}')$$

$$\times \nabla(\mu_0\psi_0 - \mu\psi)]\, ds' - \frac{i}{2\pi\omega(\mu + \mu_0)}\hat{n}$$

$$\times \int_s [\mathbf{K}_m(\mathbf{r}')(k_0^2\psi_0 - k^2\psi) + \mathbf{K}_m(\mathbf{r}')(\nabla \cdot \nabla)(\psi_0 - \psi)]\, ds' \quad (2.2\text{-}35)$$

and

$$\mathbf{K}_m(\mathbf{r}) = -\left[\frac{2\epsilon_0}{\epsilon + \epsilon_0}\right][\hat{n} \times \mathbf{E}^i(\mathbf{r})] - \frac{1}{2\pi(\epsilon + \epsilon_0)}\hat{n} \times \int_s [\mathbf{K}_m(\mathbf{r}')$$

$$\times \nabla(\epsilon_0\psi_0 - \epsilon\psi)]\, ds' + \frac{i}{2\pi\omega(\epsilon + \epsilon_0)}\hat{n}$$

$$\times \int_s [\mathbf{K}_e(\mathbf{r}')(k_0^2\psi_0 - k^2\psi) + \mathbf{K}_e(\mathbf{r}')(\nabla \cdot \nabla)(\psi_0 - \psi)]\, ds' \quad (2.2\text{-}36)$$

† Since magnetic charge has never been observed and its existence is doubtful, magnetic surface currents presumably do not exist and their introduction here is merely a mathematical convenience.

In these equations the operator $\nabla$ is taken to operate on the unprimed coordinate, and $\psi$ is defined as

$$\psi = \frac{e^{ik|\mathbf{r}-\mathbf{r}'|}}{|\mathbf{r}-\mathbf{r}'|} \tag{2.2-37}$$

with $k = \omega\sqrt{\mu\epsilon}$ the wave number of the dielectric scatterer.

In the case of dielectric scatterers, another integral equation can be formulated in terms of a volume integral over equivalent sources. However, this type of formulation is often not too useful because of the three-dimensional integrations required rather than the two-dimensional surface integrals indicated.

In general, it is much more difficult to solve an integral equation than a differential equation, so that attempts to solve these integral equations exactly are rarely made. They are, however, extremely useful as starting points for various approximations; for example, many numerical techniques are based on the integral equations for the surface currents on a perfect conductor. Another approximation is known as the Born approximation in quantum mechanics,[41] where for a general inhomogeneous Fredholm integral equation of the form

$$f(x) = g(x) + \lambda \int_a^b f(x)\, G(x, x')\, dx'$$

with $g(x)$ and $G(x, x')$ known functions, the Born approximation to $f(x)$ is given by

$$f^0(x) = g(x) + \lambda \int_a^b g(x)\, G(x, x')\, dx$$

Higher order approximations can be generated by continued iteration of this type. When the Born approximation is applied to Eqs. (2.2-24) and (2.2-25), with the unknown total field under the integral replaced by the known incident field, the result is related to the physical optics approximation discussed in Section 2.2.2.3. When the Born approximation is applied to other integral equation formulations, the physical interpretation of the results may be different. For example, low-frequency or weak scatterer approximations may result rather than the high-frequency approximation represented by physical optics.

Many integral equations other than those indicated here have been used by various researchers for scattering problems. In a given problem, special circumstances may allow an integral equation formulation other than those given. Often these are scalar integral equations, and as a result are considerably easier to deal with than the vector equations.

### 2.2.1.3. Integral Transform and Function Theoretical Techniques

In addition to boundary value and integral equation techniques, there exists a third class of methods for obtaining formally exact solutions to scattering problems. These, designated here as integral transform or function theoretical techniques, have in common the property that integral transforms, and/or the theory of analytic continuation from the theory of analytic functions of a complex variable, are used in formulating the solution.[42] Although formally exact solutions can be obtained by these methods, various approximations are required to obtain numerical results, as in the case of the boundary value or integral equation approaches previously discussed.

The first example of the use of such a technique to solve a scattering problem was Sommerfeld's solution for the fields diffracted by a perfectly conducting half-plane.[43] Although Sommerfeld's specific technique has not been successfully applied to any other problem, the general function theoretical approach can be applied to many shapes. For example, the book by Noble[44] gives a variety of applications for a function theoretic version of the Wiener–Hopf technique developed by Jones.[45]

Integral transforms have also been successfully applied to a number of scattering problems; for example, exact solutions for the field scattered by a perfectly conducting infinite cone due to an incident plane wave can be obtained by the use of the Kantorovich–Lebedev transform.[45]

Due to the great diversity of these methods, and the fact that the exact problems which are solvable with these methods, at least to date, are also solvable by separation of variables or integral equation approaches, no further discussion will be given of function theoretic or integral transform techniques.

### 2.2.2. Approximation Techniques

In the previous section the exact formulation of scattering problems was discussed. Unfortunately, as yet very few problems have an exact solution and most of the exact solutions must be evaluated approximately in order to obtain numerical answers. This section reviews the major techniques for the solution of scattering problems in an approximate manner. Approximate formulations, often based on some physical assumption, have the advantages that they are in general easier to apply than the exact formulations, and perhaps most importantly, they provide numerical answers.

#### 2.2.2.1. Geometrical Optics

When the radar target is large with respect to the observing wavelength, specific local portions of the target are essentially the major contributors to the radar cross section. These contributors are points of specular reflection (a specular point is a point on the surface where the normal to the surface bisects the angle between the incident and scattered field directions), shadow boundaries, and edges. The geometrical optics approximation assumes that the radar energy is propagated mainly along special trajectories known as ray paths. These ray paths are not, in general, straight lines. They are governed by Fermat's principle[46] which states that the ray path is one for which the optical path length is stationary, that is,

$$\delta \int_a^b m \, ds = 0 \qquad (2.2\text{-}38)$$

where $\delta$ signifies variation, $m$ is the index of refraction, $ds$ is a differential line element, and $a$ and $b$ are points on the trajectory. The classical geometrical optics approximation[47] neglects polarization and phase.

One application of Fermat's principle involves reflection from a perfectly reflecting curved surface located in a homogeneous medium. The ray bundle incident upon the surface is assumed to be parallel, corresponding to a plane wave. Since the propagation medium is homogeneous, both the incident and reflected rays travel in straight lines. The incident bundle of parallel rays of energy density $W_0$ strike the surface and are locally reflected specularly, according to Fermat's principle. If $\rho_1$ and $\rho_2$ are the principal radii of curvature of the surface at a given point, the curvature of the phase front of the reflected field at that point can be shown to have principal radii of curvature of $\rho_1/2$ and $\rho_2/2$. This arises from purely geometrical considerations.

The reflected tube of rays has the geometry shown in Figure 2-8. The energy crossing an element of the phase front at the surface is $W_0 \Delta S_1 \Delta S_2$, where $\Delta S_1$ and $\Delta S_2$ are the sides of the phase front element. The reflected field in this tube of rays now appears to come from two caustics, labeled 1–2 and 3–4 in the figure. Far down the tube, the same rays passing through $\Delta S_1 \Delta S_2$ will pass through the surface element of sides $\Delta S_{1r}$, $\Delta S_{2r}$ with energy density $W_r$. Due to conservation of energy,

$$W_0 \Delta S_1 \Delta S_2 = W_r \Delta S_{1r} \Delta S_{2r} \qquad (2.2\text{-}39)$$

If the surface element $\Delta S_{1r} \Delta S_{2r}$ is a large distance $R$ from $\Delta S_1 \Delta S_2$ (not nearby as illustrated in the figure), then $R$ is also the approximate distance from $\Delta S_{1r} \Delta S_{2r}$ to both caustics. Therefore,

$$\Delta S_{1r} = R \Delta \theta_1 = 2 \frac{R}{\rho_1} \Delta S_1$$

and

$$\Delta S_{2r} = R \Delta \theta_2 = 2 \frac{R}{\rho_2} \Delta S_2$$

Substituting these results back into Eq. (2.2-39) one obtains

$$\frac{W_r}{W_0} = \frac{\rho_1 \rho_2}{4R^2}$$

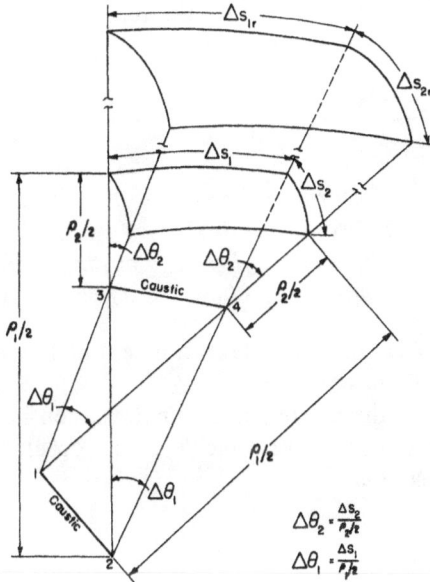

Fig. 2-8. Geometry of a ray bundle reflected geometrically from a perfectly conducting convex surface.

The backscatter cross section is then

$$\sigma(0) = 4\pi R^2 \frac{W_r}{W_0} = \pi \rho_1 \rho_2 \qquad (2.2\text{-}40)$$

where $\rho_1$, $\rho_2$ are the principle radii of curvature of the body surface at the reflection point. This, an often quoted result, is valid only for a second-degree surface, that is, a surface expressable in terms of a second-degree polynomial. A more general result is[48]

$$\sigma(0) = \pi \frac{ds}{d\Omega} \qquad (2.2\text{-}41)$$

where $ds$ is an element of area on the surface of the scatterer; and $d\Omega$ is the solid angle enclosing the normal to the surface, and is equal to $\sin\theta\, d\theta\, d\phi$ if the coordinate system of Figure 2-1 is oriented so that the normal to the surface is located along the z-axis.

Equation (2.2-40) is deceptively simple in that it may be quite difficult to evaluate $\rho_1$ and $\rho_2$. Initially, it is necessary to locate the specular points. The normal to the body must be determined, and all points where the unit normal coincides with the direction of propagation must be located. At all these points, $\rho_1$ and $\rho_2$ must be evaluated. If the equation of the surface is given in the form $f(x, y, z) = 0$, then the product of $\rho_1$ and $\rho_2$ at the specular point is given by the following expression:

$$\rho_1\rho_2 = \frac{[(\partial f/\partial x)^2 + (\partial f/\partial y)^2 + (\partial f/\partial z)^2]^2}{\varDelta}\Bigg|_{\text{specular point}} \quad (2.2\text{-}42)$$

where

$$\varDelta = - \begin{vmatrix} \partial^2 f/\partial x^2 & \partial^2 f/\partial x\,\partial y & \partial^2 f/\partial x\,\partial z & \partial f/\partial x \\ \partial^2 f/\partial x\,\partial y & \partial^2 f/\partial y^2 & \partial^2 f/\partial y\,\partial z & \partial f/\partial y \\ \partial^2 f/\partial x\,\partial z & \partial^2 f/\partial y\,\partial z & \partial^2 f/\partial z^2 & \partial f/\partial z \\ \partial f/\partial x & \partial f/\partial y & \partial f/\partial z & 0 \end{vmatrix} \quad (2.2\text{-}43)$$

Expressions for other coordinate systems can be found by the usual rules for coordinate transformations.

As an example of the use of these equations, consider the backscatter from an ellipsoid with a plane wave incident nose-on along the $x$-axis. The equation of the ellipsoid is

$$\frac{x^2}{a^2} + \frac{y^2}{b^2} + \frac{z^2}{c^2} = 1 \quad (2.2\text{-}44)$$

From Eq. (2.2-42)

$$\rho_1\rho_2 = -\frac{[4x^2/a^4 + 4y^2/b^4 + 4z^2/c^4]^2}{\begin{vmatrix} 2/a^2 & 0 & 0 & 2x/a^2 \\ 0 & 2/b^2 & 0 & 2y/b^2 \\ 0 & 0 & 2/c^2 & 2z/c^2 \\ 2x/a^2 & 2y/b^2 & 2z/c^2 & 0 \end{vmatrix}_{\substack{x=a\\y=0\\z=0}}} \quad (2.2\text{-}45)$$

and

$$\rho_1\rho_2 = \frac{b^2 c^2}{a^2} \quad (2.2\text{-}46)$$

It is thus seen that

$$\sigma(0) = \frac{\pi b^2 c^2}{a^2} \quad (2.2\text{-}47)$$

Neither Eq. (2.2-40) nor (2.2-41) is valid for a flat plate or for a cylinder, since the target must have sufficient curvature so that at least the first Fresnel zone is included on the surface. On the other hand, the equations are not valid if $\rho_1$ or $\rho_2 \ll 1$; thus the geometric optics approximation is not valid in the neighborhood of a caustic such as the tip of a cone or the edge of a wedge.

In order to overcome the lack of information as to polarization and phase, information on these is often added artificially to the geometric optics approximation. When this is done, the geometric optics approximation becomes identical to the first term of an asymptotic solution to Maxwell's equations introduced by Luneburg[49] and Kline.[50,51]

The Luneburg–Kline asymptotic expression[52,53] of the electric field for large $\omega$ splits the field into two factors. One, the amplitude factor, is a slowly varying factor which is represented by a power series in $\omega$, each term of which is slowly varying. The other is an exponential factor which is rapidly oscillating. The series representation is

$$\mathbf{E}(\mathbf{R}, \omega) = e^{ik_0\psi(\mathbf{R})} \sum_{n=0}^{\infty} \frac{\mathbf{E}_n(\mathbf{R})}{(i\omega)^n} \tag{2.2-48}$$

Substituting Eq. (2.2-48) into the wave equation $\nabla^2\mathbf{E} + k_0^2\mathbf{E} = 0$, and equating like powers of $\omega$, one obtains the eikonal equations[†]

$$|\nabla\psi| = m \tag{2.2-49}$$

$$\nabla\psi \cdot \nabla\mathbf{E}_0 + \tfrac{1}{2}(\nabla^2\psi)\,\mathbf{E}_0 = 0 \tag{2.2-50}$$

$$\nabla\psi \cdot \nabla\mathbf{E}_n + \tfrac{1}{2}(\nabla^2\psi)\,\mathbf{E}_n = \tfrac{1}{2}\nabla^2\mathbf{E}_{n-1} \tag{2.2-51}$$

where

$$n = 1, 2, 3,...$$

Equation (2.2-49) implies that traveling waves with an exponential factor $e^{ik_0(\psi-ct)}$ propagate along ray trajectories perpendicular to the wave fronts $\psi = $ constant, and that the curvatures of these trajectories and the wave fronts are determined by the local variations of the refractive index $m$. However, it should be noted that in anisotropic media, ray paths are not generally perpendicular to the wave fronts. Also, it might be noted that Eq. (2.2-38) can be derived from Eq. (2.2-49) by the application of the principle of least action.

In the high-frequency limit, only the $n = 0$ term in Eq. (2.2-48) is considered. Equation (2.2-50) can be integrated and the result combined with Eq. (2.2-48) to give

$$\mathbf{E}(\mathbf{R}) \approx \mathbf{E}_0(\mathbf{R}_0)\exp(ik_0\psi_0 + k_0\,|\,\mathbf{R} - \mathbf{R}_0\,|)\exp\left(-\frac{1}{2}\int_{\mathbf{R}_0}^{\mathbf{R}} \nabla^2\psi\,dl'\right) \tag{2.2-52}$$

where $\mathbf{R}_0$ is a reference point at which the field is known. In the above line integral, the integration is along the ray path from the reference point, $\mathbf{R}_0$, to the field point $\mathbf{R}$. In a homogeneous medium, this path is a straight line. The term $\exp(ik_0\psi_0)$ could be absorbed into $\mathbf{E}_0(\mathbf{R}_0)$, if desired. For isotropic homogeneous media

$$\exp\left[-\int_{\mathbf{R}_0}^{\mathbf{R}} \nabla^2\psi\,dl'\right] = \frac{G(\mathbf{R})}{G(\mathbf{R}_0)} \tag{2.2-53}$$

where $G$ is the Gaussian curvature of the wavefront and is equal to $1/\rho_1\rho_2$, where $\rho_1$ and $\rho_2$ are the principal radii of curvature of the wave-front surface. Substituting Eq. (2.2-53) into Eq. (2.2-52), one obtains

$$\mathbf{E}(\mathbf{R}) \approx \mathbf{E}_0(\mathbf{R}_0)[G(\mathbf{R})/G(\mathbf{R}_0)]^{1/2}\exp(ik_0\psi_0 + ik_0\,|\,\mathbf{R} - \mathbf{R}_0\,|) \tag{2.2-54}$$

[†] In these equations, $\nabla\mathbf{E}$ is interpreted as a dyad.

The presence in Eq. (2.2-54) of the Gaussian curvature signifies that the asymptotic form of the Luneberg–Kline series has the same limitations as the classical geometrical optics formulation (that is, $\rho_1$, $\rho_2 \neq 0$ or $\infty$). Now, however, phase and polarization information are included, and there is the possibility that determination of higher terms might enhance the accuracy at lower frequencies.

The Luneberg–Kline series fails to account for diffracted fields from edges, vertices, corners, or shadow boundaries. Caustics of the diffracted field are located at shadow boundaries. Overcoming these difficulties was a purpose of the development of the geometrical theory of diffraction, discussed in the following section.

Before turning away from geometrical optics, two other methods for deriving geometrical optics results should be mentioned.[54] Instead of expressing the fields in differential equation form, they can be expressed in integral equation form which may then be evaluated by saddlepoint or stationary phase approximations. This method is very illustrative for the physical understanding of ray paths, shows very clearly the significance of Fresnel zones, and may have application to many problems involving caustics.

The second method uses the elegant theory of characteristics to investigate wave fronts. Moving wave fronts associated with the propagation of field disturbances are characteristic hypersurfaces, three-dimensional domains in a four-dimensional space-time. The magnitude of the field discontinuities on these wave fronts is proportional to the geometrical–optical amplitude of monochromatic waves for which the successive portions of these moving wave fronts then constitute steady wave fronts. The rigorous derivation of the equations requires integration of Maxwell's equations over four-dimensional space-time. An introduction to this technique is given by Bremmer,[54] and a more complete treatment is given by Kline.[50]

### 2.2.2.2. Geometrical Theory of Diffraction

As discussed in the previous section, classical geometrical optics or ray-tracing techniques do not apply to edges, tips, corners, tangent points, or shadow regions. Keller[55,56] has extended geometrical optics to include a class of rays called diffracted rays. Such an extension is not unique to Keller, but his work has given practical results and constitutes what is generally known as the geometrical theory of diffraction. Consider two points, $P$ and $Q$. All direct trajectories from $P$ to $Q$ satisfying Eq. (2.2-38) make up the class of ray paths denoted $D_{000}$. The meaning of the "zero" subscripts will become clear shortly. Now, if $P$ and $Q$ lie near an opaque surface, rays can propagate from $P$ to $Q$ along the surface, or via reflection off the surface. If the object causing the reflection is penetrable, the rays can also be refracted before arriving at $Q$. All these ray paths from $P$ to $Q$ via the surface of discontinuity satisfying Eq. (2.2-38) constitute the class $D_{r00}$, where $r$ represents the number of smooth arcs and reflection points on the discontinuity surface; or in the case of refracted rays, the number of equivalent reflection points.

In order to include all the diffracted rays, one must consider a triple of nonnegative integers $r$, $s$, and $t$, where $r$ has the same meaning as before, $s$ is the number assigned to each pertinent point on the edge(s) of the discontinuity surface, and $t$ is the number assigned to points on the vertices of these surfaces. The class $D_{rst}$ consists of the ray paths through all triples, where $r$, $s$, $t$ run from zero to as many points as can be identified, or perhaps can be economically used in the calculation. Since an infinite number of rays are diffracted from each point, one, in general, considers volumes of the ray tubes and applies conservation of energy. The class $D_{rst}$ includes $D_{r00}$ and $D_{000}$ as special cases. In Keller's geometrical diffraction theory, an edge is defined to be a line on a surface along which the first or second or higher derivatives of the surface are discontinuous. Discontinuities in any derivatives of the edges must be counted as vertices. Isolated points not on edges, at which first or higher order derivatives of a surface are discontinuous, are also vertices, for example, the apex of a cone. Now the questions are: What is the behavior of the amplitude and phase of the field along these trajectories? And, is the class $D_{rst}$ sufficient to explain all diffraction behavior?

In order to consider the second question first, the types of diffracted rays are illustrated in Figure 2-9. Figure 2-9a shows the plane of diffracted rays produced by a ray normally incident on the edge of a wedge. Figure 2-9b considers oblique incidence on the same wedge. The diffracted rays in this case form a cone. Figure 2-9c gives the example of tip diffraction from a corner. The same picture would be valid for diffraction from the point of a cone. Figure 2-9d shows multiple diffraction, in this case, from the two edges of an aperture. The same type of behavior would also be true for composite structures with two or more edges. The behavior of rays which are tangent to the surface of the radar target for any incidence is illustrated in Figure 2-9e. A creeping ray travels from the tangent point around the surface of the body with some of the energy radiating off along lines tangent to the surface. The creeping ray continues around the body indefinitely, but the energy is soon negligible due to the tangential radiation. The radiation from creeping waves is an important topic in predicting radar cross sections. Besides the discussion of this which follows here, further discussion in connection with Fock's technique for predicting the fields near the shadow boundary and in the shadow region appears in Section 2.2.2.6. A general discussion of creeping waves from the viewpoint of asymptotic expressions of exact solutions is given in Section 2.3. In addition, discussions on creeping wave contributions to individual body cross sections are given in their separate chapters; note, especially, spheres in Chapter 3 and cone spheres in Chapters 6 and 8. Figure

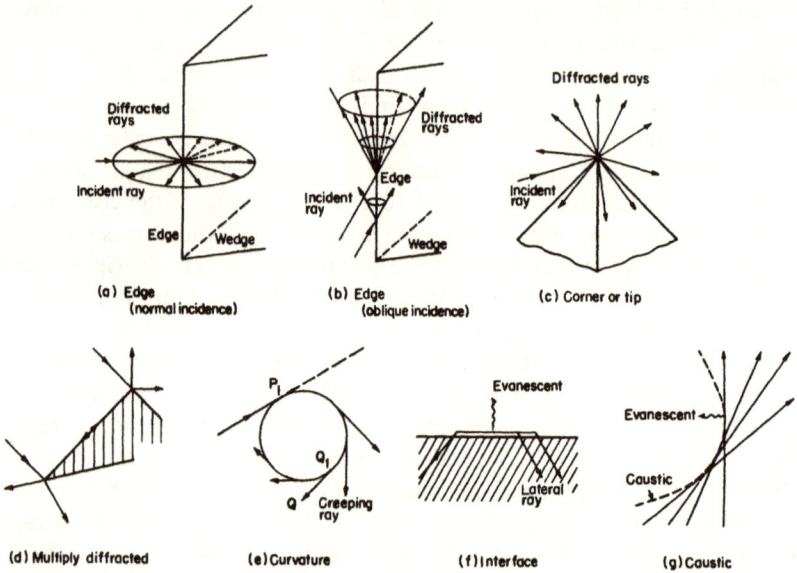

Fig. 2-9.   Types of diffracted rays considered in the geometrical theory of diffraction.

2-9f illustrates the lateral ray at an interface which is discussed further in Chapter 7, and Figure 2-9g illustrates the region shadowed by a caustic.

Not even the diffracted rays penetrate into the regions shadowed by a caustic or by a dielectric interface at angles beyond the Brewster angle. In order to account for the electromagnetic field in these regions, as illustrated in Figures 2-9f and 2-9g, Keller[55,56] has suggested a further extension of geometrical optics. The set of Eqs. (2.2-49) to (2.2-51) can have complex, as well as real solutions. These solutions are complex rays which have, in general, only one point in real space at which a physically meaningful field exists. However, each point in the shadow region is an intersection of some complex ray with real space. The field in the shadow region is proportional to $\exp(-k_0 \, \mathrm{Im} \, \psi) \exp(ik_0 \, \mathrm{Re} \, \psi)$, with $\psi$ having a positive imaginary part. Thus, the field is of higher order than in any region where it is associated with real rays.

The phase of a diffracted ray, real or complex, is determined by the optical distance which the ray has traveled from the phase reference point. The amplitude varies along the ray in a medium, as it does in classical geometrical optics. The initial value of the field on a diffracted ray is determined by use of diffraction coefficients.

In free space, the field strength $u$ in a reflected wave at a distance $R$ from the point of reflection is of the form

$$u = \Gamma \sqrt{\rho_1' \rho_2' / (\rho_1' + R)(\rho_2' + R)} \, u^i \exp(ik_0 R) \qquad (2.2\text{-}55)$$

where $\Gamma$ is the reflection coefficient of the surface, $u^i$ is incident field strength at the surface, and $\rho_1'$ and $\rho_2'$ are the principal radii of curvature of the reflected wavefront ($\frac{1}{2}$ the principal radii of curvature of the reflecting surface for normal incidence and monostatic reflection). As $\rho_2'$ becomes very small the surface has an edge. The intensity of the scattered field remains finite, and Keller modifies Eq. (2.2-55) to the form

$$u_d = D \sqrt{\rho_1'/R(\rho_1' + R)}\, u^i \exp(ik_0 R) \qquad (2.2\text{-}56)$$

where $u_d$ is the diffracted field and $D$ is the diffraction coefficient. The problem then is to find $D$. Keller[57,58] finds values for $D$ by matching his theory to the known solutions for scattering from a two-dimensional wedge and from a half-plane. The scattering from a wedge is discussed in Sections 2.3.1.1.2 and 6.5. As is discussed there, the diffraction coefficients are

$$\left.\begin{matrix} D_\perp \\ D_\parallel \end{matrix}\right\} = \frac{\exp(i\pi/4)\sin(\pi/n)}{n\sqrt{2\pi k_0}} \left\{ \left[\cos\frac{\pi}{n} - 1\right]^{-1} \pm \left[\cos\frac{\pi}{n} - \cos\frac{2\phi}{n}\right]^{-1} \right\}$$

with

$$n = \frac{\gamma}{\pi} \qquad (2.2\text{-}57)$$

In Eq. (2.2-57), the upper sign refers to $D_\perp$, the lower sign refers to $D_\parallel$, $\phi$ is the angle between the projection of the incident ray in a plane normal to the wedge apex line and the face of the wedge, and $\gamma$ is the external angle of the wedge. For the equation to hold, $\phi \neq \pi/2$. If $\gamma = 2\pi$, the wedge is a half-plane and Eq. (2.2-57) reduces to

$$\left.\begin{matrix} D_\perp \\ D_\parallel \end{matrix}\right\} = -\frac{\exp(i\pi/4)}{2\sqrt{2\pi k_0}}(1 \mp \sec\phi) \qquad (2.2\text{-}58)$$

where, again, $\phi \neq \pi/2$. The case of $\phi = \pi/2$, or normal incidence upon the face of the wedge, is discussed in Section 6.5.

For a corner or apex, both $\rho_1'$ and $\rho_2'$ are small. In this case, Eq. (2.2-55) is modified to

$$u = [D_t u^i \exp(ikR)]/R \qquad (2.2\text{-}59)$$

where $D_t$ is the coefficient of tip diffraction. As discussed in Chapter 6, the coefficient of tip diffraction is not well known, especially for directions of incidence far from the axis of the cone. Using Felsen's[59] results, discussed in more detail in Section 6.2.1.2, one obtains for a narrow angle cone with half angle $\alpha \ll \pi/2$, and the angle of incidence $\theta$, $0 < \theta \leqslant \alpha$

$$D_{t\perp} = \frac{i}{k_0}\frac{(3 + \cos^2\theta)}{4\cos^2\theta}\left(\frac{\alpha}{2}\right)^2 \qquad (2.2\text{-}60)$$

The physical optics approximation (the general method of which is discussed in the following section), not expected to be valid near edges or tips, amazingly enough, agrees quite well with Felsen's result, giving

$$D_{t\perp} = \frac{i \tan^2 \alpha}{4k_0 \cos^3 \theta (1 - \tan^2 \alpha \tan^2 \theta)^{3/2}} \qquad (2.2\text{-}61)$$

for $| \theta | < \alpha$. Equation (2.2-61) when $\theta = 0$ reduces to

$$D_{t\perp} = D_{t\parallel} = \frac{i \tan^2 \alpha}{4k_0} \qquad (2.2\text{-}62)$$

The surprisingly good agreement of the physical optics approximation with more exact results for the case of scattering from a conical tip serves as a justification for applying this approximation to a plane corner with two straight edges meeting at an angle. Keller *et al.*[58] obtain a diffraction coefficient for this case of

$$C_\perp = \frac{i}{4\pi k_0} \frac{(\cos \theta + \cos \theta') \sin \delta}{(\cos a - \cos a')(\cos b - \cos b')} \qquad (2.2\text{-}63)$$

where $a$ and $b$ are the angles between the incident field and the two edges at the corner; $a'$ and $b'$ are the angles between the diffraction field and the two edges at the corner; $\theta$ and $\theta'$ are the angles between the normal to the plane of the corner and the incident and diffracted rays, respectively; and $\delta$ is the angle between the two edges at the corner. It should be emphasized that there is no check on the validity of this equation. It is seen that the edge diffraction coefficients are proportional to $\lambda_0^{1/2}$, and that the tip or corner diffraction coefficients are proportional to $\lambda_0$. Further discussions of diffraction from edges, tips, and wedges are available in Section 2.3 and Chapter 6.

For a ray which strikes a smooth object tangentially, excites a creeping wave (see Section 2.3.2), and radiates tangentially off the body, Keller[60] and Levy and Keller [61] obtain

$$u_d(Q) = u^i(P_1) \exp\{ik_0[\phi^i(P_1) + t + s]\} \left[\frac{dw_f(P_1)}{dw_f(Q_1)}\right]^{1/2} \left[\frac{\rho_1}{s(\rho_1 + s)}\right]^{1/2}$$

$$\times \sum_m D_m(Q_1) D_m(P_1) \exp \left\{-\int_0^t \alpha_m(t)\, dt\right\} \qquad (2.2\text{-}64)$$

In this equation:

$P_1$, $Q_1$, and $Q$ are points illustrated in Figure 2-9e;
$u^i(P_1)$ is the incident amplitude at $P_1$;
$\phi^i(P_1)$ is the incident phase at $P_1$ (there is no abrupt change in phase as a ray enters onto or leaves the body);
$t$ is the geodesic arc length from $P_1$ to $Q_1$;

$s$ is the straight line distance from $Q_1$ to $Q$;

$dw_f(P_1)/dw_f(Q_1)$ is the limit of the ratio of wave-front width at $P_1$ to wave-front width at $Q_1$ of a strip of rays on the surface as the width of the strip approaches zero;

$\rho_1$ is the nonzero principal radius of curvature of the wave front at $Q_1$;

$D_m(P_1)$ is the diffraction coefficient or proportionality constant between the diffracted and incident amplitudes at $P_1$ for the $m$th mode;

$D_m(Q_1)$ is the proportionality constant between the amplitude off the body and the amplitude on the body at $Q_1$ for the $m$th mode;

$\alpha_m$ is a decay coefficient which accounts for rays leaving the body tangentially at each point of the geodesic arc.

The diffraction coefficient $D_m(P_1)$ and $D_m(Q_1)$ must have the same functional dependence to insure reciprocity. The different modes are introduced to insure agreement with asymptotic expansions of exact solutions. $D_m$ and $\alpha(t)$ are assumed to depend only on the incident field properties, on $k_0$, and on the local properties of the body at the point at which they are evaluated. Keller evaluates $D_m$ and $\alpha(t)$ for an infinite circular cylinder illuminated by an external line source parallel to the cylinder. The results are then said to be applicable to any smooth convex body. He proves[62] that for a perfectly conducting body, which is rotationally symmetric around an axis in the direction of propagation of the incident plane wave, the axial backscattered field $\mathbf{E}_s$, is given by

$$\mathbf{E}_s = \tfrac{1}{2}(u_s - u_h)\,\mathbf{E}_i \qquad (2.2\text{-}65)$$

where $u_s$ and $u_h$ are geometrically constructed scalar fields. The scalar field, $u_s$, satisfies the acoustically soft boundary condition, $u = 0$, on the body surface, and $u_h$ satisfies the acoustically hard boundary condition, $\partial u/\partial n = 0$, on the body surface, where $n$ is the normal to the body.

For the soft boundary condition on a circular cylinder

$$\alpha_m = \exp(-i\pi/6)(k_0/6a^2)^{1/3}\,q_m \qquad (2.2\text{-}66)$$

and

$$D_m = (\pi/6)\exp(i\pi/12)(2\pi/k_0)^{1/2}\,(k_0a/6)^{1/3}\,[A'(q_m)]^{-2} \qquad (2.2\text{-}67)$$

where $a$ is the radius of curvature and $A(x)$ is related to the Airy function, $\mathrm{Ai}(y)$, with

$$A(x) = \int_0^\infty \cos(\tau^3 - x\tau)\,d\tau = \pi 3^{-1/3}\mathrm{Ai}[(-3)^{-1/3}x] \qquad (2.2\text{-}68)$$

The parameter $q_m$ is the $m$th zero of $A(q)$.

For the hard boundary condition on a circular cylinder,

$$\alpha_m = \exp(-i\pi/6)[k_0/6a^2]^{1/3}q_m' \qquad (2.2\text{-}69)$$

and

$$D_m = (\pi/2) \exp\left(i\pi/12\right)(2\pi/k_0)^{1/2}(k_0 a/6)^{1/3}\left[q_m' A^2(q_m')\right]^{-1} \qquad (2.2\text{-}70)$$

where $q_m'$ is determined by the expression $A'(q_m') = 0$. These expressions obtained from the scattered fields of an infinite cylinder were applied to the scattered fields from a sphere and good agreement was obtained with the asymptotic expansion of the exact solution.[61]†

It is seen that the field diffracted around a curved surface decreases exponentially with $\lambda_0$, and is consequently weaker in the high-frequency limit than the field diffracted by a tip, which is in turn weaker than that diffracted by an edge.

At a caustic or focus of the reflected or diffracted rays, conventional geometrical diffraction theory predicts an infinite amplitude for the field. This, of course, does not correspond to physical reality, so a caustic correction factor must be introduced in order to obtain the correct finite result. There are no general techniques—valid for an arbitrary body—for finding these caustic corrections. As is the case with the diffraction coefficients, the caustic correction terms are obtained from asymptotic expansions of the exact solutions for scattering from a known body such as a circular disk, sphere, etc. The manner in which this is accomplished is discussed in a number of papers[63-65].

In summary, the general validity of Keller's techniques has never been established; nevertheless, these techniques have given good results for a variety of objects.

### 2.2.2.3. Physical Optics and Stationary Phase

2.2.2.3.1. Simplification of the Chu–Stratton Integral Equation.   The name physical optics is often used synonymously with the terms "Kirchhoff approximation," "tangent plane approximation," and "Huygens' principle." This is largely due to a protracted historical evolution and individual preferences. Huygens' principle or the Kirchhoff approximation was first formulated long before Maxwell's equations showed the true vector nature of electromagnetic fields. Consequently, the original Kirchhoff approximation is based upon representation of light waves by a scalar rather than a vector field. Since in fact electromagnetic fields satisfy vector rather than scalar boundary conditions, the use of the scalar Kirchhoff approximation can be justified only

---

† Hong [Asymptotic Theory of Electromagnetic and Acoustic Diffraction by Smooth Convex Surfaces of Variable Curvature, *J. Math. Phys.* 8:1223 (1967)] has presented a general method for obtaining successive terms in the short-wavelength asymptotic expansions of the diffracted field produced by plane acoustic and electromagnetic waves incident on an arbitrary smooth convex surface. He solves the boundary-value problem for restricted regions of a general surface by an asymptotic method. The leading terms obtained agree with those of Levy and Keller[61] and Fock[74] (see Section 2.2.5). Hong has also mathematically justified some of the physical reasoning of the geometrical theory of diffraction.

in cases where it has been clearly shown to be in agreement with the vector formulation. A vector form of the physical optics method will be developed here for scattering from finite three-dimensional objects.[†]

It has been shown previously [Eqs. (2.2-19), (2.2-20)] that in source-free space, the total electric (or magnetic) field can be represented by an integral over a closed surface. This vector integral can be obtained using a vector analog of Green's theorem directly from Maxwell's equations; it is often called the Chu–Stratton integral. The left side represents the total vector field at the observation point or receiving antenna, and the field quantities on the right side are the total fields at the scatterer surface. These total electric and magnetic fields consist of both the incident and the scattered fields. The incident field is assumed to be time harmonic and a plane wave over the entire finite scattering object, and its source may thus be considered to be at infinity. Since the desired scattered field appears on both sides of the Chu–Stratton equation, one is faced with the solution of a vector integral equation. However, when certain simplifying assumptions are made, this formidable vector integral equation reduces to a simple definite integral over the scatterer surface. In the latter form it is known as the physical optics integral and represents a powerful and useful technique for calculating approximate scattering cross sections.

The assumptions necessary to simplify the Chu–Stratton integral equation are listed here and considered individually later.

(a) The application of the integral to a nonclosed surface or finite portion of a closed surface;

(b) Removal of the observation point (or receiving antenna) to a distance far from the object in terms of wavelength and scattering object dimensions;

(c) The tangent-plane approximation, or representation of the total fields at the surface in terms of the incident fields.

The first restriction arises when the scattering object is large (all dimensions perpendicular to the incident wave direction are large); in this case it is difficult to determine the total fields on the surface of the object not directly illuminated by the incident wave. Thus, the assumption is made that the total fields at the surface on the "shadowed side" of the object are zero. This means that at a line, $\Gamma$, called the shadow boundary, the fields at the surface are assumed to drop abruptly to zero as one moves from the illuminated side to the shadowed side (from $S_1$ to $S_2$ in Figure 2-10).[‡] This assumption is valid only when $\lambda_0$ (wavelength) is very small in comparison with the dimensions of the object.

---

[†] The retention of polarization information in the vector formulation is especially convenient for bistatic scattering and nonperfectly conducting targets. It circumvents the need for artificially introducing polarization *de facto* into the scalar formulation.

[‡] The curve $\Gamma$ is the locus of points such that $\hat{n} \cdot \hat{k}_0{}^i = 0$.

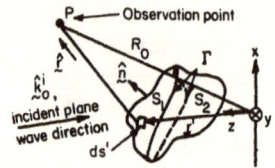

Fig. 2-10. Coordinates of the field point $P$, and points on the scatterer surface used in reduction of the Chu–Stratton equation.

For most scattering objects, this assumption implies that the fields suffer a discontinuity at the shadow boundary. The resulting Chu–Stratton integral over $S$, for which the integrand is now zero over $S_2$, becomes an integral over an open or finite surface, $S_1$. The presence of the edge discontinuity at $\Gamma$ violates one of the requirements in the application of Green's theorem, and the resulting field at the observation point no longer satisfies Maxwell's equations. This discontinuity, however, can be properly taken into account by the addition of a line integral term about the shadow boundary, $\Gamma$.[66] The resulting Chu–Stratton integral over the open surface, $S_1$, then becomes

$$\mathbf{E}^s(P) = + \frac{1}{4\pi} \int_{S_1} [i\omega\mu_0(\hat{\mathbf{n}} \times \mathbf{H}^T)\,\psi_0 + (\hat{\mathbf{n}} \times \mathbf{E}^T) \times \nabla'\psi_0 + (\hat{\mathbf{n}} \cdot \mathbf{E}^T)\nabla'\psi_0]\,ds'$$

$$+ \frac{1}{4\pi i \omega \epsilon_0} \oint_{\Gamma} \nabla'\psi_0 \mathbf{H}^T \cdot d\mathbf{l} \qquad\qquad (2.2\text{-}71)$$

$$\mathbf{H}^s(P) = - \frac{1}{4\pi} \oint_{S_1} [i\omega\epsilon_0(\hat{\mathbf{n}} \times \mathbf{E}^T)\,\psi_0 - (\hat{\mathbf{n}} \times \mathbf{H}^T) \times \nabla'\psi_0 - (\hat{\mathbf{n}} \cdot \mathbf{H}^T)\nabla'\psi_0]\,ds'$$

$$- \frac{1}{4\pi i \omega \mu_0} \oint_{\Gamma} (\nabla'\psi_0)\mathbf{E}^T \cdot d\mathbf{l} \qquad\qquad (2.2\text{-}72)$$

The symbols $\mathbf{E}^T$ and $\mathbf{H}^T$ represent the total electric and magnetic fields on the surface $S_1$; $\hat{\mathbf{n}}$ is the unit normal pointing out of $S_1$; $\psi_0$ is the free-space Green's function ($\psi_0 = \exp(ik_0\,|\,\mathbf{r}\,|)/|\,\mathbf{r}\,|$), where $\mathbf{r}$ is the vector distance from the surface area element $ds'$ to the observation point $P$; $\mathbf{E}^s(P)$ and $\mathbf{H}^s(P)$ are the scattered fields at $P$; $\mu_0$, $\epsilon_0$ are the permeability and permittivity of free space; and $\Gamma$ is the curve defined as the locus of all points where $(\hat{\mathbf{n}} \cdot \hat{\mathbf{k}}_0{}^i) = 0$.

The second assumption permits considerable simplification of the above equation. When the distance to the observation point from the object satisfies the criterion $r > 2D^2/\lambda_0$, one can approximate the unit vector $\hat{\mathbf{r}}$ from $ds'$ on the scatterer surface pointing toward $P$ by a constant vector (this distance requirement can be relaxed considerably when the surface of the object is curved; in that case, $D$ may be taken as the largest dimension of the first Fresnel zone on the scatterer surface.) This assumption permits one to evaluate $\nabla'\psi_0$ and to convert the line integral to a surface integral and simplify. The distance $r$ is considered a constant, $R_0$, when in the denominator. The resulting expressions now become[67]

$$\mathbf{E}^s(P) = \frac{+i\omega\mu_0 \exp(ikR_0)}{4\pi R_0} \int_{S_1} \left[ (\hat{\mathbf{n}} \times \mathbf{H}^T) - \hat{\mathbf{k}}_0{}^s \cdot (\hat{\mathbf{n}} \times \mathbf{H}^T) \, \hat{\mathbf{k}}_0{}^s \right.$$

$$\left. - \sqrt{\frac{\epsilon_0}{\mu_0}} (\hat{\mathbf{n}} \times \mathbf{E}^T) \times \hat{\mathbf{k}}_0{}^s \right] \exp(-ik_0 \hat{\mathbf{k}}_0{}^s \cdot \mathbf{r}') \, ds' \qquad (2.2\text{-}73)$$

$$\mathbf{H}^s(P) = \frac{-i\omega\epsilon_0 \exp(ik_0 R_0)}{4\pi R_0} \int_{S_1} \left[ (\hat{\mathbf{n}} \times \mathbf{E}^T) - \hat{\mathbf{k}}_0{}^s \cdot (\hat{\mathbf{n}} \times \mathbf{E}^T) \, \hat{\mathbf{k}}_0{}^s \right.$$

$$\left. + \sqrt{\frac{\mu_0}{\epsilon_0}} (\hat{\mathbf{n}} \times \mathbf{H}^T) \times \hat{\mathbf{k}}_0{}^s \right] \exp(-ik_0 \hat{\mathbf{k}}_0{}^s \cdot \mathbf{r}') \, ds' \qquad (2.2\text{-}74)$$

or, in terms of equivalent magnetic and electric surface currents defined in Eq. (2.2-34),

$$\mathbf{H}^s(P) = \frac{i\omega\epsilon_0 \exp(ikR_0)}{4\pi R_0} \int_{S_1} \left[ \mathbf{K}_m - (\hat{\mathbf{k}}_0{}^s \cdot \mathbf{K}_m) \, \hat{\mathbf{k}}_0{}^s \right.$$

$$\left. - \sqrt{\frac{\mu_0}{\epsilon_0}} \mathbf{K}_e \times \hat{\mathbf{k}}_0{}^s \right] \exp(-ik_0 \hat{\mathbf{k}}_0{}^s \cdot \mathbf{r}') \, ds' \qquad (2.2\text{-}75)$$

where $\mathbf{r}'$ is the distance from the coordinate origin to $ds'$ on the surface.

The preceding equations are valid in the far-zone limits previously given and when all surface dimensions perpendicular to the incidence direction are very large compared to wavelength. Edge effects are properly accounted for and $\mathbf{E}^s(P)$ and $\mathbf{H}^s(P)$ satisfy Maxwell's equations. It can be seen from the above integrands that there are no far-zone field components in the direction of scattering $\hat{\mathbf{k}}_0{}^s$; this is physically required of all far-zone scattered fields that satisfy Maxwell's equations.

2.2.2.3.2. *The Tangent-Plane Approximation.* In order to convert the above integral equations into definite integrals, one must relate the scattered field at the surface to the incident field. This is difficult in general, except when the surface is such that its radii of curvature at all points are much larger than a wavelength. When this restriction applies, one can approximate the surface at and near any point as an infinite plane tangent to the surface at that point. Then the relationship between incident and reflected fields at the surface is easily obtained from elementary considerations of plane wave reflection from an infinite plane. These relationships are given below.

(a) *Perfectly Conducting Surface*

When a plane wave is incident upon a perfectly conducting surface, the reflected wave is planar and its direction is specified by Snell's law (see Chapter 7). The relationships at the surface are

$$\hat{\mathbf{n}} \times \mathbf{E}^T = \hat{\mathbf{n}} \times (\mathbf{E}^i + \mathbf{E}^s) = 0 \qquad (2.2\text{-}76)$$

$$\hat{\mathbf{n}} \times \mathbf{H}^T = 2\hat{\mathbf{n}} \times \mathbf{H}^i \qquad (2.2\text{-}77)$$

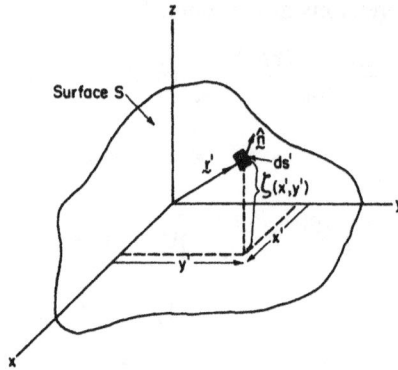

Fig. 2-11.   Coordinates used in describing
a scattering surface in terms of the surface
height $\zeta$.

Using these boundary conditions, the integrals (2.2-73), and (2.2-74) become

$$\mathbf{E}^s(P) = \frac{+i\omega\mu_0 \exp(ik_0R_0)}{2\pi R_0} \int_{S_1} (\hat{\mathbf{n}} \times \mathbf{H}^i)$$
$$- \hat{\mathbf{k}}_0{}^s \cdot (\mathbf{n} \times \mathbf{H}^i)\, \hat{\mathbf{k}}_0{}^s \exp(-ik_0\hat{\mathbf{k}}_0{}^s \cdot \mathbf{r})\, ds' \qquad (2.2\text{-}78)$$

and

$$\mathbf{H}^s(P) = \frac{-ik_0 \exp(ik_0R_0)}{2\pi R_0} \int_{S_1} (\hat{\mathbf{n}} \times \mathbf{H}^i) \times \hat{\mathbf{k}}_0{}^s \exp(-ik_0\hat{\mathbf{k}}_0{}^s \cdot \mathbf{r})\, ds' \quad (2.2\text{-}79)$$

Either of the above equations is valid for the scattered far field, but the second
is the more commonly used form. In many cases, one wishes to describe the
surface in terms of some $x$ and $y$ coordinates, as shown in Figure 2-11. Let the
surface height in the $z$ direction be $\zeta(x', y')$. The incident plane wave at $ds'$
on the surface has the form $\mathbf{H}^i = \mathbf{H}^0 \exp[ik_0{}^i \cdot \mathbf{r}]$, where $\mathbf{H}^0$ is a constant
vector perpendicular to the direction of incidence. The normal to the surface,
$\hat{\mathbf{n}}$, and the element of area, $ds'$, on the surface become

$$\hat{\mathbf{n}} = \frac{-(\partial\zeta/\partial x')\,\hat{\mathbf{x}}' - (\partial\zeta/\partial y')\,\hat{\mathbf{y}}' + \hat{\mathbf{z}}'}{\sqrt{1 + (\partial\zeta/\partial x')^2 + (\partial\zeta/\partial y')^2}},$$
$$ds' = dx'\,dy'\,\sqrt{1 + \left(\frac{\partial\zeta}{\partial x'}\right)^2 + \left(\frac{\partial\zeta}{\partial y'}\right)^2} \qquad (2.2\text{-}80)$$

and Eq. (2.2-79) can be written as

$$\mathbf{H}^s(P) = \frac{-ik_0 \exp[ik_0R_0]}{2\pi R_0}\,\hat{\mathbf{k}}_0{}^s$$
$$\times \left[ \mathbf{H}_0 \times \iint_{S_1} \left[ -\frac{d\zeta}{dx'}\,\hat{\mathbf{x}}' - \frac{\partial\zeta}{\partial y'}\,\hat{\mathbf{y}}' + \hat{\mathbf{z}}' \right] \exp[ik_0(\hat{\mathbf{k}}_0{}^i - \hat{\mathbf{k}}_0{}^s) \cdot \mathbf{r}']\, dx'\,dy' \right],$$
$$(2.2\text{-}81)$$

where $\hat{\mathbf{k}}_0{}^s$ and $\hat{\mathbf{k}}_0{}^i$ are constant unit vectors in the directions of scattering and incidence, while $\zeta$, $\partial\zeta/\partial x'$, and $\partial\zeta/\partial y'$ are functions of position on the surface. The vector $\mathbf{r}'$ is then given by $\mathbf{r}' = x'\hat{\mathbf{x}}' + y'\hat{\mathbf{y}}' + \zeta(x', y')\hat{\mathbf{z}}'$.

## (b) Nonperfectly Conducting Surface

When a plane wave is incident upon a nonperfectly conducting infinite plane, the reflected wave is also plane; however, the reflection coefficient for the reflected wave depends upon the polarization of the incident wave. As an example, consider the material beneath the surface to be homogeneous with relative constitutive constants, $\mu_r$, $\epsilon_r$ (each of which may be complex if the material is lossy), and of nearly infinite extent; thus, the wave entering this material can be considered to continue on indefinitely.[†] Then the incident field at every point on the surface may be broken up into components, $E_\parallel{}^i$, in (parallel to) the plane of incidence, and $E_\perp{}^i$, normal to the plane of incidence. The scattered field corresponding to each of these incident field components is given by [Eq. (7.1-3), Chapter 7]

$$E_\perp{}^s = R_\perp E_\perp{}^i \qquad \text{and} \qquad \hat{\mathbf{s}} \times \mathbf{E}_\parallel{}^s = R_\parallel \hat{\mathbf{k}}_0{}^i \times \mathbf{E}_\parallel{}^i \qquad (2.2\text{-}82)$$

where $\hat{\mathbf{s}}$ points in the direction of specular reflection from the surface, and the Fresnel or reflection coefficients, $R_\perp$ and $R_\parallel$, are given by Eqs. (7.1-9), and (7.1-10) of Chapter 7, or

$$R_\perp = \frac{\mu_r \cos\theta_i - \sqrt{\mu_r\epsilon_r - \sin^2\theta_i}}{\mu_r \cos\theta_i + \sqrt{\mu_r\epsilon_r - \sin^2\theta_i}}$$

$$R_\parallel = \frac{\epsilon_r \cos\theta_i - \sqrt{\mu_r\epsilon_r - \sin^2\theta_i}}{\epsilon_r \cos\theta_i + \sqrt{\mu_r\epsilon_r - \sin^2\theta_i}} \qquad (2.2\text{-}83)$$

with $\theta_i$ the local angle of incidence.

The angle of incidence and plane of incidence are functions of the position on the surface of the scatterer. One can formally resolve the incident plane waves into components in and normal to the plane of incidence; use Eqs. (2.2-82) and (2.2-83) to find the total fields on the surface; and substitute these back into Eqs. (2.2-73) or (2.2-74). However, the resulting equations are quite complex in the general case. For backscattering, results will be obtained in somewhat simpler fashion, as follows.

2.2.2.3.3. *Physical Optics Integral for Backscattering.* When one is interested in backscattering, considerable simplification of the previous equations is possible. In this case, $\hat{\mathbf{k}}_0{}^i = -\hat{\mathbf{k}}_0{}^s$. Without loss of generality, assume that the z-axis is located along the incident or scattering direction (that is, $\hat{\mathbf{k}}_0{}^s = \hat{\mathbf{z}}$).

---

† This is equivalent to requiring that the Leontovich boundary condition hold at the surface.

(a) *Perfectly Conducting Surface*

Equations (2.2-79) and (2.2-81) now become

$$\mathbf{H}^s(P) = \frac{-ik_0 \exp(ik_0 R_0)}{2\pi R_0} \mathbf{H}_0 \int_{S_1} (\hat{z} \cdot \hat{n}) \exp(-i2k_0\zeta) \, ds' \quad (2.2\text{-}84)$$

or

$$\mathbf{H}^s(P) = \frac{-ik_0 \exp(ik_0 R_0)}{2\pi R_0} \mathbf{H}_0 \iint_{S_1} \exp(-i2k_0\zeta) \, dx' \, dy' \quad (2.2\text{-}85)$$

Both equations are equivalent, of course. These equations also show that the physical optics backscattered field from an object with a perfectly conducting surface is polarized in the same direction as the incident field.

Equation (2.2-84) can be converted to another interesting and often useful form. Note that $(\hat{n} \cdot \hat{z}) \, ds' = n_z ds'$ is the projection of the element of surface area $ds'$ onto the $x' - y'$ plane; hence, $n_z ds' = ds'_z = (\partial s'_z/\partial \zeta) \, d\zeta$, where $s'_z$ is defined to be the projection of the surface area for $z > \zeta$ onto the $x' - y'$ plane. Thus, Eq. (2.2-84) becomes

$$\mathbf{H}^s(P) = \frac{+ik_0 \exp(ik_0 R_0)}{2\pi R_0} \mathbf{H}_0 \int_0^L \exp(-i2k_0\zeta) \frac{\partial s_z}{\partial \zeta} \, d\zeta \quad (2.2\text{-}86)$$

where now the $x' - y'$ (or $z = 0$) plane is taken to be behind the shadow boundary, $\Gamma$, and $L$ is the maximum length of the object in the $+z$ direction. This last equation is often useful when one knows the exact equation for the surface of the object, because then $s'_z$ as a function of $\zeta$ can be readily determined, as can $\partial s'_z/\partial \zeta$. Equation (2.2-86) shows that in the physical optics approximation, the scattered field is proportional in magnitude to the Fourier transform of $\partial s'_z/\partial \zeta$ (where one defines $s'_z = 0$ for $\zeta > L$ and $s'_z = s'_{z\max}$ for $\zeta < 0$).

(b) *Nonperfectly Conducting Surface*

Using the coordinate orientation defined above for backscattering, one can use Eqs. (2.2-73) and (2.2-74), along with the tangent plane approximation for nonperfectly conducting surfaces. In general, the reflection coefficients from an infinite plane are dependent upon polarization as expressed in Eq. (2.2-83).[†] The forms taken by Eq. (2.2-73) for the far-zone scattered electric field, when the tangent plane approximation is applied to a nonperfectly conducting surface for various incident field polarizations, are given below.[67]

---

[†] Reflection coefficients written in this manner can apply not only to a homogeneous medium extending to infinity, as considered previously, but in general to any infinite surface. $R_\perp$ and $R_\parallel$ are discussed in Chapter 7 for arbitrary isotropic media; for example, a coated perfectly conducting plane, a surface consisting of several stratified layers, and an inhomogeneous medium.

(These results are alternately written in terms of scattering matrix elements, where 1 refers to $x$-polarization and 2 refers to $y$-polarization. Note that $a_{12} = a_{21}$, as is demanded by reciprocity for backscattering.)

(a) *Incident Field is x-Polarized,* $\mathbf{E}^i = E_{0x}\hat{\mathbf{x}}\exp(-ik_0 z)$

$$\mathbf{E}^s(P) = \frac{+ik_0\exp(ik_0 R_0)}{2\pi R_0}E_{0x}$$

$$\times\left[\hat{\mathbf{x}}\int_{S_1}\frac{(\partial\zeta/\partial x')^2\,R_\parallel - (\partial\zeta/\partial y')^2\,R_\perp}{(\partial\zeta/\partial x')^2 + (\partial\zeta/\partial y')^2}\exp(-i2k_0\zeta)\,dx'\,dy'\right.$$

$$\left. + \hat{\mathbf{y}}\int_{S_1}\frac{(R_\perp + R_\parallel)(\partial\zeta/\partial x')(\partial\zeta/\partial y')}{(\partial\zeta/\partial x')^2 + (\partial\zeta/\partial y')^2}\exp(-i2k_0\zeta)\,dx'\,dy'\right]$$

$$= \frac{\exp(ik_0 R_0)}{\sqrt{4\pi}\,R_0}[a_{11}\hat{\mathbf{x}} + a_{21}\hat{\mathbf{y}}] \qquad\qquad (2.2\text{-}87)$$

(b) *Incident Field is y-Polarized,* $\mathbf{E}^i = E_{0y}\hat{\mathbf{y}}\exp(-ik_0 z)$

$$\mathbf{E}^s(P) = \frac{+ik_0\exp(ik_0 R_0)}{2\pi R_0}E_{0y}$$

$$\times\left[\hat{\mathbf{x}}\int_{S_1}\frac{(R_\parallel + R_\perp)(\partial\zeta/\partial x')(\partial\zeta/\partial y')}{(\partial\zeta/\partial x')^2 + (\partial\zeta/\partial y')^2}\exp(-i2k_0\zeta)\,dx'\,dy'\right.$$

$$\left. + \hat{\mathbf{y}}\int_{S_1}\frac{(\partial\zeta/\partial y')^2\,R_\parallel - (\partial\zeta/\partial x')^2\,R_\perp}{(\partial\zeta/\partial x')^2 + (\partial\zeta/\partial y')^2}\exp(-i2k_0\zeta)\,dx'\,dy'\right]$$

$$= \frac{\exp(ik_0 R_0)}{\sqrt{4\pi}\,R_0}[a_{12}\hat{\mathbf{x}} + a_{22}\hat{\mathbf{y}}] \qquad\qquad (2.2\text{-}88)$$

(c) *Incident Field is Right-Circularly Polarized,* $\mathbf{E}^i = E_{0R}\hat{\mathbf{R}}\exp(-ik_0 z)$

$$\mathbf{E}^s(P) = \frac{+ik_0\exp(ik_0 R_0)}{2\pi R_0}E_{0R}$$

$$\times\left[\frac{1}{2}\hat{\mathbf{R}}\int_{S_1}(R_\parallel + R_\perp)\frac{(\partial\zeta/\partial x)^2 + i2(d\zeta/\partial x')(\partial\zeta/\partial y') - (\partial\zeta/\partial y')^2}{(\partial\zeta/\partial x')^2 + (\partial\zeta/\partial y')^2}\right.$$

$$\left. \times\exp(-i2k_0\zeta)\,dx'\,dy' + \frac{1}{2}\hat{\mathbf{L}}\int_{S_1}(R_\parallel - R_\perp)\exp(-i2k_0\zeta)\,dx'\,dy'\right]$$

$$= \frac{\exp(ik_0 R_0)}{\sqrt{4\pi}\,R_0}[a_{RR}\hat{\mathbf{R}} + a_{LR}\hat{\mathbf{L}}] \qquad\qquad (2.2\text{-}89)$$

Here, $\hat{\mathbf{R}}$ and $\hat{\mathbf{L}}$ are orthogonal unit vectors representing right and left circular polarization. They are defined as $\hat{\mathbf{R}} = (\hat{\mathbf{x}} \pm i\hat{\mathbf{y}})/\sqrt{2}, \hat{\mathbf{L}} = (\hat{\mathbf{x}} \mp i\hat{\mathbf{y}})/\sqrt{2}$, where the upper sign is used for propagation in the $+z$ direction and the lower for propagation in the $-z$ direction.

(d) *Incident Field is Left-Circularly Polarized,* $\mathbf{E}^i = E_{0L}\hat{\mathbf{L}}\exp(-ik_0z)$

$$\mathbf{E}^s(P) = \frac{+ik_0\exp(ik_0R_0)}{2\pi R_0} E_{0L}\left[\frac{1}{2}\hat{\mathbf{R}}\int_{S_1}(R_{\parallel} - R_{\perp})\exp(-i2k_0\zeta)\,dx'\,dy'\right.$$

$$\times \frac{1}{2}\hat{\mathbf{L}}\int_{S_1}(R_{\perp} + R_{\parallel})\frac{(\partial\zeta/\partial x')^2 - i2(\partial\zeta/\partial x')(\partial\zeta/\partial y') - (\partial\zeta/\partial y')^2}{(\partial\zeta/\partial x')^2 + (\partial\zeta/\partial y')^2}$$

$$\left.\times \exp(-i2k_0\zeta)\,dx'\,dy'\right]$$

$$= \frac{\exp(ik_0R_0)}{\sqrt{4\pi}\,R_0}[a_{RL}\hat{\mathbf{R}} + a_{LL}\hat{\mathbf{L}}] \tag{2.2-90}$$

These expressions are written in terms of integrals over the projected area of the scatterer normal to the incident field direction. In these equations the Fresnel coefficients $R_{\parallel}$, $R_{\perp}$, given by Eq. (2.2-83), must be expressed in terms of the integration variables $x'$, $y'$. This can be done by using

$$\cos\theta_i = \hat{\mathbf{n}}\cdot\hat{\mathbf{z}} = \frac{1}{\sqrt{1 + (\partial\zeta/\partial x')^2 + (\partial\zeta/\partial y')^2}} \tag{2.2-91}$$

$$\sin\theta_i = \left\{\frac{(\partial\zeta/\partial x')^2 + (\partial\zeta/\partial y')^2}{1 + (\partial\zeta/\partial x')^2 + (\partial\zeta/\partial y')^2}\right\}^{1/2} \tag{2.2-92}$$

in Eq. (2.2-83). From the above expressions it appears that a portion of the backscattered power is depolarized, as evidenced by the existence of the elements $a_{12}$, $a_{21}$, $a_{RR}$, and $a_{LL}$. Upon application of the stationary phase principle to evaluate these physical optics integrals in the high-frequency limit, however, these elements disappear, indicating no depolarization. This is due to the fact that only those areas whose normals lie along the incidence direction contribute to the backscatter. At normal incidence, $R_{\parallel} = -R_{\perp}$, the terms in the above equations which contain the factor $(R_{\parallel} + R_{\perp})$ disappear.

Before passing, it should be noted that in addition to the previous assumptions and restrictions, physical optics neglects the effects of multiple scattering of waves from one portion of the object surface to another. This is a result of the tangent plane approximation; that is, the reflected fields at the surface are assumed to be proportional to the incident fields only at that point, and not to the scattered field from any other point on the object. Multiple scattering is undoubtedly a major contributor to depolarization for many objects, and is an important consideration when the scatterer surface is concave at many points; it is nonexistent in the high frequency limit when the object is everywhere convex, for instance, for spheres, spheroids, ellipsoids, etc.

*2.2.2.3.4. Application of Stationary Phase to Scattering from a Specular Point on the Object Surface.* It is a well-known phenomenon that waves of short wavelength, compared with object dimensions, are scattered almost entirely from those surface points which are specularly oriented. Everyone has observed this, for instance, when looking at the "glistening" areas of a lake on a windy, moonlit night. It will be shown here that when physical optics is applied to finite objects whose dimensions are much larger than wavelength, most scattering does indeed come from these specular points and the small neighborhood surrounding them. To show this formally, use will be made of an analytical technique called the stationary phase principle.

(a) *Application of Stationary Phase to Backscattering*

Let the coordinate system for scattering from an arbitrary object be oriented so that the $z$-axis lies along the backscattering direction, and passes through the point on the surface, $S_1$, where the normal to the surface, $\hat{n}$, also points in the direction of backscattering. This point will be called a specular point on the object. The position of the $x' - y'$ plane or $x' - y'$ axis orientation is immaterial and arbitrary. Consider for the moment that the surface is perfectly conducting. Then the specular point has coordinates $[0, 0, \zeta(0, 0)]$. Now the surface may be expanded in a two-variable Maclaurin series about that point as follows:

$$\zeta(x', y') = \zeta(0, 0) + \frac{\partial \zeta}{\partial x'}\bigg|_{\substack{x'=0 \\ y'=0}} x' + \frac{\partial \zeta}{\partial y'}\bigg|_{\substack{x'=0 \\ y'=0}} y'$$

$$+ \frac{\partial \zeta^2}{\partial x'^2}\bigg|_{\substack{x'=0 \\ y'=0}} \frac{x'^2}{2} + \frac{\partial^2 \zeta}{\partial x' \partial y'}\bigg|_{\substack{x'=0 \\ y'=0}} x'y' + \frac{\partial^2 \zeta}{\partial y'^2}\bigg|_{\substack{x'=0 \\ y'=0}} \frac{y'^2}{2} + \cdots$$

$$(2.2\text{-}93)$$

However, with the coordinate system as described, the slopes

$$\frac{\partial \zeta}{\partial x'}\bigg|_{\substack{x'=0 \\ y'=0}} \quad \text{and} \quad \frac{\partial \zeta}{\partial y'}\bigg|_{\substack{x'=0 \\ y'=0}}$$

are identically zero since the surface normal at the specular point lies along the $z$-axis. Hence, the second and third terms are zero. Now, using Eq. (2.2-93) for $\zeta(x', y')$ in (2.2-85), one has

$$\mathbf{H}^s(P) = \frac{-ik_0 \exp(ik_0[R_0 - 2\zeta(0, 0)])}{2\pi R_0} \mathbf{H}_0 \int\!\int \exp\left(-i2k_0 \left[\frac{\partial^2 \zeta}{\partial x'^2}\bigg|_0 \frac{x'^2}{2}\right.\right.$$

$$\left.\left. + \frac{\partial^2 \zeta}{\partial x' \partial y'}\bigg|_0 x'y' + \frac{\partial^2 \zeta}{\partial y'^2}\bigg|_0 \frac{y'^2}{2} + \cdots\right]\right) dx'\, dy'$$

$$(2.2\text{-}94)$$

where the limits on the integral depend upon the shape of the projection of the body on the $x' - y'$ plane. Now, new $x_1 - y_1$ axes will be chosen by a rotation about the $z$-axis by an angle $\alpha$ in such a way that the second or cross-product term of the exponent vanishes, and then the integral variables will be changed to $x_1$, $y_1$. Thus,

$$K_1 x_1{}^2 + K_2 y_1{}^2 = \frac{1}{2} \frac{\partial^2 \zeta}{\partial x'^2}\bigg|_0 x'^2 + \frac{\partial^2 \zeta}{\partial x' \, \partial y'}\bigg|_0 x' y' + \frac{1}{2} \frac{\partial^2 \zeta}{\partial y'^2}\bigg|_0 y'^2 \quad (2.2\text{-}95)$$

Let

$$p = \frac{\partial^2 \zeta}{\partial x'^2}\bigg|_0, \qquad q = \frac{\partial^2 \zeta}{\partial y'^2}\bigg|_0, \qquad r = \frac{\partial^2 \zeta}{\partial x' \, \partial y'}\bigg|_0 \quad (2.2\text{-}96)$$

then $\alpha$, $K_1$, and $K_2$ are given by

$$\alpha = \tan^{-1}\left[-\frac{p-q}{2r} \pm \sqrt{\left(\frac{p-q}{2r}\right)^2 + 1}\right] \qquad \begin{array}{l} \text{if } \dfrac{p-q}{2r} < 0, \text{ use upper sign} \\[2mm] \text{if } \dfrac{p-q}{2r} > 0, \text{ use lower sign} \end{array}$$

$$(2.2\text{-}97)$$

$$2K_1 = \frac{\partial^2 \zeta}{\partial x_1{}^2}\bigg|_0 = -\frac{p+q}{2} \pm \sqrt{\left(\frac{p-q}{2}\right)^2 + r^2} \qquad (2.2\text{-}98)$$

$$2K_2 = \frac{\partial^2 \zeta}{\partial y_1{}^2}\bigg|_0 = \frac{p+q}{2} \pm \sqrt{\left(\frac{p-q}{2}\right)^2 + r^2} \qquad (2.2\text{-}99)$$

These two constants $K_1$ and $K_2$ have simple geometrical interpretations. From differential geometry the quantities $\rho_1 = 1/|K_1|$ and $\rho_2 = 1/|K_2|$ represent the maximum and minimum radii of curvature of the surface $\zeta(x', y')$ at the point $x' = 0$, $y' = 0$.[†] Now 2.2-94 may be rewritten:

$$\mathbf{H}^s(P) = \frac{-ik_0 \exp(ik_0[R_0 - 2\zeta(0, 0)])}{2\pi R_0}$$

$$\times \mathbf{H}_0 \iint \exp(-i2k_0[K_1 x_1{}^2 + K_2 y_1{}^2 + \cdots]) \, dx_1 \, dy_1 \quad (2.2\text{-}100)$$

The method of stationary phase may be applied to solve the preceding equation in the limit $\lambda_0 \to 0$, such that $k_0 \to \infty$ and $k_0 K_1$, $k_0 K_2$ are very large

---

[†] Note that $K_1$ and $K_2$ may either be positive or negative. When both are of the same sign, the surface in the vicinity of the specular point is either convex or concave. But when they have opposite signs, the surface around the specular point has the shape of a "saddle." Nonetheless, this method gives valid results for all of these situations.

numbers. The requirements on the size of $k_0 \mid K_1 \mid$ and $k_0 \mid K_2 \mid$ are such that they be large enough, with respect to the constant multiplying the next series terms, so that by the time $k_0 \mid K_1 \mid x_1^2 = 2\pi$, for example, the next-higher-order term is still negligible. If higher-order terms exist, then $K_1$ and $K_2$ must be nonzero in order to apply this method. Physically, this means that the shape of the surface near the specular point must be quadric, and that its quadratic curving along the $x_1$ and $y_1$ directions must be large enough that $(1/\lambda_0)[\zeta(0, 0) - \zeta(x_1, y_1)]$ is somewhat greater than unity, in the neighborhood of $[0, 0, \zeta(0, 0)]$, before terms higher than quadratic begin to effect the shape of the surface. In this case, the first cycles of fluctuation of the exponential are due to the $x_1^2$ and $y_1^2$ terms. It is then shown in more detailed texts on the subject that these first few cycles determine the entire value of the integral. As $x_1$ and $y_1$ become larger, the integrand fluctuates increasingly more rapidly due to their quadratic effect, and the more rapid the alternate positive–negative oscillation, as a function of $x$, and $y$, the less significant the contribution to the integral. Thus, the stationary phase approximation permits one to do the following:

(a) Neglect higher order terms in the exponential, since these terms become significant only after the quadratic terms are producing rapid oscillations of the integral.

(b) Extend the limits of integration to $\pm\infty$, since the contribution for larger $x_1$ and $y_1$ is negligible.

Under these simplifications, the scattered field becomes

$$\mathbf{H}^s(P') = \frac{-ik_0 \exp(ik_0[R_0 - 2\zeta(0, 0)])}{2\pi R_0}$$

$$\times \int_{-\infty}^{\infty} \exp(-i2k_0 K_1 x_1^2)\, dx_1 \int_{-\infty}^{\infty} \exp(-i2k_0 K_2 y_1^2)\, dy_1 \quad (2.2\text{-}101)$$

These integrals are the familiar Fresnel integrals, and their values are known exactly. The resulting backscattered field after application of the stationary phase method is

$$\mathbf{H}^s(P') = \frac{\exp(ik_0[R_0 - 2\zeta(0, 0)])}{2R_0 \sqrt{K_1 K_2}} \mathbf{H}_0$$

$$(2.2\text{-}102)$$

$$= \frac{\exp(ik_0[R_0 - 2\zeta(0, 0)] - i\psi)}{2R_0} \sqrt{\rho_1 \rho_2}\, \mathbf{H}_0$$

where $\psi$ is either zero, $\pi/4$, or $\pi/2$, depending on whether $K_1$, $K_2$ are both positive, one is negative, or both are negative.

Thus the backscattering cross section for such a surface, according to Eq. (2.2-102), is

$$\sigma^B = \pi \rho_1 \rho_2 \qquad\qquad (2.2\text{-}103)$$

which is identical to Eq. (2.2-40), predicted by geometrical optics.[†]

This analysis has assumed that there is only one specular point on the scatterer. In reality, there may be several. In that case, if these points are separated (on the average) by many wavelengths, one can treat the contribution from each point separately in the manner outlined. The total scattered field will then be a sum of terms like Eq. (2.2-102) from each of the specular points.

(b) *Application of Stationary Phase to Bistatic Scattering*

Having followed through the application of stationary phase to the simplest scattering configuration, one can extend this to bistatic scattering from generally nonperfectly conducting objects. The specular point on the object is that point (or points) whose normal vector, $\hat{n}$, bisects the angle between the incidence and scattering directions, that is, $-\hat{k}_0{}^i$ and $\hat{k}_0{}^s$, and lies in the plane formed by these vectors. Stationary phase results are interpreted, in this case, as showing that almost all the scattered field comes from the immediate neighborhood of the specular point when (i) $k_0$ is sufficiently large, and (ii) the surface is quadric in nature near the specular point.

One starts in this case with Eq. (2.2-73), and finds the total fields at the nonperfectly conducting surface according to the tangent plane approximation by using Eq. (2.2-83). Then it is shown that the complicated expression in the square brackets inside the integral varies very little in the neighborhood of the specular point. On this basis, it is treated as a constant and removed from the integrand; its value is taken as that at the specular point (that is, $\hat{n} = \hat{n}_s$, where $\hat{n}_s$ is the surface normal at the specular point). Thus, the remaining integrand contains the same exponential factor as before, and this is reduced to two Fresnel integrals by invoking stationary phase.

It should be emphasized before leaving physical optics that it provides a solution correct only in the high-frequency limit. In this respect, it is similar to and no more accurate than geometrical optics. Its only advantage over geometrical optics is that it can be applied to a wider variety of targets, including flat plates, wedges, cones, etc., for which geometrical optics

---

[†] The same stationary phase analysis can also be performed for backscatter from non-perfectly conducting surfaces. One begins with Eqs. (2.2-87) and (2.2-88) instead of (2.2-85). While the algebra is somewhat tedious, one finally obtains Eq. (2.2-102), with the factor $R_{\parallel}(0)$ [or $-R_{\perp}(0)$] appearing, i.e., the Fresnel reflection coefficients for normal incidence. The backscattered field, expressed in (2.2-102), applies only to the scattering matrix elements $a_{11}$, $a_{12}$, $a_{RL}$, and $a_{LR}$, since the remaining elements (representing the depolarized component) are zero.

cannot be used. Many persons have missed this point and, in some cases, have claimed that retention of higher-order terms in $1/k_0$ will provide an improvement to the solution in the resonance region. Others have claimed that certain methods of evaluating the physical optics integral lead to better results, or that solutions provided by physical optics are inherently more accurate than those of geometrical optics. Such claims are not true. The use of the tangent-plane approximation itself involves the omission of higher-order terms in $1/k_0$. In addition, such important phenomena as creeping waves, edge diffraction, etc., contribute higher-order terms, all of which are inherently missing from the physical optics formulation. Demonstration of the failure of higher-order terms to add accuracy in the resonance region is exhibited in Figure 4-2, where exact solutions are compared with the physical optics approximation for the circular cyclinder. Even though higher-order terms were retained in the latter example, they fail to exhibit the proper behavior in the resonance region.

### 2.2.2.4. Impulse Approximation

*2.2.2.4.1. Basic Properties of the Impulse Response.* Practically all of the historical development of scattering theory and its application for computing radar cross sections has been concerned with monochromatic incident fields. Viewed from a linear system-theory standpoint, one can consider the incident monochromatic wave as an input signal and the scattered wave as the product of the incident wave and a "scattering transfer function." This scattering transfer function is a function of frequency and, of course, depends upon the size, shape, and material of the scatterer.

Discussion in this section will be restricted from the outset to back-scattering from perfectly conducting objects, because this case has received the most attention in the literature. The methods, of course, are valid for the more general situation as well. The notation and conventions established by Kennaugh[68,69] will be followed here. Assume that the scattering matrix is used to relate any far-field scattered field component to an incident field component, as defined and discussed in Section 2.1.4. Then a scattering transfer function may be defined from the scattering matrix element $a_{ij}$ for a given object as follows:

$$F(i\omega) = \frac{2r_0 a_{ij}}{c} \tag{2.2-104}$$

where

$$a_{ij} = \frac{E_i^s}{E_j^i} \tag{2.2-105}$$

$E_i^s$ refers to the $i$-polarization state component of the scattered field, $E_j^i$ is the $j$-polarization component of the incident field, $c$ is the velocity of light, $r_0$ is the distance from the leading portion of the object to the observation point, and $k_0 = 2\pi/\lambda_0 = \omega/c$. Since $E_i{}^s$ and $a_{ij}$ depend upon frequency, they are implied functions of $\omega$.

Now, instead of a monochromatic incident wave, consider an incident plane wave in the form of an impulse, that is, $E_{jl}^i = \delta[t + (r_0/c)] e_j{}^i$, which strikes the object at $t = -r_0/c$. The backscattered electric field response, $E_i{}^s$, now begins at $t = 0$. Define

$$A_{ij} = \frac{E_i{}^s}{e_j{}^i} \quad \text{and} \quad F_I(t) = \frac{2r_0}{c} A_{ij} \qquad (2.2\text{-}106)$$

with $F_I(t)$ the Fourier transform of $F(i\omega)$, or

$$F_I(t) = \frac{1}{2\pi} \int_{\infty}^{\infty} F(i\omega) e^{-i\omega t} \, d\omega \quad \text{and} \quad F(i\omega) = \int_{\infty}^{\infty} F_I(t) e^{i\omega t} \, dt \qquad (2.2\text{-}107)$$

Physically, this reveals that the impulse response of a scatterer is readily determinable if the scattered field as a function of frequency (or wavelength) is known. From this impulse response, the response of the scatterer to other incident plane-wave time functions, $f(t)$ for example, can be obtained by convolution, or

$$F(t) = \int_0^{\infty} F_I(\tau) f(t - \tau) \, d\tau \qquad (2.2\text{-}108)$$

where the incident field is $E_j{}^i = f[t + (r_0/c)] e_j{}^i$, and $F(t)$ is the scattering response of the object to this incident wave.

There are several properties of the impulse response which relate it to the properties of the monochromatic scattered field.

## (a) Relationship to Physical Optics

In the case of backscattering along the $z$-axis, one form of the physical optics backscattered field is given in Eq. (2.2-86). Let the object be located with its leading edge tangent to the $z = 0$ plane. The $z$-axis will now be reversed in direction (see Fig. 2-12). Now $\zeta$, the $z$ distance to a point on the object, can be measured as a function of the time that it takes for the incident impulsive wave to travel from the origin to a point on the object and back to the origin again at the velocity of light, $c$. Thus, $\zeta = (c/2) t$. Substituting this into Eq. (2.2-86), integrating once by parts, and expressing $H^S(P)$ and $H_0$ in terms of $F(i\omega)$ according to Eq. (2.2-104), then

$$F(i\omega) = \frac{-1}{4\pi} \int_{-\infty}^{\infty} \frac{d^2 S_z}{d\zeta^2} e^{i\omega t} \, dt \qquad (2.2\text{-}109)$$

By comparing 2.2-109 with the second equation of (2.2-107), one can see immediately that

$$F_I(t) = \frac{-1}{4\pi} \frac{d^2 S_z}{d\zeta^2} = \frac{-1}{\pi c^2} \frac{d^2 S_z}{dt^2} \qquad (2.2\text{-}110)$$

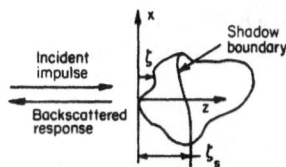

Fig. 2-12. Coordinates used in the physical optics approximation to the impulse response of a scatterer.

Physically, this means that one can obtain the impulse response predicted by physical optics by taking the second derivative of the area intercepted by the pulse moving along the object projected onto the $x - y$ plane. This projected area, $S_z$, is zero for $z < 0$, and increases monotonically to a maximum value at the trailing edge of the shadow boundary, $\zeta_S$. For $z > \zeta_S$, $S_z$ is assumed to remain at this constant value. From Fourier transform theory, if $F_S(t)$ and $F_R(t)$ are defined as the scattering response to an incident step function and ramp function, respectively, these can be simply related to the impulse response, $F_I(t)$. According to the physical optics relationship of Eq. (2.2-110),

$$F_S(t) = \frac{-1}{\pi c^2} \frac{dS_z}{dt} = \frac{-1}{2\pi c} \frac{dS_z}{d\zeta} \qquad (2.2\text{-}111)$$

and

$$F_R(t) = \frac{-1}{4\pi} S_z \qquad (2.2\text{-}112)$$

The impulse response predicted by physical optics suffers all of the deficiencies of the usual physical optics analysis. The fields are assumed to be zero on the shadowed side of the object, consequently ignoring the effects of creeping waves. In addition, multiple surface interactions are neglected. Since physical optics is basically a high-frequency technique, the impulse response derived therefrom is valid while the impulse excitation travels along the illuminated region. In this region, all of the high-frequency phenomena, such as sudden jumps and impulses in the response, are correctly predicted. As the incident impulse moves past the shadow boundary, however, physical optics no longer gives correct results. One can say in general that as the impulse moves past the object, the impulse response must die away exponentially due to resistive damping of the surface currents set up by the incident excitation.

(b) *Relationship to Low-Frequency Scattering*

The low-frequency scattered field from any object excited by a plane harmonic wave varies as $\omega^2$ (or $k_0^2$). From this fact it follows that

$$\lim_{\omega \to 0} F(i\omega) = 0, \qquad \lim_{\omega \to 0} \frac{dF(i\omega)}{d\omega} = 0, \qquad \lim_{\omega \to 0} \frac{d^2F(i\omega)}{d\omega^2} = 2K_R \qquad (2.2\text{-}113)$$

$K_R$ is proportional to the volume of the scatterer and is known if the Rayleigh cross section of the object is known. If $\sigma_R$ is the Rayleigh cross section, then

$$K_R{}^2 = \frac{\sigma_R}{\pi c^2 \omega^4} = \frac{\sigma_R}{\pi k_0{}^4 c^6}$$

From Fourier transform theory, $d^n F(i\omega)/d(i\omega)^n$ has the transform $t^n F_I(t)$. Using this fact and Eq. (2.2-113),

$$\int_{-\infty}^{\infty} F_I(t)\, dt = 0 \qquad\qquad (2.2\text{-}114)$$

$$\int_{-\infty}^{\infty} t F_I(t)\, dt = 0 \qquad\qquad (2.2\text{-}115)$$

and

$$\int_{-\infty}^{\infty} t^2 F_I(t)\, dt = -2K_R \qquad\qquad (2.2\text{-}116)$$

These are known as moment conditions on the impulse response. Higher-order moment conditions also exist and can be useful if higher-order terms in the expansion of $F(i\omega)$ are known. Note that $F_I(t)$ is zero for $t < 0$, since the scatterer cannot respond before the incident impulse strikes it.

After integration by parts, these first three moment conditions can be expressed alternatively in terms of the step and ramp response as follows:

$$\int_{0}^{\infty} F_I(t)\, dt = 0 \qquad\qquad (2.2\text{-}117)$$

$$\int_{0}^{\infty} F_S(t)\, dt = 0 \qquad\qquad (2.2\text{-}118)$$

$$\int_{0}^{\infty} F_R(t)\, dt = -K_R \qquad\qquad (2.2\text{-}119)$$

Thus, the impulse response, $F_I(t)$, can be determined exactly from Eq. (2.2–107) if the scattered field, $F(i\omega)$, is known at all frequencies. For example, $F(i\omega)$ for a sphere is known exactly, and $F_I(t)$ can be determined from it. An example of the use of this transformation to obtain the impulse response of a sphere is found in Section 3.5. The interpretation of the shape of the impulse response, as discussed there, provides valuable insight into the scattering process.

2.2.2.4.2. *Prediction of Scattering Cross Section Using the Impulse Response.* In the preceding section several properties of the impulse response and their relationship to the frequency response have been discussed. The important point to be emphasized here is that the impulse response is not only valuable for its time-domain information, but provides a simple technique for approximating the radar cross section of an object as a function of frequency. It is especially valuable in the resonance region, where target cross section prediction is extremely difficult.

To use the impulse response approximation to estimate the target cross section, Kennaugh and Moffatt[68] have established the following basic procedure. $F_I(t)$ is approximated initially according to the physical optics prediction of Eq. (2.2-110). Then this estimate is improved by adding an arbitrary function in the shadow region and forcing it to satisfy Eqs. (2.2-114) and (2.2-115). If an estimate of the Rayleigh-region cross section[70-72] is available, Eq. (2.2-116) can be used to improve this shadow region estimate further. Usually these moment equations cannot be carried further because higher-order Rayleigh-region estimates for the scattered field are non-existent. Finally, the Fourier transform of this estimate of $F_I(t)$ is taken to obtain $F(i\omega)$ according to Eq. (2.2-107). The approximate cross section then is $\sigma = \pi c^2 \mid F(i\omega)\mid^2$.

Surprisingly, this procedure has yielded extremely close agreement with observed and theoretically calculated cross sections throughout the Rayleigh, resonance, and high-frequency regions. Results of this approximation for prolate spheroids and cone-spheres are shown in later sections. The approximate answers are accurate in the high-frequency region because they are based upon the physical optics formulation. They agree in the Rayleigh region because they comply with the lowest-order moment conditions. For some reason not yet fully understood, the simultaneous enforcement of these high-frequency and low-frequency constraints on the impulse response results in a radar cross section which agrees quite well in the resonance region. The oscillatory nature of the cross section in this region is remarkably close to correct, even though physical optics alone usually yields a completely erroneous oscillatory behavior.

### 2.2.2.5. Fock Theory

In 1945 and 1946, the Russian physicist V. A. Fock published a series of papers on the electromagnetic fields induced on or near the surfaces of convex conducting bodies, for which the radius of curvature at all points is large in terms of the wavelength.[73-75] The primary contribution of this work was to demonstrate that to a first approximation the surface fields in the penumbral region surrounding the geometrical shadow boundary are local in nature, depending only on the radius of curvature and nature of the surface in the immediate vicinity of the shadow boundary. Fock used several methods to obtain equivalent or compatible results; however, the method which most clearly illustrates the physical assumptions underlying his work was the generation of the field components in the region near the shadow boundary from scalar solutions of a parabolic diffusion-type partial differential equation, which had been suggested earlier by Leontovich.[76] An essentially physical argument was used by Fock to justify the usage of a parabolic differential equation instead of an elliptical equation, such as represented by the scalar or vector Helmholtz equation.

In Fock's original work, he obtained solutions for the fields on or near the surface of a body valid only in a small region in the vicinity of the shadow boundary. However, by means of minor modifications, Fock's work can be extended to provide approximate expressions for the surface fields valid in the illuminated region and far in the shadow region, as well as near or on the shadow boundary.

To illustrate the details of Fock's technique and its connection with other approximation techniques, some simple examples will be considered.

2.2.2.5.1. *Surface Fields on a Convex, Finitely Conducting Body.* Consider a plane wave incident on a convex, finitely conducting body, with its direction of propagation along the $x$-axis of a coordinate system oriented with the $z$-axis normal to the body surface at a point on the shadow boundary. Thus, the origin of the coordinate system is a point on the shadow boundary, and the $y$-axis is chosen to form a right-handed system. These coordinates are illustrated in Figure 2-13. The equation of the surface in rectangular coordinates is

$$f(x, y, z) = 0$$

and the shadow boundary is determined from

$$\frac{\partial f}{\partial x} = 0$$

Near a given point, the surface will satisfy a parabolic equation of the form

$$z + \tfrac{1}{2}(ax^2 + 2bxy + cy^2) = 0$$

where $R = 1/a$ is the radius of curvature of the surface in the $x - z$ plane at the origin and $1/c$ is the radius of curvature in the $y - z$ plane.

In the rectangular coordinate system illustrated, each of the vector components of the total field will satisfy the scalar Helmholtz equation

$$\nabla^2 \psi + k_0^2 \psi = 0 \tag{2.2-120}$$

with the various field components being related by Maxwell's equations. Now

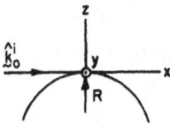

Fig. 2-13. Local coordinate system used in deriving Fock's parabolic equation.

following the derivation outlined in Fock's original paper,[75] the factor, $e^{ik_0x}$, can be separated out with

$$\psi = e^{ik_0x}\psi' \tag{2.2-121}$$

thus,

$$\nabla^2\psi' + 2ik_0\frac{\partial\psi'}{\partial x} = 0 \tag{2.2-122}$$

Similarly, $e^{ik_0x}$ can be separated out of the field components giving

$$E_j = e^{ik_0x}E_j', \qquad H_j = e^{ik_0x}H_j' \tag{2.2-123}$$

where the subscript $j$ represents any of the vector components $x$, $y$, $z$. Inserting these into Maxwell's equations then,

$$\frac{\partial E_z'}{\partial y} - \frac{\partial E_y'}{\partial z} = ik_0H_x', \qquad\qquad \frac{\partial H_z'}{\partial y} - \frac{\partial H_y'}{\partial z} = -ik_0E_x'$$

$$\frac{\partial E_x'}{\partial z} - \frac{\partial E_z'}{\partial x} - ik_0E_z' = ik_0H_y', \qquad \frac{\partial H_x'}{\partial z} - \frac{\partial H_z'}{\partial x} - ik_0H_z' = -ik_0E_y' \tag{2.2-124}$$

$$\frac{\partial E_y'}{\partial x} - \frac{\partial E_x'}{\partial y} + ik_0E_y' = ik_0H_z', \qquad \frac{\partial H_y'}{\partial x} - \frac{\partial H_x'}{\partial y} + ik_0H_y' = -ik_0E_z'$$

Now the assumption is made that the primed quantities are slowly varying functions with respect to a wavelength, and that the variation with respect to $z$ is more rapid than with respect to $x$ or $y$. This assumption can be formulated more precisely by stating

$$\frac{\partial\psi'}{\partial z} = O\left(\frac{k_0}{m}\psi'\right); \qquad \frac{\partial\psi'}{\partial x} = O\left(\frac{k_0}{M}\psi'\right); \qquad \frac{\partial\psi'}{\partial y} = O\left(\frac{k_0}{M}\psi'\right) \tag{2.2-125}$$

where the symbol $O(x)$ means "of the order of," and $m$ and $M$ are dimensionless parameters with

$$M \gg m \gg 1 \tag{2.2-126}$$

It follows from these assumptions that

$$\frac{\partial^2\psi'}{\partial z^2} + 2ik_0\frac{\partial\psi'}{\partial x} = 0 \tag{2.2-127}$$

where

$$M = m^2 \tag{2.2-128}$$

and $\psi'$ represents either $E_j'$ or $H_j'$. From Eqs. (2.2-123), (2.2-124), and the above assumptions, all the field components, neglecting terms of order $1/m$ or smaller, can be found in terms of $H_y'$ and $H_z'$ or equivalently in terms of $H_y$ and $H_z$ as

$$E_x = \frac{i}{k_0} \left[ \frac{\partial H_z}{\partial y} - \frac{\partial H_y}{\partial z} \right]$$

$$E_y = H_z$$

$$E_z = -H_y \tag{2.2-129}$$

$$H_x = \frac{i}{k_0} \left[ \frac{\partial H_y}{\partial y} + \frac{\partial H_z}{\partial z} \right]$$

Thus, if a solution to (2.2-127) which satisfies the boundary conditions and the radiation condition can be found, then all the field components can be obtained from Eq. (2.2-129).

Now, if the Leontovich boundary condition[77]

$$\mathbf{E} - (\hat{\mathbf{n}} \cdot \mathbf{E}) = \sqrt{\frac{\mu}{\epsilon}} (\hat{\mathbf{n}} \times \mathbf{H}) \tag{2.2-130}$$

is satisfied on the surface of the body, then by utilizing the ordering equations (2.2-125) and (2.2-126), the following boundary conditions must be satisfied by the magnetic field components $H_y$, $H_z$:

$$H_z = -n_y H_y \tag{2.2-131}$$

and

$$\frac{\partial H_y}{\partial z} + ik_0 \left( n_x + \sqrt{\frac{\epsilon}{\mu}} \right) H_y = \frac{\partial H_z}{\partial y} \tag{2.2-132}$$

Neglecting terms in Eqs. (2.2-131) and (2.2-132), which are of order $1/M$, even more simplified boundary conditions are

$$H_z = 0 \tag{2.2-133}$$

and

$$\frac{\partial H_y}{\partial z} + ik_0 \left( n_x + \sqrt{\frac{\epsilon}{\mu}} \right) H_y = 0 \tag{2.2-134}$$

These are completely uncoupled and, along with Eq. (2.2-127), allow $H_y$ and $H_z$ to be determined separately and uniquely to terms of order $1/m$. To obtain higher-order correction terms, the boundary conditions (2.2-131) and (2.2-132) must be used.

Considering the first order solution for $H_y$, let

$$H_y = H_y{}^i e^{ik_0 x} \psi' \tag{2.2-135}$$

where $H_y{}^i$ is the amplitude of the incident wave at infinity. The function $\psi'$

must satisfy the differential equation (2.2-127) and, from Eq. (2.2-134) and the equation of the surface, the boundary condition

$$\frac{\partial \psi'}{\partial z} + ik_0 \left(ax + by + \sqrt{\frac{\epsilon}{\mu}}\right) \psi' = 0 \qquad (2.2\text{-}136)$$

Now introduce two new variables defined by

$$\xi = m(ax + by) \qquad (2.2\text{-}137)$$

and

$$\zeta = 2aM[z + \tfrac{1}{2}(ax^2 + 2bxy + cy^2)] \qquad (2.2\text{-}138)$$

where $\xi$ and $\zeta$ are scaled so that on the shadow boundary $\zeta = \xi^2$. In terms of these variables, the differential equation (2.2-127) takes the form

$$\frac{\partial^2 \psi'}{\partial \zeta^2} + \frac{ik_0}{2m^3 a} \left(\frac{\partial \psi'}{\partial \xi} + 2\xi \frac{\partial \psi'}{\partial \zeta}\right) = 0 \qquad (2.2\text{-}139)$$

Choosing $m$ such that the coefficient in this equation is unity,

$$m = \left(\frac{k_0}{2a}\right)^{1/3} = \left(\frac{k_0 R}{2}\right)^{1/3} \qquad (2.2\text{-}140)$$

then the differential equation becomes

$$\frac{\partial^2 \psi'}{\partial \zeta^2} + i \left(\frac{\partial \psi'}{\partial \xi} + 2\xi \frac{\partial \psi'}{\partial \zeta}\right) = 0 \qquad (2.2\text{-}141)$$

Writing the boundary condition (2.2-136) in terms of $\xi$ and $\zeta$ gives

$$\frac{\partial \psi'}{\partial \zeta} + i\xi\psi' + q\psi' = 0 \qquad (2.2\text{-}142)$$

where

$$q = im\sqrt{\frac{\epsilon}{\mu}} = i\sqrt{\frac{\epsilon}{\mu}} \left(\frac{k_0 R}{2}\right)^{1/3} \qquad (2.2\text{-}143)$$

The differential equation (2.2-141) can be simplified by setting

$$\psi' = e^{i\xi\zeta + i\xi^3/3} V \qquad (2.2\text{-}144)$$

which gives as the differential equation and boundary condition for $V$;[†]

$$\frac{\partial^2 V}{\partial \zeta^2} + i \frac{\partial V}{\partial \xi} + \zeta V = 0 \qquad (2.2\text{-}145)$$

and

$$\frac{\partial V}{\partial \zeta} + qV \bigg|_{\zeta=0} = 0 \qquad (2.2\text{-}146)$$

[†] Equation (2.2-145) is often called the Fock–Leontovich parabolic equation.

Particular solutions of Eq. (2.2-145) are

$$V = e^{i\xi t}w(t - \zeta) \qquad (2.2\text{-}147)$$

where $w(t)$ is a solution of the Airy equation,

$$\frac{\partial^2 w}{\partial t^2} - tw = 0 \qquad (2.2\text{-}148)$$

Two independent solutions of the Airy equation are

$$w_1(t) = \frac{1}{\sqrt{\pi}} \int_{\Gamma_1} e^{vt - v^3/3} \, dv \qquad (2.2\text{-}149)$$

and

$$w_2(t) = \frac{1}{\sqrt{\pi}} \int_{\Gamma_2} e^{vt - v^3/3} \, dv \qquad (2.2\text{-}150)$$

where the contours $\Gamma_1$ and $\Gamma_2$ are shown in Figure 2-14.

The functions $w_1(t)$ and $w_2(t)$ can be expressed in terms of the Airy functions, $Ai(t)$, $Bi(t)$ defined and tabulated in Reference 78 as

$$w_1(t) = \sqrt{\pi}\,[Bi(t) + iAi(t)] \qquad (2.2\text{-}151)$$

and

$$w_2(t) = \sqrt{\pi}\,[Bi(t) - iAi(t)] \qquad (2.2\text{-}152)$$

The solution for $V$, which satisfies the boundary condition (2.2-146), and which for $\xi$ large and negative (corresponding to the illuminated side of the shadow boundary) correctly merges with the optical results, is

$$V = \frac{i}{2\sqrt{\pi}} \int_c e^{i\xi t} \left[ w_2(t - \zeta) - \frac{w_2'(t) - qw_2(t)}{w_1'(t) - qw_1(t)} w_1(t - \zeta) \right] dt \qquad (2.2\text{-}153)$$

Thus,

$$H_y = H_y{}^i e^{ik_0 x - i\xi\zeta + i\xi^3/3} V \qquad (2.2\text{-}154)$$

gives the $y$-component of the magnetic field. For $\zeta = 0$ (on the body surface)

$$V = \frac{1}{\sqrt{\pi}} \int_c e^{i\xi t} \frac{dt}{w_1'(t) - qw_1(t)} \qquad (2.2\text{-}155)$$

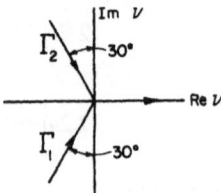

Fig. 2-14. Integration contours in the complex plane for Fock's integral solutions of the Airy equation.

and

$$H_y = H_y{}^i e^{ik_0 x} G(\xi, q) \tag{2.2-156}$$

where

$$G(\xi, q) = \frac{e^{i\xi^3/3}}{\sqrt{\pi}} \int_c e^{i\xi t} \frac{dt}{w_1'(t) - q w_1(t)} \tag{2.2-157}$$

For a perfectly conducting body $q = 0$, and

$$H_y = H_y{}^i e^{ik_0 x + i\xi^3/3} g(\xi) \tag{2.2-158}$$

with

$$g(\xi) = \frac{1}{\sqrt{\pi}} \int \frac{e^{i\xi t}}{w_1'(t)} dt \tag{2.2-159}$$

The first-order solution for $H_z$ can be obtained similarly; that is, let

$$H_z = H_z{}^i e^{ik_0 x} \phi' \tag{2.2-160}$$

where the function $\phi'$ satisfies the differential equation

$$\frac{\partial^2 \phi'}{\partial z^2} + 2ik_0 \frac{\partial \phi'}{\partial x} = 0 \tag{2.2-161}$$

and at the surface the boundary condition

$$\phi' = 0 \tag{2.2-162}$$

Again changing variables to $\xi$, $\zeta$ as previously defined in Eqs. (2.2-137) and (2.2-138), and letting

$$\phi' = e^{-i\xi\zeta + i\xi^3/3} U \tag{2.2-163}$$

then

$$\frac{\partial^2 U}{\partial \zeta^2} + i \frac{\partial U}{\partial \xi} + \zeta U = 0 \tag{2.2-164}$$

and

$$U \big|_{\zeta=0} = 0 \tag{2.2-165}$$

A solution for $U$, satisfying the boundary conditions and properly merging with geometrical results in the illuminated region, is

$$U = \frac{i}{2\sqrt{\pi}} \int_c e^{i\xi t} \left[ w_2(t - \zeta) - \frac{w_2(t)}{w_1(t)} w_1(t - \zeta) \right] dt \tag{2.2-166}$$

and

$$H_z = H_z{}^i e^{ik_0 x - i\xi\zeta + i\xi^3/3} U \tag{2.2-167}$$

To obtain second-order solutions for $H_y$, $H_z$, let

$$H_y = H_y{}^i e^{ik_0 x} \psi' + \frac{1}{m} H_z{}^i e^{ik_0 x} A \qquad (2.2\text{-}168)$$

$$H_z = H_z{}^i e^{ik_0 x} \phi' + \frac{1}{m} H_y{}^i e^{ik_0 x} B \qquad (2.2\text{-}169)$$

and use the boundary conditions (2.2-131) and (2.2-132), along with the differential equation (2.2-141), which all four functions $\psi'$, $\phi'$, $A$, $B$ satisfy. The results of such a process are

$$A = \frac{ib}{a} \frac{\partial \phi'}{\partial \zeta} + \left( \frac{ac - b^2}{a} my - \frac{ib}{a} q \right) (\phi' - \psi') \qquad (2.2\text{-}170)$$

and

$$B = -\frac{ib}{a} \frac{\partial \psi'}{\partial \zeta} + \left( \frac{ac - b^2}{a} my + \frac{ib}{a} q \right) (\phi' - \psi') \qquad (2.2\text{-}171)$$

thus giving the second-order expressions for $H_y$, $H_z$ when inserted into Eqs. (2.2-168) and (2.2-169).

The remaining field components can be obtained from Eqs. (2.2-129) and are

$$E_y = H_z, \qquad E_z = H_y \qquad (2.2\text{-}172)$$

with

$$E_x = -i \left( \frac{2}{k_0 R} \right)^{1/3} H_y{}^i e^{ik_0 x} \frac{\partial \psi'}{\partial \zeta} \qquad (2.2\text{-}173)$$

$$H_x = i \left( \frac{2}{k_0 R} \right)^{1/3} H_z{}^i e^{ik_0 x} \frac{\partial \phi'}{\partial \zeta} \qquad (2.2\text{-}174)$$

On the surface of a perfect conductor, $H_z = 0$, and $H_y$ is given by Eq. (2.2-158). Evaluating Eq. (2.2-174) for $H_x$ on the surface gives

$$H_x = i \left( \frac{2}{k_0 R} \right)^{1/3} H_z{}^i e^{ik_0 x + i \xi^3/3} f(\xi) \qquad (2.2\text{-}175)$$

with

$$f(\xi) = \frac{1}{\sqrt{\pi}} \int \frac{e^{i \xi t}}{w_1(t)} dt \qquad (2.2\text{-}176)$$

The functions $g(\xi)$ and $f(\xi)$ are thus universal functions which can be used to determine the surface currents in the vicinity of the shadow boundary of a perfectly conducting convex body which is large in terms of wavelength. Curves of $g(\xi)$ and $f(\xi)$, taken from Logan's work,[78] are shown in Figure 2-15, and in Table 2-1 a tabulation of $g(\xi)$, $f(\xi)$ is given for $-1 \leqslant \xi \leqslant 6$.

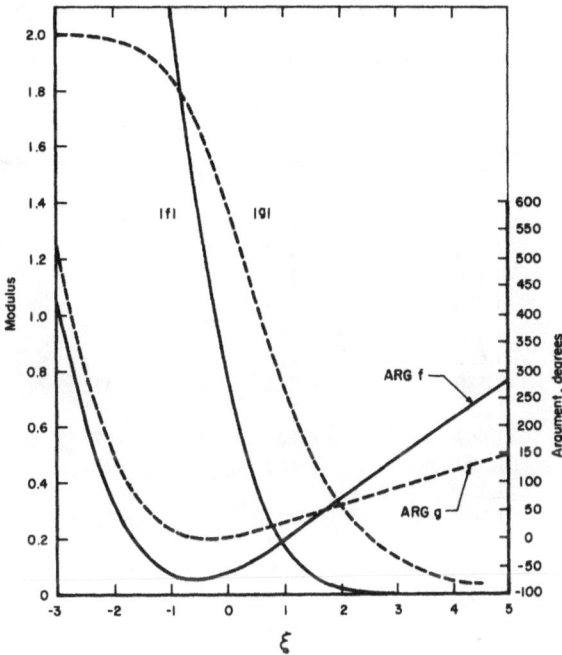

Fig. 2-15. Argument and modulus of Fock's current functions $f$ and $g$ versus the parameter $\xi$ [After Logan[78]].

For $\xi$ large and negative, the asymptotic expansions for $g(\xi)$ and $f(\xi)$ are

$$\lim_{\xi \to -\infty} f(\xi) = 2i\xi e^{-i\xi^3/3} \left[ 1 - \frac{i}{4\xi^3} + \frac{1}{2\xi^6} + O\left(\frac{1}{\xi^9}\right) \right] \qquad (2.2\text{-}177)$$

$$\lim_{\xi \to -\infty} g(\xi) = 2e^{-i\xi^3/3} \left[ 1 + \frac{i}{4\xi^3} - \frac{1}{\xi^6} + O\left(\frac{1}{\xi^9}\right) \right] \qquad (2.2\text{-}178)$$

For $\xi$ positive and greater than one, expansions for $g(\xi)$ and $f(\xi)$ are

$$f(\xi) = e^{-i\pi/3} \sum_{s=1}^{\infty} \frac{\exp[-\frac{1}{2}(\sqrt{3} - i)\,\alpha_s \xi]}{\mathrm{Ai}'(-\alpha_s)} \qquad (2.2\text{-}179)$$

and

$$g(\xi) = \sum_{s=1}^{\infty} \frac{\exp[-\frac{1}{2}(\sqrt{3} - i)\,\beta_s \xi]}{\beta_s \mathrm{Ai}(-\beta_s)} \qquad (2.2\text{-}180)$$

where the $\alpha_s$, $\beta_s$ are the roots of the Airy function and it's derivative, or $\mathrm{Ai}(-\alpha_s) = 0$, $\mathrm{Ai}'(-\beta_s) = 0$. A tabulation of the first five values of $\alpha_s$, $\beta_s$ is given in Table 2-3 of Section 2.3.2.

Table 2-1.   Fock's Current Distribution Functions, $g(\xi), f(\xi)$†

| $\xi$ | Modulus $g$ | Argument $g$, degrees | Modulus $f$ | Argument $f$, degrees |
|---|---|---|---|---|
| − 1.0 | 1.861 | 15.43 | 2.160 | 295.88 |
| − 0.9 | 1.833 | 10.07 | 1.992 | 291.82 |
| − 0.8 | 1.802 | 5.78 | 1.829 | 288.97 |
| − 0.7 | 1.766 | 2.47 | 1.672 | 287.25 |
| − 0.6 | 1.726 | 0.07 | 1.521 | 286.57 |
| − 0.5 | 1.682 | 358.49 | 1.377 | 286.85 |
| − 0.4 | 1.633 | 357.63 | 1.241 | 288.01 |
| − 0.3 | 1.580 | 357.42 | 1.112 | 289.98 |
| − 0.2 | 1.523 | 357.79 | 0.992 | 292.68 |
| − 0.1 | 1.463 | 358.67 | 0.879 | 296.04 |
| 0 | 1.399 | 360.00 | 0.776 | 300.00 |
| 0.1 | 1.334 | 1.71 | 0.681 | 304.99 |
| 0.2 | 1.266 | 3.75 | 0.594 | 309.46 |
| 0.3 | 1.197 | 6.07 | 0.516 | 314.83 |
| 0.4 | 1.113 | 8.62 | 0.446 | 320.58 |
| 0.5 | 1.059 | 11.36 | 0.383 | 326.63 |
| 0.6 | 0.991 | 14.24 | 0.327 | 332.95 |
| 0.7 | 0.925 | 17.24 | 0.279 | 339.49 |
| 0.8 | 0.859 | 20.32 | 0.236 | 346.22 |
| 0.9 | 0.797 | 23.46 | 0.199 | 353.09 |
| 1.0 | 0.738 | 26.64 | 0.167 | 0.08 |
| 1.5 | 0.488 | 42.56 | 0.0665 | 35.88 |
| 2.0 | 0.315 | 57.98 | 0.0250 | 71.61 |
| 2.5 | 0.203 | 72.90 | 0.0091 | 106.51 |
| 3.0 | 0.130 | 87.57 | 0.0033 | 140.67 |
| 4.0 | 0.054 | 116.75 | — | — |
| 5.0 | 0.022 | 145.93 | — | — |
| 6.0 | 0.009 | 175.12 | — | — |

† After Logan.[78]

2.2.2.5.2. *Application to a Perfectly Conducting Smooth Convex Cylinder.*
To examine the ability of Fock's physically based theory to correctly predict the surface fields on convex conducting surfaces, the results of the previous section will be compared with the asymptotic solutions, due to Goryainov,[79] for the surface currents on a perfectly conducting convex cylinder obtained by other techniques. The simplest such structure for which exact solutions are available is the circular cylinder, and it will be discussed initially.

2.2.2.5.2.1. *Circular Cylinder.*   Consider a plane wave normally incident on a perfectly conducting cylinder as illustrated in Figure 2-16.

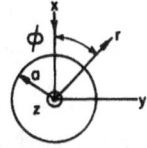

Fig. 2-16. Illustration of coordinate relations for a plane wave normally incident on a perfectly conducting cylinder of radius $a$.

For perpendicular polarization with $E$ having only a $z$-component, the exact solution for the scattered field is (see Section 4.1)[†]

$$E_z = - \sum_{n=0}^{\infty} \epsilon_n(-i)^n \frac{J_n(k_0a)}{H_n^{(1)}(k_0a)} H_n^{(1)}(k_0\rho) \cos n\phi \qquad (2.2\text{-}181)$$

and for the $z$-component of the surface current, setting $\eta = \sqrt{\mu/\epsilon} = 1$ for simplicity,

$$K_z = \frac{-2}{\pi k_0 a} \sum_{n=0}^{\infty} \epsilon_n(-i)^n \frac{\cos n\phi}{H_n^{(1)}(k_0a)} \qquad (2.2\text{-}182)$$

The first step in obtaining an asymptotic solution for $K_z$ which is valid for large $k_0a$ is to convert $K_z$ to a contour integral.[(79)] The details of this procedure will not be given here, but it is very similar to the techniques discussed in Section 2.3.2. This contour integral is split into two components, one of which is evaluated by the stationary phase method and gives the geometrical optics currents in the illuminated region. The other term of the contour integral is evaluated by conversion to a residue series and gives the diffraction component of the current. Thus, in the shadow region, $\pi \geqslant \phi \geqslant \pi/2$, the $z$-component of the surface current is[(79)]

$$K_z = \frac{i}{m} \left[ \sum_{n=0}^{\infty} \exp\left(ik_0a\left[\left(\frac{4n+3}{2}\right)\pi - \phi\right]\right) f(\xi_{2n+1}) \right.$$
$$\left. + \exp\left(ik_0a\left[\left(\frac{4n-1}{2}\right)\pi + \phi\right]\right) f(\xi_{2n+2}) \right] \qquad (2.2\text{-}183)$$

where

$$\xi_{2n+1} = m\left[\left(\frac{4n+3}{2}\right)\pi - \phi\right] \qquad (2.2\text{-}184)$$

$$\xi_{2n+2} = m\left[\left(\frac{4n-1}{2}\right)\pi + \phi\right] \qquad (2.2\text{-}185)$$

and

$$m = (k_0a/2)^{1/3} \qquad (2.2\text{-}186)$$

The $\xi_{2n+2}$ term corresponds to currents which can be thought of as beginning at the shadow boundary, $\phi = \pi/2$, and flowing to the observation point, having made $n$ successive transits around the cylinder. Similarly, the

[†] In the following equations, $\rho$ represents the distance from the cylinder axis to the observation point.

$\xi_{2n+1}$ term corresponds to currents originating at the $\phi = -\pi/2$ shadow boundary and flowing to the observation point, having made $n$ successive transits around the cylinder.

Equation (2.2-183) can be modified to represent an interpolation formula valid anywhere on the cylinder, that is, in the illuminated region as well as at the shadow boundary and in the shadow region.[79] For large $k_0a$, only the first or $n = 0$ terms of (2.2-183) contribute significantly; so for $\pi \geqslant \phi \geqslant \pi/2$,

$$K_z \approx \frac{i}{m}\{\exp[ik_0a(3\pi/2 - \phi)]f(\xi_1) + \exp[ik_0a(\phi - \pi/2)]f(\xi_2)\}$$
(2.2-187)

From this an interpolation formula can be obtained which is approximately valid anywhere on the cylinder; or $0 \leqslant \phi \leqslant \pi$,

$$K_z \approx \frac{i}{m}\{\exp[ik_0a(3\pi/2 - \phi)]f(\xi_1) + \exp[-ik_0a\cos\phi - i\xi'^3/3]f(-\xi')\}$$
(2.2-188)

with

$$\xi' = m\cos\phi$$
(2.2-189)

For $\phi \approx \pm\pi/2$, Eqs. (2.2-187) and (2.2-188) coincide, thus providing a smooth transition from Eq. (2.2-187), valid in the shadow region, to Eq. (2.2-188), valid in the illuminated region, and an approximation in the shadow region.

For parallel polarization, the only surface current component is

$$K_\phi = \frac{2i}{\pi k_0a}\sum_{n=0}^{\infty}\epsilon_n(-i)^n\frac{\cos n\phi}{H_n^{(1)'}(k_0a)}$$
(2.2-190)

where $H_n^{(1)'}$ is the derivative with respect to $k_0a$ of the Hankel function of the first kind. Following a procedure similar to that for perpendicular polarization, the current in the shadow region $\pi/2 \leqslant \phi \leqslant \pi$ is

$$K_\phi = \sum_{n=0}^{\infty}\left\{\exp\left(ik_0a\left[\left(\frac{4n+3}{2}\right)\pi - \phi\right]\right)g(\xi_{2n+1})\right.$$
$$\left. + \exp\left(ik_0a\left[\left(\frac{4n-1}{2}\right)\pi + \phi\right]\right)g(\xi_{2n+2})\right\}$$
(2.2-191)

Again for large $k_0a$, only the first or $n = 0$ terms contribute significantly, so that for $\pi/2 \leqslant \phi \leqslant \pi$,

$$K_\phi = -\{\exp[ik_0a(3\pi/2 - \phi)]g(\xi_1) + \exp[ik_0a(\phi - \pi/2)]g(\xi_2)\}$$
(2.2-192)

which converts to an interpolation formula valid for $0 \leqslant \phi \leqslant \pi$,

$$K_\phi \approx -\{\exp[ik_0a(3\pi/2 - \phi)]g(\xi_1) + \exp[-ik_0a\cos\phi - i\xi'^3/3]g(-\xi')\}$$
(2.2-193)

It is apparent from these results that for a large two-dimensional convex surface, the surface currents can be expressed in terms of Fock's universal functions $f$ and $g$, and, in fact, these functions can be used anywhere on the surface when in the proper linear combination and with the proper argument, even though originally they were intended to apply only in the vicinity of the shadow boundary. The primary difference between Fock's original development, valid near the shadow boundary, and extended usage of the function is in the argument $\xi$. In Fock's original work, $\xi$ was defined as in Eq. (2.2-137) and related to the linear distance from the shadow boundary along a straight line tangent to the surface at the boundary.[80] In the generalization of Fock's work, however, at points in the shadow region, the parameter $\xi$ must be related to the distance along a geodesic of the surface; for example, in the case of a circular cylinder, a circular arc.

The scattered fields can be obtained by integrating Eqs. (2.2-183) and (2.2-191) for the current distributions over the surface, and will result in the introduction of two new universal functions. In the region $0 < \phi < \pi$, the scattered field is, for perpendicular polarization,

$$E_z^s = -m \sqrt{\frac{2}{k_0\rho}} \exp[i(k_0\rho + \pi/4)] \sum_{n=0}^{\infty} \left[ \exp\left(\frac{ik_0a}{m}\xi_{1,n}\right) \hat{p}(\xi_{1,n}) \right.$$

$$\left. + \exp\left(\frac{ik_0a}{m}\xi_{2,n}\right) \hat{p}(\xi_{2,n}) \right] \tag{2.2-194}$$

and for parallel polarization

$$H_z^s = -m \sqrt{\frac{2}{k_0\rho}} \exp[i(k_0\rho + \pi/4)] \sum_{n=0}^{\infty} \left[ \exp\left(\frac{ik_0a}{m}\xi_{1,n}\right) \hat{q}(\xi_{1,n}) \right.$$

$$\left. + \exp\left(\frac{ik_0a}{m}\xi_{2,n}\right) \hat{q}(\xi_{2,n}) \right] \tag{2.2-195}$$

In the preceding expressions, the parameters are

$$\xi_{1,n} = m[\pi(2n+1) - \phi], \qquad \xi_{2,n} = m[\pi(2n-1) + \phi] \tag{2.2-196}$$

and the functions $\hat{p}(\xi)$, $\hat{q}(\xi)$ are defined as

$$\hat{p}(\xi) = \frac{1}{\sqrt{\pi}} \int_{-\infty}^{\infty} e^{i\xi t} \frac{\text{Ai}(t)}{\text{Bi}(t) + i\text{Ai}(t)} dt \tag{2.2-197}$$

$$\hat{q}(\xi) = \frac{1}{\sqrt{\pi}} \int_{-\infty}^{\infty} e^{i\xi t} \frac{\text{Ai}'(t)}{\text{Bi}'(t) + i\text{Ai}'(t)} dt \tag{2.2-198}$$

with $\text{Ai}(t)$, $\text{Bi}(t)$ the Airy functions.[81]

Equations (2.2-194) and (2.2-195) are valid everywhere except in the vicinity of $\phi = 0, \pi$. For $\phi \approx 0$,

$$E_z{}^s = -m \sqrt{\frac{2}{k_0\rho}} \exp[i(k_0\rho + \pi/4)]$$

$$\times \left\{ \left(\frac{m}{2}\right)^{1/2} \sqrt{\cos(\phi/2)} \exp[-i(2k_0a\cos(\phi/2) - \pi/4)] \right.$$

$$\left. + \sum_{n=0}^{\infty} \left[ \exp\left(\frac{ik_0a}{m}\xi_{1,n}\right)\hat{p}(\xi_{1,n}) + \exp\left(\frac{ik_0a}{m}\xi_{3,n}\right)\hat{p}(\xi_{3,n}) \right] \right\} \quad (2.2\text{-}199)$$

and

$$H_z{}^s = m \sqrt{\frac{2}{k_0\rho}} \exp[i(k_0\rho + \pi/4)]$$

$$\times \left\{ \left(\frac{m}{2}\right)^{1/2} \sqrt{\cos(\phi/2)} \exp[-i(2k_0a\cos(\phi/2) - \pi/4)] \right.$$

$$\left. - \sum_{n=0}^{\infty} \left[ \exp\left(\frac{ik_0a}{m}\xi_{1,n}\right)\hat{q}(\xi_{1,n}) + \exp\left(\frac{ik_0a}{m}\xi_{3,n}\right)\hat{q}(\xi_{3,n}) \right] \right\} \quad (2.2\text{-}200)$$

with

$$\xi_{3,n} = m[\pi(2n+1) + \phi] \quad (2.2\text{-}201)$$

For $\phi \approx \pi$, then,

$$E_z{}^s = -2m \frac{\exp(ik_0\rho)}{\sqrt{k_0\rho/2}} \frac{\sin[k_0a(\pi - \phi)]}{(\pi k_0a)^{1/2}(\pi - \phi)}$$

$$\times \left\{ \exp[ik_0a(\pi - \phi)]\left[\hat{p}(m\pi - m\phi) + \frac{1}{2\sqrt{\pi}m(\pi - \phi)}\right] \right.$$

$$\left. + \exp[ik_0a(\phi - \pi)]\left[\hat{p}(m\phi - m\pi) + \frac{1}{2\sqrt{\pi}m(\phi - \pi)}\right] \right\} \quad (2.2\text{-}202)$$

and

$$H_z{}^s = -2m \frac{\exp(ik_0\rho)}{\sqrt{k_0\rho/2}} \frac{\sin[k_0a(\pi - \phi)]}{(\pi k_0a)^{1/2}(\pi - \phi)}$$

$$\times \left\{ \exp[ik_0a(\pi - \phi)]\left[\hat{q}(m\pi - m\phi) + \frac{1}{2\sqrt{\pi}m(\pi - \phi)}\right] \right.$$

$$\left. + \exp[ik_0a(\phi - \pi)]\left[\hat{q}(m\phi - m\pi) + \frac{1}{2\sqrt{\pi}m(\phi - \pi)}\right] \right\} \quad (2.2\text{-}203)$$

Interpolation formulas which are approximately valid over the entire range $0 \leqslant \phi < \pi$ are

$$E_z{}^s = -m \sqrt{\frac{2}{k_0 \rho}} \exp[i(k_0 \rho + \pi/4)]\{\exp[ik_0 a(\pi - \phi)] \, \hat{p}(m\pi - m\phi)$$

$$+ \exp[-i2k_0 a \cos \phi/2 - (i/12)(2m \cos \phi/2)^3] \, \hat{p}(-2m \cos \phi/2)\}$$

$$(2.2\text{-}204)$$

$$H_z{}^s = -m \sqrt{\frac{2}{k_0 \rho}} \exp[i(k_0 \rho + \pi/4)]\{\exp[ik_0 a(\pi - \phi)] \, \hat{q}(m\pi - m\phi)$$

$$+ \exp[-i2k_0 a \cos \phi/2 - (i/12)(2m \cos \phi/2)^3] \, \hat{q}(-2m \cos \phi/2)\}$$

$$(2.2\text{-}205)$$

For large negative $\xi$, the functions $\hat{p}(\xi)$, $\hat{q}(\xi)$ have the asymptotic values[78]

$$\lim_{\xi \to -\infty} \hat{p}(\xi) = \exp[-i(\xi^3/12 + \pi/4)] \frac{\sqrt{-\xi}}{2} \left\{1 - i \frac{2}{\xi^3} + \frac{20}{\xi^6} + O\left(\frac{1}{\xi^9}\right)\right\}$$

$$(2.2\text{-}206)$$

$$\lim_{\xi \to -\infty} \hat{q}(\xi) = -\exp[-i(\xi^3/12 + \pi/4)] \frac{\sqrt{-\xi}}{2} \left\{1 + i \frac{2}{\xi^3} - \frac{28}{\xi^6} + O\left(\frac{1}{\xi^9}\right)\right\}$$

$$(2.2\text{-}207)$$

For large positive $\xi$, residue series expansions for $\hat{p}$, $\hat{q}$ are[78]

$$\hat{p}(\xi) = \frac{-\exp(-i\pi/6)}{2\sqrt{\pi}} \sum_{s=1}^{\infty} \frac{\exp[-\frac{1}{2}(\sqrt{3} - i)\,\alpha_s \xi]}{[\mathrm{Ai}'(-\alpha_s)]^2} \qquad (2.2\text{-}208)$$

$$\hat{q}(\xi) = \frac{-\exp(-i\pi/6)}{2\sqrt{\pi}} \sum_{s=1}^{\infty} \frac{\exp[-\frac{1}{2}(\sqrt{3} - i)\,\beta_s \xi]}{\beta_s [\mathrm{Ai}(-\beta_s)]^2} \qquad (2.2\text{-}209)$$

A tabulation of $\hat{p}(\xi)$, $\hat{q}(\xi)$, taken from Logan,[78] is given in Table 2-2 for $-3 \leqslant \xi \leqslant 3$.

An interpretation of these scattered field expressions can be formulated by examining Eq. (2.2-199). The first term with the $2k_0 a \cos \phi/2$ exponential dependence represents the geometrically reflected field; the other two terms represent diffracted fields with phase factors which indicate that they originate at the shadow boundary, travel around the cylinder through the shadow region, and radiate along the tangents to the cylinder passing through the observation point. This is illustrated in Figure 2-17.

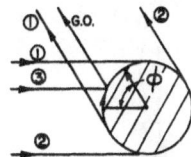

Fig. 2-17. Geometrical representations of various terms of an approximate solution for the scattered fields from a perfectly conducting cylinder using Fock functions.

Table 2-2. The Functions $\hat{p}(\xi)$, $\hat{q}(\xi)$

| $\xi$ | Real $\hat{p}$ | Imag. $\hat{p}$ | Real $\hat{q}$ | Imag. $\hat{q}$ |
|---|---|---|---|---|
| − 3.0 | 0.040 | 0.879 | − 0.135 | − 0.838 |
| − 2.9 | 0.221 | 0.840 | − 0.314 | − 0.771 |
| − 2.8 | 0.377 | 0.769 | − 0.458 | − 0.676 |
| − 2.7 | 0.503 | 0.678 | − 0.568 | − 0.562 |
| − 2.6 | 0.602 | 0.577 | − 0.646 | − 0.440 |
| − 2.5 | 0.674 | 0.469 | − 0.694 | − 0.317 |
| − 2.4 | 0.723 | 0.354 | − 0.717 | − 0.199 |
| − 2.3 | 0.754 | 0.265 | − 0.718 | − 0.090 |
| − 2.2 | 0.768 | 0.173 | − 0.703 | − 0.008 |
| − 2.1 | 0.771 | 0.091 | − 0.676 | − 0.094 |
| − 2.0 | 0.766 | 0.019 | − 0.639 | − 0.166 |
| − 1.9 | 0.756 | − 0.043 | − 0.596 | − 0.226 |
| − 1.8 | 0.742 | − 0.113 | − 0.548 | − 0.274 |
| − 1.7 | 0.726 | − 0.139 | − 0.499 | − 0.311 |
| − 1.6 | 0.712 | − 0.174 | − 0.449 | − 0.338 |
| − 1.5 | 0.699 | − 0.202 | − 0.399 | − 0.357 |
| − 1.4 | 0.689 | − 0.224 | − 0.350 | − 0.368 |
| − 1.3 | 0.682 | − 0.240 | − 0.300 | − 0.372 |
| − 1.2 | 0.679 | − 0.251 | − 0.251 | − 0.371 |
| − 1.1 | 0.682 | − 0.257 | − 0.201 | − 0.365 |
| − 1.0 | 0.690 | − 0.260 | − 0.150 | − 0.356 |
| − 0.9 | 0.707 | − 0.260 | − 0.096 | − 0.342 |
| − 0.8 | 0.732 | − 0.256 | − 0.035 | − 0.327 |
| − 0.7 | 0.770 | − 0.252 | 0.034 | − 0.309 |
| − 0.6 | 0.827 | − 0.244 | 0.119 | − 0.289 |
| − 0.5 | 0.911 | − 0.236 | 0.229 | − 0.268 |
| − 0.4 | 1.043 | − 0.225 | 0.385 | − 0.246 |
| − 0.3 | 1.270 | − 0.214 | 0.634 | − 0.223 |
| − 0.2 | 1.732 | − 0.202 | 1.118 | − 0.200 |
| − 0.1 | 3.135 | − 0.190 | 2.542 | − 0.177 |
| 0.0 | — | − 0.177 | — | − 0.154 |
| 0.1 | − 2.522 | − 0.164 | − 3.074 | − 0.131 |
| 0.2 | − 1.119 | − 0.151 | − 1.650 | − 0.109 |
| 0.3 | − 0.657 | − 0.138 | − 1.166 | − 0.088 |
| 0.4 | − 0.429 | − 0.125 | − 0.918 | − 0.067 |
| 0.5 | − 0.297 | − 0.113 | − 0.762 | − 0.048 |
| 0.6 | − 0.211 | − 0.101 | − 0.654 | − 0.031 |
| 0.7 | − 0.152 | − 0.090 | − 0.573 | − 0.014 |
| 0.8 | − 0.110 | − 0.080 | − 0.508 | − 0.0013 |
| 0.9 | − 0.080 | − 0.070 | − 0.454 | − 0.015 |
| 1.0 | − 0.058 | − 0.061 | − 0.409 | − 0.027 |
| 1.1 | − 0.041 | − 0.053 | − 0.369 | − 0.038 |
| 1.2 | − 0.029 | − 0.045 | − 0.333 | − 0.048 |
| 1.3 | − 0.019 | − 0.039 | − 0.301 | − 0.056 |
| 1.4 | − 0.012 | − 0.032 | − 0.273 | − 0.062 |

Table 2-2 *(continued)*

| $\xi$ | Real $\hat{p}$ | Imag. $\hat{p}$ | Real $\hat{q}$ | Imag. $\hat{q}$ |
|---|---|---|---|---|
| 1.5 | − 0.008 | − 0.027 | − 0.246 | − 0.068 |
| 1.6 | − 0.004 | − 0.023 | − 0.222 | − 0.072 |
| 1.7 | − 0.001 | − 0.018 | − 0.200 | − 0.075 |
| 1.8 | 0.001 | − 0.014 | − 0.180 | − 0.078 |
| 1.9 | 0.002 | − 0.011 | − 0.161 | − 0.079 |
| 2.0 | 0.003 | − 0.010 | − 0.144 | − 0.079 |
| 2.1 | — | — | − 0.128 | − 0.079 |
| 2.2 | — | — | − 0.113 | − 0.078 |
| 2.3 | — | — | − 0.100 | − 0.077 |
| 2.4 | — | — | − 0.088 | − 0.075 |
| 2.5 | — | — | − 0.077 | − 0.072 |
| 2.6 | — | — | − 0.067 | − 0.070 |
| 2.7 | — | — | − 0.058 | − 0.067 |
| 2.8 | — | — | − 0.050 | − 0.064 |
| 2.9 | — | — | − 0.042 | − 0.061 |
| 3.0 | — | — | − 0.033 | − 0.059 |

Referring to Figure 2-17, ray (1) impinges on the cylinder at the shadow boundary, travels around the cylinder a distance $k_0 a(\pi + \phi)$, and is radiated toward the observation point. Similarly, ray (2) impinges on the opposite shadow boundary, travels a distance $k_0 a(\pi - \phi)$, and is reradiated. The third ray is reflected geometrically. Thus, the summation over $n$ in Eq. (2.2-199) and the other field expressions represents rays which have made $n$ complete revolutions around the cylinder before being diffracted. This interpretation is in complete agreement with Keller's geometrical theory of diffraction.

2.2.2.5.2.2. *General Perfectly Conducting Convex Cylinder.* Now, the general case of a plane wave obliquely incident on an arbitrary perfectly conducting convex cylinder is considered. The geometry of the situation is illustrated in Figure 2-18.

The plane wave is incident from the negative $x$ region at an angle $\theta$ with respect to the $z$-axis. Two polarizations are considered; one with no $H_z$ component, designated $\perp$ polarization; and one with no $E_z$ component,

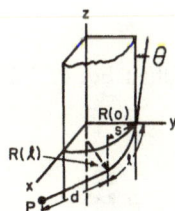

Fig. 2-18. Ray coordinates for a plane wave obliquely incident on a perfectly conducting convex cylinder.

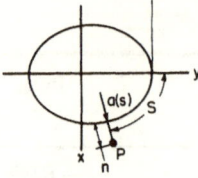

Fig. 2-19. Ray coordinates for a plane wave normally incident on a convex cylinder.

designated $\parallel$ polarization. The quantity $l$ is the length of a geodesic line of the cylinder as measured from the shadow boundary and having an angle $\theta$ with respect to the z-axis, the generator of the cylinder.[†] The quantity $R(l)$ is the radius of curvature of this geodesic line at the point $l$.

Before proceeding with the general case of oblique incidence, the two-dimensional case with $\theta = \pi/2$ will be examined. In this case two new parameters $s$ and $n$ are introduced, and are defined in Figure 2-19. The parameter $s$ is the arc length along the surface of the cylinder and $n$ is the shortest distance from a point $P$ to the surface, which for a convex surface is along a line normal to the surface. The quantity $a(s)$ is the radius of curvature of the cylinder at the point $s$.

The total field components can be obtained from a scalar, $\psi$, proportional to either $E_z$ or $H_z$ depending upon the polarization, which satisfies the wave equation

$$\nabla^2\psi + k_0^2\psi = 0 \tag{2.2-210}$$

In terms of the variables $s$ and $n$, the wave equation becomes

$$\frac{\partial}{\partial s}\left[\frac{\partial\psi/\partial s}{1 + n/a(s)}\right] + \frac{\partial}{\partial n}\left[\left(1 + \frac{n}{a(s)}\right)\frac{\partial\psi}{\partial n}\right] + k_0^2\left(1 + \frac{n}{a(s)}\right)\psi = 0 \tag{2.2-211}$$

And if $\psi = e^{ik_0 s}V'$, and $n \ll a_{min}$, then

$$2ik_0\frac{\partial V'}{\partial s} + \frac{\partial^2 V'}{\partial s^2} + \frac{\partial^2 V'}{\partial n^2} + \frac{1}{a(s)}\frac{\partial V'}{\partial n} + 2k_0^2\frac{n}{a(s)}V' = 0 \tag{2.2-212}$$

Now, introducing a new set of coordinates defined by

$$\xi = \int_0^s\left(\frac{k_0}{2a^2(s')}\right)^{1/3}ds' \tag{2.2-213}$$

and

$$\zeta = \left(\frac{2k_0^2}{a(s)}\right)^{1/3}n \tag{2.2-214}$$

then Eq. (2.2-212) becomes

$$\frac{\partial^2 V'}{\partial\zeta^2} + i\frac{\partial V'}{\partial\zeta} + \zeta V' = 0 \tag{2.2-215}$$

[†] For a circular cylinder, the geodesic curve is a spiral.

This equation is identical to Fock's parabolic equation given previously [Eq. (2.2-145)] and has the same solutions $U'$, $V'$, with

$$U' = \frac{i}{2\sqrt{\pi}} \int_\Gamma e^{i\xi t} \left[ w_2(t - \zeta) - \frac{w_2(t)}{w_1(t)} w_1(t - \zeta) \right] dt \qquad \text{for} \quad U'|_{\zeta=0} = 0$$

$$\text{(2.2-216)}$$

and

$$V' = \frac{i}{2\sqrt{\pi}} \int_\Gamma e^{i\xi t} \left[ w_2(t - \zeta) - \frac{w_2'(t)}{w_1'(t)} w_1(t - \zeta) \right] dt \qquad \text{for} \quad \frac{\partial V'}{\partial \zeta}\bigg|_{\zeta=0} = 0$$

$$\text{(2.2-217)}$$

Thus, by redefining the variables $\xi$, $\zeta$ in terms of the arc length and the normal to the surface, Fock's original results are valid for points in the shadow region such that $\zeta \ll a_{\min}$.

Returning to the oblique incidence case, it is readily apparent that $l = s \sin \theta + z \cos \theta$ is the geodesic length comparable to $s$, and that $\xi$ should be defined in terms of $l$. If this is done and $\zeta$ is defined as above, then the wave equation can again be reduced to a parabolic equation, and solutions equivalent to Eqs. (2.2-216) and (2.2-217) can be obtained.

For oblique incidence, the field close to the surface of the cylinder is given by[82]

$$\pi_z^\perp = \frac{e^{ik_0 l}}{k_0^2 \sin \theta} \left( \frac{R(0)}{R(l)} \right)^{1/6} U'(\xi, \zeta) \tag{2.2-218}$$

for $\perp$ polarization; and

$$\pi_z^\| = \frac{e^{ik_0 l}}{k_0^2 \sin \theta} \left( \frac{R(0)}{R(l)} \right)^{1/6} V'(\xi, \zeta) \tag{2.2-219}$$

for $\|$ polarization; with

$$\xi = \int_0^l \left( \frac{k_0}{2R^2(l)} \right)^{1/3} dl \tag{2.2-220}$$

$$\zeta = \left[ \frac{2k_0^2}{R(l)} \right]^{1/3} n \qquad \text{and} \quad R(l) = a(s)/\sin^2 \theta \tag{2.2-221}$$

The quantities, $\pi_z^\perp$, $\pi_z^\|$, are the $z$ components of the Hertzian potential, and are related to $\mathbf{E}$ and $\mathbf{H}$ by

$$\mathbf{E} = \nabla \times \nabla \times \hat{z}\pi_z^\perp + ik_0 \nabla \times \hat{z}\pi_z^\| \tag{2.2-222}$$

$$\mathbf{H} = \nabla \times \nabla \times \hat{z}\pi_z^\| - ik_0 \nabla \times \hat{z}\pi_z^\perp \tag{2.2-223}$$

The currents on the surface on the cylinder can be obtained from Eqs. (2.2-218), (2.2-219), and (2.2-223), giving

$$\mathbf{K}^\perp = K_z^\perp \hat{z} = \frac{ie^{ik_0 l}}{\sin \theta} \left( \frac{k_0 R(0)}{2} \right)^{1/6} \left( \frac{k_0 R(l)}{2} \right)^{-1/2} f(\xi) \hat{z} \tag{2.2-224}$$

and

$$\mathbf{K}^{\parallel} = k_0^2 \sin^2 \theta \pi_z \hat{\mathbf{e}}_s - ik_0 \cos \theta \frac{\partial \pi_z^{\parallel}}{\partial s} \hat{\mathbf{e}}_z \qquad (2.2\text{-}225)$$

In Eq. (2.2-225), $\hat{\mathbf{e}}_s$ is a unit vector tangent to the projection of the geodesic line $l$ in the $x - y$ plane, and $\hat{\mathbf{e}}_z$ is a unit vector in the $z$ direction. If in Eq. (2.2-225) terms of $O(k_0^{1/3})$ are dropped in $\partial \pi_z / \partial s$, then

$$\mathbf{K}^{\parallel} \approx e^{ik_0 l} \left( \frac{R(0)}{R(l)} \right)^{1/6} g(\xi)(\cos \theta \hat{\mathbf{e}}_z - \sin \theta \hat{\mathbf{e}}_\phi) \qquad (2.2\text{-}226)$$

where $\hat{\mathbf{e}}_s$ has been replaced by the spherical coordinate unit vector $\hat{\mathbf{e}}_\phi$.

Examining $\mathbf{K}^{\parallel}$, it is apparent that to first order for parallel polarization, the surface current in the penumbra and shadow regions is directed along the geodesic line. In addition, examination of the Poynting vector reveals that to first order it coincides with the tangent to the geodesic line at any point in the shadow or penumbral region, thus agreeing with the picture of surface diffracted rays presented by Keller in his geometrical theory of diffraction.

The far-zone scattered electric and magnetic fields in the shadow region can be obtained by integrating over the current distributions Eqs. (2.2-224) and (2.2-226). In addition, interpolation formulas valid anywhere on the cylinder can be obtained, similar to Eqs. (2.2-188) and (2.2-193), thus allowing the scattered fields to be computed for any point in the far zone.

Integrating over the surface-current distribution gives for the far-zone Hertzian potentials

$$\pi_z^{\perp} = - \frac{e^{ik_0(l_0+d_0)+i\pi/4}}{k_0^2 \sin \theta} \sqrt{\frac{2}{k_0 d_0}} \left( \frac{k_0 R(0)}{2} \right)^{1/6} \left( \frac{k_0 R(l_0)}{2} \right)^{1/6} \hat{p}(\xi_0) \qquad (2.2\text{-}227)$$

and

$$\pi_z^{\parallel} = - \frac{e^{ik_0(l_0+d_0)+i\pi/4}}{k_0^2 \sin \theta} \sqrt{\frac{2}{k_0 d_0}} \left( \frac{k_0 R(0)}{2} \right)^{1/6} \left( \frac{k_0 R(l_0)}{2} \right)^{1/6} \hat{q}(\xi_0) \qquad (2.2\text{-}228)$$

The electric and magnetic field components can be obtained from these using Eqs. (2.2-222) and (2.2-223). The parameter $\xi_0$ is defined as

$$\xi_0 = \int_0^{l_0} \left( \frac{k_0}{2R^2(l)} \right)^{1/3} dl \qquad (2.2\text{-}229)$$

where $l_0$ is the length of the geodesic line between the points of tangency of the incident ray and the diffracted ray. The distance $d_0$ is the distance from the observation point along the tangent to the geodesic.

2.2.2.5.2.3. *Elliptic Cylinder*. To illustrate the application of the techniques discussed in the previous section, the backscatter cross section of a perfectly conducting elliptic cylinder will be determined for a plane wave incident perpendicular to the cylinder axis. Letting $a$, $b$, respectively, be the semimajor and semiminor axis of the cylinder, and $\phi$ be the angle between the scattered field direction and the major axis of the ellipse, then the backscattered fields can be expressed as

$$E_z{}^s = E_z{}^0 + E_z{}^F \qquad (2.2\text{-}230)$$

and

$$H_z{}^s = H_z{}^0 + H_z{}^F \qquad (2.2\text{-}231)$$

In these expressions, $E^0$, $H^0$ are the optical contributions to the scattered fields obtained from Eqs. (4.4-17) to (4.4-20), or

$$E_z{}^0 = -\sqrt{\frac{a^2 b^2}{\rho(a^2 \cos^2\phi + b^2 \sin^2\phi)^{3/2}}} \exp\{ik_0[\rho - 2(a^2\cos^2\phi + b^2\sin^2\phi)^{1/2}]\} \qquad (2.2\text{-}232)$$

and

$$H_z{}^0 = -E_z{}^0 \qquad (2.2\text{-}233)$$

The field contributions designated $E^F$, $H^F$ are the diffraction terms emanating from the shadow boundary, and in the Fock formalism can be obtained from Eqs. (2.2-227), (2.2-228), (2.2-222), and (2.2-223), where only terms to the first order in $k_0 a$ are retained. Thus $E^F$ and $H^F$ are given by

$$E_{z,\perp}^F = -2\left(\frac{k_0 R}{2}\right)^{1/3} [e^{ik_0 s}\hat{p}(\xi)] \frac{e^{i(k_0\rho + \pi/4)}}{\sqrt{k_0\rho/2}} \qquad (2.2\text{-}234)$$

and

$$H_{z,\parallel}^F = -2\left(\frac{k_0 R}{2}\right)^{1/3} [e^{ik_0 s}\hat{q}(\xi)] \frac{e^{i(k_0\rho + \pi/4)}}{\sqrt{k_0\rho/2}} \qquad (2.2\text{-}235)$$

The parameters $R$, $s$, $\xi$ are defined as

$$R = \frac{a^2 b^2}{[b^2 \cos^2\phi + a^2 \sin^2\phi]^{3/2}} \qquad (2.2\text{-}236)$$

the radius of curvature of the ellipse at the shadow boundary;

$$s = \int_0^s ds = \int_0^{\pi/2} 2(a^2 \cos^2\gamma + b^2 \sin^2\gamma)^{1/2}\, dy \qquad (2.2\text{-}237)$$

$$= 2aE(\epsilon^2), \quad \epsilon^2 = \frac{a^2 - b^2}{a^2} \qquad (2.2\text{-}238)$$

or $s$ is one-half the circumference of the ellipse;

$$\xi = 2\left(\frac{k_0 b^2}{2a}\right)^{1/3} \int_0^{\pi/2} (1 - \epsilon^2 \sin^2\gamma)^{-1/2}\, dy \qquad (2.2\text{-}239)$$

or

$$\xi = 2 \left( \frac{k_0 b^2}{2a} \right)^{1/3} K(\epsilon^2), \qquad \epsilon^2 = \frac{a^2 - b^2}{a^2} \tag{2.2-240}$$

and is the parameter defined by Eq. (2.2-213). In the above equations, $E$ and $K$ are complete elliptic integrals of the first and second kinds, respectively.[81]

For incidence along the major or minor axes, respectively, the field expressions simplify considerably. Thus for incidence along the major axis

$$E_{z,\perp}^s(0) = -\sqrt{\frac{b^2}{2a\rho}}\, e^{ik_0(\rho - 2a)}$$

$$- 2 \left( \frac{k_0 a^2}{2b} \right)^{1/3} \left\{ e^{i2k_0 aE(\epsilon^2)} \hat{p} \left[ 2 \left( \frac{k_0 b^2}{2a} \right)^{1/3} K(\epsilon^2) \right] \right\} \frac{e^{i(k_0\rho + \pi/4)}}{\sqrt{k_0\rho/2}}$$

$$\tag{2.2-241}$$

and

$$H_{z,\parallel}^s(0) = \sqrt{\frac{b^2}{2a\rho}}\, e^{ik_0(\rho - 2a)}$$

$$- 2 \left( \frac{k_0 a^2}{2b} \right)^{1/3} \left\{ e^{i2k_0 aE(\epsilon^2)} \hat{q} \left[ 2 \left( \frac{k_0 b^2}{2a} \right)^{1/3} K(\epsilon^2) \right] \right\} \frac{e^{i(k_0\rho + \pi/4)}}{\sqrt{k_0\rho/2}}$$

$$\tag{2.2-242}$$

For incidence along the minor axis, then,

$$E_{z,\perp}^s(\pi/2) = -\sqrt{\frac{a^2}{2b\rho}}\, e^{ik_0(\rho - 2b)}$$

$$- 2 \left( \frac{k_0 b^2}{2a} \right)^{1/3} \left\{ e^{i2k_0 aE(\epsilon^2)} \hat{p} \left[ 2 \left( \frac{k_0 b^2}{2a} \right)^{1/3} K(\epsilon^2) \right] \right\} \frac{e^{i(k_0\rho + \pi/4)}}{\sqrt{k_0\rho/2}}$$

$$\tag{2.2-243}$$

and

$$H_{z,\parallel}^s(\pi/2) = \sqrt{\frac{a^2}{2b\rho}}\, e^{ik_0(\rho - 2b)}$$

$$- 2 \left( \frac{k_0 b^2}{2a} \right)^{1/3} \left\{ e^{i2k_0 aE(\epsilon^2)} \hat{q} \left[ 2 \left( \frac{k_0 b^2}{2a} \right)^{1/3} K(\epsilon^2) \right] \right\} \frac{e^{i(k_0\rho + \pi/4)}}{\sqrt{k_0\rho/2}}$$

$$\tag{2.2-244}$$

Expressions for the bistatic scattered fields can be obtained in a similar fashion, following the reasoning outlined in the previous sections.

2.2.2.5.3. *Application to a Perfectly Conducting Sphere.* Consider a plane wave incident on a perfectly conducting sphere as illustrated in

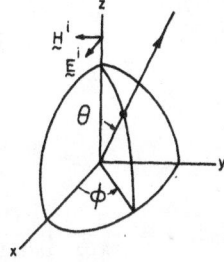

Fig. 2-20.   Coordinates for scattering from a sphere.

Figure 2-20. The field components can be obtained in terms of two potentials, $P$ and $Q$, as follows:[83]

$$
\left.\begin{aligned}
E_r &= \cos\phi\left(\frac{\partial^2}{\partial r^2} + k_0^2\right)\left(r\,\frac{\partial P}{\partial\theta}\right) \\[2mm]
E_\theta &= \cos\phi\left[\frac{1}{r}\frac{\partial}{\partial r}\left(r\,\frac{\partial^2 P}{\partial\theta^2}\right) - \frac{ik_0}{\sin\theta}\frac{\partial Q}{\partial\theta}\right] \\[2mm]
E_\phi &= \sin\phi\left[ik_0\,\frac{\partial^2 Q}{\partial\theta^2} - \frac{1}{r\sin\theta}\frac{\partial}{\partial r}\left(r\,\frac{\partial P}{\partial\theta}\right)\right] \\[2mm]
H_r &= -\sin\phi\left(\frac{\partial^2}{\partial r^2} + k_0^2\right)\left(r\,\frac{\partial Q}{\partial\theta}\right) \\[2mm]
H_\theta &= -\sin\phi\left[\frac{1}{r}\frac{\partial}{\partial r}\left(r\,\frac{\partial^2 Q}{\partial\theta^2}\right) - \frac{ik_0}{\sin\theta}\frac{\partial P}{\partial\theta}\right] \\[2mm]
H_\phi &- -\cos\phi\left[\frac{1}{r\sin\theta}\frac{\partial}{\partial r}\left(r\,\frac{\partial Q}{\partial\theta} - ik_0\,\frac{\partial^2 P}{\partial\theta^2}\right)\right]
\end{aligned}\right\} \qquad (2.2\text{-}245)
$$

where

$$
P = \frac{1}{ik_0^2 r}\sum_{n=1}^{\infty}\frac{2n+1}{n(n+1)}(-1)^n\left[j_n(k_0 r) - \frac{j_n'(k_0 a)}{h_n^{(1)\prime}(k_0 a)}\,h_n^{(1)}(k_0 r)\right]P_n(\cos\theta)
$$
$$
(2.2\text{-}246)
$$

$$
Q = \frac{1}{ik_0^2 r}\sum_{n=1}^{\infty}\frac{2n+1}{n(n+1)}(-1)^n\left[j_n(k_0 r) - \frac{j_n(k_0 a)}{h_n^{(1)}(k_0 a)}\,h_n^{(1)}(k_0 r)\right]P_n(\cos\theta)
$$
$$
(2.2\text{-}247)
$$

The functions $j_n$, $h_n^{(1)}$ are spherical Bessel functions, and the primes denote differentiation with respect to the argument; while $P_n$ is the Legendre poly-nominal of order $n$.

On the surface of a perfectly conducting sphere, the field components $E_\theta$, $E_\phi$, $H_r$ are zero, and

$$
E_r\,|_{r=a} = \frac{\cos\phi}{k_0^2 a^2}\sum_{n=1}^{\infty}(2n+1)(-1)^n\,\frac{1}{h_n^{(1)\prime}(k_0 a)}\,\frac{dP_n(\cos\theta)}{d\theta} \qquad (2.2\text{-}248)
$$

$$H_\theta \,|_{r=a} = \frac{\sin\phi}{k_0 a} \left[ \frac{i}{\sin\theta} \sum_{n=1}^{\infty} \frac{2n+1}{n(n+1)} (-1)^n \frac{1}{h_n^{(1)\prime}(k_0 a)} \frac{dP_n(\cos\theta)}{d\theta} \right.$$

$$\left. + \sum_{n=1}^{\infty} \frac{2n+1}{n(n+1)} (-1)^n \frac{1}{h_n^{(1)}(k_0 a)} \frac{d^2 P_n(\cos\theta)}{d\theta^2} \right] \qquad (2.2\text{-}249)$$

$$H_\phi \,|_{r=a} = \frac{\cos\phi}{k_0 a} \left[ \frac{1}{\sin\theta} \sum_{n=1}^{\infty} \frac{2n+1}{n(n+1)} (-1)^n \frac{1}{h_n^{(1)}(k_0 a)} \frac{dP_n(\cos\theta)}{d\theta} \right.$$

$$\left. + i \sum_{n=1}^{\infty} \frac{2n+1}{n(n+1)} (-1)^n \frac{1}{h_n^{(1)\prime}(k_0 a)} \frac{d^2 P_n(\cos\theta)}{d\theta^2} \right] \qquad (2.2\text{-}250)$$

Asymptotic expressions, valid for $k_0 a$ large, in terms of Fock functions can be derived for these field components. Following Belkina and Vaynshtein,[83] there are three regions of interest; the illuminated region, the shadow region for $\theta < \pi$, and the vicinity of the caustic, $\theta \approx \pi$. In the illuminated region, $\theta \leqslant \pi/2$,

$$E_r \,|_{r=a} = \left[ \cos\phi \sin\theta \, e^{-ik_0 a \cos\theta + i\xi'^3/3} g(\xi') + i\cos\phi \, \frac{e^{ik_0 a(3\pi/2-\theta)}}{\sqrt{\sin\theta}} g(\xi_1) \right]$$
$$(2.2\text{-}251)$$

$$H_\theta \,|_{r=a} = - \left[ \frac{i\sin\phi}{m} e^{-ik_0 a \cos\theta + i\xi'^3/3} f(\xi') + \frac{\sin\phi \, e^{ik_0 a(3\pi/2-\theta)}}{m\sqrt{\sin\theta}} f(\xi_1) \right]$$
$$(2.2\text{-}252)$$

$$H_\phi \,|_{r=a} = - \left[ \cos\phi \, e^{-ik_0 a \cos\theta + i\xi'^3/3} g(\xi') - \frac{i\cos\phi}{\sqrt{\sin\theta}} e^{ik_0 a(3\pi/2-\theta)} g(\xi_1) \right]$$
$$(2.2\text{-}253)$$

In the shadow region, away from the caustic at $\theta = \pi$, or for $\pi/2 \leqslant \theta < \pi$,

$$E_r \,|_{r=a} = \frac{\cos\phi}{\sqrt{\sin\theta}} \left[ e^{ik_0 a(\theta-\pi/2)} g(\xi) + i e^{ik_0 a(3\pi/2-\theta)} g(\xi_1) \right] \qquad (2.2\text{-}254)$$

$$H_\theta \,|_{r=a} = -\sin\phi \left[ \frac{i}{m\sqrt{\sin\theta}} [e^{ik_0 a(\theta-\pi/2)} f(\xi) - i e^{ik_0 a(3\pi/2-\theta)} f(\xi_1)] \right.$$

$$\left. - \frac{i}{k_0 a \sin^{3/2}\theta} [e^{ik_0 a(\theta-\pi/2)} g(\xi) + i e^{ik_0 a(3\pi/2-\theta)} g(\xi_1)] \right] \qquad (2.2\text{-}255)$$

$$H_\phi \,|_{r=a} = -\cos\phi \left[ \frac{1}{\sqrt{\sin\theta}} [e^{ik_0 a(\theta-\pi/2)} g(\xi) - i e^{ik_0 a(3\pi/2-\theta)} g(\xi_1)] \right.$$

$$\left. - \frac{1}{k_0 am \sin^{3/2}\theta} [e^{ik_0 a(\theta-\pi/2)} f(\xi) + i e^{ik_0 a(3\pi/2-\theta)} f(\xi_1)] \right] \qquad (2.2\text{-}256)$$

In the shadow region in the vicinity of the caustic, $\theta \approx \pi$;

$$E_r \big|_{r=a} = 2 \cos \phi \sqrt{\pi} \, m^{3/2} e^{i(k_0 a + 1/2)\pi/2} \sqrt{\frac{\pi - \theta}{\sin \theta}} \, J_1[\nu_1'(\pi - \theta)] \, g(m\pi/2)$$

$$\text{(2.2-257)}$$

$$H_\theta \big|_{r=a} = 2 \sin \phi \sqrt{m\pi} \sqrt{\frac{\pi - \theta}{\sin \theta}} \left[ e^{i(k_0 a + 1/2)\pi/2} J_0''[\nu_1{}^0(\pi - \theta)] f(m\pi/2) \right.$$

$$\left. - \frac{i e^{i(k_0 a + 1/2)\pi/2} m}{k_0 a \sin \theta} J_1[\nu_1'(\pi - \theta)] \, g(m\pi/2) \right]$$

$$\text{(2.2-258)}$$

$$H_\phi \big|_{r=a} = 2 \cos \phi \sqrt{\pi} \, m^{3/2} \sqrt{\frac{\pi - \theta}{\sin \theta}} \left[ e^{i(k_0 a - 1/2)\pi/2} J_0''[\nu_1'(\pi - \theta)] \, g(m\pi/2) \right.$$

$$\left. - \frac{e^{i(k_0 a + 1/2)\pi/2}}{k_0 a m \sin \theta} J_1[\nu_1{}^0(\pi - \theta)] f(m\pi/2) \right]$$

$$\text{(2.2-259)}$$

In all these equations, the parameters are

$$\xi = m(\theta - \pi/2); \qquad \xi_1 = m(3\pi/2 - \theta); \qquad \xi' = -m \cos \theta \quad \text{(2.2-260)}$$

and

$$\nu_1{}^0 = k_0 a + m t_1{}^0; \qquad \nu_1' = k_0 a + m t_1' \qquad \text{(2.2-261)}$$

where $t_1{}^0$ and $t_1'$ are related to the zero's of the Airy function, and are given by

$$\left. \begin{aligned} t_1{}^0 &= 2.3381 e^{i\pi/3} \\ t_1' &= 1.0188 e^{i\pi/3} \end{aligned} \right\} \qquad \text{(2.2-262)}$$

In their 1957 article[83], Belkina and Vaynshtein compared the surface fields, as computed by the exact series Eqs. (2.2-248) to (2.2-250), with those computed by the asymptotic expressions, Eqs. (2.2-251) to (2.2-259), for $k_0 a = 10$. These curves are shown in Figures 2-21 and 2-22; it is evident that for $k_0 a = 10$, the asymptotic solutions agree very well with the exact results, even at the caustic, $\theta = \pi$.

In order to determine the scattered fields, the surface currents can be integrated over the sphere; however, the currents derived from Eqs. (2.2-251) to (2.2-259) are given by complex expressions not easily integrated. Since the far-zone scattered fields are sensitive primarily only to the currents near the shadow boundary and in the illuminated region, adequate results can be obtained by using simpler expressions which are poorer approximations to the exact surface currents. These approximate surface fields are given by

$$\left. \begin{aligned} E_r &\approx \cos \phi \sin \theta \, e^{-ik_0 a \cos \theta + i\xi'^3/3} g(\xi') \\ H_\theta &\approx \frac{-i \sin \phi}{m} e^{-ik_0 a \cos \theta + i\xi'^3/3} f(\xi') \\ H_\phi &\approx -\cos \phi \, e^{-ik_0 a \cos \theta + i\xi'^3/3} g(\xi') \end{aligned} \right\} \qquad \text{(2.2-263)}$$

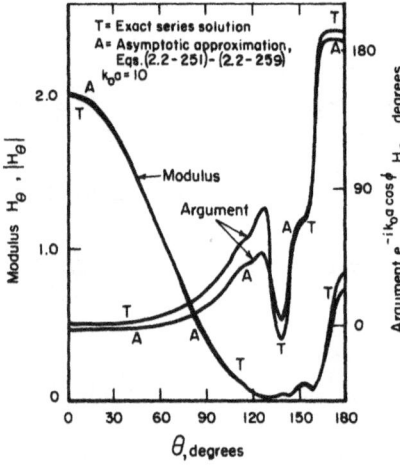

Fig. 2-21. Exact and approximate solutions for the tangential magnetic field, $H_\theta$, on the surface of a perfectly conducting sphere, $\phi = \pi/2$ plane [After Belkina and Vaynshtein[83]].

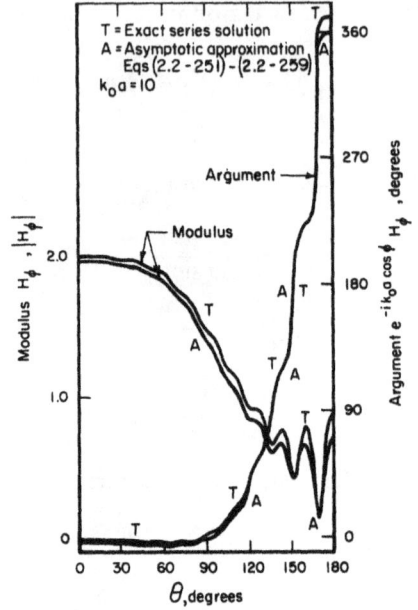

Fig. 2-22. Exact and approximate solutions for the tangential magnetic field, $H_\phi$, on the surface of a perfectly conducting sphere, $\phi = 0$ plane [After Belkina and Vaynshtein[83]].

Using these in the integral for the vector potential,

$$\mathbf{A} = \frac{1}{4\pi} \int_s \mathbf{J}\, \frac{e^{ik_0 r}}{r}\, ds$$

approximate expressions for the scattered fields, obtained by Federov[84], agree well with the exact results for $k_0 a = 5$ and above. As in the case of the cylinder, the scattered fields are expressed in terms of the universal functions $\hat{p}(\xi)$, $\hat{q}(\xi)$ for $0 < \theta \leqslant \pi$;

$$E_\theta \approx \frac{2e^{ik_0 r}}{k_0 r}\, m^4\, \sqrt{\pi}\, \cos\phi\, \Big[ [J_1'(2k_0 a \cos\theta/2) + iJ_1''(2k_0 a \cos\theta/2)]\, e^{i\tau'^{\,3}/12}\hat{q}(\tau')$$

$$+ \sqrt{\frac{\pi}{\sin\theta}\frac{\theta}{\,}}\, [J_1'(k_0 a(\pi - \theta)) - iJ_1''(k_0 a(\pi - \theta))]\, \hat{q}(\tau)$$

$$+ \frac{1}{k_0 a \sin\theta}\, [J_1(2k_0 a \cos\theta/2) - iJ_2(2k_0 a \cos\theta/2)]\, e^{i\tau'^{\,3}/12}\hat{p}(\tau')$$

$$+ \frac{\sqrt{\pi - \theta}}{k_0 a \sin^{3/2}\theta}\, [J_1(2k_0 a(\pi - \theta)) + iJ_2(k_0 a(\pi - \theta))]\, \hat{p}(\tau) \Big] \qquad (2.2\text{-}264)$$

and

$$E_\phi \approx \frac{2e^{ik_0 r}}{k_0 r} m^4 \sqrt{\pi} \sin \phi \left[ [J_1'(2k_0 a \cos \theta/2) + iJ_1''(2k_0 a \cos \theta/2)] e^{i\tau'^3/12} \hat{p}(\tau') \right.$$

$$+ \sqrt{\frac{\pi - \theta}{\sin \theta}} [J_1'(k_0 a(\pi - \theta)) - iJ_1''(k_0 a(\pi - \theta))] \hat{p}(\tau)$$

$$+ \frac{1}{k_0 a \sin \theta} [J_1(2k_0 a \cos \theta/2) - iJ_2(2k_0 a \cos \theta/2)] e^{i\tau'^3/12} \hat{q}(\tau')$$

$$\left. + \frac{\sqrt{\pi - \theta}}{k_0 a \sin^{3/2} \theta} [J_1(k_0 a(\pi - \theta)) + iJ_2(k_0 a(\pi - \theta))] \hat{q}(\tau) \right] \qquad (2.2\text{-}265)$$

while for $0 \leqslant \theta < \pi$,

$$E_\theta \approx - \frac{e^{ik_0 r}}{r} \left[ \frac{a}{2} \cos \phi\, e^{-2ik_0 a \cos \theta/2} + \frac{2im^4 \sqrt{\pi} \cos \phi}{k_0} e^{ik_0 a \pi} \sqrt{\frac{\theta}{\sin \theta}} \right.$$

$$\times \left( [J_1'(k_0 a\theta) + iJ_1''(k_0 a\theta)] \hat{q}(\tau) + [J_1'(k_0 a\theta) - iJ_1''(k_0 a\theta)] \hat{q}(\tau'') \right.$$

$$\left. \left. - [J_1(k_0 a\theta) - iJ_2(k_0 a\theta)] \frac{\hat{p}(\tau)}{k_0 a \sin \theta} - [J_1(k_0 a\theta) + iJ_2(k_0 a\theta)] \frac{\hat{p}(\tau'')}{k_0 a \sin \theta} \right) \right]$$

$$(2.2\text{-}266)$$

$$E_\phi \approx \frac{e^{ik_0 r}}{r} \left[ (a/2) \sin \phi\, e^{-2ik_0 a \cos \theta/2} - \frac{2im^4 \sqrt{\pi} \sin \phi}{k_0} e^{ik_0 a \pi} \sqrt{\frac{\theta}{\sin \theta}} \right.$$

$$\times \left( [J_1'(k_0 a\theta) + J_1''(k_0 a\theta)] \hat{p}(\tau) + [J_1'(k_0 a\theta) - iJ_1''(k_0 a\theta)] \hat{p}(\tau'') \right.$$

$$\left. \left. - [J_1(k_0 a\theta) - iJ_2(k_0 a\theta)] \frac{\hat{q}(\tau)}{k_0 a \sin \theta} - [J_1(k_0 a\theta) + iJ_2(k_0 a\theta)] \frac{\hat{q}(\tau'')}{k_0 a \sin \theta} \right) \right]$$

$$(2.2\text{-}267)$$

The parameters $\tau$, $\tau'$, $\tau''$ are defined as

$$\tau = m(\pi - \theta), \qquad \tau' = -2m \cos \theta/2, \qquad \tau'' = m(\pi + \theta) \qquad (2.2\text{-}268)$$

Table 2-2 of $\hat{p}$ and $\hat{q}$, along with a table of Bessel functions, allows the computation of $E_\theta$, $E_\phi$ for any value of $k_0 a$ and $\theta$. In the illuminated region for $\tau'$ large and negative, the quantities $e^{i\tau'^3/12}\hat{q}(\tau')$ and $e^{i\tau'^3/12}\hat{p}(\tau')$ can be replaced by their asymptotic expressions.

Comparing Eqs. (2.2-254) to (2.2-256) for the surface fields near the shadow boundary with the results predicted by Fock's original work, several differences are apparent. First, as in the case of a cylindrical structure, the phase factors associated with $f$ or $g$, as well as the arguments of $f$ and $g$, are related to the arc lengths along a geodesic of the surface. In addition, however, unlike the two-dimensional case, a factor of $(\sin \theta)^{-1/2}$ is included in each field term. This $(\sin \theta)^{-1/2}$ is a convergence factor due to the geometrical structure of the sphere forcing the fields moving into the shadow region from the shadow boundary to converge on the caustic at $\theta = \pi$. This convergence factor can be deduced for an arbitrary three-dimensional body by considering the total energy flow between two geodesic curves separated an infinitesimal amount[85]. When this is done, the field functions on a finite three-dimensional body are required to be multiplied by a factor $\Gamma$, where

$$\Gamma = \frac{1}{\sqrt{\dfrac{dr(s, x)}{dx}}} \qquad (2.2\text{-}269)$$

The function, $r(s, x)$, is the distance between two equiphase points on any two geodesic lines of the surface which are separated at the shadow boundary a distance $x$. The distance along the geodesics from the shadow boundary to the equiphase points is $s$. This is illustrated in Figure 2-23. For the sphere, it is apparent that $dr = a \sin \theta \, d\phi$ and $dx = a \, d\phi$, so

$$\frac{dr}{dx} = \sin \theta$$

and

$$\frac{1}{\sqrt{\dfrac{dr}{dx}}} = (\sin \theta)^{-1/2} \qquad (2.2\text{-}270)$$

### 2.2.2.6. Variational Techniques

A technique which may in some cases be useful for obtaining approximate expressions valid in the resonance region, improvements to physical optics, or Rayleigh results, is the variational technique. The common factor in all variational techniques is the formulation of an expression for the desired function, such that first-order errors in approximating an unknown quantity result in only second-order errors in the approximation to the desired function. As an example, consider the integral equation

$$Lu = F \qquad (2.2\text{-}271)$$

where $u$ is an unknown function, $L$ is a linear integral or integrodifferential operator, and $F$ is a known function. Assume that it is desired to find $(u, H)$,

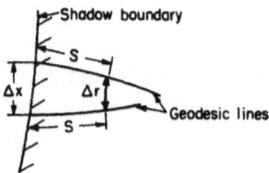

Fig. 2-23. Illustration of the differential orthogonal distance along a convex surface between two equiphase points located on two geodesic lines separated an infinitesimal amount at the shadow boundary.

a scalar product of $u$ and some known function $H$, defined such that $(u, v) = (v, u)$. An example of such a scalar product might be

$$(u, H) = \int_a^b u(x)\, H(x)\, dx \qquad (2.2\text{-}272)$$

It can be shown[86] that stationary functionals for $(u, H)$ are

$$(u, H) + (v, F) - (Lu, v) \qquad (2.2\text{-}273)$$

and

$$\frac{(u, H)(v, F)}{(Lu, v)} \qquad (2.2\text{-}274)$$

where[†]

$$\tilde{L}v = H \qquad (2.2\text{-}275)$$

This means that if the approximations to $u$ and $v$, $u'$ and $v'$, where $u' = u + \varDelta$ and $v' = v + \varDelta$ are inserted into the expressions (2.2-273) and (2.2-274), the error in the resultant approximation to $(u, H)$ will be of order $\varDelta^2$.

Equation (2.2-271) is a homogeneous integral equation. For an inhomogeneous equation of the type

$$u = F - Lu \qquad (2.2\text{-}276)$$

stationary functionals other than (2.2-273) and (2.2-274) result. These can be obtained from Eqs. (2.2-273) and (2.2-274) by substituting for $L$, the operator

$$Q = 1 + L \qquad (2.2\text{-}277)$$

which converts Eq. (2.2-276) into a homogeneous equation, or

$$Qu = F \qquad (2.2\text{-}278)$$

When this is done the resulting stationary functionals are

$$(u, H) + (v, F) - (u, v) - (Lu, v) \qquad (2.2\text{-}279)$$

and

$$\frac{(u, H)(v, F)}{(u, v) + (Lu, v)} \qquad (2.2\text{-}280)$$

In the preceding discussion, the functions $u, v, F, H$ are considered scalar functions of one or more variables. Stationary functionals, equivalent to Eqs. (2.2-273), (2.2-274) and (2.2-279), (2.2-280) result when the integral equations (2.2-271) and (2.2-276) are vector equations, in which case the operator $L$ is a dyadic integrodifferential operator. For example, Kodis[87] has given a stationary expression for the far zone scattered field amplitude in

[†] The symbol $\tilde{L}$ is used to designate the adjoint operator to the operator $L$ (see Reference 86).

terms of the surface currents on a perfectly conducting finite scatterer. Starting with an integral equation for the surface current of the form

$$\hat{n} \times \mathbf{E}^i(\rho) = i\omega\mu \int_{s'} \mathbf{K}(\hat{\mathbf{k}}_0{}^i, \rho') \cdot \tilde{\tilde{\Gamma}}_0(\rho, \rho') \, ds' \qquad (2.2\text{-}281)$$

where $\rho$, $\rho'$ are position vectors from an arbitrary origin to points on the scatterer surface; $\mathbf{K}(\hat{\mathbf{k}}_0{}^i, \rho')$ is the surface current on the scatterer due to a plane wave incident in the direction indicated by the unit vector $\hat{\mathbf{k}}_0{}^i$. The symbol $\tilde{\tilde{\Gamma}}_0$ represents the free-space dyadic Green's function defined as

$$\tilde{\tilde{\Gamma}}_0(\mathbf{r}, \mathbf{r}') = \left(\tilde{\tilde{I}} + \frac{1}{k_0{}^2} \nabla^2\right) \psi_0(\mathbf{r}, \mathbf{r}') \qquad (2.2\text{-}282)$$

with $\tilde{\tilde{I}}$ the unit dyad or tensor and $\psi_0$ the free-space Green's function defined by Eq. (2.2-21).

In the far zone, the scattered electric field is given by

$$\mathbf{E}^s(\mathbf{r}) \approx -\frac{i\omega\mu}{4\pi} \frac{\exp(ik_0 r)}{r} \left[\hat{\mathbf{k}}_0{}^s \times \left(\hat{\mathbf{k}}_0{}^s \times \int_{s'} \mathbf{K}(\hat{\mathbf{k}}_0{}^i, \rho') \exp[-ik_0(\hat{\mathbf{k}}_0{}^s \cdot \rho')] \, ds'\right)\right]$$

$$(2.2\text{-}283)$$

Defining the scattering amplitude as

$$\mathbf{A}(\hat{\mathbf{k}}_0{}^i, \hat{\mathbf{k}}_0{}^s) = \frac{i\omega\mu}{4\pi} \int_{s'} \mathbf{K}(\hat{\mathbf{k}}_0{}^i, \rho') \exp[-ik_0(\hat{\mathbf{k}}_0{}^s \cdot \rho')] \, ds' \qquad (2.2\text{-}284)$$

then a stationary expression for $\mathbf{A}$ is

$$\hat{e}^i \cdot \mathbf{A}(\hat{\mathbf{k}}_0{}^i, -\hat{\mathbf{k}}_0{}^s) = -\frac{1}{4\pi} \left[\hat{e}^s \cdot \int_{s'} \mathbf{K}(\hat{\mathbf{k}}_0{}^i, \rho') \exp[ik_0(\hat{\mathbf{k}}_0{}^s \cdot \rho')] \, ds'\right]$$

$$\times \left[\frac{\hat{e}^i \cdot \int_{s'} \mathbf{K}(\hat{\mathbf{k}}_0{}^s, \rho') \exp[ik_0(\hat{\mathbf{k}}_0{}^s \cdot \rho')] \, ds'}{\int_{s'}\int_s \mathbf{K}(\hat{\mathbf{k}}_0{}^i, \rho') \cdot \tilde{\tilde{\Gamma}}_0(\rho', \rho) \cdot \mathbf{K}(\hat{\mathbf{k}}_0{}^s, \rho) \, ds \, ds'}\right] \qquad (2.2\text{-}285)$$

which is equivalent to Eq. (2.2-274). The unit vectors $\hat{e}^i$, $\hat{e}^s$ in this equation specify the direction of the incident and scattered electric fields, respectively.

Although stationary vector functionals, such as Eq. (2.2-285) can be formulated starting from the general vector integral equations for the surface currents or total fields, in most applications the general vector problem is reduced to a scalar problem prior to the utilization of any variational principle. Variational techniques have been applied to only a few scattering problems. Notable are the applications by Levine and Schwinger,[88,89] and Huang[90] to the computation of the transmission coefficient of circular, elliptic, and rectangular apertures in a perfectly conducting screen. Variational methods have also been applied by Kouyoumjian[91] to calculate the backscatter cross section of circular loops and rectangular plates, by Tai[92] to compute the backscattered fields from a thin wire, and by Cohen[93] to the backscattered field from a dielectric cylinder.

In all of these applications, relatively simple approximations to the unknown fields or surface currents have been used in the stationary functional. This is primarily due to the difficulty of using more complex expressions. For objects such as discs, square plates, loops, etc., where at normal incidence resonance effects are minimal, variational techniques have yielded excellent agreement with experiment (see Figure 7-29). For more complex objects, such as thin wires, the agreement is still good; however, the frequency range over which good agreement is attainable is considerably smaller.

The major difficulty with variational approaches, other than their analytical complexity, is that, in general, no estimates are available of the degree to which the approximate solution deviates from the exact solution.

### 2.2.2.7. Low-Frequency Approximations

For bodies whose ratio of the maximum dimension of the body to the radar wavelength is very small, the scattering is termed low-frequency scattering, or the body is said to be in the Rayleigh region. (Since Rayleigh made many contributions to scattering problems other than low-frequency approximations, there is a possibility of confusion in designating the low-frequency region as the Rayleigh region. However, since it is a customary designation, it is used throughout this book.) Rayleigh or low-frequency scattering should be distinguished from Rayleigh–Gans scattering. Rayleigh–Gans scattering occurs with objects whose dielectric constants are close to that of the ambient medium, that is, the total field in the presence of a dielectric body for which $\mu = \mu_0$ is given by

$$\mathbf{E}^T = \mathbf{E}^i + \frac{1}{4\pi\epsilon_0}(\nabla\nabla\cdot + k_0^2)\int_V (\epsilon - \epsilon_0)\,\mathbf{E}^T\psi_0(\mathbf{r}',\mathbf{r})\,dV \quad (2.2\text{-}286)$$

where $V$ is the volume of the radar target and $\psi_0$ is the free space Green's function. For the Rayleigh–Gans approximation, since $\epsilon - \epsilon_0$ is very small, the second term is small, and an error in the assumption of the value of $\mathbf{E}^T$ inside the integral is not so significant. In the Rayleigh–Gans approximation, $\mathbf{E}^T$ inside the integral is assumed to be $\mathbf{E}^i$, thus, it is a form of the Born approximation discussed in Section 2.2.1.2. The Rayleigh–Gans approximation gives good results for all size objects (measured in wavelengths) for relative dielectric constants close to unity. Rayleigh scattering is useful for all dielectric constants, but only for small objects.

There are several techniques available to determine the cross section for a body in the Rayleigh region.[94] If the exact expression for the radar cross section as a function of frequency for a body is known, the cross section at low frequencies may be obtained by expanding the known solution in a power series in frequency and using the first or the first and second terms. Since a formal solution is available for only a few bodies, this technique is not very useful and other approximations are needed. In an 1897 paper, Rayleigh[95] showed that a first approximation could be obtained by solving the problem for the desired body in static electric and magnetic fields. This technique has the advantage that it is Laplace's equation which must be solved, rather than the vector Helmholtz equation. Thus it is reduced to a problem in potential theory. The basic equation of Rayleigh scattering for a dielectric body is Eq. (2.2-286). The scattered field at a large distance $R$ in the direction of the unit vector $\hat{e}$ when the target is very small is ($\mu = \mu_0$)

$$E^s \approx \frac{k_0^2 e^{ik_0 R}}{4\pi\epsilon_0 R} \{\mathbf{p} - (\mathbf{p} \cdot \hat{e})\,\hat{e}\} \qquad (2.2\text{-}287)$$

where

$$\mathbf{p} = \int_V (\epsilon - \epsilon_0)\, \mathbf{E}^T \, dV'$$

The target has a scattered field similar to an electric dipole of moment $\mathbf{p}$. For a perfectly conducting target the basic equations are, from Eqs. (2.2-26) to (2.2-29),

$$\mathbf{E}^T = \mathbf{E}^i + \frac{1}{4\pi} \int_s \{(\hat{n}' \cdot \mathbf{E}^T)\,\nabla'\psi_0 + i\omega\mu(\hat{n}' \times \mathbf{H}^T)\,\psi_0\}\, ds' \quad (2.2\text{-}288)$$

and

$$\mathbf{H}^T = \mathbf{H}^i + \frac{1}{4\pi} \int_s (\hat{n}' \times \mathbf{H}^T) \times \nabla'\psi_0\, ds' \qquad (2.2\text{-}289)$$

where $S$ is the surface of the obstacle and $\hat{n}'$ a unit vector normal to the surface pointing from the interior of $S$ to the exterior. Making the same far field assumptions as were made for the dielectric case, one obtains

$$E^s \approx \frac{k_0^2 e^{ik_0 R}}{4\pi\epsilon_0 R} \left[ \mathbf{p} - (\mathbf{p} \cdot \hat{e})\,\hat{e} + \frac{1}{\eta_0}(\mathbf{m} \times \hat{e}) \right] \qquad (2.2\text{-}290)$$

where

$$\mathbf{m} = \frac{1}{2}\mu_0 \int_s \mathbf{r}' \times (\hat{e} \times \mathbf{H}^T)\, ds', \qquad \mathbf{p} = \epsilon_0 \int_s (\hat{n}' \cdot \mathbf{E}^T)\,\mathbf{r}'\, ds' \quad (2.2\text{-}291)$$

Thus, for a perfectly conducting target, the first approximation for the scattered field is that it comes from an electric dipole of moment $\mathbf{p}$ and a magnetic dipole of moment $\mathbf{m}$.

Stevenson[96] extended Rayleigh's work and observed that a formal solution could be obtained as a power series in the ratio of the maximum dimension of the scatterer to the wavelength by solving problems in potential theory for each term in the series. Stevenson has a special technique for finding the first three terms of this series which he used in his worked examples.[97] Stevenson's general technique is applicable to any isotropic object either metallic or dielectric. However, Kleinman[98] reports that the general technique needs clarification in order to prevent errors, and presents an improved version which is applicable only to perfectly conducting objects.

Another technique for determining Rayleigh scattering has been derived by Darling.[99] He expressed the first term of the Rayleigh series for the spherically capped cone in terms of infinite matrices. His method could be applied to more general bodies, since he has shown that potential problems can be solved for any region which is the intersection of regions for each of which the exterior electrostatic Green's function is known, provided the region is axially symmetric with the same axes of symmetry as the source. The technique is discussed primarily for the scalar case in a paper by Darling and Senior.[100]

An approximate method of determining the Rayleigh backscatter cross section for axial incidence on a perfectly conducting body of revolution has been given by Siegel.[101,102] The exact series for the cross section can be written as

$$\sigma = 4\pi k_0^4 \sum_{l=0}^{\infty} |f_l|^2 \qquad (2.2\text{-}292)$$

Siegel expressed $f_0$ as the product of a constant, the volume of the target, and a dimensionless correction factor $F$, that is,

$$f_0 = \frac{1}{\pi} VF \qquad (2.2\text{-}293)$$

He then found $F$ by approximating the target by a spheroid of semimajor and semiminor axes $a$ and $b$, respectively. From this he deduced an $F$ of the form

$$F = 1 + \frac{1}{\pi \tau} e^{-\tau} \qquad (2.2\text{-}294)$$

where $\tau = a/b$.

For a sphere this method gives a Rayleigh cross section of $1.5889\, k_0^4 V^2$, as compared to the correct result (see Chapter 3) of $1.6114\, k_0^4 V^2$. A generalization of Siegel's method, using the exact Rayleigh results for general spheroids and ellipsoids, is given in Chapter 8. Also in Chapter 8 is a listing of some Rayleigh cross sections for various objects obtained using this technique. Further discussion of the determination of the Rayleigh cross section for various standard shapes such as spheres, cones, disks, etc., is included in the chapters devoted to these shapes.

### 2.2.2.8. *Numerical Solutions*

The term "numerical solution" implies the formulation of a problem in a style suitable for machine evaluation and final solution. The advent of the high-speed digital computer and its nearly universal availability has emphasized and accelerated the evolution of numerical techniques and their application to the solution of physical problems. The great speed of elementary calculations has enhanced the use of simple iterative techniques for the solution of problems that were formerly intractable in their exact form.

There are many areas of electromagnetic scattering theory where computer and numerical techniques can be used. For instance, they can be, and are, used in the subsequent chapters on spheres and cylinders to sum the exact series solution and generate the Bessel functions involved in each series term. Or they can be used to solve the potential problem of a body immersed in static fields in order to obtain the low-frequency electromagnetic scattered fields from the object. As a third example, one could use the computer along with a numerical integration program to evaluate a physical optics integral for scattering from a complex body in the high-frequency limit. In all of these cases, the computer and numerical techniques are used to solve the problem. However, in the first example given, the exact series solution is already known, and the computer is merely used as an aid in performing the summation. In the second and third examples, one has already reduced and restricted the scattering problem to the low- or high-frequency region. Hence, one is using the computer to solve an approximation in the first place, and the numerical solution is only as valid as the original approximation. Therefore, in these examples the computer is used as a secondary tool in the solution, since either the exact solution is already known, or else one is starting with an approximate solution in the first place. Nonetheless, the computer is an invaluable and indispensable tool in evaluation of this type of problems.

The general usage of the term "numerical solution" of scattering problems, however, does not include the preceding examples in which the computer is used as a secondary tool in the problem solution. Rather, the term applies to direct solution of the exact Maxwell's equations or their integral forms by numerical techniques. The bodies or targets for which the scattered fields are sought are of sufficient complexity in shape that an exact analytical solution cannot be derived as it can, for instance, for the sphere. Hence, the solution is not readily obtainable by classical boundary value or integral equation techniques. In beginning with an exact formulation upon which the numerical solution is to be based, rather than an approximation, the solution for the scattered field is not restricted to certain frequency regions or body sizes. In fact, the main value of a numercial solution proceeding from an exact formulation is its validity in the resonance region where neither high- nor low-frequency approximations are generally applicable.

Several investigators, somewhat simultaneously and independently, began looking at such solutions toward the end of the 1950's and early 1960's. The techniques of these investigators differ on some points but possess many similar characteristics. As of now, several of these techniques have been reduced to working computer programs.[†] The general similar properties of these numerical techniques and computer programs are the following: (a) All of the programs use as input functions the size and shape of the scattering target, the frequency, the incidence and scattering angles of interest, and the incident wave polarization direction. (b) The output of such programs can be the far-zone scattered field strength, phase angle, and polarization directions. The scattering cross section is also available. Most of the programs can also give the surface currents on the scatterer (assuming the scatterer is perfectly conducting). (c) Although the techniques used to obtain the numerical solution for the scattered field from perfectly conducting targets can be extended to dielectric or coated targets in a straightforward manner, up to now the only investigator who has obtained results for dielectric objects (cylinders of arbitrary cross section) has been Richmond of Ohio State University.[108] (d) All of the solutions, in some manner, reduce to an infinite set of linear equations in an infinite number of unknowns by enforcing boundary or constraint conditions at an infinite number of points on the surface of the body. This infinite set is truncated in a suitable fashion resulting in a finite set of equations. The number of equations (or number of points of enforcement on the scatterer) is chosen large enough so that the spacing between points is generally smaller than a wavelength. Hence, the number of equations to be solved is proportional to the size of the target as compared to the wavelength. Thus, the computation time varies at least as the fourth power (for rotationally symmetric targets), and, in most cases, as the sixth power of the object size (or frequency). (e) Once the unknowns have been solved in the aforementioned simultaneous equations, they must be either summed in some manner or numerically integrated (which is a type of summation process) over the surface of the body to obtain the scattered far-zone field.

As can be seen from the preceding discussion, numerical solutions of scattering problems are practical in the low-frequency and resonance regions. In the high-frequency region where the object is greater than, say, 500 square wavelengths in its surface area, the computation time becomes quite long and costly. In this high-frequency region, one would logically resort to various approximations such as physical or geometrical optics to solve the scattering problem.

The numerical solutions discussed here are exact, inasmuch as they arise

---

[†] Among these are Andreasen and his group, formerly of TRG, Inc.; Waterman, formerly of AVCO; Richmond and his colleagues at the Ohio State University; and Oshiro and Mitzner at Northrup Norair. There are, undoubtedly, several more groups with such programs, as this list is by no means comprehensive.

from an exact formulation of the scattering problem. The only approximation arises in truncating the set of equations in the solution of the problem.
However, as one increases the number of points or equations, the solution
can be made to approach the true solution to within any specified accuracy.
In fact, in some of the programs available, one specifies the desired accuracy
and the computer chooses the number of points or equations large enough so
that the specified accuracy is obtained.

The main advantage of numerical solutions lies in the ability to obtain
an immediate estimate of the cross section of a given scatterer and the cross-
section variation as a function of aspect angle, frequency, and polarization.
The estimate is exact since it was obtained from an exact formulation. As
such, it may provide the only possible answers (short of measurements) in
the low-frequency and resonance regions where many analytical approximations are not valid.

The main disadvantage of numerical solutions is the lack of insight they
provide into the nature of the scattering process. The geometrical theory of
diffraction, for example, while only a high-frequency approximation,
not only provides solutions but relates scattered field behavior to certain
definite areas on the scatterer surface. Hence, one obtains an understanding
of the scattering process and can use the knowledge to shape the surface
better in order to either enhance or conceal the object at certain frequencies
and aspects. Conversely, one can hope to relate the shape of an unknown
target to its cross-section signature through the knowledge gained from many
of the analytical theories and approximations. On the other hand, a numerical
solution provides as an answer a simple number; this number cannot be
easily related to different parts of the scatterer. Nor can the numerical
solution for one given shape be readily extrapolated to other shapes.

Thus, numerical solutions can be invaluable in situations where one
needs a rapid, relatively accurate estimate of the cross section. They will not,
however, replace the analytical theories and approximations heretofore
developed. In most situations, both approaches to scattering problems
complement each other and can possibly accelerate the development of
hybrid approaches which combine the two.

Very briefly, an attempt will be made below to categorize and summarize
the highlights of these numerical techniques. Investigators and pertinent
references will be mentioned where possible.[†] A somewhat more complete
bibliography and summary is given in reference 104.

The first formulation of a numerical solution to scattering problems
employed the cylindrical eigenfunction series to represent the scattered field
in two-dimensional scattering geometries, and the spherical eigenfunction

[†] Starting in 1965, there has been a large increase in the research in this area, and the
techniques are being developed and improved rapidly. This summary represents the
state of the art as reported in the literature up to mid-1967.

series to represent the scattered field in three-dimensional situations. Alternately, these eigenfunctions, when normalized differently, are called multipoles[105] or modal functions.[104] At any rate, the unknown coefficients in the series solution for the scattered field, which must be determined, are functions of the scatterer shape and size, and depend upon the direction of incidence and the polarization of the incident wave.

The first type of scattering problems formulated in the preceding manner had the scatterer surface as portions of the natural coordinates of the spherical or cylindrical system. For example, Sommerfeld[106] formulates the problem of scattering from a portion of a cylinder in this manner, and Rogers and Schultz[107] formulate a numerical solution for scattering from a cone–sphere where both the cone and sphere parts of the scatterer are portions of the "constant $r$" and "constant $\theta$" coordinate surfaces. When such is the case, the method of least-square error can be used to determine the coefficients.[106] An infinite set of linear equations in the unknown series coefficients results; this is truncated and solved directly. If the surface of the scatterer cannot be represented by portions of the natural coordinate surfaces of the system, then the unknown coefficients of the series solution must be determined differently. Kennaugh[105] treats and solves this problem for three-dimensional scattering from spheroids. Mullin, Sandburg, and Velline[108] use essentially the same procedure for two-dimensional scattering from elliptic cylinders.† The procedure consists basically of (a) truncating the scattered field series after a finite number of terms, say $N$, in $N$ unknown coefficients; and (b) enforcing the boundary conditions, involving this series, at a finite number of points on the object, say at $M$ points. Thus, one has $N$ equations in $M$ unknowns. If $M = N$, these coefficients are solved for by standard algebraic methods.[108] If $M$ is greater than $N$, one can use instead the boundary conditions at all $M$ points, along with the method of least square error at all of these points, to determine the coefficients.[105] Once the scattered-field, series-solution coefficients are known, the series are summed to obtain the scattered field.

A second method starts with the Chu–Stratton integral equation (alternatively referred to as the Green's functional formulation). For perfectly conducting surfaces, this has the form shown in Eq. (2.2-27). Waterman[109] uses a slightly different integral equation which avoids the singularity of the Green's function. At any rate, the integral equation over the scatterer surface contains the total field in the integrand. The integral can then be reduced to a summation, for numerical purposes, involving the scattered and incident fields (or, alternatively, the surface currents) at $N$ points on the scatterer. By allowing the nonintegral term of the equation, involving the fields, also to be

---

† The method of Mullin *et al.* is not exact. They employ the eigenfunctions for the circular cylinder everywhere. The error becomes intolerable for axial ratios less than about one-half.

specified at $N$ different points on the scatterer surface, one has $N$ equations in $N$ unknowns. Such is basically the technique of Richmond[104] and Andreasen.[110]

There are certain variations to this second approach which are employed. Andreasen, for example, expresses the incident wave in cylindrical modes or eigenfunctions. Harrington et al.[111] choose other functions to represent the fields; these functions are chosen to be convenient for the given problem.

The manner in which the $N$ points on the body surface are chosen and spaced follows the same pattern as those for the series-solution formulation. For both formulations, there are several possibilities:

(a) The $N$ points can be equally distributed over the surface of the body with spacing somewhat less than a wavelength. (A convenient figure is $\frac{1}{8}$ wavelength, which is mentioned in Reference 110.)

(b) The $N$ points can be unevenly distributed with spacing varying with local body shape or radius of curvature in some predetermined manner.

(c) Instead of enforcing the surface constraints only at $N$ points, a least-squares error approach can be used. This generally involves another integration over the body surface with the integrals now appearing in the system of coupled linear equations. Evaluation of the integrals increases the computation time as $N^2$.

Of these three approaches, the first is the least time-consuming, but probably the least accurate. The third is the most time-consuming, but accuracy increases correspondingly. For a given accuracy, it may be possible to reduce the number of equations used for the third approach and hence the computation time.

When the scatterer is a very thin wire of finite length, the integral equation takes on a special form.[104] Solution of this integral equation must be enforced only at points along the length of the object, since the fields on the thin wire do not vary appreciably around the wire in azimuth. This reduces the scattering "surface" to a line, and the number of points needed in this case varies directly with scatterer length rather than with its square.

It is obvious that the two basic formulations of the scattering problem for numerical solution bear strong resemblances. Essentially, the same numerical operations are performed, but each starts with a different initial problem formulation. As such, both approaches are of nearly equal complexity for a given degree of accuracy. Any variations or improvements providing shortcuts for one of the two approaches should also be applicable to the other approach.

The preceding discussion in no way covers the details of the numerical solutions mentioned in the given references, nor does it include all of the references pertinent to the subject (see Reference 104 for a more extensive bibliography). In presenting only the highlights, one cannot hope to do justice to the powerful techniques and variations developed by the individual investigators. The discussion here is only intended to give the reader who is unfamiliar with numerical solution techniques a feeling for what is involved in such solutions.

## 2.3. EDGE DIFFRACTION, CREEPING WAVES, AND TRAVELING WAVES

In estimating the high-frequency or upper resonance-region cross section of both simple and complex bodies, the observed cross section often can be attributed to a combination of various scattering components. One such component which has already been encountered is the specular component discussed in Section 2.2.2.1. Three of the more important remaining possible contributions to the radar cross section of a body are those listed in the title of this section. The term "analytic components" will be used throughout this book to collectively describe scattering components of this type, and was chosen because these components appear as a result of mathematical manipulations of the analytic expressions for the scattered fields from various bodies. The manner in which this is accomplished is discussed in the following sections.

### 2.3.1. Edge and Tip Diffraction

For bodies which are not smoothly curved and contain sharp edges, tips, corners, etc., that is, points or regions where the radii of curvature in one or more directions approach zero, or are much less than a wavelength, techniques such as geometrical optics, physical optics, etc., fail to predict correctly the scattering properties of the body. In the vicinity of points of this nature, the surface currents differ vastly from those predicted by geometrical or physical optics.

#### 2.3.1.1. *Edge Diffraction*

To gain some insight into the influence of an edge on the fields scattered by a body, the simplest body having a sharp edge for which exact results are available will be examined. This is the perfectly conducting, infinite half-plane.

2.3.1.1.1. *Infinite Perfectly Conducting Half-Plane.* The geometry of the situation is shown in Figure 2-24. The infinite half-plane lies in the $x - z$ plane, with its edge along the $z$-axis and extending to infinity in the positive $x$-direction. A plane wave is incident from a direction specified by the spherical coordinates $\theta$, $\phi$. Any arbitrary polarization can be specified by a linear combination of two polarizations; one designated perpendicular ($\perp$), where the incident magnetic field has no $z$ component; and the other parallel ($\parallel$), where the incident electric field has no $z$ component. Using the extension

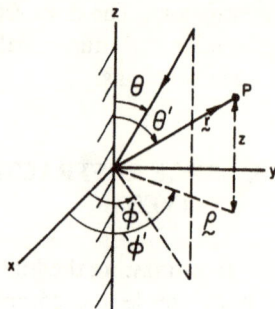

Fig. 2-24. A plane wave obliquely incident on a perfectly conducting semi-infinite plane.

to three dimensions of Sommerfeld's exact solution for the total fields,[112] the currents on the surface of the half-plane can be obtained from

$$\mathbf{K} = (\hat{n} \times \mathbf{H})|_{\phi'=0} - (\hat{n} \times \mathbf{H})|_{\phi'=2\pi} \qquad (2.3\text{-}1)$$

where $\hat{n} = \hat{y}$ is a unit vector normal to the surface and $\mathbf{H}$ is the total magnetic field at the surface.

For perpendicular polarization, the incident fields are

$$E_{\perp}{}^i = (-\cos \phi \cos \theta; \ -\sin \phi \cos \theta; \ \sin \theta) \sin \theta$$
$$\times \exp[-ik_0(x \cos \phi \sin \theta + y \sin \phi \sin \theta + z \cos \theta)] \quad (2.3\text{-}2)$$

$$H_{\perp}{}^i = 1/\eta(-\sin \phi; \ \cos \phi; \ 0) \sin \theta$$
$$\times \exp[-ik_0(x \cos \phi \sin \theta + y \sin \phi \sin \theta + z \cos \theta)] \quad (2.3\text{-}3)$$

with the terms within the first set of parenthesis being the direction cosines of the fields with respect to the $(x, y, z)$ axes.[†] The total fields are[113]

$$E_{z_\perp} = \frac{e^{-i\pi/4}}{\sqrt{\pi}} \sin^2 \theta \ \exp[ik_0(\rho \sin \theta - z \cos \theta)][G(p) - G(q)]$$

$$H_{x_\perp} = \frac{-e^{-i\pi/4}}{\eta \sqrt{\pi}} \sin \theta \ \exp[ik_0(\rho \sin \theta - z \cos \theta)] \left[ \sin \phi[G(p) - G(q)] \right.$$

$$\left. + i \sqrt{\frac{2}{k_0\rho \sin \theta}} \sin \frac{\phi}{2} \cos \frac{\phi'}{2} \right]$$

$$(2.3\text{-}4)$$

$$H_{y_\perp} = \frac{e^{-i\pi/4}}{\eta \sqrt{\pi}} \sin \theta \ \exp[ik_0(\rho \sin \theta - z \cos \theta)] \left[ \cos \phi[G(p) - G(q)] \right.$$

$$\left. - i \sqrt{\frac{2}{k_0\rho \sin \theta}} \sin \frac{\phi}{2} \sin \frac{\phi'}{2} \right]$$

$$E_{x_\perp} = -\eta H_{y_\perp} \cos \theta, \qquad E_{y_\perp} = \eta H_{x_\perp} \cos \theta, \qquad H_{z_\perp} = 0$$

[†] The notation $(a; b; c)$ used in (2.3-2) and (2.3-3) is equivalent to $(a\hat{x} + b\hat{y} + c\hat{z})$.

The function $G$ is defined as

$$G(p) = e^{-ip^2} \sqrt{\frac{\pi}{2}} \left[ \frac{e^{i\pi/4}}{\sqrt{2}} \mp C\left(\pm\sqrt{\frac{2}{\pi}}\,p\right) \mp iS\left(\pm\sqrt{\frac{2}{\pi}}\,p\right)\right] \qquad (2.3\text{-}5)$$

for $p$ positive or negative, respectively; while $p$ and $q$ are

$$p = -\sqrt{2k_0\rho}\,\sin\theta\cos\frac{(\phi-\phi')}{2}, \qquad q = -\sqrt{2k_0\rho}\,\sin\theta\cos\frac{(\phi+\phi')}{2}$$
$$(2.3\text{-}6)$$

with $C(\tau)$, $S(\tau)$, the Fresnel cosine and sine integrals defined and tabulated in reference 114. Using Eqs. (2.3-1) and (2.3-4), the surface currents on the half-plane for perpendicular polarization are

$$K_{z\perp} = \frac{4e^{-i\pi/4}}{\eta\,\sqrt{\pi}}\sin\theta\sin\phi\,\exp[-ik_0(x\sin\theta\cos\phi + z\cos\theta)]$$

$$\times\left[\sqrt{\frac{\pi}{2}}\left(C\left[2\sqrt{\frac{k_0 x\sin\theta}{\pi}}\cos\frac{\phi}{2}\right] + iS\left[2\sqrt{\frac{k_0 x\sin\theta}{\pi}}\cos\frac{\phi}{2}\right]\right)\right.$$

$$\left. + \frac{i\exp[2ik_0 x\sin\theta\cos^2(\phi/2)]\sin(\phi/2)}{4\sin\phi\,\sqrt{(k_0 x\sin\theta)/2}}\right] \qquad (2.3\text{-}7)$$

and

$$K_{x\perp} = K_{y\perp} = 0$$

Similarly, for parallel polarization, the incident fields are

$$E_\parallel{}^i = \eta(\sin\phi;\,-\cos\phi;\,0)\sin\theta$$
$$\times\exp[-ik_0(x\sin\theta\cos\phi + y\sin\theta\sin\phi + z\cos\theta)] \quad (2.3\text{-}8)$$

$$H_\parallel{}^i = (-\cos\phi\cos\phi;\,-\sin\phi\cos\theta;\,\sin\theta)\sin\theta$$
$$\times\exp[-ik_0(x\sin\theta\cos\phi + y\sin\theta\sin\phi + z\cos\theta)] \quad (2.3\text{-}9)$$

The total fields are

$$H_{z\parallel} = \frac{e^{i\pi/4}}{\sqrt{\pi}}\sin^2\theta\,\exp[ik_0(\rho\sin\theta - z\cos\theta)][G(p) + G(q)]$$

$$H_{y\parallel} = \frac{-e^{i\pi/4}}{\sqrt{\pi}}\sin\theta\cos\theta\,\exp[ik_0(\rho\sin\theta - z\cos\theta)]\left[\sin\phi[G(p) - G(q)]\right.$$

$$\left. - i\sqrt{\frac{2}{\pi k_0\rho\sin\theta}}\cos\frac{\phi}{2}\sin\frac{\phi'}{2}\right]$$
$$(2.3\text{-}10)$$

$$H_{x\parallel} = \frac{-e^{i\pi/4}}{\sqrt{\pi}}\sin\theta\cos\theta\,\exp[ik_0(\rho\sin\theta - z\cos\theta)]\left[\cos\phi[G(p) + G(q)]\right.$$

$$\left. - i\sqrt{\frac{2}{\pi k_0\rho\sin\theta}}\cos\frac{\phi}{2}\cos\frac{\phi'}{2}\right]$$

$$E_{x\parallel} = \frac{-\eta H_y}{\cos\theta}, \qquad E_y = \frac{\eta H_x}{\cos\theta}, \qquad E_z = 0$$

Using Eqs. (2.3-1) and (2.3-10), the surface currents for parallel polarization are then

$$K_{x_\parallel} = \frac{4e^{-i\pi/4}}{\sqrt{\pi}} \sin^2 \theta \exp[-ik_0(x \sin \theta \cos \phi + z \cos \theta)]$$

$$\times \sqrt{\frac{\pi}{2}} \left[ C \left[ 2 \sqrt{\frac{k_0 x \sin \theta}{\pi}} \cos \frac{\phi}{2} \right] + iS \left[ 2 \sqrt{\frac{k_0 x \sin \theta}{\pi}} \cos \frac{\phi}{2} \right] \right]$$

$$K_{y_\parallel} = 0 \qquad\qquad\qquad\qquad\qquad\qquad\qquad\qquad\qquad\qquad (2.3\text{-}11)$$

and

$$K_{z_\parallel} = \frac{4e^{-i\pi/4}}{\sqrt{\pi}} \sin \theta \cos \phi \exp[-ik_0(x \sin \theta \cos \phi + z \cos \theta)]$$

$$\times \left[ \sqrt{\frac{\pi}{2}} \left( C \left[ 2 \sqrt{\frac{k_0 x \sin \theta}{\pi}} \cos \frac{\phi}{2} \right] + iS \left[ 2 \sqrt{\frac{k_0 x \sin \theta}{\pi}} \cos \frac{\phi}{2} \right] \right) \right.$$

$$\left. - \frac{i \exp[2ik_0 x \sin \theta \cos^2(\phi/2)] \cos^2(\phi/2)}{4 \cos \phi \sqrt{(k_0 x \sin \theta)/2}} \right] \qquad (2.3\text{-}12)$$

Examining these currents as $x$ approaches zero, or near the edge of the half-plane, it is apparent that $K_{z_\perp}$ and $K_{z_\parallel}$ approach infinity as $(x)^{-1/2}$, since the Fresnel integrals are finite for small $x$. The normal current component, $K_{x_\parallel}$, approaches zero as $(x)^{1/2}$, since $C(\tau) \approx \tau$ for $\tau \ll 1$ and $S(\tau) \approx (\pi/6) \tau^3$ for $\tau \ll 1$.

At the other extreme, for $x$ large or far from the edge, $C(\tau) \approx S(\tau) \approx \frac{1}{2}$ for $\tau \gg 1$, so the currents become for $x \gg 1$,

$$K_{z_\perp} \approx \frac{2}{\eta} \sin \theta \sin \phi \exp[-ik_0(x \sin \theta \cos \phi + z \cos \theta)] \qquad (2.3\text{-}13)$$

$$K_{x_\parallel} \approx 2 \sin^2 \theta \exp[-ik_0(x \sin \theta \cos \phi + z \cos \theta)] \qquad (2.3\text{-}14)$$

$$K_{z_\parallel} \approx 2 \sin \theta \cos \theta \cos \phi \exp[-ik_0(x \sin \theta \cos \phi + z \cos \theta)] \qquad (2.3\text{-}15)$$

and by referring to Eqs. (2.3-3) and (2.3-9), are seen to equal $2(\hat{n} \times \mathbf{H}^i)$. Thus, far from the edge, the surface currents are those predicted by physical optics. In fact, it is not necessary for $x$ to be very large for the current distribution to closely approximate the physical optics distribution. Even for $x \approx \lambda_0/4$, the difference is small. Thus, the deviations of the exact current distributions from the physical optics distribution, which are designated the edge currents, are local in nature and confined very closely to the edge. These so-called edge currents can be written as

$$\mathbf{K}^e = \mathbf{K} - 2(\hat{n} \times \mathbf{H}^i)$$

or

$$K^e_{z_\perp} = \frac{4e^{-i\pi/4}}{\eta \sqrt{\pi}} \sin\theta \sin\phi \exp[-ik_0(x\sin\theta\cos\phi + z\cos\theta)]$$

$$\times \left[ \sqrt{\frac{\pi}{2}} \left( C\left[2\sqrt{\frac{k_0 x \sin\theta}{\pi}} \cos\frac{\phi}{2}\right] + iS\left[2\sqrt{\frac{k_0 x \sin\theta}{\pi}} \cos\frac{\phi}{2}\right] \right. \right.$$

$$\left. \left. - \frac{e^{i\pi/4}}{\sqrt{2}} \right) + \frac{i\exp[2ik_0 x \sin\theta\cos^2(\phi/2)]\sin\phi/2}{4\sin\phi\sqrt{(k_0 x \sin\theta)/2}} \right] \qquad (2.3\text{-}16)$$

$$K_{x_\parallel} = \frac{4e^{-i\pi/4}}{\sqrt{\pi}} \sin^2\theta \exp[-ik_0(x\sin\theta\cos\phi + z\cos\theta)]$$

$$\times \left[ \sqrt{\frac{\pi}{2}} \left( C\left[2\sqrt{\frac{k_0 x \sin\theta}{\pi}} \cos\frac{\phi}{2}\right] + iS\left[2\sqrt{\frac{k_0 x \sin\theta}{\pi}} \cos\frac{\phi}{2}\right] \right. \right.$$

$$\left. \left. - \frac{e^{i\pi/4}}{\sqrt{2}} \right) \right] \qquad (2.3\text{-}17)$$

$$K^e_{z_\parallel} = \frac{4e^{-i\pi/4}}{\sqrt{\pi}} \sin\theta\cos\theta\cos\phi \exp[-ik_0(x\sin\theta\cos\phi + z\cos\theta)]$$

$$\times \left[ \sqrt{\frac{\pi}{2}} \left( C\left[2\sqrt{\frac{k_0 x \sin\theta}{\pi}} \cos\frac{\phi}{2}\right] + iS\left[2\sqrt{\frac{k_0 x \sin\theta}{\pi}} \cos\frac{\phi}{2}\right] \right. \right.$$

$$\left. \left. - \frac{e^{i\pi/4}}{\sqrt{2}} \right) - \frac{i\exp[2ik_0 x \sin\theta\cos^2(\phi/2)]}{4\cos\phi\sqrt{(k_0 x \sin\theta)/2}} \right] \qquad (2.3\text{-}18)$$

If the total electric and magnetic fields are examined in the limit of large $k_0\rho$, that is far from the half-plane, they can be separated into three components: the incident field, the geometrically reflected field, and a third component which will be designated the diffraction component. The geometrically reflected portion of the total field can be attributed to radiation by the physical optics currents flowing on the half-plane, while the diffraction component can be attributed to the edge currents. To illustrate the form of this decomposition, the component of the total electric field along the z-axis for perpendicular polarization will be examined. From Eq. (2.3-4),

$$E_{z_\perp} = \frac{e^{-i\pi/4}}{\sqrt{\pi}} \sin^2\theta \exp[ik_0(\rho\sin\theta - z\cos\theta)][G(p) - G(q)]$$

Now to clearly separate $E_z$ into incident, reflected, and diffracted fields, the quantity $[G(p) - G(q)]$ must be evaluated for large $\rho$ or in the far-field region. In order to do this, three angular regions must be distinguished. These are

(i) $0 \leqslant \phi' < \pi - \phi$, where the incident, reflected, and diffracted fields are present;

(ii) $\pi - \phi < \phi' < \pi + \phi$, where the incident and diffracted fields are present;

(iii) $\pi + \phi < \phi' \leqslant 2\pi$, where only the diffracted field is present.

This distinction comes about analytically, due to the requirement for different asymptotic forms for $G(p)$ and $G(q)$ depending upon whether $p$ and $q$ are positive or negative. Thus, in evaluating the quantity $[G(p) - G(q)]$, there are four possibilities:

(i) $p < 0, q < 0$, then

$$[G(p) - G(q)] = \sqrt{\pi}\, e^{i\pi/4}(e^{-ip^2} - e^{-iq^2}) + \frac{i}{2}\left[\frac{1}{p} - \frac{1}{q}\right] \qquad (2.3\text{-}19)$$

(ii) $p < 0, q > 0$ or $p > 0, q < 0$, then

$$[G(p) - G(q)] = \sqrt{\pi}\, e^{-ip^2 + i\pi/4} + \frac{i}{2}\left[\frac{1}{p} - \frac{1}{q}\right]$$

or

$$[G(p) - G(q)] = \sqrt{\pi}\, e^{-iq^2 + i\pi/4} + \frac{i}{2}\left[\frac{1}{q} - \frac{1}{p}\right] \qquad (2.3\text{-}20)$$

(iii) $p > 0, q > 0$, then

$$[G(p) - G(q)] = \frac{i}{2}\left[\frac{1}{p} - \frac{1}{q}\right] \qquad (2.3\text{-}21)$$

For case i), where $[G(p) - G(q)]$ is given by Eq. (2.3-19), the total field $E_{z_\perp}$ is

$$E_{z_\perp} = \sin^2 \theta \exp[ik_0(\rho \sin \theta - z \cos \theta)]\left[\exp\left[-i2k_0\rho \sin \theta \cos\left(\frac{\phi - \phi'}{2}\right)\right]\right.$$

$$\left. - \exp\left[-i2k_0\rho \sin \theta \cos\left(\frac{\phi + \phi'}{2}\right)\right]\right]$$

$$- \frac{ie^{-i\pi/4}}{2\sqrt{\pi}}\sin^2 \theta \frac{\exp[ik_0(\rho \sin \theta - z \cos \theta)]}{\sqrt{2k_0\rho \sin \theta}}$$

$$\times \left[\frac{\cos[(\phi + \phi')/2] - \cos[(\phi - \phi')/2]}{\cos[(\phi + \phi')/2]\cos[(\phi - \phi')/2]}\right] \qquad (2.3\text{-}22)$$

or, simplifying,

$$E_{z_\perp} = \sin^2 \theta \exp\left[-ik_0\left(\rho \sin \theta \cos\left(\frac{\phi - \phi'}{2}\right) + z \cos \theta\right)\right]$$

$$- \sin^2 \theta \exp\left[-ik_0\left(\rho \sin \theta \cos\left(\frac{\phi + \phi'}{2}\right) + z \cos \theta\right)\right]$$

$$+ \frac{e^{i\pi/4}\sin^2 \theta\, e^{ik_0(\rho \sin \theta - z \cos \theta)}}{\sqrt{(\pi k_0\rho \sin \theta)/2}}\left[\frac{\sin(\phi/2)\sin(\phi'/2)}{\cos \phi + \cos \phi'}\right] \qquad (2.3\text{-}23)$$

The first term of this is the incident field; the second term, the geometrically reflected field; and the third term, the diffracted field, which as discussed previously, can be attributed to the edge currents.

Similarly, the fields can be written for cases (ii) and (iii) using Eqs. (2.3-4), (2.3-20), and (2.3-21), as

$$E_{z_\perp} = \sin^2 \theta \exp\left[-ik_0\left(\rho \sin \theta \cos \left(\frac{\phi - \phi'}{2}\right) + z \cos \theta\right)\right]$$

$$+ \left[\frac{\sin(\phi/2) \sin(\phi'/2)}{\cos \phi + \cos \phi'}\right] \frac{e^{i\pi/4} \sin^2 \theta \exp[ik_0(\rho \sin \theta - z \cos \theta)]}{\sqrt{(\pi k_0 \rho \sin \theta)/2}} \qquad (2.3\text{-}24)$$

for case (ii) and

$$E_{z_\perp} = \frac{e^{i\pi/4} \sin^2 \theta \exp[ik_0(\rho \sin \theta - z \cos \theta)]}{\sqrt{(\pi k_0 \rho \sin \theta)/2}} \left[\frac{\sin(\phi/2) \sin(\phi'/2)}{\cos \phi + \cos \phi'}\right] \qquad (2.3\text{-}25)$$

for case (iii).

For parallel polarization, the results of a decomposition of $H_z$ into incident, reflected, and diffracted fields are

$$H_{z_\parallel} = \sin^2 \theta \exp\left[-ik_0\left(\rho \sin \theta \cos \left(\frac{\phi - \phi'}{2}\right) + z \cos \theta\right)\right] + \sin^2 \theta$$

$$\times \exp\left[-ik_0\left(\rho \sin \theta \cos \left(\frac{\phi + \phi'}{2}\right) + z \cos \theta\right)\right]$$

$$- \frac{e^{i\pi/4} \sin^2 \theta \exp[ik_0(\rho \sin \theta - z \cos \theta)]}{\sqrt{(\pi k_0 \rho \sin \theta)/2}} \left[\frac{\cos(\phi/2) \cos(\phi'/2)}{\cos \phi + \cos \phi'}\right]$$

$$(2.3\text{-}26)$$

for case (i);

$$H_{z_\parallel} = \sin^2 \theta \exp\left[-ik_0\left(\rho \sin \theta \cos \left(\frac{\phi - \phi'}{2}\right) + z \cos \theta\right)\right]$$

$$- \frac{e^{i\pi/4} \sin^2 \theta \exp[ik_0(\rho \sin \theta - z \cos \theta)]}{\sqrt{(\pi k_0 \rho \sin \theta)/2}} \left[\frac{\cos(\phi/2) \cos(\phi'/2)}{\cos \phi + \cos \phi'}\right]$$

$$(2.3\text{-}27)$$

for case (ii); and

$$H_{z_\parallel} = \frac{e^{i\pi/4} \sin^2 \theta \exp[ik_0(\rho \sin \theta - z \cos \theta)]}{\sqrt{(\pi k_0 \rho \sin \theta)/2}} \left[\frac{\cos(\phi/2) \cos(\phi'/2)}{\cos \phi + \cos \phi'}\right] \qquad (2.3\text{-}28)$$

for case (iii).

Equivalent decompositions can be carried out for all the total field components for both perpendicular and parallel polarizations. Decomposing the fields into incident, reflected, and diffracted components is valid for

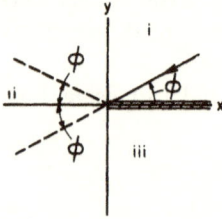

Fig. 2-25. Boundaries along which $p$ or $q$ are zero, and regions in which different analytical solutions are valid for a plane wave normally incident on a perfectly conducting half-plane.

observation points in the far field and not too near the boundaries of the regions (i), (ii), and (iii), shown in Figure 2-25, where $p$ or $q$ or both are zero, thus invalidating the asymptotic expansion of $G(p)$ and $G(q)$.

If the expression (2.3-25) for the diffracted field is examined, the phase factor $\exp[ik_0(\rho \sin \theta - z \cos \theta)]$ is observed to be independent of $\phi'$ and varies as $\rho \sin \theta$; thus, the diffracted field at any point appears to be traveling along a cone of semi-angle $\theta$, whose axis is the edge of the half-plane. This, of course, agrees with the picture presented by Keller in his geometrical theory of diffraction, where edge diffracted rays are assumed to lie on a cone with a half-angle equal to the angle between the incident ray and the tangent to the edge, and whose axis coincides with the tangent to the edge.

2.3.1.1.2. *Infinite Perfectly Conducting Wedge.* To obtain an insight into the influence of edges on finite bodies, a more useful model than the half-plane will be an infinite perfectly conducting wedge. For simplicity, the two-dimensional problem of a plane cylindrical wave incident on a perfectly conducting wedge of exterior angle $\gamma$ will be examined. The coordinates of the system are shown in Figure 2-26.

For an incident plane wave perpendicularly polarized, the incident field is

$$\mathbf{E}^i = E_z{}^i \hat{\mathbf{z}} = e^{-ik_0\rho \cos(\phi-\phi')}\hat{\mathbf{z}} \tag{2.3-29}$$

and the resultant total fields are[115]

$$\mathbf{E}_\perp = E_z\hat{\mathbf{z}} = \frac{\pi}{\gamma} \sum_{n=0}^{\infty} \epsilon_n e^{-i\nu_n\pi/2} J_{\nu_n}(k_0\rho)[\cos \nu_n(\phi - \phi') - \cos \nu_n(\phi + \phi')]\,\hat{\mathbf{z}} \tag{2.3-30}$$

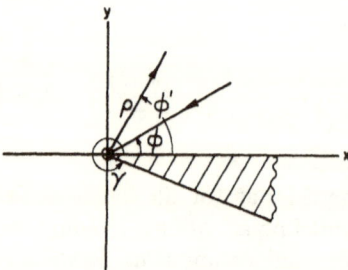

Fig. 2-26. A plane wave incident normally on a perfectly conducting wedge.

with

$$H_\rho = \frac{1}{i\omega\mu\rho}\frac{\partial E_z}{\partial \phi'} \quad \text{and} \quad H_\phi = \frac{-1}{i\omega\mu}\frac{\partial E_z}{\partial \rho}$$

In Eq. (2.3-30),

$$\epsilon_0 = 1, \qquad \epsilon_{n\neq0} = 2, \qquad \nu_n = \frac{n\pi}{\gamma}$$

Similarly, for parallel polarization the incident field is

$$\mathbf{H}^i = H_z^i \hat{\mathbf{z}} = e^{-ik_0\rho\,\cos\,(\phi-\phi')}\hat{\mathbf{z}} \tag{2.3-31}$$

and the resultant total fields are

$$\mathbf{H}_{\parallel} = H_z\hat{\mathbf{z}} = \frac{\pi}{\gamma}\sum_{n=0}^{\infty}\epsilon_n e^{-i\nu_n\pi/2}J_{\nu_n}(k_0\rho)[\cos\nu_n(\phi-\phi') + \cos\nu_n(\phi+\phi')]\,\hat{\mathbf{z}} \tag{2.3-32}$$

with

$$E_\rho = \frac{-1}{i\omega\epsilon\rho}\frac{\partial H_z}{\partial \phi'} \quad \text{and} \quad E_\phi = \frac{1}{i\omega\epsilon}\frac{\partial H_z}{\partial \rho}$$

By using an integral representation for $J_{\nu_n}(k_0\rho)$, and suitably deforming the contour of integration, the infinite series in Eqs. (2.3-30), (2.3-32) can be summed. Evaluating the resultant integrals by stationary-phase techniques yields solutions which are asymptotically valid for large $k_0\rho$, and in which the incident, geometrically reflected, and diffracted components are separated.[116] The result of such a process is as follows:

$$E_{z_\perp} = [U(\rho, \phi-\phi') - U(\rho, \phi+\phi')] \tag{2.3-33}$$

and

$$H_{z_\parallel} = [U(\rho, \phi-\phi') + U(\rho, \phi+\phi')] \tag{2.3-34}$$

The function $U$ has various forms depending upon the relative angles of incidence and observation, that is, whether the incident, reflected, and diffracted waves are observable simultaneously or only in certain combinations. Thus the $\rho$, $\phi$ plane must initially be split into two regions; one corresponding to that where only one face of the wedge is exposed to the incident field; and the other, where both faces are exposed. These are illustrated in Figure 2-27.

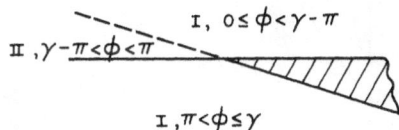

Fig. 2-27. Regions in which the function $U$ has different forms.

In region I where only one face is exposed, $U(\rho, \phi \pm \phi')$ has the form

$$U(\rho, \phi \pm \phi') = D(\rho, \phi \pm \phi') + e^{-ik_0\rho\cos(\phi\pm\phi')}, \qquad 0 < \phi' < \pi - \phi$$
$$U(\rho, \phi - \phi') = D(\rho, \phi - \phi') + e^{-ik_0\rho\cos(\phi-\phi')} \quad \text{and}$$
$$U(\rho, \phi + \phi') = D(\rho, \phi + \phi'), \qquad \text{for} \quad \pi - \phi < \phi' < \pi + \phi$$
$$U(\rho, \phi \pm \phi') = D(\rho, \phi \pm \phi'), \qquad \pi + \phi < \phi' < \gamma$$

(2.3-35)

In region II where both faces are exposed, $U$ has the form

$$U(\rho, \phi \pm \phi') = D(\rho, \phi \pm \phi') + e^{-ik_0\rho\cos(\phi\pm\phi')}, \qquad 0 < \phi' < \pi - \phi$$
$$U(\rho, \phi - \phi') = D(\rho, \phi - \phi') + e^{-ik_0\rho\cos(\phi-\phi')} \quad \text{and}$$
$$U(\rho, \phi + \phi') = D(\rho, \phi + \phi'), \qquad \text{for} \quad \pi - \phi < \phi < 2\gamma - \pi - \phi$$
$$U(\rho, \phi - \phi') = D(\rho, \phi - \phi') + e^{-ik_0\rho\cos(\phi-\phi')} \quad \text{and}$$
$$U(\rho, \phi - \phi') = D(\rho, \phi + \phi') + e^{-ik_0\rho\cos(2\gamma-\phi-\phi')},$$
$$\text{for} \quad 2\gamma - \pi - \phi < \phi' < 2\gamma$$

(2.3-36)

The diffraction term, $D(\rho, \phi \pm \phi')$, is given by

$$D(\rho, \phi \pm \phi') = \frac{e^{i(k_0\rho+\pi/4)}}{\sqrt{2\pi k_0\rho}} \left[\frac{(\pi/\gamma)\sin\pi^2/\gamma}{\cos\pi^2/\gamma - \cos[(\pi/\gamma)(\phi \pm \phi')]}\right] \quad (2.3\text{-}37)$$

In the preceding expressions for $U$, the terms in $e^{-ik_0\rho\cos(\phi-\phi')}$ correspond to the incident wave, while the $e^{-ik_0\rho\cos(\phi+\phi')}$ terms correspond to the geometrically reflected field from the face of the wedge parallel to the $x$-axis, and the $e^{-ik_0\rho\cos(2\gamma-\phi-\phi')}$ term corresponds to the field geometrically reflected from the other face.

The results just given can easily be generalized to the three-dimensional situation of a plane wave incident on a perfectly conducting infinite wedge at an arbitrary oblique angle. This is illustrated in Figure 2-28.

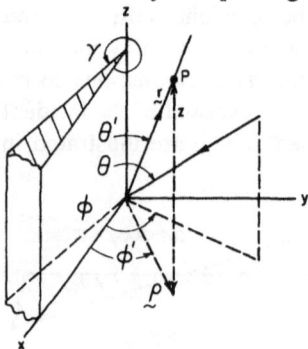

Fig. 2-28. A plane wave incident obliquely on an infinite perfectly conducting wedge.

For perpendicular polarization the incident field is specified by Eqs. (2.3-2) and (2.3-3); and for parallel polarizations, by Eqs. (2.3-8) and (2.3-9). The resulting total field components are

$$E_{z_\perp} = \frac{\pi}{\gamma} \sin^2 \theta \, e^{-ik_0 z \cos \theta} \sum_{n=0}^{\infty} \epsilon_n e^{-i\nu_n \pi/2} J_{\nu_n}(k_0 \rho \sin \theta)$$

$$\times \, [\cos \nu_n(\phi - \phi') - \cos \nu_n(\phi + \phi')] \qquad (2.3\text{-}38)$$

and

$$H_{z_\parallel} = \frac{\pi}{\gamma} \sin^2 \theta \, e^{-ik_0 z \cos \theta} \sum_{n=0}^{\infty} \epsilon_n e^{-i\nu_n \pi/2} J_{\nu_n}(k_0 \rho \sin \theta)$$

$$\times \, [\cos \nu_n(\phi - \phi') + \cos \nu_n(\phi + \phi')] \qquad (2.3\text{-}39)$$

with the remainder of the field components obtainable by differentiation of the above two using Maxwell's equations.

The far-zone asymptotic field solutions would then be given as in the two-dimensional case by

$$E_{z_\perp} = [U(\rho, \theta, \phi - \phi') - U(\rho, \theta, \phi + \phi')] \qquad (2.3\text{-}40)$$

and

$$H_{z_\parallel} = [U(\rho, \theta, \phi - \phi') + U(\rho, \theta, \phi + \phi')] \qquad (2.3\text{-}41)$$

where $U(\rho, \theta, \phi \pm \phi')$ is obtained from the two-dimensional values given in Eqs. (2.3-35) and (2.3-36) by replacing $k_0$ by $k_0 \sin \theta$, and multiplying by $\sin^2 \theta \, e^{-ik_0 z \cos \theta}$. Thus, the diffraction term becomes

$$D(\rho, \theta, \phi \pm \phi') = \frac{\sin^2 \theta \, \exp[ik_0(\rho \sin \theta) - z \cos \theta) + i\pi/4]}{\sqrt{2\pi k_0 \rho \sin \theta}}$$

$$\times \left[ \frac{(\pi/\gamma) \sin \pi^2/\gamma}{\cos \pi^2/\gamma - \cos[(\pi/\gamma)(\phi \pm \phi')]} \right] \qquad (2.3\text{-}42)$$

If the wedge is allowed to degenerate into a half-plane by letting $\gamma = 2\pi$, then it is easily seen that the diffracted field, given by

$$D(\rho, \theta, \phi - \phi') \pm D(\rho, \theta, \phi + \phi')$$

using Eq. (2.3-42), is the same as that given by Eqs. (2.3-25) and (2.3-28) for the half-plane.

To examine the surface currents in the vicinity of the edge, the series solutions (2.3-38) and (2.3-39) must be used. The surface currents are obtained from the tangential components of the magnetic field at the surface; thus, in the case of perpendicular polarization, only the $H_x$ component is of interest; while for parallel polarization, both $H_x$ and $H_z$ contribute to the surface currents. From Eq. (2.3-38) and Maxwell's equations

$$H_{x_\perp} = \frac{-i}{\omega\mu\sin^2\theta}\frac{\partial E_z}{\partial y} = \frac{-i\pi\sin\theta}{\eta\gamma}e^{-ik_0z\cos\theta}\sum_{n=0}^{\infty}\epsilon_n e^{-i\nu_n\pi/2}$$

$$\times\left[\sin\phi'J'_{\nu_n}[\cos\nu_n(\phi-\phi')-\cos\nu_n(\phi+\phi')]+\frac{\cos\phi'\nu_n J_{\nu_n}}{k_0\rho\sin\theta}\right.$$

$$\left.\times[\sin\nu_n(\phi-\phi')+\sin\nu_n(\phi+\phi')]\right] \tag{2.3-43}$$

with the argument of $J_{\nu_n}$ being $k_0\rho\sin\theta$. Similarly,

$$H_{x_\parallel} = \frac{-i\cos\theta}{\omega\epsilon\sin^2\theta}\frac{\partial H_z}{\partial y} = -\eta\frac{i\pi}{\gamma}\cos\theta\sin\theta\, e^{-ik_0z\cos\theta}\sum_{n=0}^{\infty}\epsilon_n e^{-i\nu_n\pi/2}$$

$$\times\left[\cos\phi'J'_{\nu_n}[\cos\nu_n(\phi-\phi')+\cos\nu_n(\phi+\phi')]-\frac{\sin\phi'\nu_n J_{\nu_n}}{k_0\rho\sin\theta}\right.$$

$$\left.\times[\sin\nu_n(\phi-\phi')-\sin\nu_n(\phi+\phi')]\right] \tag{2.3-44}$$

and $H_{z_\parallel}$ is given by Eq. (2.3-39). On the front surface of the wedge $\phi' = 0$, so that

$$H_{x_\perp} = \frac{-i2\pi}{\eta k_0 x}e^{-ik_0z\cos\theta}\sum_{n=0}^{\infty}\epsilon_n\, e^{-i\nu_n\pi/2}\nu_n J_{\nu_n}(k_0 x\sin\theta)\sin\nu_n\phi \tag{2.3-45}$$

$$H_{x_\parallel} = \frac{\eta 2\pi}{i\gamma}\cos\theta\sin\theta\, e^{-ik_0z\cos\theta}\sum_{n=0}^{\infty}\epsilon_n\, e^{-i\nu_n\pi/2}J'_{\nu_n}(k_0 x\sin\theta)\cos\nu_n\phi \tag{2.3-46}$$

$$H_{z_\parallel} = \frac{2\pi}{\gamma}\sin^2\theta\, e^{-ik_0z\cos\theta}\sum_{n=0}^{\infty}\epsilon_n\, e^{-i\nu_n\pi/2}J_{\nu_n}(k_0 x\sin\theta)\cos\nu_n\phi \tag{2.3-47}$$

For small $x$, the Bessel functions can be expanded in a power series, retaining only the first terms. Thus, the surface currents near the edge on the face $\phi' = 0$ become

$$K_{z_\perp} \approx \frac{2i}{\eta k_0}\left(\frac{\pi}{\gamma}\right)^2\frac{[(k_0 x\sin\theta)/2]^{(\pi/\gamma)-1}}{\Gamma(1+\pi/\gamma)}\exp(-ik_0z\cos\theta - i\pi^2/2\gamma)\sin\pi\phi/\gamma \tag{2.3-48}$$

$$K_{z_\parallel} \approx 2i\eta\left(\frac{\pi}{\gamma}\right)^2\frac{[(k_0 x\sin\theta)/2]^{(\pi/\gamma)-1}}{\Gamma(1+\pi/\gamma)}\exp(-ik_0z\cos\theta - i\pi^2/2\gamma)\cos\pi\phi/\gamma \tag{2.3-49}$$

and

$$K_{x_\parallel} \approx \frac{-2\pi}{\gamma}\sin^2\theta\exp(-ik_0z\cos\theta) \tag{2.3-50}$$

As for the half-plane, the surface current components parallel to the edge go to infinity provided $\pi < \gamma \leqslant 2\pi$, while the current normal to the edge is finite.

2.3.1.1.3. *Arbitrary Perfectly Conducting Edges.* The results just obtained are representative of some general principles regarding the fields in the vicinity of an edge on a perfectly conducting body. Consider an edge with a continuously varying tangent, and choosing a point on the edge, set up a coordinate system with the $z$-axis along the tangent, and the $x$-axis tangent to one sheet of the surface. The tangent to the other sheet of the surface then lies at an angle $\phi = 2\pi - \gamma$, where $\gamma$ is the wedge angle formed by the tangent planes to the surface. The situation is illustrated in Figure 2-29. In a region around the point $P$, which can be taken to be much smaller than the wavelength, the problem of the fields in the vicinity of the point is essentially a Rayleigh problem. Detailed examination of the situation reveals that, with respect to the distance, $\rho$, from the edge in the $x - y$ plane, the field variation is

$$E_\rho, E_\phi, H_\rho, H_\phi \propto \rho^{(\pi/\gamma)-1}$$

$$E_z \propto \rho^{(\pi/\gamma)}$$

and

$$H_z \propto \text{constant} \tag{2.3-51}$$

Thus, the field components parallel to the edge are bounded, while for $\gamma > \pi$ the components perpendicular to the edge are singular; however, the total field energy in any finite region is always finite.

The results obtained for the wedge, then, are locally valid even for a finite curved edge, and provide a means for estimating the contribution of an edge to the high-frequency scattering cross section of a finite body. To utilize the wedge results, it is necessary that they be modified to represent spherical waves emanating from points on the edge in the far field, rather than a cylindrical wave. The value of this technique depends, to a large extent, upon how this modification is accomplished. For a plane wave incident in a plane perpendicular to the tangent to the edge, the proper modification

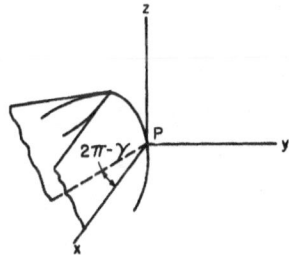

Fig. 2-29. Coordinates used for describing the local fields on an arbitrary convex perfectly conducting edge.

consists in a multiplication of the infinite-wedge diffraction coefficient D by

$$\frac{e^{-i\pi/4}}{\sqrt{\rho\lambda_0}}\delta l \qquad (2.3\text{-}52)$$

where $\delta l$ is the length of an infinitesimal segment of the edge. Thus, the contribution from edges lying wholly perpendicular to the plane of incidence can be obtained by multiplying the infinite-wedge results by the above factor, and integrating along the entire edge. Such a technique is called the Sommerfeld–MacDonald technique in this book, although it has been applied in variously modified forms by a large number of authors, often without being specifically named. Notable among the recent applications are those by Clemmow and Millar, to convex apertures in infinite perfectly conducting surfaces;[117–119] by Siegel,[120] in determining the contribution of the base of a finite cone to the nose-on backscatter cross section; and by Ufimtsev,[121–123] to various scattering problems.

For a plane wave incident arbitrarily upon a finite edge, a method of accounting for the edge effects which bypasses the problem of converting the infinite-edge cylindrical waves into a spherical wave in the far field would be to assume that near the edge the current distributions on a curved edge are locally the same as that on the infinite wedge tangent to the edge, and integrate along the edge over a region near the edge to obtain the scattered field contributions. If the optical contributions to the infinite-wedge currents have been subtracted out and only the edge currents contributing to the diffraction fields are used, these will decay rapidly away from the edge and the integrations can be carried out over the entire surface, or only over a region near the edge, with essentially the same results. The difficulty in applying this technique lies in the complexity of the integrations, and the fact that for the general wedge the surface currents are expressed in terms of the series solutions (2.3-38) and (2.3-39), which converge slowly for distances $x$ far from the edge.

### 2.3.1.2. Tip Diffraction

Consider a semi-infinite circular cone oriented as shown in Figure 2-30 with a plane wave incident at an angle $\theta$. In order to determine the contribu-

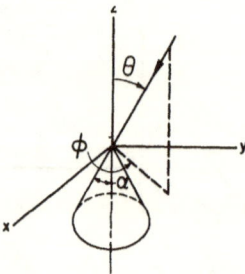

Fig. 2-30.   Coordinates for a plane wave incident on the tip of a semi-infinite cone of half angle $\alpha$.

tion of a tip, such as that for the cone, to the total backscattered field of a body, the results for the semi-infinite cone will be examined for angles of incidence such that the tip is the only contributor, that is, angles such that $\theta < \alpha$.

2.3.1.2.1. *Narrow-Angle Cones.* For narrow-angle cones, $\alpha \ll \pi/2$, at an arbitrary angle of incidence, $\theta \leqslant \alpha$, Felson[124] has obtained the expression

$$\mathbf{E}_\perp^s \approx \frac{ie^{ik_0 r}}{k_0 r} \left(\frac{\alpha}{2}\right)^2 \left[\frac{3 + \cos^2 \theta}{4 \cos^3 \theta}\right] [\hat{\mathbf{x}} \sin \phi - \hat{\mathbf{y}} \cos \phi] \qquad (2.3\text{-}53)$$

for the backscattered electric field where the incident field has its electric field vector lying in the $x - y$ plane. For parallel polarization, similar results should be obtained when $\theta$ is near zero and $\alpha \ll \pi/2$.

From physical optics, the solution is

$$\mathbf{E}^s \approx \frac{e^{ik_0 r}}{r} \left[\frac{i \tan^2 \alpha}{4k_0 \cos^3 \theta (1 - \tan^2 \alpha \tan^2 \theta)^{3/2}}\right] [\hat{\mathbf{x}} \sin \phi - \hat{\mathbf{y}} \cos \phi] \qquad (2.3\text{-}54)$$

for $\theta < \alpha$. Amazingly enough, for small $\theta$ and $\alpha$, the physical optics result agrees with Felson's result; thus either one can be used to estimate the contribution of a tip to the scattered fields. Since it is somewhat simpler, Eq. (2.3-53) (Felson's result) will be examined in more detail. For a plane wave incident along the axis, the backscattered field becomes

$$\mathbf{E}^s = \frac{ie^{ik_0 r}}{k_0 r} \left(\frac{\alpha}{2}\right)^2 [\hat{\mathbf{x}} \sin \phi + \hat{\mathbf{y}} \cos \phi] \qquad (2.3\text{-}55)$$

and assuming a receiver polarization the same as for the incident field, the backscatter cross section is

$$\sigma = \frac{\pi \alpha^4}{4k_0^2} \qquad \text{or} \qquad \frac{\sigma}{\lambda_0^2} = \frac{\alpha^4}{16\pi} \qquad (2.3\text{-}56)$$

It is evident from the above equation that the backscatter due to a small angle tip is quite small. For example, if $\alpha = 10°$, $\sigma/\lambda_0^2$ is $1.3 \times 10^{-5}$ and the contribution from the tip to the total backscatter cross section of most objects will be insignificant.

Returning to the case $\theta \neq 0$, the backscatter cross section from Eq. (2.3-53) is

$$\frac{\sigma}{\lambda_0^2} = \frac{\alpha^4}{16\pi} \left[\frac{3 + \cos^2 \theta}{4 \cos^3 \theta}\right]^2 \qquad (2.3\text{-}57)$$

From this it is apparent that the change in cross section with the angle $\theta$ is small; for example, only an 8 % increase in $\sigma/\lambda_0^2$ occurs for $\theta$ varying from

0 to 10°. Thus, the tip contribution can be considered to be essentially that given by Eq. (2.3-56), or the physical optics equivalent

$$\frac{\sigma}{\lambda_0^2} \approx \frac{\tan^4 \alpha}{16\pi} \qquad (2.3\text{-}58)$$

and in most cases can be neglected in comparison with other contributions to the cross section.

2.3.1.2.2. *Wide-Angle Cones.* For cones where $0 \ll \alpha \leqslant \pi/2$ and $\theta = 0$, the backscatter cross section is given by Felson as

$$\frac{\sigma}{\lambda_0^2} \approx \frac{1 - 2\cos^2(\pi - \alpha)}{16\pi \cos^4(\pi - \alpha)} \qquad (2.3\text{-}59)$$

which agrees closely with the physical optics result, Eq. (2.3-58), for $\alpha \approx \pi/2$. Goryainov[125] has obtained an expression for the scattered fields with a plane wave incident along the $z$-axis, and in terms of his results the axially incident backscatter cross section is

$$\frac{\sigma}{\lambda_0^2} = \frac{L^2}{\pi} \qquad (2.3\text{-}60)$$

where $L$ is plotted in Figure 2-31 for $0 < \alpha < \pi/2$. For narrow-angle cones with $\alpha \ll \pi/2$, then

$$L \approx \left(\frac{\alpha}{2}\right)^2 \qquad (2.3\text{-}61)$$

and for wide-angle cones with $0 \ll \alpha < \pi/2$, then

$$L \approx \frac{1}{(\pi - 2\alpha)^2} \qquad (2.3\text{-}62)$$

essentially in agreement with both physical optics and Felson's results.

## 2.3.2. Creeping Waves

In succeeding chapters of the book, creeping waves will often be encountered in connection with high-frequency approximations to the scattered fields from specific shapes, and in fact have already appeared in Section 2.2, although not explicitly discussed. They are generally characterized as waves which are launched at the shadow boundary of an object and then propagate along the body surface in the shadow region, emerging at the opposite shadow boundary. They are of course closely related to the surface-diffracted rays of Keller's geometrical theory of diffraction and to the theoretical results of Fock. The mathematical details of creeping wave theory are quite complex

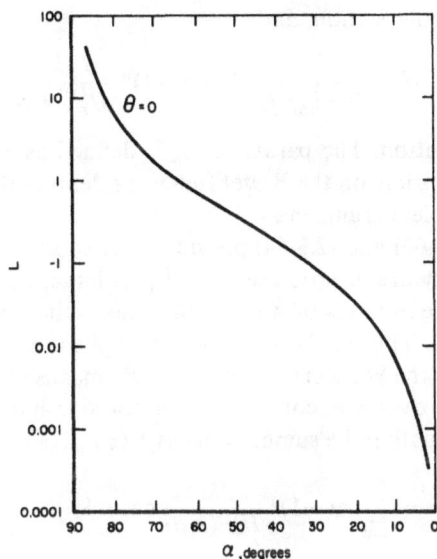

Fig. 2-31. Goryainov's $L$ function for axial
incidence backscatter from a perfectly conducting cone [after Goryainov[125]].

and often tend to obscure the physical picture of the phenomena. Perhaps the
best way to arrive at an understanding of the creeping wave formalism is to
follow the historical development of the subject.

The term "creeping waves" was first applied by Franz and Depperman[126]
to some mathematical results they obtained while searching for high-frequency approximations to the fields scattered or diffracted by an infinite perfectly
conducting cylinder. Similar mathematical techniques had been applied
previously to problems involving both cylinders and spheres; however, the
physical interpretation and the term "creeping waves" were provided by
Franz and Depperman.

### 2.3.2.1. Perfectly Conducting Cylinder

The historical starting point for creeping wave theory is the eigenfunction-
series solution for the scattered fields from an infinite perfectly conducting
cylinder at normal incidence. This can be obtained by the techniques discussed
in Section 2.2.1.1, and is considered in some detail in Chapter 4. From
Eqs. (4.1-5) and (4.1-6), the far-zone fields scattered by an infinite perfectly
conducting cylinder are

$$E_z^s = -E_z^0 \frac{e^{ik_0\rho}}{\sqrt{(i\pi k_0\rho)/2}} \sum_{n=0}^{\infty} \epsilon_n(-1)^n \frac{J_n(k_0a)}{H_n^{(1)}(k_0a)} \cos n\phi \qquad (2.3\text{-}63)$$

for perpendicular polarization, and

$$H_z^s = -H_z^0 \frac{e^{ik_0\rho}}{\sqrt{(i\pi k_0\rho)/2}} \sum_{n=0}^{\infty} \epsilon_n(-1)^n \frac{J_n'(k_0a)}{H_n^{(1)'}(k_0a)} \cos n\phi \quad (2.3\text{-}64)$$

for parallel polarization. The parameter $\epsilon_n$ is defined as $\epsilon_0 = 1$ and $\epsilon_n = 2$ for $n > 0$, and the prime on the Bessel functions denotes differentiation with respect to the complete argument.

Equations (2.3-63) and (2.3-64) provide exact solutions for the scattered fields; however, they are of little use when $k_0a$ is large, corresponding to the cylinder radius large in terms of wavelength, due to the slowness with which the series converges. This can be remedied by application of a mathematical technique known as the Watson transform.[127] By means of this transform the infinite summation over $n$ is converted to a contour integral in a complex $\nu$-plane. Thus designating the summations as $f_\perp(\phi), f_\parallel(\phi)$, they become[†]

$$f_\perp(\phi) = \sum_{n=0}^{\infty} \epsilon_n(-1)^n \frac{J_n(k_0a)}{H_n^{(1)}(k_0a)} \cos n\phi$$

$$= -i \int_c e^{-i\nu\pi} \frac{\cos \nu(\pi - \phi)}{\sin \nu\pi} \frac{J_\nu(k_0a)}{H_\nu^{(1)}(k_0a)} d\nu \quad (2.3\text{-}65)$$

and

$$f_\parallel(\phi) = -i \int_c e^{-i\nu\pi} \frac{\cos \nu(\pi - \phi)}{\sin \nu\pi} \frac{J_\nu'(k_0a)}{H_\nu^{(1)'}(k_0a)} d\nu \quad (2.3\text{-}66)$$

where the contour $c$ is shown in Figure 2-32. This contour encloses the zeros of $\sin \nu\pi$ along the real axis. Now by expanding $\cos \nu(\pi - \phi)$ as

$$\cos \nu(\pi - \phi) = e^{i\nu\pi} \cos \nu\phi - ie^{i\nu\phi} \sin \nu\pi \quad (2.3\text{-}67)$$

$f_\perp, f_\parallel$ become

$$f_\perp(\phi) = -i \int_c \frac{\cos \nu\phi}{\sin \nu\pi} \frac{J_\nu(k_0a)}{H_\nu^{(1)}(k_0a)} d\nu - \int_c e^{i\nu(\phi-\pi)} \frac{J_\nu(k_0a)}{H_\nu^{(1)}(k_0a)} d\nu \quad (2.3\text{-}68)$$

$$f_\parallel(\phi) = -i \int_c \frac{\cos \nu\phi}{\sin \nu\pi} \frac{J_\nu'(k_0a)}{H_\nu^{(1)'}(k_0a)} d\nu - \int_c e^{i\nu(\phi-\pi)} \frac{J_\nu'(k_0a)}{H_\nu^{(1)'}(k_0a)} d\nu \quad (2.3\text{-}69)$$

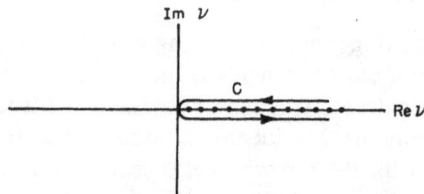

Fig. 2-32. Integration contour for the Watson transform, and the positive zero's of $\sin \nu\pi$.

[†] The polarization notations, $\perp$ and $\parallel$, used here are opposite those used in Chapter 4.

The last terms of $f_\perp$ and $f_\parallel$ can be evaluated by saddle-point integrations, and give the specular point contributions to the scattered fields exactly as given by the geometrical optics series (Luneberg series). The first terms contain the diffraction effects, and are evaluated by deforming the contour to enclose the zeros of $H_\nu^{(1)}(k_0 a)$ and $H_\nu^{(1)'}(k_0 a)$. These lie in the first quadrant of the $\nu$-plane as shown in Figure 2-33, when $k_0 a \gg 1$.

In deforming the contours $c$ of Figure 2-32 into the contours $D$ of Figure 2-33, it is necessary to insure that no poles other than those shown are enclosed and that there are no contributions to the integrals from the arcs at infinity. This is the case for the backscattered fields with $\phi < \pi/2$, and the first terms of Eqs. (2.3-68) and (2.3-69) can be evaluated by a residue series yielding

$$f_\perp^{(1)}(\phi) = -2\pi \sum_{s=1}^{\infty} \frac{\cos \nu_s \phi}{\sin \nu_s \pi} \left[ \frac{J_\nu(k_0 a)}{(\partial/\partial\nu)\, H_\nu^{(1)}(k_0 a)} \right]_{\nu=\nu_s} \qquad (2.3\text{-}70)$$

and

$$f_\parallel^{(1)}(\phi) = -2\pi \sum_{s=1}^{\infty} \frac{\cos \bar\nu_s \phi}{\sin \bar\nu_s \pi} \left[ \frac{J_\nu'(k_0 a)}{(\partial/\partial\nu)\, H_\nu^{(1)'}(k_0 a)} \right]_{\nu=\bar\nu_s} \qquad (2.3\text{-}71)$$

The parameters, $\nu_s$ and $\bar\nu_s$, are the complex zeros of $H_\nu^{(1)}(k_0 a)$ and $H_\nu^{(1)'}(k_0 a)$, respectively. As shown in Figure 2-33, they lie along a line oriented at 60° with respect to the positive real axis with $\nu_s$ and $\bar\nu_s$ alternating, and $\bar\nu_1$ closest to the real axis. To determine these as a function of $k_0 a$ is a difficult mathematical problem; however, for large $k_0 a$, they can be expressed in terms of the zeros of the Airy function Ai$(x)$ and its derivative, of which the first 56 have been tabulated by Logan.[128] Thus, the expressions for $\nu_s$ and $\bar\nu_s$ are

$$\nu_s = k_0 a + \alpha_s\, e^{i\pi/3} \left[ \frac{k_0 a}{2} \right]^{1/3} + \frac{\alpha_s^2}{60}\, e^{i2\pi/3} \left[ \frac{2}{k_0 a} \right]^{1/3} + O(k_0 a)^{-4/3} \qquad (2.3\text{-}72)$$

$$\bar\nu_s = k_0 a + \beta_s\, e^{i\pi/3} \left[ \frac{k_0 a}{2} \right]^{1/3} + \left[ \frac{\beta_s^2}{60} + \frac{1}{10\beta_s} \right] e^{i2\pi/3} \left[ \frac{2}{k_0 a} \right]^{1/3} + O(k_0 a)^{-4/3}$$
$$(2.3\text{-}73)$$

where Ai$(-\alpha_s) = 0$, Ai$'(-\beta_s) = 0$. Expressions for the roots $\nu_s$ and $\bar\nu_s$, in somewhat different notation, are given for powers of $(k_0 a)$ up to and including $(k_0 a)^{-5/3}$ by Franz and Galle.[129]

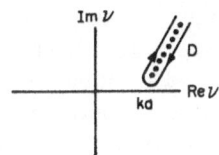

Fig. 2-33. Integration contour for a residue series evaluation and the complex zero's of the Hankel function of the first kind and its derivative.

Now returning to Eqs. (2.3-70) and (2.3-71), the ratio $\cos \nu\phi/\sin \nu\pi$ can be expanded by setting

$$\cos \nu\phi = \frac{e^{i\nu\phi} + e^{-i\nu\phi}}{2}$$

and since $\operatorname{Im} \nu_s, \bar{\nu}_s > 0$,

$$\frac{1}{\sin \nu\pi} = -2i\, e^{i\nu\pi} \sum_{m=0}^{\infty} e^{i\nu 2\pi m}$$

Using these results in Eqs. (2.3-70) and (2.3-71),

$$f_{\perp}^{(1)}(\phi) = 2\pi i \sum_{s=1}^{\infty} \sum_{m=0}^{\infty} \left[ \frac{J_{\nu}(k_0 a)}{(\partial/\partial \nu)\, H_{\nu}^{(1)}(k_0 a)} \right]_{\nu=\nu_s} [e^{i\nu_s[(2m+1)\pi+\phi]} + e^{i\nu_s[(2m+1)\pi-\phi]}]$$

(2.3-74)

and

$$f_{\parallel}^{(1)}(\phi) = 2\pi i \sum_{s=1}^{\infty} \sum_{m=0}^{\infty} \left[ \frac{J_{\nu}'(k_0 a)}{(\partial/\partial \nu)\, H_{\nu}^{(1)'}(k_0 a)} \right]_{\nu=\bar{\nu}_s} [e^{i\nu_s[(2m+1)\pi+\phi]} + e^{i\nu_s[(2m+1)\pi-\phi]}]$$

(2.3-75)

From these equations, recalling that the leading term in both $\nu_s$ and $\bar{\nu}_s$ is $k_0 a$, the creeping-wave interpretation is clearly apparent. The $\phi$ dependence of $f_{\perp}$ and $f_{\parallel}$ appears in terms of the form

$$e^{i\nu[(2m+1)\pi+\phi]} \approx e^{ik_0 a[(2m+1)\pi+\phi]}$$

and

$$e^{i\nu[(2m+1)-\phi]} \approx e^{ik_0 a[(2m+1)\pi-\phi]}$$

which are the phase factors corresponding to waves which have encircled the cylinder $m + \frac{1}{2}$ times in both a clockwise and counterclockwise sense, where $m$ is an integer. If instead of the scattered fields, the eigenfunction expansions for the surface currents on the cylinder are similarly transformed, then similar terms result with phase factors corresponding to waves propagating around the shadowed portion of the cylinder.

For each root $\nu_s$, $\bar{\nu}_s$ there is an infinity of such creeping wave terms, each integral change in $m$ corresponding to a complete traverse around the cylinder circumference. The imaginary parts of $\nu_s$ and $\bar{\nu}_s$ determine the rates at which these waves are attenuated as they move around the cylinder. From Eqs. (2.3-72) and (2.3-73), the imaginary parts of $\nu_s$ and $\bar{\nu}_s$ are proportional to $(k_0 a)^{1/3}$; thus for large $k_0 a$, considerable attenuation can occur, and the waves are strongly damped after only one traversal around the back of the cylinder.

The amplitude with which the creeping waves contribute to the scattered fields is determined by evaluating the terms in the square brackets of Eqs. (2.3-74) and (2.3-75). This is accomplished by first eliminating $J_{\nu}(k_0 a)$ and

$J_\nu'(k_0 a)$ by means of the Wronskian relationship between $J_\nu(x)$ and $H_\nu^{(1)}(x)$, that is,

$$J_\nu H_\nu' - J_\nu' H_\nu = \frac{2i}{\pi k_0 a}$$

which results in

$$\frac{J_\nu(k_0 a)}{(\partial/\partial\nu)\, H_\nu^{(1)}(k_0 a)} = \frac{2i}{\pi k_0 a H_\nu^{(1)\prime}(\partial/\partial\nu)\, H_\nu^{(1)}(k_0 a)}$$

and

$$\frac{J_\nu'(k_0 a)}{(\partial/\partial\nu)\, H_\nu^{(1)\prime}(k_0 a)} = \frac{-2i}{\pi k_0 a H_\nu^{(1)\prime}(\partial/\partial\nu)\, H_\nu^{(1)\prime}(k_0 a)}$$

when the fact that $H_{\nu_s}^{(1)}(k_0 a) = H_{\bar\nu_s}^{(1)\prime}(k_0 a) = 0$ is also used.

These can then be evaluated by using a uniformly valid asymptotic expansion of the Hankel function and its derivative, again giving results expressible in terms of the Airy function $\mathrm{Ai}(x)$ and its derivative. The details of this evaluation will not be given; however, the results are

$$\left[\frac{J_\nu(k_0 a)}{(\partial/\partial\nu)\, H_\nu^{(1)}(k_0 a)}\right]_{\nu=\nu_s} \approx \frac{e^{i5\pi/6}[k_0 a/2]^{1/3}}{4\pi[\mathrm{Ai}'(-\alpha_s)]^2}\left[1 + \frac{e^{i\pi/3}\alpha_s}{30[k_0 a/2]^{2/3}} + O(k_0 a)^{-4/3}\right] \tag{2.3-76}$$

$$\left[\frac{J_\nu'(k_0 a)}{(\partial/\partial\nu)\, H_\nu^{(1)\prime}(k_0 a)}\right]_{\nu=\nu_s} \approx \frac{e^{i5\pi/6}[k_0 a/2]^{1/3}}{4\pi\beta_s[\mathrm{Ai}(-\beta_s)]^2}\left[1 + \frac{e^{i\pi/3}}{[k_0 a/2]^{2/3}}\left[\frac{1}{10\beta_s^2} - \frac{\beta_s}{30}\right]\right.$$
$$\left. + O(k_0 a)^{-1/3}\right] \tag{2.3-77}$$

The contributions to the fields scattered from the cylinder due to creeping waves, using Eqs. (2.3-76), (2.3-77), (2.3-74), and (2.3-75), are

$$E_z^c = E_z^{\,0}\frac{e^{ik_0\rho}}{\sqrt{2k_0\rho}}\sum_{s=1}^{\infty}\sum_{m=0}^{\infty} D_s[e^{i\bar\nu_s[(2m+1)\pi+\phi]} + e^{i\bar\nu_s[(2m+1)\pi-\phi]}] \tag{2.3-78}$$

and

$$H_z^c = H_z^{\,0}\frac{e^{ik_0\rho}}{\sqrt{2k_0\rho}}\sum_{s=1}^{\infty}\sum_{m=0}^{\infty} \bar D_s[e^{i\nu_s[(2m+1)\pi+\phi]} + e^{i\nu_s[(2m+1)\pi-\phi]}] \tag{2.3-79}$$

where

$$D_s = \frac{e^{i\pi/12}[k_0 a/2]^{1/3}}{\pi^{1/2}[\mathrm{Ai}'(-\alpha_s)]^2}\left[1 + \frac{e^{i\pi/3}\alpha_s}{30[k_0 a/2]^{2/3}} + O(k_0 a)^{-4/3}\right] \tag{2.3-80}$$

and

$$\bar D_s = \frac{e^{i\pi/12}[k_0 a/2]^{1/3}}{\pi^{1/2}\beta_s[\mathrm{Ai}(-\beta_s)]^2}\left[1 - \frac{e^{i\pi/3}}{[k_0 a/2]^{2/3}}\left[\frac{1}{10\beta_s^2} - \frac{\beta_s}{30}\right] + O(k_0 a)^{-4/3}\right] \tag{2.3-81}$$

Franz and Galle,[129] using a different notation, give expressions from which $D_s$ and $\bar{D}_s$ can be computed up to and including terms of order $(k_0 a)^{-2}$.

In Table 2-3, values of $\alpha_s$, $\beta_s$, Ai$'(-\alpha_s)$, Ai$(-\beta_s)$ are given for $s$ up to five.

**Table 2-3**

| $s$ | $\alpha_s$ | Ai$'(-\alpha_s)$ | $\beta_s$ | Ai$(-\beta_s)$ |
|---|---|---|---|---|
| 1 | 2.33810 | + 0.70121 | 1.01879 | + 0.53565 |
| 2 | 4.08794 | − 0.80311 | 3.24819 | − 0.41901 |
| 3 | 5.52055 | + 0.86520 | 4.82009 | + 0.38040 |
| 4 | 6.78670 | − 0.91085 | 6.16330 | − 0.35790 |
| 5 | 7.94413 | + 0.94733 | 7.37217 | + 0.34230 |

From the table it is apparent that $\alpha_s$ and $\beta_s$ rise rapidly as a function of $s$, and since $\nu_s$ and $\bar{\nu}_s$ are to first-order linear functions of $\alpha_s$ and $\beta_s$, then for $k_0 a$ moderately large, only the $s = 1$ term contributes significantly to the scattered field. Thus, the primary contributions to the scattered fields come from the $s = 1$, $m = 0$ terms. To this order of approximation, the scattered fields are

$$E_{z,1}^c = E_z^0 \frac{e^{ik_0\rho}}{\sqrt{2k_0\rho}} D_1[e^{i\nu_1(\pi+\phi)} + e^{i\nu_1(\pi-\phi)}]$$

and

$$H_{z,1}^c = H_z^0 \frac{e^{ik_0\rho}}{\sqrt{2k_0\rho}} \bar{D}_1[e^{i\bar{\nu}_1(\pi+\phi)} + e^{i\bar{\nu}_1(\pi-\phi)}]$$

which for backscatter reduce to

$$E_{z,1}^c(0) = E_z^0 \frac{e^{ik_0\rho}}{\sqrt{k_0\rho/2}} D_1 e^{i\nu_1\pi} \qquad (2.3\text{-}82)$$

and

$$H_{z,1}^c(0) = H_z^0 \frac{e^{ik_0\rho}}{\sqrt{k_0\rho/2}} \bar{D}_1 e^{i\bar{\nu}_1\pi} \qquad (2.3\text{-}83)$$

Curves of $|D_1 e^{i\nu_1\pi}|$ and $|\bar{D}_1 e^{i\bar{\nu}_1\pi}|$ versus $k_0 a$ are given in Figure 2-34. For $k_0 a = 5$, neglecting the terms for $s > 1$ and $m > 0$ results in an error of $<1\%$, which decreases for larger $k_0 a$.

### 2.3.2.2. Perfectly Conducting Sphere

Similar creeping-wave contributions to the scattered fields can be obtained for other smoothly curving shapes. The sphere will be considered as

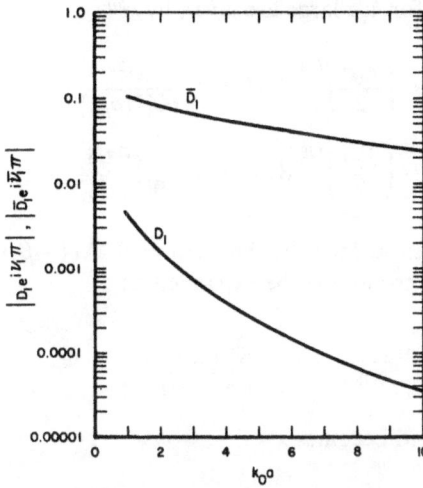

Fig. 2-34. Backscatter creeping wave magnitudes for a perfectly conducting cylinder, parallel and perpendicular polarizations.

an additional example. The eigenfunction series or so-called Mie-series solution for the backscattered fields from a perfectly conducting sphere can be transformed into a residue series similar to that for the cylinder (see Chapter 3). For a plane wave incident along the $z$-axis, the diffraction component of the backscattered field, after the specular contribution has been subtracted out, is[130]

$$E^c = -i\,\frac{e^{ik_0\rho}}{k_0\rho}\left(\frac{\pi}{2}\sum_s\left[\frac{\nu L^{(2)}_{\nu-1/2}(k_0a)}{(\partial/\partial\nu)\,L^{(1)}_{\nu-1/2}(k_0a)}\right]_{\nu=\zeta_s}(\cos\zeta_s\pi)^{-1}\right.$$

$$\left.-\frac{\pi}{2}\sum_s\left[\frac{\nu L^{(2)'}_{\nu-1/2}(k_0a)}{(\partial/\partial\nu)\,L^{(1)'}_{\nu-1/2}(k_0a)}\right]_{\nu=\zeta_s}(\cos\zeta_s\pi)^{-1}\right) \qquad (2.3\text{-}84)$$

where $L^{(1)}_{\zeta_s-1/2}(k_0a) = 0$ and $L^{(1)'}_{\zeta_s-1/2}(k_0a) = 0$. These functions are defined in terms of the spherical Hankel functions as

$$L^{(1,2)}_{\nu-1/2}(k_0a) = k_0ah^{(1,2)}_{\nu-1/2}(k_0a)$$

$$L^{(1,2)'}_{\nu-1/2}(k_0a) = \frac{\partial}{\partial k_0a}\,k_0ah^{(1,2)}_{\nu-1/2}(k_0a)$$

The roots $\zeta_s$ and $\bar{\zeta}_s$ for $k_0 a$ large are given by[130]

$$\zeta_s = k_0 a + \left[\frac{k_0 a}{2}\right]^{1/3} \alpha_s\, e^{i\pi/3} + \frac{e^{i2\pi/3}}{60[k_0 a/2]^{1/3}}\, \alpha_s{}^2 \qquad (2.3\text{-}85)$$

$$\bar{\zeta}_s = k_0 a + \left[\frac{k_0 a}{2}\right]^{1/3} \beta_s\, e^{i\pi/3} + \frac{e^{i2\pi/3}}{60[k_0 a/2]^{1/3}} \left[\beta_s{}^2 - \frac{9}{\beta_s}\right] \qquad (2.3\text{-}86)$$

where again $\alpha_s$, $\beta_s$ are defined by $\mathrm{Ai}(-\alpha_s) = 0$, $\mathrm{Ai}'(-\beta_s) = 0$. Now in Eq. (2.3-84) the factor $1/\cos \nu\pi$ can be expanded as

$$\frac{1}{\cos \nu\pi} = 2 \sum_{l=0}^{\infty} (-1)^l\, e^{i\nu(2m+1)\pi}$$

thus the scattered field becomes

$$E^c = -i\, \frac{\pi\, e^{ik_0\rho}}{k_0\rho} \sum_{s=1}^{\infty} \sum_{m=0}^{\infty} (-1)^m [C_s\, e^{i\zeta_s(2m+1)\pi} - \bar{C}_s\, e^{i\bar{\zeta}_s(2m+1)\pi}] \qquad (2.3\text{-}87)$$

with

$$C_s = \left[\frac{\nu L_{\nu-1/2}^{(2)}(k_0 a)}{(\partial/\partial\nu)\, L_{\nu-1/2}^{(1)}(k_0 a)}\right]_{\nu=\zeta_s}$$

$$\bar{C}_s = \left[\frac{\nu L_{\nu-1/2}^{(2)'}(k_0 a)}{(\partial/\partial\nu)\, L_{\nu-1/2}^{(1)'}(k_0 a)}\right]_{\nu=\bar{\zeta}_s}$$

The amplitude factors or launch coefficients $C_s$, $\bar{C}_s$ can be evaluated similarly to those for cylinders, giving[130]

$$C_s = \frac{(k_0 a)^{4/3}\, e^{i5\pi/6}}{\pi[\mathrm{Ai}'(-\alpha_s)]^2} \left[1 + e^{i\pi/3}\, \frac{8\alpha_s}{15[k_0 a/2]^{2/3}} + O(k_0 a)^{-4/3}\right] \qquad (2.3\text{-}88)$$

and

$$\bar{C}_s = \frac{(k_0 a)^{4/3}\, e^{i5\pi/6}}{\pi\beta_s[\mathrm{Ai}(-\beta_s)]^2} \left[1 + \frac{e^{i\pi/3}}{60[k_0 a/2]^{2/3}} \left[32\beta_s + \frac{9}{\beta_s{}^2}\right] + O(k_0 a)^{-4/3}\right] \qquad (2.3\text{-}89)$$

The creeping-wave interpretation is again readily apparent from Eq. (2.3-87): since the leading term of both $\zeta_s$ and $\bar{\zeta}_s$ is $k_0 a$, the summation over $m$ represents waves which have circled the sphere $m + \frac{1}{2}$ times. Like the cylinder solution, the terms for $m > 0$ decay rapidly as a function of $k_0 a$, as do the terms for $s > 1$. Thus dominant terms correspond to $m = 0$, $s = 1$. As a result of the difference between $\zeta_s$, $\bar{\zeta}_s$, the term in Eq. (2.3-87) involving $\bar{C}_s$, $\bar{\zeta}_s$ is larger by several orders of magnitude for $k_0 a = 5$ or greater. Those terms involving $\bar{C}_s$, $\bar{\zeta}_s$ are often called "$E$" waves because at the shadow boundary the electric field vector is normal to the surface. Those terms involving $C_s$, $\zeta_s$ are designated "$H$" waves from the fact that their magnetic field vectors are normal to the surface at the shadow boundary.

Although the mathematical details vary slightly, similar decompositions of the scattered fields can be carried out for directions other than backscatter, as well as for the surface-current distributions. Geometrical and creeping-wave contributions are obtained in the illuminated region, while in the shadow region only the creeping wave terms appear.

Comparing Eqs. (2.3-72) and (2.3-73) for the creeping-wave propagation coefficients on a cylinder with Eqs. (2.3-85) and (2.3-86) for the creeping-wave propagation coefficients on a sphere, it appears that they differ only in the higher-order terms, that is,

$$\nu_s \approx \zeta_s \approx k_0 a + \left[\frac{k_0 a}{2}\right]^{1/3} \alpha_s\, e^{i\pi/3}$$

$$\bar{\nu}_s \approx \bar{\zeta}_s \approx k_0 a + \left[\frac{k_0 a}{2}\right]^{1/3} \beta_s\, e^{i\pi/3}$$

The launch coefficients $C$, $\bar{C}$, $D$, $\bar{D}$ differ, of course, since the shadow boundary for the sphere is a circle, whereas it is a pair of parallel straight lines for the cylinder. However, the equivalence of the propagation coefficients for sufficiently large $k_0 a$ indicates that the creeping-wave propagation rates for smooth perfect conductors are determined solely by the frequency and the curvature along the geodesics of the surface. Thus, additional justification is provided for the assumptions made regarding surface-diffracted rays in the geometrical theory of diffraction.

### 2.3.2.3. *Relationship Between Creeping Waves and Fock Theory*

The connection between creeping-wave theory and Fock theory is clearly apparent if the asymptotic expansions [Eqs. (2.2-208) and (2.2-209)] for the Fock functions $\hat{p}$ and $\hat{q}$ are substituted in Eqs. (2.2-199) and (2.2-200) for the scattered fields from a perfectly conducting cylinder. Carrying out this process in the case of Eq. (2.2-199) for the axial electric field where

$$E_z{}^s = - \left(\frac{k_0 a}{2}\right)^{1/3} \sqrt{\frac{2}{k_0 \rho}}\, e^{i(k_0 \rho + \pi/4)} \left[ e^{i k_0 a(\phi - \pi)} \hat{p}\left[\left(\frac{k_0 a}{2}\right)^{1/3}(\phi - \pi)\right]\right.$$

$$+ \sum_{n=0}^{\infty} \hat{p}\left[\left(\frac{k_0 a}{2}\right)^{1/3}(2n\pi + \pi + \phi)\right] e^{i k_0 a(2n\pi + \pi + \phi)}$$

$$\left. + \sum_{n=0}^{\infty} \hat{p}\left[\left(\frac{k_0 a}{2}\right)^{1/3}(2n\pi + \pi - \phi)\right] e^{i k_0 a(2n\pi + \pi - \phi)}\right] \qquad (2.3\text{-}90)$$

and substituting the asymptotic expansion for $\hat{p}$ in the last two terms, then

$$E_z{}^s = -\left(\frac{k_0 a}{2}\right)^{1/3} \sqrt{\frac{2}{k_0 \rho}}\, e^{i(k_0 \rho + \pi/4)} \left[ e^{ik_0 a(\phi - \pi)} \hat{p} \left[ \left(\frac{k_0 a}{2}\right)^{1/3} (\phi - \pi) \right] \right]$$

$$+ \sum_{n=0}^{\infty} \sum_{s=1}^{\infty} \left( e^{ik_0 a[(2n+1)\pi + \phi]} \right.$$

$$\times \left[ \frac{-\exp(-i\pi/6)\exp(-\sqrt{3}/2 + i/2)\,\alpha_s(k_0 a/2)^{1/3}\,[(2n+1)\pi + \phi]}{2\sqrt{\pi}\,[\mathrm{Ai}'(-\alpha_s)]^2} \right]$$

$$+ e^{ik_0 a[(2n+1)\pi - \phi]}$$

$$\left. \times \left[ \frac{-\exp(-i\pi/6)\exp(-\sqrt{3}/2 + i/2)\,\alpha_s(k_0 a/2)^{1/3}\,[(2n+1)\pi - \phi]}{2\sqrt{\pi}\,[\mathrm{Ai}'(-\alpha_s)]^2} \right] \right) \right]$$

$$(2.3\text{-}91)$$

Neglecting the first term which corresponds to the geometrical optics field and simplifying the remaining terms gives

$$E_z^{s,cw} = \frac{e^{ik_0 \rho}}{\sqrt{2k_0 \rho}} \sum_{n=0}^{\infty} \sum_{s=1}^{\infty} D_s{}^F \Big[ \exp i\nu_s{}^F[(2n+1)\pi + \phi]$$

$$+ \exp i\nu_s{}^F[(2n+1)\pi - \phi] \Big] \qquad (2.3\text{-}92)$$

where

$$D_s{}^F = \frac{e^{i\pi/12}(k_0 a/2)^{1/3}}{\pi^{1/2}[\mathrm{Ai}'(-\alpha_s)]^2} \qquad (2.3\text{-}93)$$

$$\nu_s{}^F = k_0 a + e^{i\pi/3}(k_0 a/2)^{1/3}\,\alpha_s \qquad (2.3\text{-}94)$$

and it is apparent that these are identical to the first terms of $D_s$ and $\nu_s$ given by Eqs. (2.3-72) and (2.3-80).

If the scattered magnetic field predicted by Fock theory [Eq. (2.2-200)] is similarly expanded by using the asymptotic expansion for $\hat{q}$, then the first terms of $\bar{D}_s$ and $\bar{\nu}_s$ are obtained.

### 2.3.3. Traveling Waves

#### 2.3.3.1. Perfectly Conducting Bodies

For long thin bodies, such as wires, ogives, prolate spheroids, etc., illuminated near nose-on incidence, the backscatter cross section exhibits large peaks for parallel polarization, with the backscattered energy appearing to emanate from the rear of the body. Peters, by considering such bodies to act as traveling-wave antennas excited by the incident field, obtained analytical expressions for the backscattered cross section which agreed remarkably well with experiment.[131] The backscatter cross section of a long, thin, perfectly conducting body due to a traveling wave mode is

$$\sigma = \frac{\lambda_0^2 \gamma^2}{\pi} Q^2 \left[ \left[ \frac{\sin \theta}{1 - p \cos \theta} \right] \left[ \sin \frac{k_0 l}{2} (1 - p \cos \theta) \right] \right]^4 \qquad (2.3\text{-}95)$$

where $p = v/c$ is the relative phase velocity of the traveling wave, $\gamma$ is a reflection coefficient determined by the rear shape of the scatterer, and $Q$ is a constant related to the maximum directivity of the body when considered as an antenna. In the case of a thin traveling-wave antenna radiating in the lowest order $TM$ mode, which is dominant for smooth highly conducting bodies, $Q$ is given by[131]

$$\frac{1}{Q} = \frac{-2}{p^2} + \frac{\text{Cin}[k_0 l(1 + p)] - \text{Cin}[k_0 l(1 - p)]}{p^3}$$
$$+ \frac{1}{2p^3} \Big[ (p - 1) \cos[k_0 l(1 + p)] + (p + 1) \cos[k_0 l(1 - p)]$$
$$+ (p^2 - 1) \, k_0 l \Big( \text{Si}[k_0 l(1 + p)] - \text{Si}[k_0 l(1 - p)] \Big) \Big] \qquad (2.3\text{-}96)$$

where $\text{Si}(x)$ is a sine integral tabulated in several places,[132,133] and $\text{Cin}(x)$ is a modified cosine integral tabulated by Kraus.[134]

Using $p = 1$, $\gamma = 0.32$, and evaluating $Q$ for a wire 39 wavelengths long gives

$$\frac{\sigma}{\lambda_0^2} = 0.00085 \left[ \frac{\sin \theta}{1 - \cos \theta} \sin[124.5(1 - \cos \theta)] \right]^4 \qquad (2.3\text{-}97)$$

as the traveling-wave contribution to the cross section of a long thin wire. This is compared in Figure 2-35 with measurements made by Peters on a $39\lambda_0$ silver rod, $\lambda_0/4$ in diameter.[135] The agreement is seen to be excellent.

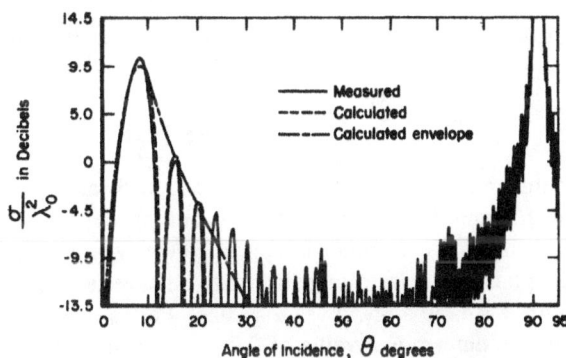

Fig. 2-35. Calculated traveling wave backscatter cross section of a $39\lambda_0$ rod of radius $\lambda_0/8$ along with the measured cross section [after Peters[135]].

For $k_0 l \gg 1$ and $p = 1$, Eqs. (2.3-95) and (2.3-96) can be simplified. Eq. (2.3-95) becomes

$$\sigma = \frac{\gamma^2 \lambda_0^2}{\pi} Q^2 [\cot^2 \theta/2 \sin^2(k_0 l \sin^2 \theta/2)]^2 \qquad (2.3\text{-}98)$$

while for $p = 1$, Eq. (2.3-96) for $Q$ becomes

$$\frac{1}{Q} = \text{Cin}(2k_0 l) - 1$$

Now,

$$\text{Cin}(x) = \ln \zeta x - \text{Ci}(x)$$

and for $x \gg 1$

$$\text{Cin}(x) \approx \ln \zeta x - \frac{\sin x}{x} \approx \ln \zeta x$$

where $\zeta$ equals Euler's constant 1.78. Thus, the expression for $1/Q$ becomes

$$\frac{1}{Q} \approx \ln \zeta 2 k_0 l - 1 \approx \ln 2 k_0 l + 0.5772 - 1$$

$$\approx \ln 2 k_0 l - 0.4228 \qquad (2.3\text{-}99)$$

From Eqs. (2.3-98) and (2.3-99),

$$\sigma \approx \frac{\gamma^2 \lambda_0^2}{\pi} \left[ \frac{\cot^2(\theta/2) \sin^2(k_0 l \sin \theta/2)}{\ln 2 k_0 l - 0.4228} \right]^2 \qquad (2.3\text{-}100)$$

this expression being valid for $k_0 l \gg 1$, and $p = 1$. From Eq. (2.3-100), the angle at which the maximum cross section occurs is

$$\theta_{\max} \approx 49.35 (\lambda_0/l)^{1/2} \qquad (2.3\text{-}101)$$

measured with respect to the axis, and the value of the cross section at this angle is

$$\sigma_{\max} \approx \frac{\lambda_0^2}{\pi} \gamma^2 \left[ \frac{0.725 k_0 l}{\ln 2 k_0 l - 0.4228} \right]^2 \qquad (2.3\text{-}102)$$

For long thin bodies other than cylindrical, such as ogives, spheroids, etc., $p$ will not in general equal one, and can be approximated by assuming that $p$ equals the average normalized phase velocity along the axis when the normalized phase velocity along the body surface is one. For example, a $39\lambda_0$ ogive with a $30°$ tip angle gives $p = 0.99$. Peters using this value of $p$, and the empirically determined value of 0.7 for $\gamma$, obtained

$$\frac{\sigma}{\lambda_0^2} = 0.0032 \left[ \frac{\sin \theta}{1 - 0.99 \cos \theta} \sin[124.5(1 - 0.99 \cos \theta)] \right]^4 \qquad (2.3\text{-}103)$$

for the traveling-wave contribution to the backscatter cross section for this ogive.[131] Good agreement with experimental measurements resulted.

The primary difficulty in applying this traveling-wave theory to the analysis of the radar cross sections of different shapes is that no analytical means exist for estimating the reflection coefficient $\gamma$. For thin cylindrical wires, Peters has determined empirically that $\gamma \approx 0.32$; while for a 30° ogive 39 wavelengths long, empirical results give $\gamma \approx 0.7$. Various authors have given differing opinions regarding the variation in $\gamma$ with the radius of curvature of the body at the reflection point. Seigel[136] indicates that $\gamma$ should be proportional to the radius of curvature of the body at the reflection point, with $\gamma$ approaching one for large radii of curvature. Others[137] have suggested that $\gamma$ is proportional to $\sin^2 \theta$, with a proportionality constant which is unity for a sharp termination and goes to zero as the radius of curvature at the reflection point approaches several wavelengths or greater. These two viewpoints are rather widely divergent, and about all that can be said is that if the termination is sharp, $\gamma$ is to first order independent of $\lambda_0$ and reasonably large, although shape-dependent to some extent. If the termination is smooth, $\gamma$ is considerably smaller and probably dependent upon both $\lambda_0$ and the shape at the termination. These conclusions appear to agree with the observed traveling wave components of the backscatter cross sections of cone spheres (see Chapters 6 and 8).

### 2.3.3.2. Dielectric and Lossy Bodies

For long thin bodies which are rough or not perfectly conducting, higher-order traveling-wave modes can become dominant. As a result, unlike the perfectly conducting case, large end-fire echoes can occur, that is, the traveling-wave cross section may have a maximum along the axis. The first case in which this effect was observed and analytically estimated was the backscattering cross section of a polyrod antenna.[131,135] Recently, Peters

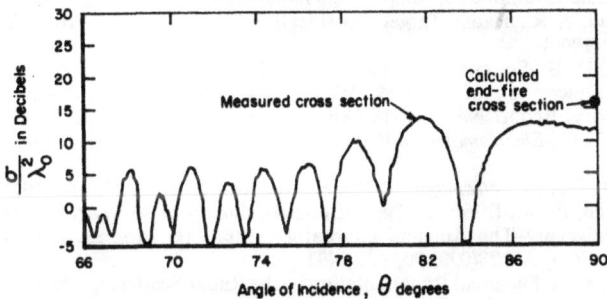

Fig. 2-36. Measured backscatter cross section of a threaded steel rod along with the calculated end-fire cross section [after Peters[139]].

measured the backscatter cross section of a $19\frac{1}{2}$-inch-long piece of 0.375-inch O. D. threaded steel rod at $K_a$ band, terminated with a $\frac{5}{16}$-inch-diameter reflecting disc to make $\gamma = 1$. In addition, he calculated the end-fire cross section using $c/v = 1.041$. The results are shown in Figure 2-36, and indicate relatively good agreement considering the simplicity of the theoretical model used.[138]

## 2.4.  REFERENCES

1. Ridenour, L. N., *Radar System Engineering*, McGraw-Hill, New York (1947).
2. Skolnik, M. I., *Introduction to Radar Systems*, McGraw-Hill, New York (1962).
3. Saybel, A. G., *Fundamentals of Radar*, "Soviet Radio" Publishing House, Moscow (1961).
4. Barton, D. K., *Radar System Analysis*, Prentice-Hall, Englewood Cliffs, New Jersey (1964).
5. Berkowitz, R. S., *Modern Radar Analysis, Evaluation, and System Design*, John Wiley and Sons, New York (1965).
6. Van Bladel J., *Electromagnetic Fields*, McGraw-Hill, New York (1964), pp. 254-258.
7. Hiatt, R. E., Siegel, K. M., and Weil, H., Forward Scattering by Coated Objects Illuminated by Short Wavelength Radar, *Proc. IRE* **48**:1630 (1960).
8. Siegel, K. M., Bistatic Radars and Forward Scattering, *Proc. National Conference on Aeronautical Electronics*, Dayton, Ohio, (1958), p. 286.
9. Crispin, J. W., Jr., Goodrich, R. F., and Siegel, K. M., A Theoretical Method for the Calculation of the Radar Cross Sections of Aircraft and Missiles, University of Michigan, Report No 2591-1-M (July 1959), AF 19(604)-1949, AFCRC-TN-59-774.
10. Kell, R. E., On the Derivation of Bistatic RCS from Monostatic Measurements, *Proc. IEEE* **53**:983 (1965).
11. IRE Standards on Radio Wave Propagation (definition of terms), *Supplement to Proc. IRE* **30**:2 (1942).
12. Rumsey, V. H., Deschamps, G. A., Kales, M. L., and Bohnert, J. I., Techniques for Handling Elliptically Polarized Waves with Special Reference to Antennas, *Proc. IRE* **39**:533 (1951).
13. Morgan, M. G., and Evans, W., Jr., Synthesis and Analysis of Elliptic Polarization Loci in Terms of Space-Quadrature Sinusoidal Components, *Proc. IRE* **39**:552 (1951).
14. Clayton, L., and Hollis, S., Antenna Polarization Analysis by Amplitude Measurement of Multiple Components, *Microwave Journal* **8**:35 (1965).
15. Copeland, J. R., Radar Target Classification by Polarization Properties, *Proc. IRE* **48**:1290 (1960).
16. Barrick, D. E., Summary of Concepts and Transformations Commonly Used in the Matrix Description of Polarized Waves, The Ohio State University–The Antenna Laboratory, Report No. 1388-16 (April 1965), NASA Grant No. NSG-213-61.
17. Jordan, E. C., *Electromagnetic Waves and Radiating Systems*, Prentice-Hall, New York (1950).
18. Stratton, J. A., *Electromagnetic Theory*, McGraw-Hill, New York (1941).
19. Kennaugh, E. M., Effects of Type of Polarization on Echo Characteristics, The Ohio State University–The Antenna Laboratory, Quarterly Progress Report No. 389-13 (March 1962), AF 28(099)-90, AD 2493.
20. Borison, S. L., Diagonal Representation of the Radar Scattering Matrix for an Axially Symmetric Body, *IEEE Trans. Antennas and Propagation* **AP-13**:176 (1965).
21. Mikulski, J. J., The Scattering Matrix, Polarization, Power and Periodic Bodies from the Viewpoint of Matrix Theory, Lincoln Laboratory-MIT, Group Report No. 47-38 (March 1960).

22. Tsu, R., Representations of a Scattering Pattern by a Collection of Induced Dipole Sources, The Ohio University–The Antenna Laboratory, Report No. 827-2 (June 1959), AF 19(604)-3501.
23. Kraus, J. D., *Antennas*, McGraw-Hill, New York (1950).
24. Kouyoumjian, R. G., and Peters, L., Jr., Range Requirements in Radar Cross-Section Measurements, *Proc. IRE* **53**:920 (1965).
25. Hansen, R. C., *Microwave Scanning Antennas*, Vol. 1, Academic Press, New York (1964), p. 1.
26. Brysk, H., The Radar Cross Section of a Semi-Infinite Body, *Canad. J. Phys.* **38**:48 (1960).
27. RAT SCAT Cross Section Measurements of the Telstar Vehicle, Air Force Missile Development Center, Holloman Air Force Base, FRXM-021-2 (March 1965).
28. Pettengill, G. H., Kraft, L. G., Earth Satellite Observations Made with the Millstone Hill Radar. In *Avionics Research: Satellites and Problems of Long Range Detection and Tracking*, Pergamon Press, New York (1960), p. 125.
29. Radar Properties of Aircraft, Naval Research Laboratory, Interim Report No. C-3460-143/52 (December 1952), AD 4531.
30. Swerling, P., Detection of Fluctuating Pulsed Signals in the Presence of Noise, *IRE Trans.* **IT-3**:175 (1957).
31. Weinstock, W. W., Target Cross Section Models for Radar System Analysis, Ph. D. Dissertation, University of Pennsylvania, 1964.
32. Huynen, J. R., Dynamic Radar Cross Section Predictions, *Proc. 7th Military Electronics Conference*, 1963.
33. Pederson, N., Halsey, H., Toney, J., and Clements, J., A New Method of Correlation of Down-Range Measurements with Static Radar Cross Section Measurements. In *Radar Reflectivity Measurements Symposium*, Vol. II, RADC-TDR-64-25 (1964), p. 177.
34. Stratton, J. A., *Electromagnetic Theory*, McGraw-Hill, New York, (1941), p. 392.
35. Morse, P. M., and Feshbach, H., *Methods of Theoretical Physics*, McGraw-Hill, New York (1953), p. 492.
36. Rogers, C. C., and Schultz, F. V., The Scattering of a Plane Electromagnetic Wave by a Finite Cone, Purdue University, Report No. 1 (August 1960).
37. Stratton, J. A., *Electromagnetic Theory*, McGraw-Hill, New York (1941), p. 464.
38. Jones, D. S., *The Theory of Electromagnetism*, Macmillan, New York (1964), p. 53.
39. Maue, A. W., Zur Formulierung eines allgemeinen Beugungsproblems durch eine Integralgleichung, *Z. Physik* **126**:601 (1949).
40. Muller, C., Über die Beugung elektromagnetischer Schwingungen an endlichen homogenen Körpern, *Math. Ann.* **123**:345 (1951).
41. Schiff, L. I., *Quantum Mechanics*, McGraw-Hill, New York (1955), p. 159.
42. Heins, A. E., Function-Theoretic Aspects of Diffraction Theory. In Langer, R. E., *Electromagnetic Waves*, University of Wisconsin Press, Madison, Wisconsin (1962), p. 99.
43. Sommerfeld, A., Mathematische Theorie der Diffraction, *Math. Ann.* **47**:317 (1896).
44. Noble, B., *The Wiener–Hopf Technique*, Pergamon Press, New York (1958).
45. Jones, D. S., *The Theory of Electromagnetism*, Macmillan, New York (1964), p. 573.
46. Morse, P. M., and Feshbach, H., *Methods of Theoretical Physics*, Vol. II, McGraw-Hill, New York (1953), p. 1106.
47. Born, M., and Wolf, E., *Principles of Optics*, Pergamon Press, New York (1959), p. 108.
48. Kerr, D. E., *Propagation of Short Radio Waves*, McGraw-Hill, New York (1951), p. 728.
49. Luneberg, R. K., *Mathematical Theory of Optics*, Stanford University Press, Stanford, California (1965).
50. Kline, M., An Asymptotic Solution of Maxwell's Equations. In *The Theory of Electromagnetic Waves*, John Wiley and Sons, Interscience Div. New York (1951), p. 225.
51. Kline, M., Electromagnetic Theory and Geometrical Optics. In *Electromagnetic Waves*, Langer, L. E., University of Wisconsin Press, Madison, Wisconsin (1962), p. 3.

52. Kline, M., and Kay, I. W., *Electromagnetic Theory and Geometrical Optics*, John Wiley and Sons, Interscience Div., New York (1965).
53. Kouyoumjian, R. G., An Introduction to Geometrical Optics and the Geometrical Theory of Diffraction. In *Recent Advances in Antenna and Scattering Theory*, short course notes, The Ohio State University (1965).
54. Bremmer, H., Propagation of Electromagnetic Waves, *Handbuch der Physik*. In Flugge, S. (Ed.), *Electric Fields and Waves*, Vol. 16, Springer-Verlag, Berlin (1958), p. 423.
55. Keller, J. B., A Geometrical Theory of Diffraction. In *Symposium on the Calculus of Variations and Its Application*, McGraw-Hill, New York (1958), p. 27, and a discussion at the end of Chapter IX.
56. Keller, J. B., Geometrical Theory of Diffraction, *J. Opt. Soc. Am.* **52**:116 (1962).
57. Keller, J. B., Diffraction by an Aperture, *J. Appl. Phys.* **28**: 426 (1957).
58. Keller, J. B., Lewis, R. M., and Seckler, B. D., Diffraction by an Aperture (II), *J. Appl. Phys.* **28**:570 (1957).
59. Felsen, L. D., Plane-Wave Scattering by Small-Angle Cones, *IRE Trans.* **AP-5**:211 (1957).
60. Keller, J. B., Diffraction by a Convex Cylinder, *IRE Trans.* **AP-4**:312 (1956).
61. Levy, B. R., and Keller, J. B., Diffraction by a Smooth Object, *Comm. Pure Appl. Math.* **12**:159 (1959).
62. Levy, B. R., and Keller, J. B., Diffraction by a Spheroid, *Canad. J. Phys.* **38**:128 (1960).
63. Buchal, R. N., and Keller, J. B., Boundary Layer Problems in Diffraction Theory, *Comm. Pure Appl. Math* **13**:85 (1960).
64. Kravtsov, Yu. A., Asymptotic Solution of the Maxwell Equations Near the Caustic Surface, *Izv. Vuz., Radiofizika* **7**:1049 (1964).
65. Ludwig, D., Uniform Asymptotic Expansions at a Caustic, *Comm. Pure Appl. Math* **19**:215 (1966).
66. Stratton, J. A., *Electromagnetic Theory*, McGraw-Hill, New York (1941), p. 468.
67. Barrick, D. E., A More Exact Theory for the Scattering of Electromagnetic Waves from Statistically Rough Surfaces, Dissertation, The Ohio State University, 1965.
68. Kennaugh, E. M., and Cosgriff, R. L., The Use of Impulse Response in Electromagnetic Scattering Problems, *IRE National Convention Record*, Part 1 (1958), p. 72.
69. Kennaugh, E. M., and Moffatt, D. L., Transient and Impulse Response Approximations, *Proc. IEEE* **53**:893 (1965).
70. Kennaugh, E. M., and Moffatt, D. L., On the Axial Echo Area of the Cone-Sphere Shape, *Proc. IRE* **50**:199 (1962), and **51**:231 (1963).
71. Siegel, K. M., Schultz, F. V., Gere, B. H., and Sleator, F. B., The Theoretical and Numerical Determination of the Radar Cross Section of a Prolate Spheroid, *IRE Trans. Ant. Prop.* **AP-4**:266 (1956).
72. Moffatt, D. L., and Kennaugh, E. M., The Axial Echo Area of a Perfectly Conducting Prolate Spheroid, *IEEE Trans. Ant. Prop.* **AP-13**:401 (1965).
73. Fock, V. A., Diffraction of Radio Waves Around the Earth's Surface, *J. Phys. USSR* **9**:255 (1945).
74. Fock, V. A., The Distribution of Currents Induced by a Plane Wave on the Surface of a Conductor, *J. Phys. USSR* **10**:130 (1946).
75. Fock, V. A., The Field of a Plane Wave Near the Surface of a Conducting Body, *J. Phys. USSR* **10**:399 (1946).
76. Leontovich, M. A., A Method of Solution of Problems of Electromagnetic Wave Propagation Along the Earth's Surface, *Izv. Akad. Nauk SSSR, Ser. Fiz.* **8**:16 (1944).
77. Leontovich, M. A., Approximate Boundary Conditions for the Electromagnetic Field on the Surface of a Good Conductor. In Logan, N. A., *Diffraction, Refraction and Reflection of Radio Waves* (June, 1957), AFCRL-TN-57-102, AD 117276, p. 383.
78. Logan, N. A., General Research in Diffraction Theory, Vol. 2, Lockheed Aircraft Corporation, Report No. LMSD 288088 (December 1959), AD 243 182.
79. Goryainov, A. S., An Asymptotic Solution of the Problem of Diffraction by a Cylinder, *Radiotekhnika i Elektronika* **3**, 603 (1958).
80. Fock, V. A., The Field of a Plane Wave Near the Surface of a Conducting Body, *Izv. Akad. Nauk SSSR, Ser. Fiz.* **10**:171 (1946).

81. Abramowitz, M., and Stegun, I. A., *Handbook of Mathematical Functions*, National Bureau of Standards of the Department of Commerce (March 1965).
82. Ivanov, V. I., Diffraction of Short Plane Electromagnetic Waves With Oblique Incidence on a Smooth Convex Cylinder, *Radiotekhnika i Elektronika* **5**, 524 (1960).
83. Belkina, M. G., and Vaynshteyn, L. A., Radiation Characteristics of Spherical Surface Antennas. In *Diffraction of Electromagnetic Waves on Certain Bodies of Revolution* Fock, Belkina, Vaynshteyn, Sovetskoye Radio, Moscow (1957), p. 63.
84. Fedorov, A. A., Asymptotic Solution of the Problem of Diffraction of Plane Electromagnetic Waves on Ideally Conducting Spheres, *Radiotekhnika i Elektronika* **3**:1451 (1958).
85. Goodrich, R. F., Studies in Radar Cross Sections XXVI—Fock Theory, University of Michigan, Report No. 2591-3-T (July 1958), AFCRL-TN-58-350, AD-160790.
86. Fel'd, Ya. N., Variational Method for the Calculation of Parameters Which are Linear Functionals of the Electrodynamic Integral Equations, *Radiotekhnika i Elektronika* **7**, 46 (1962).
87. Kodis, R. D., Variational Principles in High Frequency Scattering, *Proc. Cambridge Phil. Soc.* **54**, 512 (1958).
88. Levine, H., and Schwinger, J., On the Theory of Diffraction by an Aperture in an Infinite Plane Screen. I, *Phys. Rev.* **74**:958 (1948).
89. Levine, H., and Schwinger, J., On the Theory of Diffraction by an Aperture in an Infinite Plane Screen. II, *Phys. Rev.* **75**:1423 (1949).
90. Huang, C., Kodis, R. D., and Levine, H., Diffraction by Apertures, *J. Appl. Phys.* **26**:151 (1958).
91. Kouyoumjian, R. G., The Calculation of the Echo Areas of Perfectly Conducting Objects by the Variational Method, The Ohio State University Research Foundation, Report No. 444-13 (November 1953), AD 48214, DA 36-039 sc-5506.
92. Tai, C. T., Electromagnetic Backscattering Fom Cylindrical Wires, *J. Appl. Phys.* **23**:909 (1952).
93. Cohen, M. H., Application of the Reaction Concept to Scattering Problems, *IRE Trans.* **AP-3**:193 (1955).
94. Kleinman, R. E., The Rayleigh Region, *Proc. IEEE* **53**:848 (1965).
95. Lord Rayleigh, On the Incidence of Aerial and Electric Waves Upon Small Obstacles in the Form of Ellipsoids or Elliptic Cylinders and on the Passage of Electric Waves Through a Circular Aperture in a Conducting Screen, *Philosophical Magazine*XLIV: 28 (1897), and paper 230 in Vol. IV *Scientific Papers by Lord Rayleigh*, Dover Publications, New York (1964), p. 305.
96. Stevenson, A. F., Solution of Electromagnetic Scattering Problems as Power Series in the Ratio (Dimension of Scatterer)/Wavelength, *J. Appl. Phys.* **24**:1134 (1953).
97. Stevenson, A. F., Electromagnetic Scattering by an Ellipsoid in the Third Approximation, *J. Appl. Phys.* **24**:1143 (1953).
98. Kleinman, R. E., Low Frequency Solution of Three-Dimensional Scattering Problems, University of Michigan, Radiation Laboratory, Report No. 7133-4-T, AFCRL-65-639, AF 19(628)-4328.
99. Darling, D. A., Some Relations Between Potential Theory and the Wave Equation, University of Michigan, Radiation Laboratory, Report No. 2871-S-T (December 1960), AF 19(604)-4993, AD 258307.
100. Darling, D. A., and Senior, T. B. A., Low Frequency Expansions for Scattering By Separable and Nonseparable Bodies, *J. Acoust. Soc. Am.* **37**:228 (1965).
101. Siegel, K. M., Low-Frequency Radar Cross Section Computations, *Proc. IEEE* **51**:232 (1963).
102. Siegel, K. M., The Quasi-Static Radar Cross Sections of Complex Bodies of Revolution. In Langer, R., *Electromagnetic Waves*, University of Wisconsin Press, Madison, Wisconsin (1962), p. 181.
103. Richmond, J. H., Scattering By a Dielectric Cylinder of Arbitrary Cross-Section Shape, *IEEE Trans. Ant. Prop.* **AP-13**:334 (1965).
104. Richmond, J. H., Digital-Computer Solutions of the Rigorous Equations for

Scattering Problems. In *Recent Advances in Antennas and Scattering Theory* (Short Course Notes), The Ohio State University (1965).

105. Kennaugh, E. M., Multipole Field Expansions and Their Use in Approximate Solutions of Electromagnetic Scattering Problems, Ph.D. Dissertation, The Ohio State University (December 1959), also Report No. 827-5 of the Antenna Laboratory, The Ohio State University (November 1959).

106. Sommerfeld, A., *Partial Differential Equations in Physics*, Academic Press, New York (1949), pp. 29 and 159.

107. Rogers, C. C., and Schultz, F. V., The Scattering of a Plane Electromagnetic Wave by a Finite Cone, Purdue University (August 1960).

108. Mullin, C. R., Sandburg, R., and Velline, C. O., A Numerical Technique for the Determination of Scattering Cross Sections of Infinite Cylinders of Arbitrary Geometrical Cross Section, *IEEE Trans. Ant. Prop.* AP-13:141 (1965).

109. Waterman, P. C., Scattering of Electromagnetic Waves by Conducting Surfaces, AVCO Corporation, Technical Memorandum No. RAD-TM-62-94 (December 1962).

110. Andreason, M. G., Scattering from Bodies of Revolution, *IEEE Trans. Ant. Prop.* AP-13:303 (1965).

111. Harrington, R. F., Chang, L. F., Adams, A. T., *et al.*, Matrix Methods for Solving Field Problems, Syracuse University (August 1966), AD 639744.

112. Sommerfeld, A., *Optics*, Academic Press, New York (1954), p. 245.

113. Born, M., and Wolf, E., *Principles of Optics*, Pergamon Press, New York (1959), 562.

114. Pearcy, T., *Table of the Fresnel Integral*, Cambridge University Press, London (1956).

115. Ufimtsev, P. Ya., *The Edge Wave Method in Physical Diffraction Theory*, Sovetskoye Radio, Moscow (1962).

116. Ufimtsev, P. Ya., Approximate Computation of the Diffraction of Plane Electromagnetic Waves at Certain Metal Bodies, I, *Zh. Tekh. Fiz.* 27:1840 (1957).

117. Clemmow, P. C., Edge Currents in Diffraction Theory, *IRE Trans.* AP-4:282 (1956).

118. Millar, R. F., An Approximate Theory of the Diffraction of an Electromagnetic Wave by an Aperture in a Plane Screen, *Proc. IEE (GB)* 103:177 (1955).

119. Millar, R. F., The Diffraction of an Electromagnetic Wave by a Large Aperture, *Proc. IEE (GB)* 104: 240 (1956).

120. Siegel, K. M., Far Field Scattering From Bodies of Revolution, *Appl. Sci. Res.* 7:293 (1959).

121. Ufimtsev, P. Ya., Approximate Calculations of the Diffraction of Plane Electromagnetic Waves by Certain Metal Objects, *Zh. Tekh. Fiz.* 28:2604 (1958).

122. Ufimtsev, P. Ya., Secondary Diffraction of Electromagnetic Waves by a Strip, *Zh. Tekh. Fiz.* 28:569 (1958).

123. Ufimtsev, P. Ya., Secondary Diffraction of Electromagnetic Waves by a Disk, *Zh. Tekh. Fiz.* 28:583 (1958).

124. Felsen, L. B., Asymptotic Expansion of the Diffracted Wave for a Semi-Infinite Cone, *Trans. IRE* AP-5:402 (1957).

125. Goryainov, A. S., The Diffraction of a Plane Electromagnetic Wave Propagating Along the Axis of a Cone, *Radiotekhnika i Elektronika* 6:47 (1961).

126. Franz, W., and Deppermann, K., Theorie der Beugung am Zylinder unter Berücksichtigung der Kriechwelle, *Ann. Phys.* 10, 361 (1952).

127. Bremmer, H., *Terrestrial Radio Waves*, Elsevier, New York (1949), p. 31.

128. Logan, N. A., General Research in Diffraction Theory, Lockheed Aircraft Corporation, Vol. I, Report No. LMSD 288087 (December 1959), AD 241288.

129. Franz, W., and Galle, R., Semiasymptotische Reihen für die Beugung einer ebenen Welle am Zylinder, *Z. Naturforsch.* 10a:374 (1955).

130. Senior, T. B. A., and Goodrich, R. F., Scattering by a Sphere, *Proc. IEEE (GB)* 111:907, (1964).

131. Peters, L., End-Fire Echo Area of Long Thin Bodies, The Ohio State University Research Foundation, Report No. 601-9 (June 1956), AF 33(616)2546, AD 100980.

132. Jahnke, E., and Emde, F., *Tables of Functions*, Fourth Ed., Dover, New York (1945), p. 6.

133. Abramowitz, M., and Stegun, I. A., *Handbook of Mathematical Functions*, National Bureau of Standards of the Department of Commerce, (March 1965), p. 238.
134. Kraus, J. D., *Antennas*, McGraw-Hill, New York (1950), p. 539.
135. Peters, L., End-Fire Echo Area of Long Thin Bodies, *Trans. IRE* **AP-6**:133 (1958).
136. Siegel, K. M., Far-Field Scattering From Bodies of Revolution, *Appl. Sci. Res.* **7**, 293 (1959).
137. Kleinman, R. E., and Senior, T. B. A., Studies in Radar Cross Sections XLVIII —Diffraction and Scattering by Regular Bodies—II: The Cone, University of Michigan Radiation Laboratory, Report No. 3648-2-T (January 1963), AF 19(604)-6655.
138. Peters, L., Echo Area Properties of Bodies Due to Certain Traveling Wave Modes, The Ohio State University, Research Foundation, Report No. 777-19 (May 1960).

# Chapter 3

# SPHERES

D. E. Barrick

## 3.1. GENERAL THEORY OF THE EXACT SOLUTION

The only existing exact solution for scattering from a sphere is attributed to Mie.[1] This solution employs vector wave functions defined in a spherical coordinate system, and uses boundary value techniques to find the terms of the resultant series for the scattered field. The formulation can, therefore, be used regardless of the composition of the sphere because the surface of the sphere is a natural coordinate surface in the spherical coordinate system.

The geometry of the problem is shown in Figure 3-1. For this analysis the following conventions will be established:

(a) A plane wave is incident moving in the $-z$ direction.

(b) Harmonic time dependence, where the factor $e^{-i\omega t}$ will be suppressed.

(c) The electric field vector of the incident field is polarized in the $x$ direction.

(d) The point, $P$, with coordinates $(r', \theta', \phi')$ is the point outside the sphere $(r > a)$ at which the scattered field is sought. Hence, the incident field is

$$\mathbf{E}^i = E_0 \hat{\mathbf{x}} \, e^{-ik_0 z}$$

$$\mathbf{H}^i = -\sqrt{\frac{\epsilon_0}{\mu_0}} \, E_0 \hat{\mathbf{y}} \, e^{-ik_0 z} = -\frac{E_0}{\eta} \, \hat{\mathbf{y}} \, e^{-ik_0 z}$$

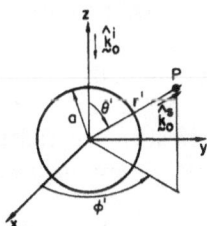

Fig. 3-1.  Scattering geometry for a sphere.

**141**

### 3.1.1.   General Solution

Following Reference 2, the Mie solution for the scattered field at point, $P$, and at angular frequency $\omega$ is

$$\mathbf{E}^s(P, \omega) = E_0 \sum_{n=1}^{\infty} (A_n \mathbf{M}_{o1n} + B_n \mathbf{N}_{e1n}) \tag{3.1-1}$$

$$\mathbf{H}^s(P, \omega) = -i \sqrt{\frac{\epsilon_0}{\mu_0}} E_0 \sum_{n=1}^{\infty} (B_n \mathbf{M}_{e1n} + A_n \mathbf{N}_{o1n}) \tag{3.1-2}$$

In the above Mie solution, the quantities, $\mathbf{M}_{\substack{e\\o}1n}$ and $\mathbf{N}_{\substack{e\\o}1n}$, are spherical vector wave functions, and depend only upon the position of $P$, or $(r', \theta', \phi')$. The constants, $A_n$ and $B_n$, are determined solely by the properties of the sphere and are independent of the scattering direction; they will be given separately for each class of spheres discussed under the appropriate section. The vector wave functions are defined here and remain unchanged for the remainder of this chapter.

$$\mathbf{M}_{\substack{e\\o}1n} = \mp \frac{1}{\sin \theta} h_n^{(1)}(k_0 r') P_n{}^1(\cos \theta') \begin{bmatrix} \sin \phi' \\ \cos \phi' \end{bmatrix} \hat{\theta}$$

$$- h_n^{(1)}(k_0 r') \frac{d}{d\theta} P_n{}^1(\cos \theta') \begin{bmatrix} \cos \phi' \\ \sin \phi' \end{bmatrix} \hat{\phi} \tag{3.1-3}$$

$$\mathbf{N}_{\substack{e\\o}1n} = \frac{n(n+1)}{k_0 r'} h_n^{(1)}(k_0 r') P_n{}^1(\cos \theta') \begin{bmatrix} \cos \phi' \\ \sin \phi' \end{bmatrix} \hat{r}$$

$$+ \frac{1}{k_0 r'} \frac{d}{d(k_0 r')} [k_0 r' h_n^{(1)}(k_0 r')] \frac{d}{d\theta} P_n{}^1(\cos \theta') \begin{bmatrix} \cos \phi' \\ \sin \phi' \end{bmatrix} \hat{\theta}$$

$$\mp \frac{1}{k_0 r' \sin \theta'} \frac{d}{d(k_0 r')} [k_0 r' h_n^{(1)}(k_0 r')] P_n{}^1(\cos \theta') \begin{bmatrix} \sin \phi' \\ \cos \phi' \end{bmatrix} \hat{\phi} \tag{3.1-4}$$

Either the upper or lower first subscript of these vector wave functions is used depending upon which is called for in Eq. (3.1-1) or (3.1-2). If, for example, the upper subscript is used, then the upper sign in front of each term is used (where two signs appear), and the upper trigonometric factor in the brackets [ ] is used. The spherical Hankel functions, $h_n{}^{(1)}(k_0 r')$, and associated Legendre functions, $P_n{}^1(\cos \theta')$, are defined and partially tabulated in references 3 and 4.

The infinite series of (3.1-1) and (3.1-2) are exact for scattering in any direction and at any distance from the sphere (i.e., near field or far field). For the far field, where $r' \gg a$, the solution simplifies considerably, and this region is discussed in Section 3.1.2. For the near field the solutions (3.1-1) and (3.1-2) do not simplify.

From a practical standpoint, one cannot use an infinite number of terms in the solution. The number of terms required for a given accuracy is discussed in Section 3.1.3, along with other practical considerations involved in summing the Mie series numerically to obtain a result.

### 3.1.2. Far-Field Approximations

If the point $P$ at which the scattered field is sought is far enough from the sphere that $r' \gg a$, then the spherical vector wave functions of Eqs. (3.1-3) and (3.1-4) simplify considerably. Thus, the scattered electric field at $P$ becomes[2]

$$\mathbf{E}^s(P, \omega) = E_0 \frac{e^{ik_0r'}}{k_0r'} [\cos \phi' S_1(\theta') \,\hat{\theta} - \sin \phi' S_2(\theta') \,\hat{\phi}] \qquad (3.1\text{-}5)$$

where

$$S_1(\theta') = \sum_{n=1}^{\infty} (-i)^{n+1} \left[ A_n \frac{P_n^1(\cos \theta')}{\sin \theta'} + iB_n \frac{d}{d\theta'} P_n^1(\cos \theta') \right] \qquad (3.1\text{-}6)$$

and

$$S_2(\theta') = \sum_{n=1}^{\infty} (-i)^{n+1} \left[ A_n \frac{d}{d\theta'} P_n^1(\cos \theta') + iB_n \frac{P_n^1(\cos \theta')}{\sin \theta'} \right] \qquad (3.1\text{-}7)$$

$S_1(\theta')$ and $S_2(\theta')$ are called the complex far-field amplitudes of the scattered radiation for the $\theta$ and $\phi$ polarization directions. Their magnitudes squared are proportional to the intensities of the scattered radiation at $P$ along these polarization directions.

In the far-field solutions above, the local scattered wave is planar, and, therefore, only the $E$ field is given. The $H$ field is simply related to it by the factor $\sqrt{\epsilon_0/\mu_0}$, or $|H| = \sqrt{\epsilon_0/\mu_0} |E|$.

The expression in the square brackets of Eq. (3.1-5) is often written in terms of $F(\theta', \phi')$, the scattering function, as follows:

$$F(\theta', \phi') \,\hat{\tau} = \cos \phi' S_1(\theta') \,\hat{\theta} - \sin \phi' S_2(\theta') \,\hat{\phi} \qquad (3.1\text{-}8)$$

where $\hat{\tau}$ is a unit vector in the direction of polarization of the scattered field at $P$. The scattering cross section in any arbitrary polarization direction, $\hat{\eta}$, for an incident wave polarized in the $x$ direction is then

$$\sigma_\eta(\theta', \phi') = \frac{4\pi}{k_0^2} |F(\theta', \phi')|^2 |\hat{\tau} \cdot \hat{\eta}|^2 \qquad (3.1\text{-}9)$$

The particular scattering cross sections for the $\theta$ and $\phi$ polarization states are then found by setting $\hat{\eta} = \hat{\theta}$ and $\hat{\phi}$, respectively.

$$\sigma_\theta(\theta', \phi') = \frac{4\pi}{k_0^2} S_1(\theta')^2 \cos^2 \phi' \qquad (3.1\text{-}10)$$

$$\sigma_\phi(\theta', \phi') = \frac{4\pi}{k_0^2} S_2(\theta')^2 \sin^2 \phi' \qquad (3.1\text{-}11)$$

The above quantities are sometimes referred to as the $E$- and $H$-plane cross sections, respectively.

### 3.1.2.1. *Backscattering Cross Section*

In the backscattering direction, Eq. (3.1-8) simplifies considerably because $\theta' = 0$. One commonly defines $F(0, \phi') = F(0)$, since there is no longer a dependence upon $\phi'$. In this case $S_1(0) = S_2(0) = F(0)$, and

$$F(0)\,\hat{\tau} = F(0)\,\hat{x} = -\hat{x} \sum_{n=1}^{\infty} (-i)^{n-1} \frac{n(n+1)}{2} (A_n + iB_n) \qquad (3.1\text{-}12)$$

Thus, the backscattered field is polarized in the same direction as the incident field when the incident polarization is linear. The backscattering cross section for a sphere in the direction of incident polarization is then

$$\sigma(0) = \frac{4\pi}{k_0^2} \, | \, F(0)|^2 \qquad (3.1\text{-}13)$$

### 3.1.2.2. *Forward Scattering Cross Section*

In this case, $\theta' = \pi$, and the applicable scattering function is

$$F(\pi, \phi') = F(\pi) = S_1(\pi) = -S_2(\pi)$$

where

$$F(\pi)\,\hat{\tau} = F(\pi)\,\hat{x} = \hat{x} \sum_{n=1}^{\infty} (i)^{n-1} \left[ \frac{n(n+1)}{2} (A_n - iB_n) \right] \qquad (3.1\text{-}14)$$

The above equation indicates that the forward scattered field is again polarized in exactly the same direction as the incident field. In general, the forward scattered field is always polarized in the same direction as the incident field, regardless of whether the incident polarization is linear or elliptical.

The forward scattering cross section in the incident polarization direction is then

$$\sigma(\pi) = \frac{4\pi}{k_0^2} \, | \, F(\pi)|^2 \qquad (3.1\text{-}15)$$

### 3.1.3. **Numerical Computation with the Mie Series**

As stated previously, the Mie solution [Eqs. (3.1-1) and (3.1-2)] is the only known exact solution to the problem of scattering from a sphere, and

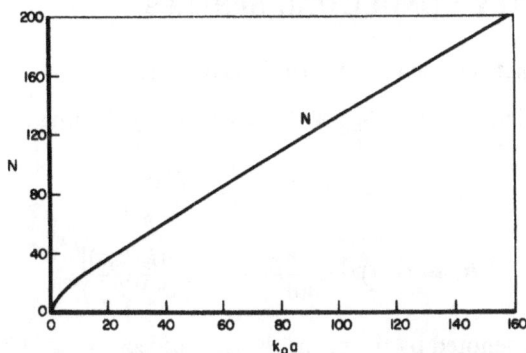

Fig. 3-2. Number of terms of Mie series solution versus $k_0 a$ necessary for no more than 4% error [after Weil et al. [5]].

in some cases, no approximate solution even exists. On the other hand, the almost universal availability of the modern digital computer suggests that a program for the solution of the Mie series might, in many cases, be desirable.

Obviously, in practice, an infinite series such as the Mie solution is unattainable and must be terminated after $N$ terms. The number of terms that must be retained, $N$, is of the same order of magnitude as $k_0 a$. Hence, use of the Mie series is more efficient for smaller values of $k_0 a$ than for large values. This dependence of $N$ upon $k_0 a$ is illustrated graphically in Figure 3-2 (from reference 5), where choosing $N$ from the curve shown guarantees no greater than 4% error. Making $N$ even larger reduces the error still more.

Generation of Bessel and Legendre functions and their derivatives on a computer for varying order, $n$, at constant argument is quite simply done by the use of recursion formulas.[†] One need only know these functions for two orders, e.g., $P_1^1(\cos \theta')$ and $P_2^1(\cos \theta')$, to generate all the others and their derivatives. The lower-order Legendre and spherical Bessel functions are expressible as simple trigonometric functions.

Since the constants $A$ and $B$ of the Mie solution are peculiar only to the properties of the sphere being considered, one could easily generate these coefficients within different subroutines, each subroutine applying to a particular class of spheres (e.g., perfectly conducting, coated, lossy dielectric, etc.) Then the appropriate subroutine would be used with the main program which generates the remainder of the terms of the Mie series and sums the series.

Suggestions for numerical generation of the coefficient $A_n$ and $B_n$ will be presented under the appropriate section of this chapter.

---

† See formulas (10.1.19) through (10.1.22) of Reference 3 for spherical Bessel functions, and formulas (8.5.3) and (8.5.4) of Reference 3, for the associated Legendre function.

## 3.2. PERFECTLY CONDUCTING SPHERES

### 3.2.1. The Exact Solution—Numerical Summation

The coefficients $A_n$ and $B_n$ for the perfectly conducting sphere are

$$A_n = -(-i)^n \frac{2n+1}{n(n+1)} \frac{j_n(k_0 a)}{h_n^{(1)}(k_0 a)} \qquad (3.2\text{-}1)$$

$$B_n = (-i)^{n+1} \frac{2n+1}{n(n+1)} \frac{[k_0 a j_n(k_0 a)]'}{[k_0 a h_n^{(1)}(k_0 a)]'} \qquad (3.2\text{-}2)$$

Differentiation, denoted by the prime, is with respect to $k_0 a$. These constants, $A_n$ and $B_n$, may then be used with whichever of the series (3.1-1), (3.1-2), (3.1-6), (3.1-7), (3.1-12), or (3.1-13) is appropriate to the scattering configuration desired.

The numerators and denominators of these coefficients may be generated by recursion techniques for machine computation, as discussed in the preceding section. One should, however, note that forward recursion generation of $j_n(x)$ (i.e., starting with $j_0(x)$ and $j_1(x)$ to find higher-order functions) is extremely inaccurate and unstable. One must start with $j_N(x)$ and $j_{N-1}(x)$ and generate the lower orders. On the other hand, in the generation of $y_n(x)$, the imaginary term of $h^{(1)}(x)$, just the opposite is true, and one must use a forward recursion. For a discussion of these computer recurrence techniques and resulting error propagation and instabilities, see reference 3, p. XIII and pp. 697–699.

### 3.2.2. Approximate Solutions—Far Field ($r \gg a$)

Approximate solutions and curves for the scattered field and scattering cross sections for the cases of backscattering, bistatic scattering, and forward scattering are presented separately. Each of three frequency regions is discussed for the above cases; resonance region, low-frequency region, and high-frequency region, in that order.

#### 3.2.2.1. Backscattering

A sphere scatters incident waves into the back direction as a perfect polarization reflector, in the same manner as a perfectly conducting infinite plane at normal incidence. Thus, if the incident polarization is linear, the backscattered polarization is linear in the same direction; if the incident polarization is right circular, the backscatter polarization is left circular. Hence, only a scalar field will be used and assumed to represent the field intensity in the polarization state of the scattered wave. The backscattering cross section used represents all of the power in the polarization state of the scattered wave.

Table 3-1

| Equation (3.2-5) | First term of (3.2-5) |
|---|---|
| $k_0a = 1.0$,  38% error | $k_0a = 1.0$,  150% error |
| $k_0a = 0.8$,  4% error | $k_0a = 0.8$,  42% error |
| $k_0a < 0.7$,  less than 0.2% error | $k_0a = 0.7$,  27% error |
|  | $k_0a = 0.6$,  10% error |
|  | $k_0a < 0.5$,  less than 1% error |

A curve of the exact backscattering cross section as a function of $k_0a$ is shown in Figure 3-3. It is normalized with respect to the geometrical cross section of the sphere $\pi a^2$. This curve was obtained from the exact Mie solution, and is often used for calibrating radar sets when a conducting sphere is used as a standard target.

3.2.2.1.1. *Resonance Region* $(0.4 \leqslant k_0a \leqslant 20)$. The resonance region for the range of $k_0a$, specified above, is the region of Figure 3-3 containing the oscillations. In this region, simple approximations, such as physical or geometrical optics, cannot correctly predict backscattering cross section including the detailed oscillatory nature. If one desires only the average value of $\sigma(0)/\pi a^2$ for $k_0a > 1$ (denoted by $\langle \sigma(0)/\pi a^2 \rangle$), it is apparent from the curve that

$$\left\langle \frac{\sigma(0)}{\pi a^2} \right\rangle = 1 \qquad (3.2\text{-}3)$$

The error involved in this approximation becomes less for larger $k_0a$, so that for $k_0a > 10$, the error in (3.2-3) is never greater than 25 % or 1 decibel.

An approximation valid in the lower portion of the resonance region and below is obtained by expanding the Mie solution into a series in $k_0a$. This results in a backscattering function and backscattering cross section of[†]

$$F(0) = \frac{3}{2}(k_0a)^3 \left[1 - \frac{5}{54}(k_0a)^2 + \frac{17}{900}(k_0a)^4 - \frac{6{,}651{,}923}{11{,}907{,}000}(k_0a)^6\right]$$

$$+ i\frac{1}{2}(k_0a)^6 \left[1 + \frac{6}{5}(k_0a)^2\right] + O[(k_0a)^{10}] \qquad (3.2\text{-}4)$$

and

$$\sigma(0) = \pi a^2 9(k_0a)^4 \left[1 - \frac{5}{27}(k_0a)^2 + \frac{3{,}379}{72{,}900}(k_0a)^4 - \frac{1{,}502{,}812}{1{,}488{,}375}(k_0a)^6\right]$$

$$+ O[(k_0a)^{12}] \qquad (3.2\text{-}5)$$

The above series are valid only for $k_0a < 1$. Table 3-1 demonstrates the error incurred in the use of Eq. (3.2-5) and only the first term of (3.2-5).

[†] Typographical errors appearing in several of the terms of Eqs. (3-3) and (3-4) in the original source (Reference 2) have been corrected here.

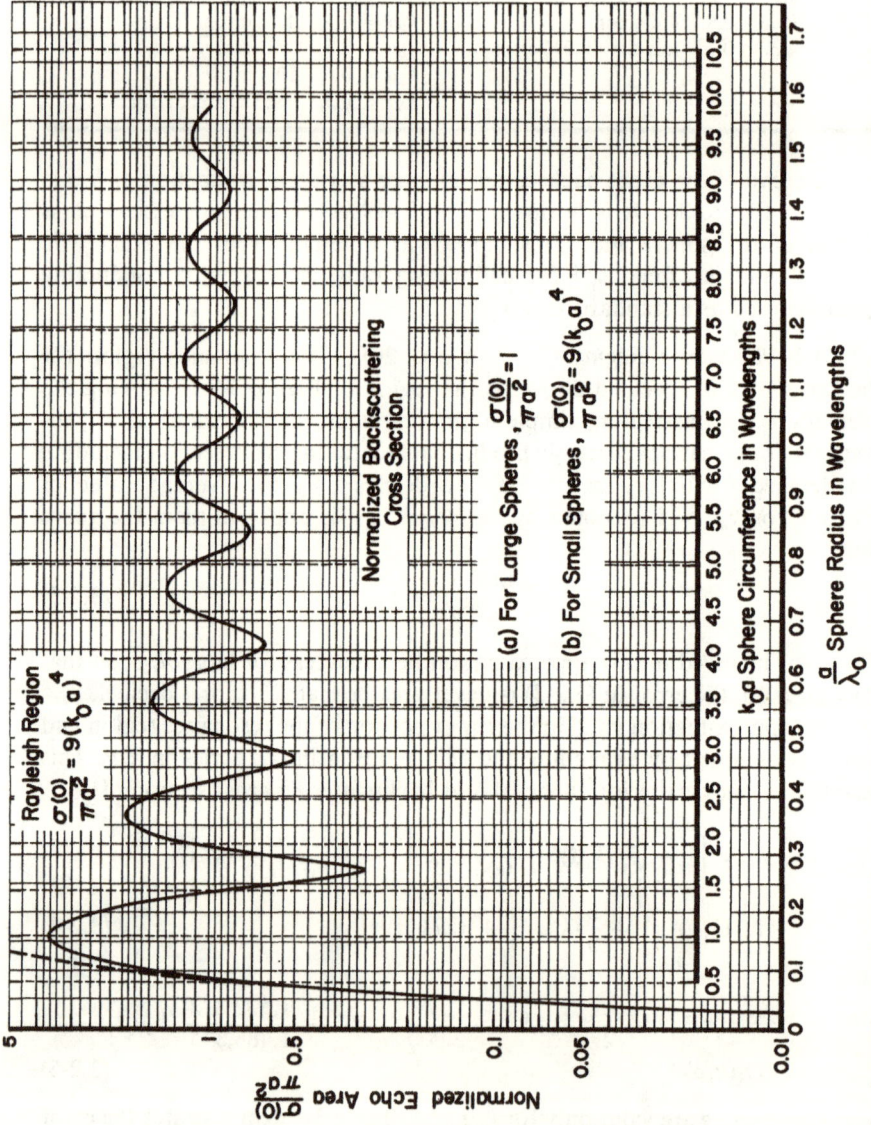

Fig. 3-3. Exact solution for normalized backscattering cross section of perfectly conducting sphere as a function of sphere size relative to wavelength.

At the upper end of the resonance region, the backscattering function may be broken up into an optics term $F^0(0)$, and a creeping wave term, $F^c(0)$. These components are obtained from the Mie solution using the Watson transformation. Thus, $F^0(0)$ accounts for the component of the backscattered field specularly reflected from the sphere, while $F^c(0)$ takes into account the waves excited by the incident field which become attached to the sphere at the shadow boundary, travel around the sphere one or more times, and then are launched into the backscattering direction as discussed in Section 2.3.2. Then,

$$F(0) = F^0(0) + F^c(0) \tag{3.2-6}$$

where[6]

$$F^0(0) = -\frac{x}{2}\, e^{-i2x}\left[1 - \frac{i}{2x} + O(x^{-3})\right], \qquad x = k_0 a \tag{3.2-7}$$

and

$$F^c(0) = -ix^{4/3}\, e^{i\pi(x-1/6)}[\{[1.357588 + (0.741196 + i1.283788)\, x^{-2/3}]$$

$$\times\ \exp[-x^{1/3}(2.200002 - i1.270172) + x^{-1/3}(0.445396 + i0.257150)]\}_{E_1}$$

$$+\ \{[0.695864 + (0.964654 + i1.670829)\, x^{-2/3}]$$

$$\times\ \exp[-x^{1/3}(7.014224 - i4.049663) - x^{-1/3}(0.444477 + i0.256619)]\}_{E_2}$$

$$-\ \{[0.807104 + (0.798821 + i1.383598)\, x^{-2/3}]$$

$$\times\ \exp[-x^{1/3}(5.048956 - i2.915016) - x^{-1/3}(0.312321 + i0.180319)]\}_{H_1}]$$

$$\tag{3.2-8}$$

The creeping wave contributions are broken up into terms representing the first and second $E$ waves and the first $H$ wave, and these terms are designated accordingly by the subscript on the brackets $\{\cdots\}_{E_1}$, $\{\cdots\}_{E_2}$, and $\{\cdots\}_{H_1}$. They represent waves which have traversed the sphere only once. The magnitude of each of these terms can be estimated from the negative exponential factor in each term. Generally, only the first $E$ wave is significant. For example, at $k_0 a = 2$, the magnitude of the first $E$ wave of $F^c(0)$ is about two orders of magnitude greater than that of the first $H$ wave and about two and a half orders greater than that of the second $E$ wave. This difference becomes much larger for larger $k_0 a$. All numbers in these equations are correct to six digits to the right of the decimal point. Terms in $x^{-2}$, $x^{-4/3}$ have been retained until they affect the magnitude of the answer by less than 1 % at $x = 5$.

For the value of $k_0 a$ chosen, the user can decide how many creeping wave terms of (3.2-8) he wishes to retain compared to the optics contribution, $F^0(0)$, and then find the backscattering cross section from $F(0)$ according to (3.1-13). The error involved in using the above equations with only the first creeping wave term is shown in Figure 3-4(*a*). It is always less than 15 % for $k_0 a > 1.5$. However, the first four creeping wave terms are used (i.e., the first two *H*-waves), the agreement is extremely close even down to about $k_0 a = 0.6$ in the Rayleigh region; this curve is shown in Figure 3-4(*b*) (from Reference 7).

3.2.2.1.2. *Low-Frequency Region* ($k_0 a < 0.4$). This region is alternatively referred to as the Rayleigh region. A solution valid in this region is obtained by retaining only the first term of (3.2-4) and (3.2-5). Thus, the low-frequency backscattering function and cross section are

$$F(0) = \tfrac{3}{2}(k_0 a)^3 \tag{3.2-9}$$

$$\sigma(0) = \pi a^2 9(k_0 a)^4 \tag{3.2-10}$$

Equation (3.2-10) is called the Rayleigh backscattering cross section for a perfectly conducting sphere, and the errors incurred in its use are given in the second column of Table 3-1.

3.2.2.1.3. *High-Frequency Region* ($k_0 a > 20$). This region is alternatively referred to as the optics region because scattering results can be accurately predicted by the optical laws of reflection as $k_0$a becomes very large.

As $k_0 a$ becomes large, $F^c(0)$ of Eq. (3.2-8) becomes negligible due to the negative exponential, and only the first term of $F^0(0)$ in (3.2-7) is significant. Thus, the backscattering function and cross section in the high-frequency region become

$$F(0) = F^0(0) = -\tfrac{1}{2}k_0 a \, e^{-i2k_0 a} \tag{3.2-11}$$

$$\sigma(0) = \pi a^2 \tag{3.2-12}$$

The above cross section can also be derived using geometrical or physical optics methods. It is equivalent to the geometrical optics backscattering cross section from any curved surface segment derived in Section 2.2.2.1, but specialized to the sphere where the radii of curvature are equal, $a_1 = a_2$.

3.2.2.2. *Bistatic Scattering*

The polarization direction of the bistatically scattered far field is not the same generally as that of the incident field. Hence, the quantities sought in this case are $S_1(\theta)$ and $S_2(\theta)$. From these complex scattering amplitudes, the scattering cross section at any point in the far field and for any arbitrary polarization direction of the receiving antenna, $\hat{n}$, may be obtained using Eqs. (3.1-8) and (3.1-9).

Often for purposes of illustration, one specializes Eqs. (3.1-10) and (3.1-11) to the case where $\phi' = 0$ and $\pi/2$, respectively. These quantities are

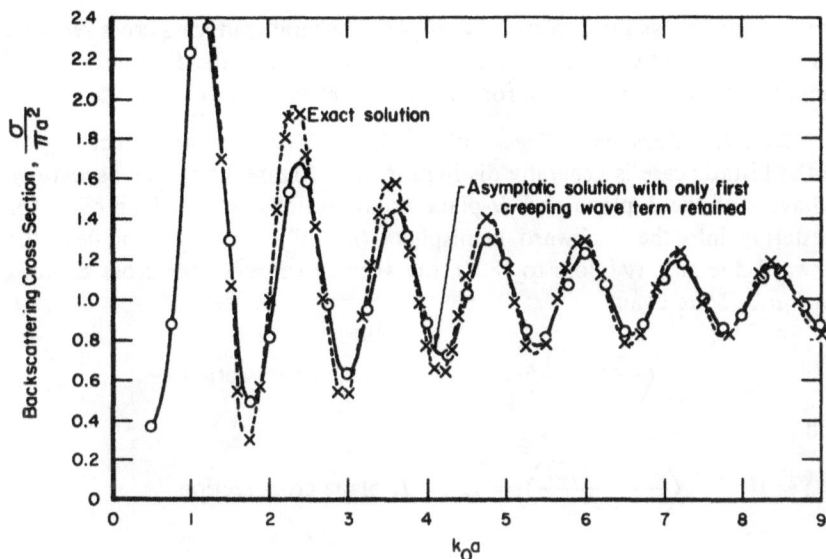

(a) One creeping wave term.

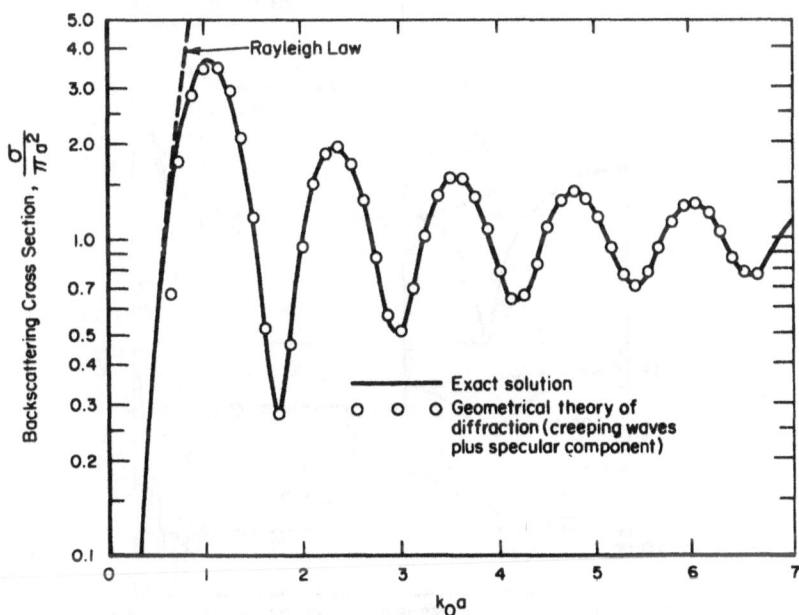

(b) Four creeping wave terms.

Fig. 3-4.  Comparison of exact solution and creeping wave approximation for back-scattering cross section of perfectly conducting sphere as a function of relative sphere size [after Kouyoumjian[7]].

then referred to as the *E*-plane and *H*-plane bistatic scattering cross sections when the incident wave has its electric field polarized in the *E* plane. Plots of these functions[2,8] are shown for various values of $k_0 a$ in Figure 3-5.

3.2.2.2.1. *Resonance Region* $(0.4 < k_0 a < 20)$. The resonance region in the bistatic case is generally distinguished in Figure 3.5 by the oscillatory behavior of the *E*-plane and *H*-plane cross sections as functions of $\theta'$ for scattering into the backward hemisphere (i.e., $\theta' < 90°$). Except near the forward direction ($\theta'$ close to 180°), the average values of the cross sections for $k_0 a > 1$ are again

$$\left\langle \frac{\sigma_\theta(\theta', 0)}{\pi a^2} \right\rangle = 1, \qquad E\text{-plane cross section}$$

and

$$\left\langle \frac{\sigma_\phi(\theta', \pi/2)}{\pi a^2} \right\rangle = 1, \qquad H\text{-plane cross section}$$

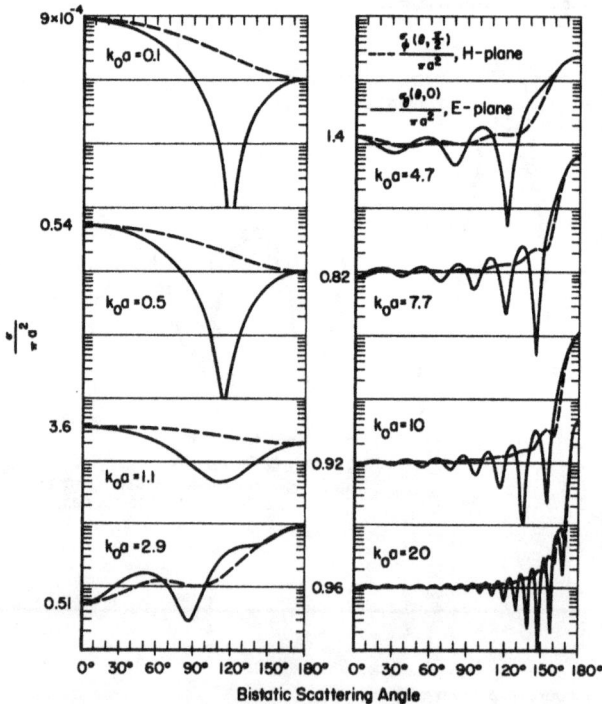

Fig. 3-5. Exact solutions for the bistatic scattering cross sections of a perfectly conducting sphere as a function of the bistatic angle for various values of relative sphere size.

The complex scattering amplitudes near the lower end of the resonance region (say, $0.4 < k_0 a < 1$) may be approximated by

$$S_1(\theta') = (k_0 a)^3 \left[ \left( \tfrac{1}{2} + \cos \theta' \right) - \left( \tfrac{3}{10} - \tfrac{11}{45} \cos \theta' + \tfrac{1}{12} \cos 2\theta' \right) (k_0 a)^2 \right.$$

$$\left. + \tfrac{1}{120} \left( \tfrac{1807}{70} - \tfrac{2531}{105} \cos \theta' + \tfrac{57}{42} \cos 2\theta' + \tfrac{1}{3} \cos 3\theta' \right) (k_0 a)^4 \right]$$

$$+ i (k_0 a)^6 \left[ \tfrac{1}{6} (4 \cos \theta' - 1) + \tfrac{1}{5} (1 + 2 \cos \theta')(k_0 a)^2 \right] + O[(k_0 a)^9]$$

$$(3.2\text{-}13)$$

and

$$S_2(\theta') = (k_0 a)^3 \left[ \left( \tfrac{1}{2} \cos \theta' + 1 \right) + \left( \tfrac{3}{10} - \tfrac{23}{60} \cos \theta' - \tfrac{1}{18} \cos 2\theta' \right) (k_0 a)^2 \right.$$

$$\left. + \tfrac{1}{60} \left( - \tfrac{1343}{105} + \tfrac{3769}{280} \cos \theta' + \tfrac{57}{63} \cos 2\theta' + \tfrac{1}{8} \cos 3\theta' \right) (k_0 a)^4 \right]$$

$$+ i (k_0 a)^6 \left[ \tfrac{1}{6} (4 - \cos \cdot \theta') + \tfrac{1}{5} (\cos \theta' + 2)(k_0 a)^2 \right] + O[(k_0 a)^9]$$

$$(3.2\text{-}14)$$

for the E-plane and H-plane cases, respectively. The errors involved in the use of these two equations are given again by the first column of Table 3-1 as a function of $k_0 a$. These equations are not valid at all for $k_0 a > 1$. They are valid for all $\theta'$.

At the upper end of the resonance region the optics contribution, $S_1^0(\theta')$ and $S_2^0(\theta')$, as well as the creeping wave contribution, $S_1^c(\theta')$ and $S_2^c(\theta')$, can be obtained by a Watson transformation of the Mie series.[6] Thus;

$$S_1(\theta') = S_1^0(\theta') + S_1^c(\theta')$$

$$S_2(\theta') = S_2^0(\theta') + S_2^c(\theta')$$

$$(3.2\text{-}15)$$

where

$$S_1^0(\theta') = \frac{k_0 a}{2} \exp[-i 2 k_0 a \cos(\theta'/2)] \left[ 1 - \frac{i}{2 k_0 a \cos^3(\theta'/2)} \right.$$

$$\left. - \frac{7}{4(k_0 a)^2} \frac{\sin^2(\theta'/2)}{\cos^6(\theta'/2)} + O[(k_0 a)^{-3}] \right]$$

$$(3.2\text{-}16)$$

and

$$S_2^0(\theta') = \frac{k_0 a}{2} \exp[-i 2 k_0 a \cos(\theta'/2)] \left[ 1 - \frac{i \cos \theta'}{2 k_0 a \cos^3(\theta/2)} \right.$$

$$\left. + \frac{1}{4(k_0 a)^2} \frac{(6 + \cos \theta') \sin^2(\theta'/2)}{\cos^6(\theta'/2)} + O[(k_0 a)^{-3}] \right]$$

$$(3.2\text{-}17)$$

$$S_1{}^c(\theta') = -ix^{1/3}\sqrt{\frac{x}{2\pi\sin\theta'}}\,e^{i\pi(x+1/12)}$$

$$\times\left[[2.715175 + x^{-2/3}(0.933508 + i1.616911)]\right.$$

$$\times\left\{\left[1 - \frac{i0.8750}{x\tan\theta'}\right]\exp[-ix\theta' - (\pi-\theta')\,x^{1/3}(0.700283 - i0.404308)\right.$$

$$+ (\pi-\theta')\,x^{-1/3}(0.141774 + i0.081853)]$$

$$- i\left[1 + \frac{i0.8750}{x\tan\theta'}\right]\exp[ix\theta' - (\pi+\theta')\,x^{1/3}(0.700283 - i0.404308)$$

$$\left.+ (\pi+\theta')\,x^{-1/3}(0.141774 + i0.081853)]\right\}_{E_1}$$

$$+ [1.391727 + x^{-2/3}(1.032306 + i1.788036)]$$

$$\times\left\{\left[1 - \frac{i0.8750}{x\tan\theta'}\right]\exp[-ix\theta' - (\pi-\theta')\,x^{1/3}(2.232697\right.$$

$$- i1.289048) - (\pi-\theta')\,x^{-1/3}(0.141482 + i0.081684)]$$

$$- i\left[1 + \frac{i0.8750}{x\tan\theta'}\right]\exp[ix\theta' - (\pi+\theta')\,x^{1/3}(2.232697 - i1.289048)$$

$$\left.- (\pi+\theta')\,x^{-1/3}(0.141482 + i0.081684)]\right\}_{E_2}$$

$$+ \frac{0.201776}{x\sin\theta'}\{\exp[-ix\theta' - (\pi-\theta')\,x^{1/3}(1.607133 - i0.927879)$$

$$- (\pi-\theta')\,x^{-1/3}(0.099415 + i0.057397)]$$

$$- i\exp[ix\theta' - (\pi+\theta')\,x^{-1/3}(1.607133 - i0.927879)$$

$$\left.- (\pi+\theta')\,x^{-1/3}(0.099415 + i0.057397)]\}_{H_1}\right] \qquad (3.2\text{-}18)$$

$$S_2^c(\theta') = ix^{1/3} \sqrt{\frac{x}{2\pi \sin \theta'}}\, e^{i\pi(x+1/12)}$$

$$\times \left[ \frac{0.339397}{x \sin \theta'} \{\exp[-ix\theta' - (\pi - \theta')\, x^{1/3}(0.700283 - i0.404308) \right.$$

$$+ (\pi - \theta')\, x^{-1/3}(0.141774 + i0.081853)]$$

$$- i \exp[ix\theta' - (\pi + \theta')\, x^{1/3}(0.700283 - i0.404308)$$

$$+ (\pi + \theta')\, x^{-1/3}(0.141774 + i0.081853)]\}_{E_1}$$

$$+ \frac{0.173966}{x \sin \theta'} \{\exp[-ix\theta' - (\pi - \theta')\, x^{1/3}(2.232697 - i1.289048)$$

$$- (\pi - \theta')\, x^{-1/3}(0.141482 + i0.081684)]$$

$$- i \exp[ix\theta' - (\pi + \theta')\, x^{1/3}(2.232697 - i1.289048)$$

$$- (\pi + \theta')\, x^{-1/3}(0.141482 + i0.081684)]\}_E$$

$$+ [1.614208 + x^{-2/3}(0.848747 + i1.470073)] \left\{ \left[ 1 - \frac{i0.8750}{x \tan \theta'} \right] \right.$$

$$\times \exp[-ix\theta' - (\pi - \theta')\, x^{1/3}(1.607133 - i0.927879)$$

$$- (\pi - \theta')\, x^{-1/3}(0.099415 + i0.057397)] - i \left[ 1 + \frac{i0.8750}{x \tan \theta'} \right]$$

$$\times \exp[ix\theta' - (\pi + \theta')\, x^{1/3}(1.607133 - i0.927879)$$

$$\left. - (\pi + \theta')\, x^{-1/3}(0.099415 + i0.057397)] \right\}_{H_1} \right]$$

and

$$x = k_0 a \qquad (3.2\text{-}19)$$

Equations (3.2-16) and (3.2-17) are valid for any value of $\theta'$, except near the forward direction, such that $\theta' < \pi - \delta$, where $\delta = O(1/k_0 a)$. However, Eqs. (3.2-18) and (3.2-19) are not valid near either the backscattering or forward scattering directions, but are valid in an intermediate region, such that $\delta < \theta' < \pi - \delta$, where $\delta$ is given above. The error involved in the use of the above expressions for $k_0 a > 2$ is generally less than 15%.

3.2.2.2.2. *Low-Frequency Region* ($k_0 a < 0.4$). The low-frequency or Rayleigh-region complex scattering amplitudes are obtained by taking the first term involving $(k_0 a)^3$ of Eqs. (3.2-13) and (3.2-14).

$$S_1(\theta') = (k_0 a)^3 [\tfrac{1}{2} + \cos \theta'] \qquad (3.2\text{-}20)$$

$$S_2(\theta') = (k_0 a)^3 [\tfrac{1}{2} \cos \theta' + 1] \qquad (3.2\text{-}21)$$

The $E$-plane and $H$-plane scattering cross sections are obtained by squaring (3.2-20) and (3.2-21) respectively, and multiplying by $4\pi/k_0{}^2$ according to Eqs. (3.1-10) and (3.1-11):

$$\sigma_\theta(\theta', 0) = \pi a^2 \cdot 4(k_0 a)^4[\tfrac{1}{2} + \cos\theta']^2 \qquad (3.2\text{-}22)$$

$$\sigma_\phi\left(\theta', \frac{\pi}{2}\right) = \pi a^2 4(k_0 a)^4[\tfrac{1}{2}\cos\theta' + 1]^2 \qquad (3.2\text{-}23)$$

The error involved in the use of the above equations is less than $1\%$ at $k_0 a = 0.5$, and decreases rapidly for smaller $k_0 a$. They are valid for all $\theta'$.

3.2.2.2.3. *High-Frequency Region* $(k_0 a > 20)$.  As $k_0 a$ increases beyond 20, the creeping waves of (3.2-18) and (3.2-19) become small compared with the optics contribution [Eqs. (3.2-15) and (3.2-16)]. The only term of the optics contribution which is important is the first. Hence, the complex scattering amplitudes and $E$-plane and $H$-plane cross sections are

$$S_1(\theta') = S_2(\theta') = -\tfrac{1}{2}k_0 a \exp[-i2k_0 a \cos\theta'/2] \qquad (3.2\text{-}24)$$

$$\sigma_\theta(\theta', 0) = \sigma_\phi\left(\theta', \frac{\pi}{2}\right) = \pi a^2 \qquad (3.2\text{-}25)$$

These are valid everywhere except near the forward direction, i.e., for $0 \leqslant \theta' \leqslant \pi - \delta$, where $\delta = O(1/k_0 a)$. This region, near forward scattering at high frequencies, will be treated in the forward scattering section. The above formulas are in error by about $5\%$ at $k_0 a = 20$, and the error decreases rapidly for increasing $k_0 a$.

From (3.2-25), it is evident that the bistatic cross section of a perfectly conducting sphere at high frequencies is the same as the backscattering cross section independent of $\theta'$.

### 3.2.2.3. *Forward Scattering*

In the forward direction, the scattered field is always polarized in the same direction as the incident field. Hence, the cross section to be discussed here is assumed to be a measure of the scattered power in the same polarization state as the incident wave.[†] The scattering function and complex scattering amplitudes are related by $F(\pi) = S_1(\pi) = -S_2(\pi)$.

The exact forward scattering cross section as a function of $k_0 a$ is plotted in Figure 3-6. This curve has been constructed from several sources.[8–10] Most noticeable are the lack of oscillations in the resonance region and the monotonically increasing character of the curve as compared to backscattering.

---

[†] The forward scattering cross section is not proportional to the total power propagating into the forward direction, as is the case in the bistatic and backward directions. The incident wave also propagates into the forward direction.

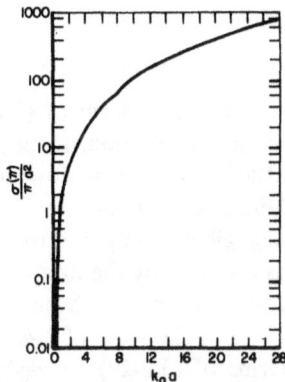

Fig. 3-6. Exact solution for the forward scattering cross section of a perfectly conducting sphere as a function of sphere size relative to wavelength.

3.2.2.3.1. *Resonance Region* $(0.4 < k_0a < 20)$. Equations applicable near the lower end of this range for the scattering and forward scattering cross section are

$$F(\pi) = -\frac{1}{2}(k_0a)^3 \left[1 + \frac{113}{90}(k_0a)^2 - \frac{1783}{2100}(k_0a)^4 - \frac{670{,}057}{396{,}900}(k_0a)^6\right]$$

$$- i\frac{5}{6}(k_0a)^6 \left[1 + \frac{6}{25}(k_0a)^3\right] + O[(k_0a)^{10}] \tag{3.2-26}$$

and

$$\sigma(\pi) = \pi a^2(k_0a)^4 \left[1 + \frac{113}{45}(k_0a)^2 - \frac{6{,}899}{56{,}700}(k_0a)^4 - \frac{5{,}419{,}129}{1{,}984{,}500}(k_0a)^6\right]$$

$$+ O[(k_0a)^{12}] \tag{3.2-27}$$

As before, these equations are only valid for $k_0a < 1$. The error incurred in their use is as shown in the left side of Table 3-1.

At the upper end of the resonance region, the equations for the scattering function are obtained from a Watson transformation of the Mie solution.[6] Expressed in terms of the optics and creeping wave contributions, these are

$$F^0(\pi) = -i\frac{1}{2}\left[(k_0a)^2 - \frac{11}{12}\right] \tag{3.2-28}$$

and

$$F^c(\pi) = -ix^{4/3}[[0.032927 + i0.057154 + x^{-2/3}(0.242679 - i0.710672)$$
$$- x^{-4/3}(0.001846 + i0.008027)]] - ix^{4/3}\, e^{i2\pi(x+1/6)}[\{[1.357588$$
$$+ x^{-2/3}(0.741196 + i1.283788)] \exp[-x^{1/3}(4.400004 - i2.540343)$$
$$+ x^{-1/3}(0.890792 + i0.514299)]\}_{E_1} + \{[0.807104$$
$$+ x^{-2/3}(0.798821 + i1.383598)] \exp[-x^{1/3}(10.097912 - i5.830032)$$
$$- x^{-1/3}(0.624641 + i0.360637)]\}_{H_1}] \tag{3.2-29}$$

where

$$F(\pi) = F^0(\pi) + F^c(\pi) \qquad (3.2\text{-}30)$$

The first term of (3.2-29) in curly brackets represents a quasicreeping wave contribution. It may be thought of as a creeping wave which becomes attached at the shadow boundary, travels zero distance, and is immediately launched again into the forward direction. As can be seen, it is not necessarily negligible. Compared to $F^0(\pi)$ for $k_0 a = 10$, this first term is still 6 % as large as $F^0(\pi)$. Only the first $E$ and $H$ waves are retained of the true creeping wave contributions which have completed one entire circuit around the sphere.

For the value of $k_0 a$ chosen, the user can decide how many creeping wave terms of (3.2-29) he wishes to retain compared to the optics contribution $F^0(\pi)$. The error involved in the use of the above equations is less than 15 % at $k_0 a = 2$, and decreases rapidly for $k_0 a > 2$.

3.2.2.3.2. *Low-Frequency Region* ($k_0 a < 0.4$). The low-frequency or Rayleigh approximation for the forward scattering function and cross section is found by including only the lowest-order term of Eqs. (3.2-26) and (3.2-27). The results are

$$F(\pi) = -\tfrac{1}{2}(k_0 a)^3 \qquad (3.2\text{-}31)$$

$$\sigma(\pi) = \pi a^2 (k_0 a)^4 \qquad (3.2\text{-}32)$$

The error involved in the use of these equations is less than 1 % for $k_0 a < 0.5$.

3.2.2.3.3. *High-Frequency Region—Forward-Scattering Function and Cross Section* ($k_0 a > 20$). The forward-scattering function and cross section for $k_0 a$ very large can be found from (3.2-28); $F^0(\pi)$, the optics contribution for large $k_0 a$, is much larger than $F^c(\pi)$, the creeping wave contribution. Hence, from the first term of (3.2-28),

$$F(\pi) = -i\tfrac{1}{2}(k_0 a)^2 \qquad (3.2\text{-}33)$$

and

$$\sigma(\pi) = \pi a^2 (k_0 a)^2 \qquad (3.2\text{-}34)$$

Error involved in the use of these equations is less than 5 % at $k_0 a = 20$, and decreases rapidly for $k_0 a > 20$.

Semi-empirical formulas for the scattering function and forward scattering cross section, valid over a wider range of $k_0 a$, were obtained in Reference 9, and are given by

$$F(\pi) = -i\tfrac{1}{2}k_0 a[k_0 a + \tfrac{2}{5}]$$

$$\sigma(\pi) = \pi a^2 [k_0 a + \tfrac{2}{5}]^2 \qquad (3.2\text{-}35)$$

These formulas are in error by 10 % at $k_0 a = 1$, and by 2 % at $k_0 a = 2$. It can be seen that these equations become identical to (3.2-33) and (3.2-34) as $k_0a$ becomes very large.

*3.2.2.3.4. High-Frequency Region—Forward Scattering Pattern* $(k_0a > 100)$. In and near the forward direction, the bistatic scattered far field in the high-frequency limit is defined to be the diffracted field. The diffracted field pattern is not a function of the properties of the sphere (i.e., whether it is a perfect conductor, dielectric, etc.), and does not even depend upon the spherical shape. It depends only upon the size and shape of the cross section of the object in a plane passing through the shadow boundary. By Babinet's principle, this diffraction pattern is the same as would be produced by waves passing through a perfectly conducting screen with a hole in it of the same cross sectional area as that of the object at the shadow boundary (circular for the case under consideration).

One can see from Figure 3-5 that as frequency increases (or $k_0a$ increases), the scattering into the backward hemisphere becomes constant, while the oscillations in the forward direction become larger in amplitude but confined closer to $\theta' = 180°$. The value at $\theta' = 180°$ increases as $(k_0a)^2$. This indicates a focusing effect at higher frequencies in the forward direction. The $E$-plane and $H$-plane cross sections become nearly the same at $k_0a > 100$, indicating that at high frequencies the forward diffraction pattern is both polarization independent and independent of $\phi'$, the azimuth angle. The polarization is the same as that of the incident field.

Since this forward diffraction pattern is independent of $\phi'$, let $F(\pi - \alpha) = F(\theta')$ represent the high-frequency scattering function, where $\alpha$ is very much less than $\pi$. Then this scattering pattern is[10]

$$F(\pi - \alpha) = -i(k_0a)^2 \left[ \frac{J_1(k_0a \sin \alpha)}{k_0a \sin \alpha} \right] \tag{3.2-36}$$

and

$$\sigma(\pi - \alpha) = \frac{4\pi}{k_0^2} |F(\pi - \alpha)|^2 = 4\pi a^2 (k_0a)^2 \left[ \frac{J_1(k_0a \sin \alpha)}{k_0a \sin \alpha} \right]^2 \tag{3.2-37}$$

which, by comparison with Eq. (7.5-11), is seen to be the high-frequency, near-normal-incidence backscatter cross section for a perfectly conducting disc.

This expression converges to the proper result at the forward scattering direction (as $\alpha \to 0$), since the quantity in the brackets reduces to $\frac{1}{2}$.

The above diffraction patterns can be used to find the beamwidth of the main lobe in the forward direction. The half-power point of the cross section of (3.2-37) occurs at $\sin \alpha_h = 1.61/k_0a$, which gives the beamwidth, $\alpha_h$. The angle, $\alpha_n$, at which the first null or dark ring occurs is defined by $\sin \alpha_n = 3.83/k_0a$. The intensity of any of the fringes or bright rings is much less than that of the main lobe, as for example, the intensity of the first bright ring is less than 2 % that of the main lobe maximum, independent of $k_0a$.

## 3.3. DIELECTRIC SPHERES—LOSSY AND LOSSLESS

In this section, the more general situation of nonperfectly conducting homogeneous spheres will be considered, and results will be presented where they exist. The medium of the sphere is now denoted as Medium 1, with a wave number

$$k_1 = \omega \sqrt{\epsilon_1 \mu_1} = \omega \sqrt{(\epsilon_1' + i\epsilon_1'')(\mu_1' + i\mu_1'')} = k_0 m_1 \qquad (3.3\text{-}1)$$

where $\epsilon_1 = \epsilon_0 \cdot \epsilon_{r1}$ and $\mu_1 = \mu_0 \cdot \mu_{r1}$ are the effective permittivity and permeability of the sphere material, and $m_1$, as defined above, is called the refractive index of the medium and can be expressed in terms of its real and imaginary parts,

$$m_1 = m_{R1} + im_{I1} \qquad (3.3\text{-}2)$$

The perfectly conducting sphere treated in the preceding section is a special case where $\epsilon_1'' = \infty$, since the conductivity $\sigma_1$ is assumed to be infinite. The perfect dielectric sphere to be discussed here also is a special case, where $\epsilon_1'' = 0$ and $\mu_{r1} = 1$. Although the lossy dielectric sphere seems more difficult to handle, because of the complex refractive index, than the perfect dielectric sphere, it turns out that approximate solutions for the scattering cross section at higher frequencies are more readily obtained for the former case. Besides, almost all materials in nature exhibit some finite nonzero conductivity, $\sigma_1$; hence, $\epsilon_1''$ may be small but is not identically zero.

Since the historical treatment of the results for a nonmetallic sphere with permeability, $\mu_1$ different from $\mu_0$, are limited, this section will be restricted solely to dielectric spheres where $\mu_1 = \mu_0$ (or $\mu_{r1} = 1$).

### 3.3.1. The Exact Solution—Numerical Summation

The coefficients of the Mie solution for this case become

$$A_n = (-i)^n \frac{2n+1}{n(n+1)} \left[ \frac{j_n(k_0a)[k_1aj_n(k_1a)]' - j_n(k_1a)[k_0aj_n(k_0a)]'}{j_n(k_1a)[k_0ah_n^{(1)}(k_0a)]' - h_n^{(1)}(k_0a)[k_1aj_n(k_1a)]'} \right] \quad (3.3\text{-}3)$$

$$B_n = (-i)^{n+1} \frac{2n+1}{n(n+1)} \left[ \frac{j_n(k_0a)[k_1aj_n(k_1a)]' - m_1^2 j_n(k_1a)[k_0aj_n(k_0a)]'}{h_n^{(1)}(k_0a)[k_1aj_n(k_1a)]' - m_1^2 j_n(k_1a)[k_0ah_n^{(1)}(k_0a)]'} \right]$$

$$(3.3\text{-}4)$$

As before, the primes, $[\ \ ]'$, denote differentiation with respect to either $k_0a$ or $k_1a$, whichever appears within the brackets. In general, for a lossy dielectric, $k_1 = k_0 m_1$ is complex.

The spherical Bessel functions, $j_n(z)$ and $h_n^{(1)}(z)$, in these constants can be generated using recursion techniques starting with $j_0(z)$, $j_1(z)$, and $y_0(z)$, $y_1(z)$, as discussed in Section 3.2.1. These recursion techniques are equally valid when $z$ is complex; since the lowest order spherical Bessel and Neuman functions are simple trigonometric expressions, complex arguments merely change them to combinations of trigonometric and hyperbolic functions.[†]

### 3.2.2. Approximate Solutions—Far Field—Arbitrary Refractive Index $m_1$

Since the scattering mechanism is different between perfectly conducting spheres and dielectric spheres, the approximations employed and their regions of validity are different.

Curves from exact solutions for several parameters are available in the literature,[13–15] and several representative curves are shown in Figures 3-7 through 3-10. Figure 3-7(a), (b) shows computed exact backscatter cross sections for two dielectric spheres; one lossless with $m_1 = 1.61$, and one with a slight amount of loss where $m_1 = 1.61 + i0.0025$. Figure 3-7(a) shows the actual cross section for values of $k_0 a$ up to 20, and 3-7(b) shows only the average cross section for $k_0 a$ ranging from 20 to 50. Figure 3-8 shows the backscatter cross section for a lossy sphere with $m_1 = 1.78 + i0.0024$. Curve 3-8(a) gives the complete cross section for $k_0 a$, ranging from 0 to 30; and curve 3-8(b) gives the average cross section for $0 \leqslant k_0 a \leqslant 500$. In Figure 3-9, bistatic cross sections are given for a dielectric sphere with $m_1 = 2.105$ and several values of $k_0 a$. Figure 3-10 gives backscatter and forward scatter cross sections for the same sphere ($m_1 = 2.105$) versus $k_0 a$, with curve (a) showing backscatter and curve (b) forward scatter.

### 3.3.2.1. Low-Frequency Region ($k_1 a < 0.8$)

In this region, the exact Mie series solution can be expanded into a series in powers of $k_0 a$ in which only the first few terms of the series are used. In order to do this, one must expand the coefficients $A_n$ and $B_n$ of Eqs. (3.3-3) and (3.3-4) into a power series in $k_0 a$. Stratton (Reference 16, p. 571) has shown that only $A_1$, $A_2$, $B_1$, and $B_2$ are necessary if one wishes only to preserve terms up to and including $(k_0 a)^5$. Furthermore, when $\mu_1 = \mu_0$, as assumed here, these coefficients become[‡]

$$A_1 = \frac{1}{30}(m_1{}^2 - 1)(k_0 a)^5$$

$$A_2 = 0$$

$$B_1 = i\left[\frac{m_1{}^2 - 1}{m_1{}^2 + 2}(k_0 a)^3 + \frac{3}{5}\frac{(m_1{}^2 - 1)(m_1{}^2 - 2)}{(m_1{}^2 + 2)^2}(k_0 a)^5\right]$$

$$B_2 = \frac{1}{18}\frac{m_1{}^2 - 1}{2m_1{}^2 + 3}(k_0 a)^5 \qquad\qquad (3.3\text{-}5)$$

---

[†] An alternative recursion technique can be employed to generate the Mie coefficients by using logarithmic derivative functions. This is discussed in References 11 and 12.
[‡] Errors in Stratton's coefficients have been corrected here.

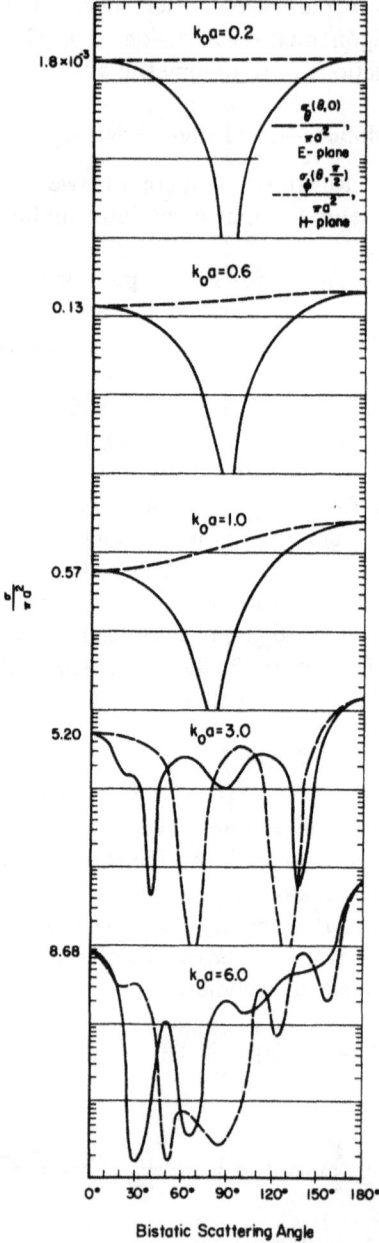

Fig. 3-7. Exact solutions for the normalized backscattering cross
Sections of lossless and slightly absorbing homogeneous spheres as
a function of sphere size relative to wavelength. Refractive indices of
spheres are $m_1 = 1.61$ and $1.61 + i0.0025$ [after Atlas *et al.*[13]].

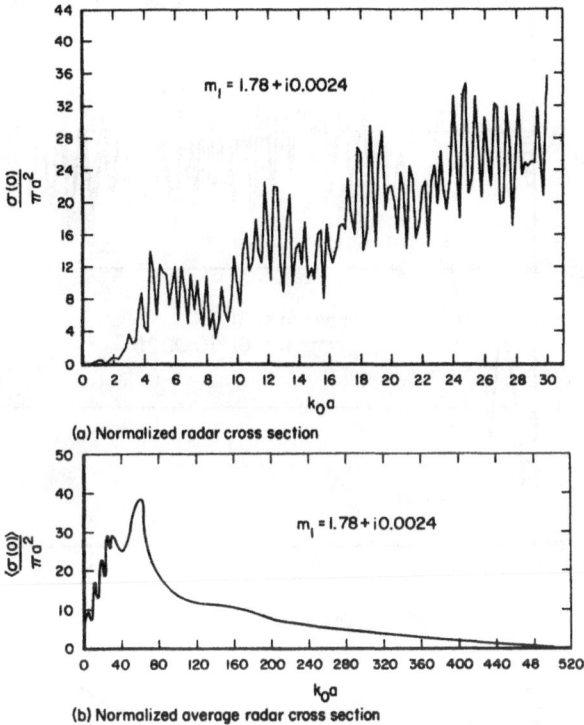

(a) Normalized radar cross section

(b) Normalized average radar cross section

Fig. 3-8. Exact solutions for the normalized backscattering cross sections of a slightly absorbing homogeneous sphere as a function of sphere size relative to wavelength. Refractive index of the sphere is $m_1 = 1.78 + 0.0024$ [after Herman and Battan[14]].

3.3.2.1.1. *Low-Frequency Backscattering.* When one uses the above coefficients in the case of far-field backscattering, the scattering function becomes

$$F(0) = \frac{m_1^2 - 1}{m_1^2 + 2}(k_0 a)^3 + \left[\frac{3}{5}\frac{(m_1^2 - 1)(m_1^2 - 2)}{(m_1^2 + 2)^2}\right.$$
$$\left. - \frac{1}{30}(m_1^2 - 1) - \frac{1}{6}\frac{(m_1^2 - 1)}{(2m_1^2 + 3)}\right](k_0 a)^5 \qquad (3.3\text{-}6)$$

For $k_0 a$ small enough, only the first term is singificant. Preserving only the first term, one obtains the following well-known Rayleigh backscattering cross section for dielectric spheres[†]:

$$\sigma(0) = 4\pi a^2 \left|\frac{m_1^2 - 1}{m_1^2 + 2}\right|^2 (k_0 a)^4 \qquad (3.3\text{-}7)$$

[†] Notice should be taken that Eqs. (3.3-6) and (3.3-7) do not reduce to those for a perfectly conducting sphere when $m_1$ approaches infinity. This fact is noted and explained in Reference 5 (pp. 8 and 9). The reason for this is that the method of derivation of the coefficients of Eq. (3.3-5) is not valid in this limiting case.

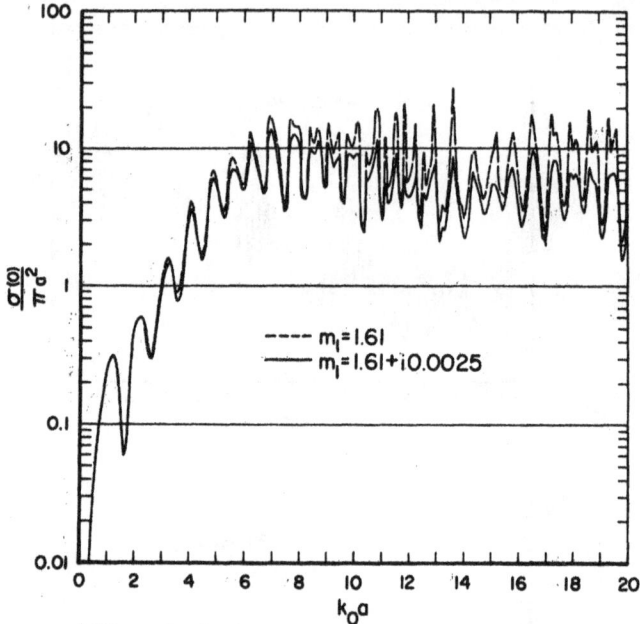

(a) Normalized radar cross section.

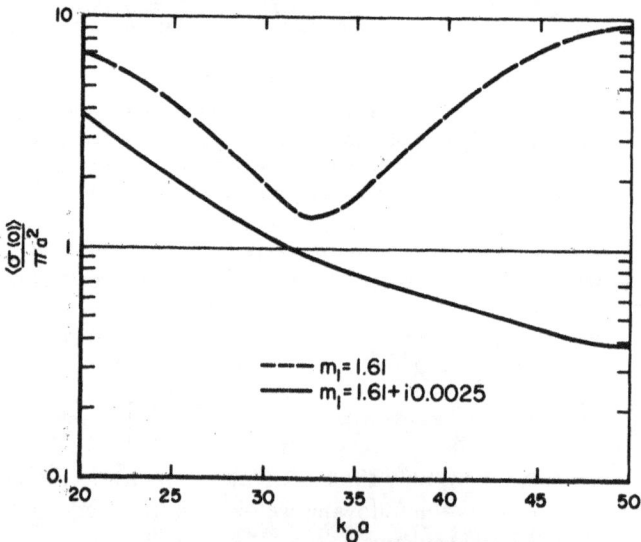

(b) Normalized average radar cross section.

Fig. 3-9.   Exact solutions for the bistatic
cross sections of a homogeneous sphere
as a function of bistatic angle for various
values of relative sphere size. Refractive
index of the sphere is $m_1 = 2.105$
[after Kerker and Matievic[15]].

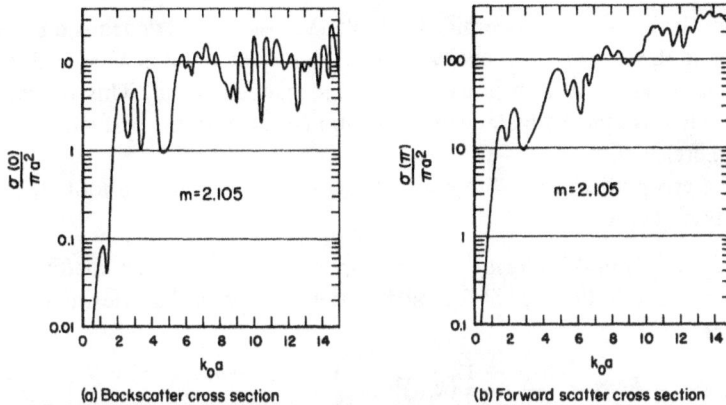

(a) Backscatter cross section          (b) Forward scatter cross section

Fig. 3-10.  Exact Solutions for back and forward scattering cross sections of a homogeneous sphere as a function of sphere size relative to wavelength. Refractive index of the sphere is $m_1 = 2.105$. (Below $k_0a = 6$, computed points are not close enough to show exact oscillatory detail of curve.) [After Kerker and Matievic[15].]

This expression is valid for $m_1$ real or complex [i.e., for lossless or lossy spheres provided $|k_1a| < 0.8$]. As a measure of accuracy, for $\epsilon_1'' = 0$ and $\epsilon_{r1} = 4$, use of (3.3-7) results in 4% error at $k_0a = 0.5$. These errors decrease rapidly with decreasing $k_0a$.

3.3.2.1.2.  *Low-Frequency Bistatic Scattering.*  When the coefficients of (3.3-5) are used in the case of bistatic scattering, the complex scattering amplitudes become

$$S_1(\theta') = \frac{m_1^2 - 1}{m_1^2 + 2} \cos \theta' (k_0a)^3 + \left[\frac{3}{5} \frac{(m_1^2 - 1)(m_1^2 - 2)}{(m_1^2 + 2)^2} \cos \theta' \right.$$
$$\left. - \frac{1}{6} \frac{(m_1^2 - 1)}{(2m_1^2 + 3)} \cos 2\theta' - \frac{1}{30} (m_1^2 - 1)\right] (k_0a)^5 \qquad (3.3\text{-}8)$$

$$S_2(\theta') = \frac{m_1^2 - 1}{m_1^2 + 2} (k_0a)^3 + \left[\frac{3}{5} \frac{(m_1^2 - 1)(m_1^2 - 2)}{(m_1^2 + 2)^2} \right.$$
$$\left. - \frac{1}{6} \frac{(m_1^2 - 1)}{(2m_1^2 + 3)} \cos \theta' - \frac{1}{30} (m_1^2 - 1) \cos \theta'\right] (k_0a)^5 \qquad (3.3\text{-}9)$$

Hence, the $E$-plane and $H$-plane cross sections become (retaining only the first term of the above expressions)

$$\sigma_\theta(\theta', 0) = 4\pi a^2 \left|\frac{m_1^2 - 1}{m_1^2 + 2}\right|^2 \cos^2 \theta' (k_0a)^4 \qquad (3.3\text{-}10)$$

$$\sigma_\phi\left(\theta', \frac{\pi}{2}\right) = 4\pi a^2 \left|\frac{m_1^2 - 1}{m_1^2 + 2}\right|^2 (k_0a)^4 \qquad (3.3\text{-}11)$$

Thus, in the low-frequency limit, the $E$-plane cross section depends on $\theta'$ with maxima in the forward and backward directions and a null at $\theta' = \pi/2$, while the $H$-plane cross section is completely independent of $\theta'$. This is different from the low-frequency bistatic cross section behavior of the perfectly conducting sphere.

The error incurred in the use of the above equations is about the same as for backscattering.

3.3.2.1.3. *Low-Frequency Forward Scattering.* When the coefficients of (3.3-5) are used in the case of forward scatter, the forward scattering function becomes

$$F(\pi) = \frac{m_1^2 - 1}{m_1^2 + 2} (k_0 a)^3 + \left[ \frac{3}{5} \frac{(m_1^2 - 1)(m_1^2 - 2)}{(m_1^2 + 2)^2} \right.$$

$$\left. + \frac{1}{6} \frac{(m_1^2 - 1)}{(2m_1^2 + 3)} + \frac{1}{30} (m_1^2 - 1) \right] (k_0 a)^5 \qquad (3.3\text{-}12)$$

where $F(\pi) = S_1(\pi) = -S_2(\pi)$. For $k_0 a$ sufficiently small that only the first term is important, one observes that the forward scattering function is equal in magnitude to the backward scattering function, but 180° out of phase with it. Hence, retaining only this first term, the forward and backscattering cross sections are identical:

$$\sigma(\pi) = 4\pi a^2 \left| \frac{m_1^2 - 1}{m_1^2 + 2} \right|^2 (k_0 a)^4 \qquad (3.3\text{-}13)$$

Again, this behavior is different from that of the perfectly conducting sphere. The error involved in the used of these equations is about the same as for backscattering.

3.3.2.2. *Refractive Region* ($k_0 a > 2$, $k_0 a < 1.15/m_{I1}$)

This region does not correspond to the resonance region for the perfectly conducting sphere. The phenomena producing the scattered field for the dielectric sphere are different from that of the perfectly conducting sphere. One can treat the scattered field as having arisen from rays of the incident field which enter the sphere and are then multiplied, reflected and refracted, and finally emerge in the scattering direction of interest. These internal rays and their behavior can be determined entirely by modifications of geometrical and/or physical optics. As the frequency increases (or sphere radius increases), these various groups of rays interfere with each other, producing maxima and minima in the scattered field strength. In a lossless dielectric medium, this interference phenomena continues as $k_0 a$ increases because the internal rays never die out. It is only when the material is lossy enough and the internal ray paths are long enough, ($k_0 a < 1.15/m_{I1}$), that the internal rays are attenuated and diminish with $k_0 a$. Hence, in the absence of loss, there is no asymptotic or limiting value for the sphere cross section, as attested to in Figures 3-7 and 3-8. In fact, as shown in these figures, the dielectric-sphere scattering cross section can be as much as an order of magnitude greater than that for the perfectly conducting sphere in this region. Moreover, the oscillations of the curves for backscattering cross section versus $k_0 a$ are not regular in their periodicity. For most practical purposes, all dielectric materials have some losses, however small, and eventually the internal rays begin to be damped (especially those reflected a large number of times internally). As an example, notice Figure 3-8b. In that case the loss term (imaginary part of the refractive index) is very small, but when $k_0 a$ has reached 500, the backscattering cross section has passed out of this refractive region and toward its asymptotic limit, due to internal ray attenuation.

The approximations employed in this section are those of modified geometrical optics (References 17, 18, 19, 20, 21), which denote geometrical optics modified by use of physical optics for certain special cases where geometrical-optics methods fail. This approach has the advantage that it gives meaningful physical insight into the nature of the scattering process which is hard to obtain from the exact eigenfunction (Mie) solution. Only rays reflected internally a minimal number of times will be considered here because (a) they are generally the most important contributors, (b) they are easier to handle mathematically, and (c) they indicate the general trend of the scattering cross section, even though the fine detail of the curves may not be accurate.

There are five types of rays to be considered in this section. They are described briefly here and shown for backscattering in Table 3-2.

(1) *The External Ray.* This is the ray which never enters the sphere. It is reflected from each point of the sphere surface in the specular direction, and it is proportional to the Fresnel reflection coefficient of the sphere material at that angle of incidence. For backscattering, this ray is reflected from the forward-most cap, as shown in Table 3-2. The phase at this cap is taken as zero and used as a reference for all the other rays.

(2) *The Axial Rays.* These are rays which enter the sphere along the axis, are reflected one or more times, and then exit in either the back or forward scattering direction. Only the ray with one internal reflection (Table 3-2) is considered here for backscattering, and those rays with zero and two internal reflections are considered for forward scattering.

(3) *The Glory Rays.* These are rays which exist in the back and forward scattering directions only. Since the sphere is an axially symmetric body, these rays enter and leave the sphere on rings at constant incidence and refraction angles. If one were to view a sphere in the backscattering direction, he would "see" a ring of light emerging, rather than a spot near the cap as produced by the axial and external rays. This ring of light has a toroidal phase front, and physical optics is used to determine the contribution from such a front. The

## Table 3-2. Modified Geometrical-Optics

| Type of ray and diagram | Range of $m_1$ for which ray exists | Angle of incidence of ray, $\psi$ |
|---|---|---|
| <br>(1) Externally reflected ray | All $m_1$ | $\psi = 0$ |
| <br>(2) Axial ray (one internal reflection) | All $m_1{}^*$ | $\psi = 0$ |
| <br>(3a) Glory ray (one internal reflection) | $\sqrt{2} < m_{R1} \leq 2$ | $\cos\psi = \dfrac{m_{R1}^2 - 2}{2}$ |
| <br>(3b) Glory ray (two internal reflections) | $0 < m_{R1} \leq 2^{1/4\dagger}$ | $\sin^2\psi = \dfrac{m_{R1}^2}{2}\left[1 - \sqrt{\dfrac{3 - m_{R1}^2}{3}}\cos\dfrac{\beta}{3} - 120°\right]$<br>where $\beta = + \cos^{-1}\left[\dfrac{3m_{R1}^2 - 3}{3 - m_{R1}^2}\sqrt{\dfrac{3}{3 - m_{R1}^2}}\right]$ |
| <br>(4) Stationary ray (one internal reflection) | $m_{R1} < 2$ | $\sin\psi = \sqrt{\dfrac{4 - m_{R1}^2}{3}}$,<br>$\alpha = 4\sin^{-1}\left[\dfrac{\sin\psi}{m_{R1}}\right] - 2\psi$,<br>where $\alpha$ is the angle of the emergent stationary ray |

\* For $m_1$ near 2, the divergence factor $D(0)$ for the axial ray becomes infinite when geometrical-optics methods are used. In order to obtain a true expression for $D(0)$ in this neighborhood, physical optics must be used instead.

† Actually, for $\frac{2}{3} < m_{R1} < 2^{1/4}$, there are two such twice-reflected glory rays, one entering and leaving at $\psi < 77.2°$ and the other at $77.2° < \psi < 90°$. At $m_{R1} = \frac{2}{3}$, for example, there are two such glory rays, one at $\psi = 65°$ and the other at $\psi = 90°$. However,

## Approach for Backscattering

| $D(0)$ | $R(0)$ | $A(0)$ | $\phi$ |
|---|---|---|---|
| $1$ | $-\dfrac{m_1-1}{m_1+1}$ | $1$ | $0$ |
| $\dfrac{m_1}{2-m_1}$ | $\dfrac{4m_1(m_1-1)}{(m_1+1)^2}$ | $\exp\left[-4k_0 a m_{I1}\right]$ | $4k_0 a m_{R1} - \pi$ |
| $\sqrt{\dfrac{\pi}{2}\,k_0 a}\,\dfrac{m_1^2(m_1^2-2)^{1/2}}{4-m_1^2}$ | $\dfrac{(5m_1^2-12)(m_1^2-2)}{2(m_1^2-1)^3}$ | $\exp\left[-2k_0 a m_{I1}\right]$ | $(m_{R1}^2+4)k_0 a - \pi$ |
| $\sqrt{2\pi k_0 a\,\cos\psi\,\sin^2\psi}\times\sqrt{m_{R1}^2-\sin^2\psi}$ $\times\dfrac{\mid 6\cos\psi - \sqrt{m_{R1}^2-\sin^2\psi}\mid}{\mid 3\cos\psi - \sqrt{m_{R1}^2}\ \sin^2\psi\mid}$ | $-\dfrac{R_\parallel^2(\psi)[1-R_\parallel^2(\psi)]}{2}$ $+\dfrac{R_\perp^2(\psi)[1-R_\perp^2(\psi)]}{2}$ | $\exp\left[-6k_0 a\,\dfrac{m_{I1}}{m_{R1}}\right.$ $\left.\times\sqrt{m_{R1}^2-\sin^2\psi}\right]$ | $2k_0 a\,(1-\cos\psi) + 6k_0 a$ $\times\sqrt{m_{R1}^2-\sin^2\psi} - \pi$ |
| $4(1+\cos\alpha)\sqrt{\dfrac{r_0}{a}\,\sin\psi}\left[\dfrac{m_{R1}^2-1}{18\cos\alpha}\right.$ $\times\left.\sqrt{\dfrac{m_{R1}^2-1}{4-m_{R1}^2}}\right]^{1/3}(k_0 a)^{2/3}\,\mathrm{Ai}(-z),^{\dagger\dagger}$ where $z=\left[\dfrac{4(m_{R1}^2-1)}{9\cos\alpha}\sqrt{\dfrac{m_{R1}^2-1}{4-m_{R1}^2}}\right]^{1/3}$ $\times (k_0 a)^{2/3}\,\sin\alpha,$ and $\dfrac{r_0}{a} = \sin(\alpha+\psi) + [4(m_{R1}$ $-\sqrt{m_{R1}^2-\sin^2\psi}) - (1-\cos\psi)]\sin\alpha$ | $-\dfrac{R_\parallel(\psi)[1-R_\parallel^2(\psi)]}{2}$ $+\dfrac{R_\perp(\psi)[1-R_\perp^2(\psi)]}{2}$ | $\exp\left[-4k_0 a\,\dfrac{m_{I1}}{m_{R1}}\right.$ $\left.\times\sqrt{m_{R1}^2-\sin^2\psi}\right]$ | $k_0 a\,\{\cos(\alpha+\psi) + (1-\cos\psi)$ $-4\sqrt{m_{R1}^2-\sin^2\psi}$ $-(1-\cos\alpha)$ $\times[4(m_{R1}-\sqrt{m_{R1}^2-\sin^2\psi})$ $-(1-\cos\psi)]\} + 3\pi/2$ |

the second set of rays closer to grazing is neglected here with respect to the first set because (a) the range of $m_{R1}$ in which the second set occurs is very small, and (b) the amount of energy transmitted into the sphere in rays in this second set is very small since the transmission coefficient approaches zero at grazing.

‡ The Airy function $\mathrm{Ai}(z)$ is defined and its properties listed in Reference 3, pp. 446–455.

contribution of this ray to the backscattering cross section is not constant in magnitude as a function of $k_0 a$, as are those for the axial and external rays obtained from geometric optics; the contribution of this ray increases in magnitude in proportion to $k_0 a$.

The angle of the ring, $\psi$, at which the rays enter and leave the sphere from the axis (see Table 3-2) depends upon the dielectric constant and the number of internal reflections. In many cases, glory rays exist only for certain limited ranges of dielectric constants. These ranges are given in Table 3-2, as well as the equations for computing the angle of incidence, $\psi$, for the glory ray ring.

(4) *The Stationary or Rainbow Rays.* These rays produce significant contributions in both the back and bistatic directions. They are strongest at a bistatic angle $\alpha$, as shown in Table 3-2. $\alpha$ is the angle at which the rate of change of scattering angle, $\theta'$, of the emerging ray as a function of the incident ray angle, $\psi$, is equal to zero. As shown in Table 3-2, the ray $b$ which enters at $\psi$ exits at $\alpha$; all other rays entering above and below $\psi$ exit at angles less than $\alpha$. Hence, the rays pass through a "stationary" point at direction $\alpha$. Consequently, at angles near and less than $\alpha$, this ray produces a strong contribution. If $\alpha$ is near zero (in general, less than 20°), this ray can contribute in the backscattering direction. In that case, the contribution is a ring of light, as with the glory ray, due to axial symmetry. In the general bistatic case, however, the light emerges as a spot rather than a ring; although the intensity of the scattered field is strongest at bistatic angle $\alpha$, this ray contributes at bistatic angles, $\theta'$ near $\alpha$, especially less than $\alpha$.

The phase front near the stationary direction is cubic in nature. In order to take this effect into account, physical optics is used. The resulting contribution, hence, contains Airy functions. The magnitude of this component in the backscattering direction is not a constant, but increases with $k_0 a$, as does the glory-ray contribution. As with the glory ray, stationary rays exist only for certain ranges of dielectric constant. These ranges are given in Table 3-2. Even though a stationary ray exists, it may not contribute at backscattering; only when $\alpha < 20°$ will its contribution be significant.

The stationary-ray theory explains the existence of the rainbow. The first rainbow occurs at $\alpha \approx 42°$ from backscattering. This corresponds to one internal reflection of light rays inside the raindrop. Higher-order internal reflections account for weaker rainbows at larger values of $\alpha$, but these are rarely visible.

(5) *Bistatic Geometric Optic Rays.* These are rays emerging in a given direction $\theta'$, $\phi'$, whose behavior can be predicted entirely by geometric optics. Their phase fronts are regular (unlike the "ring" or toroidal fronts of the glory and stationary rays in the backscattering direction), and, consequently, ray geometry can be used to find the radii of curvature of these phase fronts. Singly reflected internal bounce rays are shown and considered in Table 3-3.

The scattering cross section of a sphere, when $N$ rays are included, can be written as

$$\sigma_{\substack{\parallel \\ \perp}}(\theta') = \frac{4\pi}{k_0^2} \left| S_{\substack{1 \\ 2}}^{(1)}(\theta')\, e^{j\phi_1} + S_{\substack{1 \\ 2}}^{(2)}(\theta')\, e^{j\phi_2} + \cdots + S_{\substack{1 \\ 2}}^{(N)}(\theta')\, e^{j\phi_N} \right|^2 \quad (3.3\text{-}14)$$

where $\sigma_{\substack{\parallel \\ \perp}}(\theta')$ represents either the $E$-plane (upper subscript) or $H$-plane (lower subscript) cross section. $S_{\substack{1 \\ 2}}^{(i)}(\theta')$ is the scattering amplitude for the $E$ or $H$ plane, while $\phi_i$ is its phase angle with respect to the externally reflected ray. For backscattering ($\theta' = 0$) and forward scattering ($\theta' = \pi$), the $E$- and $H$-plane cross sections are identical. For bistatic scattering in any other direction at any other polarization, a scattering cross section can be obtained by vector addition [see Eq. (3.1-8) and (3.1-9)].

If one is interested merely in the average cross section and not the oscillatory behavior of the curve, Eq. (3.3-14) may be approximated by

$$\left\langle \sigma_{\substack{\parallel \\ \perp}}(\theta') \right\rangle = \frac{4\pi}{k_0^2} \left[ \left| S_{\substack{1 \\ 2}}^{(1)}(\theta') \right|^2 + \left| S_{\substack{1 \\ 2}}^{(2)}(\theta') \right|^2 + \cdots + \left| S_{\substack{1 \\ 2}}^{(N)}(\theta') \right|^2 \right] \quad (3.3\text{-}15)$$

The scattering amplitude factors, $S_{\substack{1 \\ 2}}^{(i)}(\theta')$, are further broken up in the tables into four quantities, as follows:

$$S_{\substack{1 \\ 2}}^{(i)}(\theta') = \left( \frac{k_0 a}{2} \right) D^{(i)}(\theta')\, R_{\substack{1 \\ 2}}^{(i)}(\theta')\, A^{(i)}(\theta') \quad (3.3\text{-}16)$$

These quantities are explained as follows:

$D^{(i)}(\theta')$ accounts for the divergence of scattered power in the scattering direction normalized by the divergence of rays scattered externally from a sphere. It depends upon the curvature and nature of the phase fronts of the scattered field, and is determined either by geometric or physical optics for the various rays.

$R_{\substack{1 \\ 2}}^{(i)}(\theta')$ includes all of the reflection and transmission coefficients of the rays upon striking, being reflecting by, and being transmitted through the sphere surface. For bistatic scattering in the $E$ plane, the $E$ field is always polarized parallel to the plane of incidence, and the Fresnel reflection and transmission coefficients for parallel polarization are used. In the $H$ plane, the $E$ field is always perpendicular to the incidence plane, and the Fresnel coefficients for perpendicular polarization are used. For backscattering and forward scattering in the case of the glory and stationary rays, the incident rays enter and leave the sphere on a ring; the polarization varies on this ring between parallel and perpendicular. In these cases, an average is made around the ring of the two sets of Fresnel coefficients (see References 17, 19, 20, 21 for elaboration). These reflection coefficients in the tables are written explicitly for the simpler cases, while for the more complex cases, they are

### Table 3-3.　Modified Geometrical-Optics

| Type of ray and diagram | Range of $m_1$ and $\theta$ for which ray exists | Angle of incidence, $\psi$ |
|---|---|---|
| <br>(1) Externally reflected ray | All $m_1$ | $\psi = \dfrac{\theta'}{2}$ |
| <br>(2) Bistatic transmitted ray | $2\sin^{-1}\left[\dfrac{1}{m_{R1}}\right] < \theta' < \pi^*$ for $m_{R1} > 1$ | $\sin\psi = \dfrac{m_{R1}\cos(\theta'/2)}{\sqrt{m_{R1}^2 - 2m_{R1}\sin(\theta/2) + 1}}$ |
| <br>(3) Bistatic single-reflected ray | $0 < \theta' < \pi - 4\sin^{-1}\left[\dfrac{1}{m_{R1}}\right]^*$ for $m_{R1} > 1$ | One must solve the following transcendental equation for $\psi$ in terms of $\theta$<br>$\theta' = 4\sin^{-1}\left[\dfrac{1}{m_{R1}}\sin\psi\right] - 2\psi$ |
| <br>(4) Bistatic stationary ray | $m_{R1} < 2$ | $\sin\psi = \sqrt{\dfrac{4 - m_{R1}^2}{3}}$,<br>$\alpha = 4\sin^{-1}\left[\dfrac{\sin\psi}{m_{R1}}\right] - 2\psi$,　where<br>$\alpha$ is the angle of the emergent stationary ray. |

* This approach is valid also for $|m_{R1}| < 1$ (e.g., inside a plasma sphere). In that case, see Reference 19 for alternate equations for $|m_{R1}| < 1$.

## Approach for Bistatic Scattering

| $D(\theta')$ | $R(\theta')$ | $A(\theta')$ | $\phi$ |
|---|---|---|---|
| $1$ | $\mp R_1\!\left(\dfrac{\theta'}{2}\right)$ | $1$ | $2k_0 a \cos(\theta'/2)$ |
| $\sqrt{\dfrac{\sin 2\psi}{\sin \theta'}} \cdot \dfrac{\sqrt{m^2_{R1} - \sin^2\psi}}{\lvert \cos\psi - \sqrt{m^2_{R1} - \sin^2\psi}\rvert}$ | $\mp \cdot [1 - R^2_\perp(\psi)]$ | $\exp\left[-2k_0 a \dfrac{m_{I1}}{m_{R1}} \times \sqrt{m^2_{R1} - \sin^2\psi}\right]$ | $2k_0 a (\cos\psi - \sqrt{m^2_{R1} - \sin^2\psi})$ |
| $\sqrt{\dfrac{\sin 2\psi}{\sin \theta'}} \dfrac{\sqrt{m^2_{R1} - \sin^2\psi}}{\lvert 2\cos\psi - \sqrt{m^2_{R1} - \sin^2\psi}\rvert}$ | $\mp R_\perp(\psi)[1 - R^2_\perp(\psi)]$ | $\exp\left[-4k_0 a \dfrac{m_{I1}}{m_{R1}} \times \sqrt{m^2_{R1} - \sin^2\psi}\right]$ | $2k_0 a (\cos\psi - 2\sqrt{m^2_{R1} - \sin^2\psi})$ |
| $\sqrt{\dfrac{\sin \psi}{\sin \theta'}} \cdot \dfrac{\sqrt{m^2_{R1} - \sin^2\psi}}{\lvert 4\cos\psi - \sqrt{m^2_{R1} - \sin^2\psi}\rvert}$ $\times 2\sqrt{\dfrac{2}{\pi}}(1 + \cos\alpha)\left[\dfrac{m^2_{R1} - 1}{9\cos(\alpha - \theta')}\right.$ $\times \left.\sqrt{\dfrac{m^2_{R1} - 1}{4 - m^2_{R1}}}\,\right]^{1/3} (k_0 a)^{1/6}\, \mathrm{Ai}(-z),\dagger$ where $z = \left[\dfrac{4(m^2_{R1} - 1)}{9\cos(\alpha - \theta')} \sqrt{\dfrac{m^2_{R1} - 1}{4 - m^2_{R1}}}\right]^{1/3}$ $\times (k_0 a)^{2/3}\, \sin(\alpha - \theta')$ | $\mp R_1(\psi)[1 - R^2_\perp(\psi)]$ | $\exp\left[-4k_0 a \dfrac{m_{I1}}{m_{R1}} \times \sqrt{m^2_{R1} - \sin^2\psi}\right]$ | $k_0 a \{\cos(\alpha + \psi) + (1 - \cos\psi)$ $- 4\sqrt{m^2_{R1} - \sin^2\psi}$ $- (1 - \cos\alpha)$ $\times [4(m_{R1} - \sqrt{m^2_{R1} - \sin^2\psi})$ $- (1 - \cos\psi)]\} + 3\pi/2$ |

† The Airy function $\mathrm{Ai}(z)$ is defined and its properties listed in Reference 3, pp. 446–455.

given in terms of $R_{\parallel}(\theta')$ and $R_{\perp}(\theta')$. The Fresnel reflection coefficients are defined and discussed in Chapter 7.

$A^{(i)}(\theta')$ takes into account the attenuation of the rays inside the sphere due to finite conductivity of the sphere material. For completely lossless material, this factor is unity. For slightly lossy spheres, this factor depends exponentially upon total path length of the ray inside the sphere and upon the imaginary part of the refractive index $m_{I1}$.

3.3.2.2.1. *Backscattering.* Table 3-2 gives the amplitude factors and phase angles of the various rays for backscattering. These, along with Eqs. (3.3-14) and (3.3-16), determine the cross section. Only rays reflected internally a minimal number of times are considered here.

In order to show the behavior and magnitudes of four of the various backscattered rays, Figure 3-11 has been plotted.[19] The dielectric material is lossless with $\epsilon_1 = 2.592$ ($m_{R1} = 1.61$). It is evident that for large $k_0 a$, the stationary ray and the glory ray contribute much more than the constant-magnitude external and axial rays. From Figure 3-11 and Eq. (3.3-15), one can compute an average total backscattering cross section versus $k_0 a$ for this dielectric sphere.

3.3.2.2.2. *Bistatic Scattering.* Table 3-3 gives the amplitude factors and phase angle of the various rays at bistatic angle $\theta'$ in either the $E$ plane (upper subscripts and signs) or in the $H$ plane (lower subscripts and signs). Near the

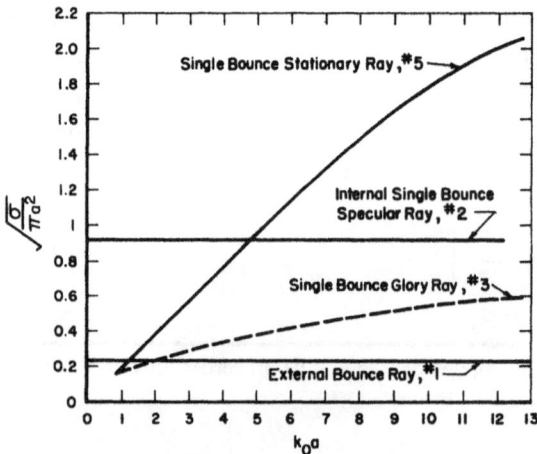

Fig. 3-11. Relative magnitudes of the various first-order rays predicted by geometrical diffraction theory contributing to the backscattered field as a function of relative sphere size for a homogeneous sphere with index of refraction $m_1 = 1.61$ [after Thomas[19]].

forward direction ($\theta' \to \pi$), however, all of the contributions from these rays will be dwarfed by the near-forward diffraction pattern for large $k_0 a$. This diffracted field pattern is given by Eqs. (3.2-36) and (3.2-37).

*3.3.2.2.3. Forward Scattering.* Table 3-4 gives the amplitude factors and phase angles of the various rays in the forward direction. The choice of subscripts is not present since the polarization of the rays is the same as that of the incident field. The first four rays in the table are the rays which penetrate the sphere and reemerge in the forward direction; these are treated by the modified geometrical optics approach. Also, there must be included two other components. These are listed in the table as (5) and (6).

Ray (3) shows two possible double-bounce glory rays; one exists for $2 < m_{R1} < 3$ and the other only for $m_{R1} < 1$. Ray (2) exists only for $m_{R1} < 1$.

Ray (5) is an edge-reflected component. This component is different from and in addition to the diffraction component. It cannot be predicted on the basis of ray optics as were rays (1) through (4). Reference 10 (pp. 362–364) discusses this component and derives it in a semi-empirical manner.

Ray (6) is the diffraction component. It dominates all of the others for large $k_0 a$, regardless of the sphere material.

### 3.3.2.3. Large Sphere Limit, $k_0 a > 1.15/m_{I1}$ (Reflective Region)

In this region, the paths through the sphere are sufficiently long that the refracted internal rays are attenuated due to the loss, or imaginary part of the refractive index, $m_{I1}$. Thus all rays which pass through the sphere are negligible, and only those which are externally reflected or diffracted are of importance. Hence, one usually notes a marked drop in scattering cross section when he passes into this region for any other than the forward direction.

*3.3.2.3.1. Backscattering.* In this case, ray (1) is the reflected ray (the only one which does not pass into the sphere). Therefore, the limiting value of the cross section is found from Eq. (3.3-14):

$$\sigma(0) = \pi a^2 \, | \, R(0)|^2 = \pi a^2 \left| \frac{m_1 - 1}{m_1 + 1} \right|^2 \qquad (3.3\text{-}17)$$

It is constant in value. For large refractive index, the cross section becomes $\pi a^2$, as seen from (3.3-17).

*3.3.2.3.2. Bistatic Scattering.* Again, only ray (1), the externally reflected ray is unattenuated since it does not pass into the sphere. Hence, the limiting value of the scattering amplitudes are (the phase angle is the same for both)

$$S_1(\theta') = \frac{k_0 a}{2} R_{\parallel} \left( \frac{\theta'}{2} \right) \qquad (3.3\text{-}18)$$

Table 3-4.   Modified Geometrical-Optics

| Type of ray and diagram | Range of $m_1$ for which ray exists | Angle of incidence of ray, $\psi$ |
|---|---|---|
| (1) Directly transmitted axial ray | All $m_1$ | $\psi = 0$ |
| (2) Single internally reflected glory ray | $m_{R1} < 1$ | $\psi = \sin^{-1}\left[\dfrac{m_{R1}^2}{4}\left(1 + \sqrt{\dfrac{8}{m_{R1}^2} + 1}\right)\right]$ |
| (3) Double internally reflected glory ray | $m_{R1} < 1$<br>$2 < m_{R1} < 3$ | $\sin\psi = \dfrac{m_{R1}}{2}\sqrt{3 + m_{R1}}$, for $m_{R1} < 1$<br><br>$\sin\psi = \dfrac{m_{R1}}{2}\sqrt{3 - m_{R1}}$, for $2 < m_{R1} < 3$ |
| (4) Double internally reflected axial ray | All $m_1$* | $\psi = 0$ |
| (5) Externally reflected component | All $m_1$ | Reflected from shadow boundary |
| (6) Diffracted component | All $m_1$ | Diffracted at shadow boundary |

* For $m_1$ near 3, the divergence factor $D(\pi)$ for this axial ray becomes infinite when geometric optics methods are used. In order to obtain a true expression for $D(\pi)$ in this neighborhood, physical optics must be used instead.

## Approach for Forward Scattering

| $D(\pi)$ | $R(\pi)$ | $A(\pi)$ | $\phi$ |
|---|---|---|---|
| $\dfrac{m_{R1}}{1 - m_{R1}}$ | $\dfrac{4m_{R1}}{m_{R1}^2 + 1}$ | $\exp[-2k_0 a m_{I1}]$ | $2k_0 a(m_{R1} - 1)$ |
| $\sqrt{4\pi k_0 a \sin^2\psi \cos\psi \cdot} \\ \times \dfrac{\sqrt{m_{R1}^2 - \sin^2\psi}}{\mid 2\cos\psi - \sqrt{m_{R1}^2 - \sin^2\psi}\mid}$ | $-\dfrac{R_\parallel(\psi)\,[1 - R_\parallel^2(\psi)]}{2} \\ -\dfrac{R_\perp(\psi)\,[1 - R_\perp^2(\psi)]}{2}$ | $\exp\left[-4k_0 a \dfrac{m_{I1}}{m_{R1}} \\ \times \sqrt{m_{R1}^2 - \sin^2\psi}\right]$ | $2k_0 a\left[2\sqrt{m_{R1}^2 - \sin^2\psi} - \cos\psi\right] + \pi/4$ |
| $\sqrt{4\pi k_0 a \sin^2\psi \cos\psi \cdot} \\ \times \dfrac{\sqrt{m_{R1}^2 - \sin^2\psi}}{\mid 3\cos\psi - \sqrt{m_{R1}^2 - \sin^2\psi}\mid}$ | $\dfrac{R_\parallel^2(\psi)\,[1 - R_\parallel^2(\psi)]}{2} \\ + \dfrac{R_\perp^2(\psi)\,[1 - R_\perp^2(\psi)]}{2}$ | $\exp\left[-6k_0 a \dfrac{m_{I1}}{m_{R1}} \\ \times \sqrt{m_{R1}^2 - \sin^2\psi}\right]$ | $2k_0 a\left[3\sqrt{m_{R1}^2 - \sin^2\psi} - \cos\psi\right] - \dfrac{5\pi}{4}$, <br> for $m_{R1} < 1$ <br> $2k_0 a\left[3\sqrt{m_{R1}^2 - \sin^2\psi} - \cos\psi\right] - \dfrac{7\pi}{4}$, <br> for $2 < m_{R1} < 3$ |
| $\dfrac{m_{R1}}{3 - m_{R1}}$ | $\dfrac{4m_{R1}(m_{R1} - 1)^2}{(m_{R1} + 1)^4}$ | $\exp[-6k_0 a m_{I1}]$ | $2k_0 a(3m_{R1} - 1)$ |
| $\dfrac{3}{2}(k_0 a)^{1/3}$ | $1$ | $1$ | $-\dfrac{5\pi}{6}$ |
| $k_0 a$ | $1$ | $1$ | $-\dfrac{\pi}{2}$ |

As with (3.3-17), for large values of refractive index, the Fresnel coefficients both approach unity in absolute value, and the bistatic cross section is the same as that for backscattering, $\pi a^2$.

3.3.2.3.3. *Forward Scattering.* In this case, all of the rays considered pass through the sphere and are attenuated. Only the reflective contribution, ray (5), and the diffraction contribution, ray (6), remain. They are, respectively,

$$\sigma_R(\pi) = \pi a^2 (9/4)(k_0 a)^{2/3} \qquad (3.3\text{-}19)$$

and

$$\sigma_D(\pi) = \pi a^2 (k_0 a)^2. \qquad (3.3\text{-}20)$$

For large $k_0 a$, only $\sigma_D(\pi)$ is important.

### 3.3.3. Approximate Solutions—Far Field—Rayleigh–Gans Region ($|m_1 - 1| \ll 1, 2k_0 a\,|m_1 - 1| \ll 1$)

The significant difference between this region and the regular Rayleigh region for dielectric spheres is that here, so long as the above two conditions are satisfied (to a reasonable degree), the parameter $k_0 a$ does not have to be much less than unity, as was necessary in the Rayleigh region; as a matter of fact, it may be greater than unity.

### 3.3.3.1. *Bistatic Scattering*

The bistatic scattering amplitude functions are derived in Reference 10 and are given here for the Rayleigh–Gans region,

$$S_1(\theta') = 2(k_0 a)^3 (m_1 - 1) \left[ \frac{j_1[2k_0 a\,\cos(\theta'/2)]}{2k_0 a\,\cos(\theta/2)} \right] \begin{bmatrix} \cos\theta' \\ 1 \end{bmatrix} \quad (3.3\text{-}21)$$

or in terms of trigonometric functions,

$$S_1(\theta') = 2(k_0 a)^3 (m_1 - 1)$$
$$\times \frac{\sin[2k_0 a\,\cos(\theta'/2)]/[2k_0 a\,\cos(\theta'/2)] - \cos[2k_0 a\,\cos(\theta'/2)]}{[2k_0 a\,\cos(\theta'/2)]^2} \begin{bmatrix} \cos\theta' \\ 1 \end{bmatrix}$$
$$(3.3\text{-}22)$$

From the above results, it can be seen that the larger the value of $k_0 a$, the more the energy is concentrated in the forward direction as circular fringes of scattered radiation.

### 3.3.3.2. *Backscattering*

When only the far-field backscattering cross section is sought, probably the simplest and least restrictive solution is that of Reference 22. This result has been derived directly from the Mie series solution by means of appropriate simplifications. The result is

$$\sigma(0) = \pi a^2 \left| \frac{m_1^2 - 1}{2} \right|^2 \left| \frac{2k_0 m_1 a \cos (2k_0 m_1 a) - \sin (2k_0 m_1 a)}{2k_0 m_1 a} \right|^2 \quad (3.3\text{-}23)$$

When $k_0 m_1 a$ is extremely large (i.e., greater than 100), only the cosine term is important. The answer then becomes

$$\sigma(0) = \pi a^2 \left| \frac{m_1^2 - 1}{2} \right|^2 \cos^2 (2k_0 m_1 a) = \pi a^2 \,|\, m_1 - 1\,|^2 \cos^2 (2k_0 m_1 a) \quad (3.3\text{-}24)$$

where $m_1^2 - 1 \approx 2(m_1 - 1)$ for $|\, m_1 - 1\,| \ll 1$.

### 3.3.3.3. Forward Scattering

The forward scattering cross section is obtained by allowing $\theta'$ to approach $\pi$ in (3.3-21) or (3.3-22), and noting that $S_1(\pi) = -S_2(\pi)$ and $\sigma(\pi) = (4\pi/k_0^2)\,|\, S_1(\pi)|^2$, where

$$S_1(\pi) = \tfrac{2}{3}(m_1 - 1)(k_0 a)^3 = -S_2(\pi) \quad (3.3\text{-}25)$$

$$\sigma(\pi) = \pi a^2 (k_0 a)^4 \cdot (16/9)(m_1 - 1)^2 \quad (3.3\text{-}26)$$

## 3.4. RADIALLY STRATIFIED AND INHOMOGENEOUS SPHERES

The following form of the Mie coefficients will be used throughout this section:[†]

$$A_n = -(-i)^n \frac{2n + 1}{n(n + 1)} \frac{k_0 a_0 j_n(k_0 a_0) - i Z_n(k_0 a_0)[k_0 a_0 j_n(k_0 a_0)]'}{k_0 a_0 h_n(k_0 a_0) - i Z_n(k_0 a_0)[k_0 a_0 h_n(k_0 a_0)]'} \quad (3.4\text{-}1)$$

$$B_n = (i)^n \frac{2n + 1}{n(n + 1)} \frac{k_0 a_0 j_n(k_0 a_0) - i Y_n(k_0 a_0)[k_0 a_0 j_n(k_0 a_0)]'}{k_0 a_0 h_n(k_0 a_0) - i Y_n(k_0 a_0)[k_0 a_0 h_n(k_0 a_0)]'} \quad (3.4\text{-}2)$$

where $a_0$ is the outermost radius of the sphere (see Figure 3-12 or Figure 3-15). The quantities, $Z_n(k_0 a_0)$ and $Y_n(k_0 a_0)$, are defined as the normalized $n$th-order modal surface impedance and admittance.[23,24] These impedances and admittances will be given and discussed for each type of sphere considered.[‡]

---

[†] In the following equations, $h_n(x)$ will be used to represent the spherical Hankel function of the first kind, $h_n^{(1)}(x)$.

[‡] This form of the Mie coefficients could have been used for the homogeneous dielectric sphere of the preceding section. In that case, $a_0 = a$, and the impedances and admittances are

$$Z_n(k_0 a_0) = - i \frac{\mu_1 k_0}{\mu_0 k_1} \frac{k_1 a_0 j_n(k_1 a_0)}{[k_1 a_0 j_n(k_0 a_0)]'} \quad \text{and} \quad Y_n(k_0 a_0) = - i \frac{\mu_0 k_1}{\mu_1 k_0} \frac{k_1 a_0 j_n(k_1 a_0)}{[k_1 a_0 j_n(k_1 a_0)]'}$$

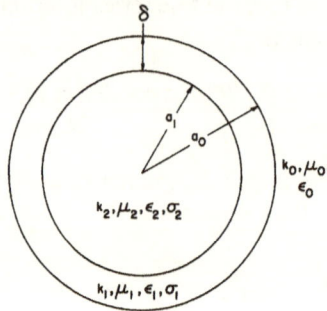

Fig. 3-12.  Scattering geometry for two-layer radially stratified sphere.

### 3.4.1.  Two Concentric Spheres

#### 3.4.1.1.  *Exact Solution*

The normalized modal surface impedances and admittances, $Z_n(k_0a_0)$ and $Y_n(k_0a_0)$, will be given for use in the Mie series solution. Although the forms of the coefficients are somewhat more complex than for homogeneous spheres, their numerical solution is nonetheless straightforward, using recursion techniques. The amount of computer time is increased over that of the homogeneous dielectric sphere only by a factor of about 2 [i.e., the time necessary to generate a new set of Bessel functions, $j_n(k_2a_1)$, $j_n(k_1a_1)$, $k_n(k_2a_1)$, $k_n(k_1a_1)$]. As will be seen, approximate solutions to this problem are for the most part nonexistent, and, hence, numerical techniques may well be the only method of arriving at any solution at all.

Two classes of problems will be discussed in this section; in the first case, the inner sphere is a dielectric; while in the second, it is a perfect conductor.[†] The latter problem is also referred to as a coated conducting sphere.

The surface impedances and admittances for the dielectric center sphere are[2]

$$Z_n(k_0a_0) = ik_0a_0\left(\frac{f_2}{f_1}\right), \qquad Y_n(k_0a_0) = ik_0a_0\left(\frac{f_4}{f_3}\right) \qquad (3.4\text{-}3)$$

where

$$f_2 = \frac{j_n(k_2a_1)}{\mu_0\mu_1}\, [[k_1a_1j_n(k_1a_1)]'\, h_n(k_1a_0) - [k_1a_1h_n(k_1a_1)]'\, j_n(k_1a_0)]$$

$$+ \frac{[k_2a_1j_n(k_2a_1)]'}{\mu_0\mu_2}\, [j_n(k_1a_0)\, h_n(k_1a_1) - j_n(k_1a_1)\, h_n(k_1a_0)] \qquad (3.4\text{-}4)$$

---

[†] The situation where the outer layer or coating is radially and continuously inhomogeneous (having a power-law variation of refractive index) is treated in Reference 24. Results are involved and not presented here for lack of space.

$$f_1 = \frac{j_n(k_2 a_1)}{\mu_1^2} \left[ [k_1 a_0 j_n(k_1 a_0)]' \, [k_1 a_1 h_n(k_1 a_1)]' - [k_1 a_1 j_n(k_1 a_1)]' \, [k_1 a_0 h_n(k_1 a_0)]' \right]$$

$$+ \frac{[k_2 a_1 j_n(k_2 a_1)]'}{\mu_1 \mu_2} \left[ j_n(k_1 a_1)[k_1 a_0 h_n(k_1 a_0)]' - h_n(k_1 a_1)[k_1 a_0 j_n(k_1 a_0)]' \right]$$

$$(3.4\text{-}5)$$

$$f_4 = \frac{k_1^2 [k_2 a_1 j_n(k_2 a_1)]'}{\mu_1^2 k_2^2} \left[ j_n(k_1 a_0) \, h_n(k_1 a_1) - j_n(k_1 a_1) \, h_n(k_1 a_0) \right]$$

$$+ \frac{j_n(k_2 a_1)}{\mu_1 \mu_2} \left[ [k_1 a_1 j_n(k_1 a_1)]' \, h_n(k_1 a_0) - [k_1 a_1 h_n(k_1 a_1)]' \, j_n(k_1 a_0) \right] \qquad (3.4\text{-}6)$$

$$f_3 = \frac{k_0^2 [k_2 a_1 j_n(k_2 a_1)]'}{\mu_0 \mu_1 k_2^2} \left[ j_n(k_1 a_1)[k_1 a_0 h_n(k_1 a_0)]' - h_n(k_1 a_1)[k_1 a_0 j_n(k_1 a_0)]' \right]$$

$$+ \frac{k^2 j_n(k_2 a_1)}{\mu_0 \mu_2 k_1^2} \left[ [k_1 a_0 j_n(k_1 a_0)]' \, [k_1 a_1 h_n(k_1 a_1)]' \right.$$

$$\left. - [k_1 a_1 j_n(k_1 a_1)]' \, [k_1 a_0 h_n(k_1 a_0)]' \right]$$

$$(3.4\text{-}7)$$

The above expressions simplify when the inner sphere, or core, is perfectly conducting. In that case, $\sigma_2 \to \infty$ and $k_2 \to +\infty$. Then the surface impedances and admittances for a perfectly conducing core are

$$Z_n(k_0 a_0) = i k_0 a_0 \frac{\mu_1}{\mu_0} \frac{j_n(k_1 a_0) \, h_n(k_1 a_1) - j_n(k_1 a_1) \, h_n(k_1 a_0)}{j_n(k_1 a_1)[k_1 a_0 h_n(k_1 a_0)]' - h_n(k_1 a_1)[k_1 a_0 j_n(k_1 a_0)]'} \qquad (3.4\text{-}8)$$

$$Y_n(k_0 a_0) = i k_0 a_0 \frac{\mu_0 k_1^2}{\mu_1 k_0^2}$$

$$\times \frac{h_n(k_1 a_0)[k_1 a_1 j_n(k_1 a_1)]' - j_n(k_1 a_0)[k_1 a_1 h_n(k_1 a_1)]'}{[k_1 a_0 j_n(k_1 a_0)]' \, [k_1 a_1 h_n(k_0 a_1)]' - [k_1 a_1 j_n(k_1 a_1)]' \, [k_1 a_0 h_n(k_1 a_0)]'}$$

$$(3.4\text{-}9)$$

3.4.1.2. *Approximate Solutions*

3.4.1.2.1. *Low-Frequency Region* ($k_0 a_0 < 0.5$). In this region (Rayleigh region), the Mie coefficients [Eqs. (3.4-1) and (3.4-2)] can be expanded in terms of $k_0 a_0$, and when this quantity is sufficiently small, only the lowest power need be considered. Then only $A_1$ and $B_1$ contain this lowest power, $(k_0 a_0)^3$, and consequently are the only coefficients of importance in this region. They are

$$A_1 = -\frac{(k_0 a_0)^3}{2} \left[ \frac{1 - 2[i Z_1(k_0 a_0)/k_0 a_0]}{1 + [i Z_1(k_0 a_0)/k_0 a_0]} \right] \qquad (3.4\text{-}10)$$

$$B_1 = -i \frac{(k_0 a_0)^3}{2} \left[ \frac{1 - 2[i Y_1(k_0 a_0)/k_0 a_0]}{1 + [i Y_1(k_0 a_0)/k_0 a_0]} \right] \qquad (3.4\text{-}11)$$

Normally, if the other arguments of the Bessel functions appearing in $Z_1(k_0 a_0)$ and $Y_1(k_0 a_0)$ are not small compared to unity (i.e., $k_1 a_0$ and $k_2 a_1$), then the functions $Z_1(k_0a_0)$ and $Y_1(k_0a_0)$ cannot be simplified, but must be computed from their defining equations [Eqs. (3.4-3) through (3.4-9)]. In that case, the various cross sections are

(i) *Backscattering Cross Section*

$$\sigma(0) = \frac{4\pi}{k_0^2} \mid A_1 + iB_1 \mid^2$$

$$= \pi a^2 (k_0 a_0)^4 \left| \frac{1 - 2[iZ_1(k_0a_0)/k_0a_0]}{1 + [iZ_1(k_0a_0)/k_0a_0]} - \frac{1 - 2[iY_1(k_0a_0)/k_0a_0]}{1 + [iY_1(k_0a_0)/k_0a_0]} \right|^2 \quad (3.4\text{-}12)$$

(ii) *Bistatic Scattering Amplitudes*[†]

$$S_1(\theta') = -(A_1 + iB_1 \cos \theta')$$

$$= \frac{(k_0a_0)^3}{2} \left[ \frac{1 - 2[iZ_1(k_0a_0)/k_0a_0]}{1 + [iZ_1(k_0a_0)/k_0a_0]} - \frac{1 - 2[iY_1(k_0a_0)/k_0a_0]}{1 + [iY_1(k_0a_0)/k_0a_0]} \cos \theta' \right]$$

$$(3.4\text{-}13)$$

$$S_2(\theta') = -(A_1 \cos \theta' + iB_1)$$

$$= \frac{(k_0a_0)^3}{2} \left[ \frac{1 - 2[iZ_1(k_0a_0)/k_0a_0]}{1 + [iZ_1(k_0a_0)/k_0a_0]} \cos \theta' - \frac{1 - 2[iY_1(k_0a_0)/k_0a_0]}{1 + [iY_1(k_0a_0)/k_0a_0]} \right]$$

$$(3.4\text{-}14)$$

(iii) *Forward Scattering Cross Section*

$$\sigma(\pi) = \frac{4\pi}{k_0^2} \mid A_1 - iB_1 \mid^2$$

$$= \pi a^2 (k_0 a_0)^4 \left| \frac{1 - 2[iZ_1(k_0a_0)/k_0a_0]}{1 + [iZ_1(k_0a_0)/k_0a_0]} + \frac{1 - 2[iY_1(k_0a_0)/k_0a_0]}{1 + [iY_1(k_0a_0)/k_0a_0]} \right|^2 \quad (3.4\text{-}15)$$

3.4.1.2.2. *Small Argument Simplifications* ($\mid k_1a_0 \mid < 0.5$, $\mid k_2a_1 \mid < 0.5$) —*Dielectric Core.* When the above conditions are met, as well as the condition $k_0a_0 < 0.5$, then the modal surface impedances and admittances of Eqs. (3.4-3) through (3.4-7) simplify. They are[‡]

$$Z_1(k_0a_0) = -i \frac{k_0a_0}{2} \frac{\mu_1}{\mu_0} \frac{\mu_2\gamma + 2\mu_1\alpha}{\mu_1\beta + \mu_2\alpha}$$

$$Y_1(k_0a_0) = -i \frac{k_0a_0}{2} \frac{k_1^2\mu_0}{k_0^2\mu_1} \frac{\mu_1(k_2^2/k_1^2)\gamma + 2\mu_2\alpha}{\mu_2\beta + \mu_1(k_2^2/k_1^2)\alpha} \quad (3.4\text{-}16)$$

where here

$$\alpha = 1 - \left(\frac{a_1}{a_0}\right)^3, \quad \beta = 2 + \left(\frac{a_1}{a_0}\right)^3, \quad \gamma = 2\left(\frac{a_1}{a_0}\right)^2 + 1 \quad (3.4\text{-}17)$$

---

[†] The *E*-plane and *H*-plane scattering cross sections are then obtained from $S_1(\theta')$ and $S_2(\theta')$, respectively, by using (3.1-10) with $\phi' = 0$ and (3.1-11) with $\phi' = \pi/2$.

[‡] These results are valid for any arbitrary sphere media. Either medium may be lossy ($\sigma_1$ or $\sigma_2$ different from zero) or have different permeabilities, $\mu_1$ and $\mu_2$, as well as permittivities, $\epsilon_1$ and $\epsilon_2$. The outer layer may be a plasma such that $\mid k_1 \mid < k_0$.

When $a_1/a_0$ approaches unity, i.e., the outer layer or coating is very thin, then $\alpha \to 0$, $\beta \to 3$, $\gamma \to 3$, and $Z_1(k_0a_0)$ and $Y_1(k_0a_0)$ are the same as for the dielectric sphere of wave number $k_2$ (the coating produces no effect on the low-frequency scattering properties).

3.4.1.2.3. *Small Argument Simplification* ($|\, k_1a_0\,| < 0.5$, $|\, k_2a_1\,| > 20$)— *Conducting or Highly Reflecting Core.* In this case the modal impedance and admittance is

$$Z_1(k_0a_0) = -ik_0a_0\frac{\mu_1}{\mu_0}\frac{\alpha}{\beta}, \qquad Y_1(k_0a_0) = -i\frac{k_0a_0}{2}\cdot\frac{k_1^2\mu_0}{k_0^2\mu_1}\cdot\frac{\gamma}{\alpha} \qquad (3.4\text{-}18)$$

where $\alpha$, $\beta$, and $\gamma$ are given in (3.4-17). Again, when the outer coating becomes vanishingly thin, $a_1/a_0 \to 1$, $\alpha \to 0$, $\beta \to 3$, $\gamma \to 3$, $Z_1(k_0a_0) \to 0$, and $Y_1(k_0a_0) \to \infty$, as they should for a perfectly conducting sphere.

For both of the above situations it can be seen that in the low-frequency limit it is possible to make the backscattering or forward scattering cross section zero, independent of frequency, by a proper choice of coating parameters ($a_0 - a_1$, $k_1$, $\mu_1$, $\epsilon_1$, $\sigma_1$).[†]

3.4.1.2.4. *Intermediate Region* ($k_0a_0 > 0.5$, $k_0\delta < 1.15/m_{I1}$, $k_0a_1 < 1.15/m_{I2}$). In this region, scattering can be thought of as produced by several mechanisms: external specular reflection, multiple internally reflected, and refracted rays, creeping waves, glory-type rays, stationary rays, etc. It was difficult to obtain reasonable approximations in this region for the homogeneous dielectric sphere of Section 3.3; it is beyond hope in the case at hand to present reasonably simple approximations which are valid over a wide range of parameters. Although it is possible to attack a given problem by a modified geometrical optics approach similar to that used in the preceding section, the complexity is generally great. Most people needing results prefer machine summation of the Mie series, using coefficients $A_n$ and $B_n$, as defined in Eqs. (3.4-1) through (3.4-9).

Several curves of backscattering cross section as a function of $a_0/\lambda$ (or $k_0a_0$) are shown here in Figure 3-13. They are obtained from the exact Mie solution. Figure 3-13 was obtained from Reference 26 and represents the backscattering cross section from a perfectly conducting core coated with a thin, pure dielectric layer (i.e., $\mu_1 = \mu_0$, $\sigma_1 = 0$). Here, the dielectric coating has a constant thickness ratio such that $\delta/a_0$ remains constant as $a_0$ or $k_0$ increases.

Figure 3-14 presents the backscattering cross section from melting ice spheres (hail stones), represented by an ice core of refractive index $m_2 = 1.78\,(1 + i0.001357)$, surrounded by a water layer which is 0.01 cm

---

[†] This assumes that $\mu_1$, $\epsilon_1$, and $\sigma_1$ are independent parameters, and are frequency independent. See Reference 25 for a detailed discussion of the effect of coatings on the low-frequency cross section of spheres and other bodies.

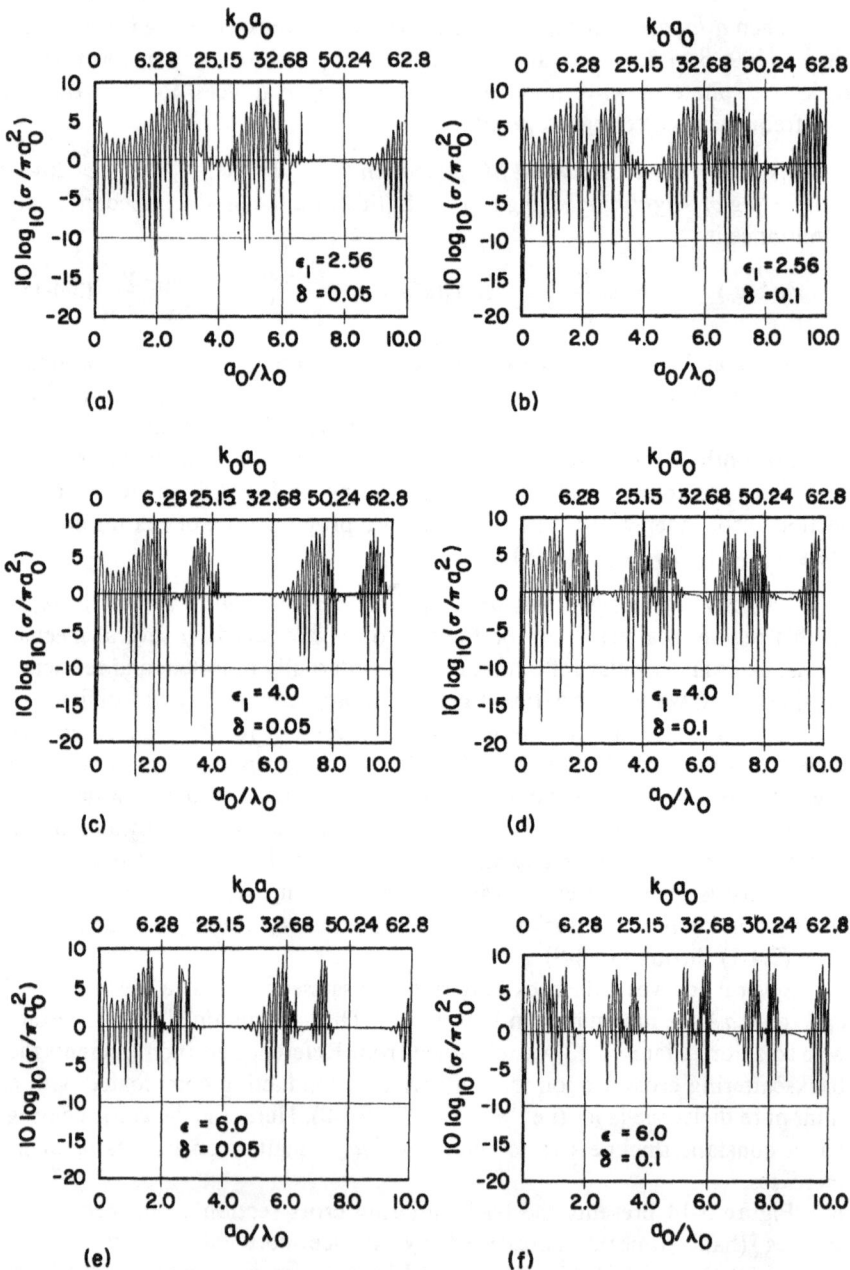

Fig. 3-13. Exact backscattering cross section of a perfectly conducting sphere coated with a lossless dielectric of relative permittivity, $\epsilon_1$, as a function of sphere size for various relative coating thicknesses and coating permittivities [after Rheinstein[26]].

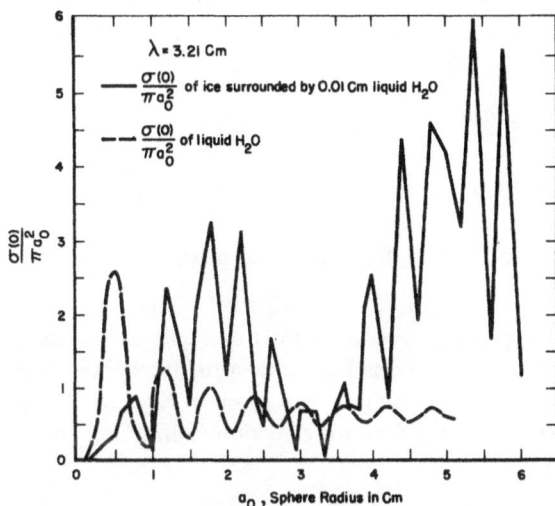

Fig. 3-14. Exact backscattering cross section for ice spheres (hail stones) surrounded by a layer of water 0.01 cm thick at wavelength $\lambda_0 = 3.21$ cm as a function of the inner sphere radius [after Herman and Battan[27]].

thick, having a refractive index $m_1 = 7.14 \ (1 + i0.4051)$ at wavelength $\lambda_0 = 3.21$ cm.[27] It is compared with the backscattering cross section of a homogeneous water drop or sphere of the same radius, $a_0$, with the above refractive index, $m_1$. Here, the layer thickness, $\delta$, is kept constant as the radius, $a_0$, increases.

3.4.1.2.5. *High-Frequency Limit—Scattering from the Core* ($k_0 a_0 > 20$, $k_0 a_1 > 1.15/m_{I2}$ *or Perfectly Conducting Core*). The above condition ensures that the center core is large enough so that all rays entering it are sufficiently attenuated, due to nonzero conductivity or loss, that they are negligible in magnitude when they leave the core. In this range, then, the scattered waves may be thought of as due to plane waves reflected from an infinite plane slab of refractive index $m_1$, thickness $\delta$, in front of a semi-infinite region of refractive index $m_2$; this slab is tangent to the sphere at the specular point (i.e., if the bistatic scattering angle of interest is $\theta'$, then the angle of incidence on the sphere at the specular point is $\theta'/2$). In this case, the plane wave reflection coefficients from a single-layered medium derived in Chapter 7 are used.

(i) *Backscattering Cross Section*

$$\sigma(0) = \pi a^2 \mid R_{\perp}(0)\mid^2 = \pi a^2 \mid R_{\parallel}(0)\mid^2 \tag{3.4-19}$$

where $R_{\perp}(\theta')$ and $R_{\parallel}(\theta')$ are obtained from Eqs. (7.1-33), (7.1-34), and (7.1-42). When the center core is perfectly conducting, the reflection coefficient at $\theta' = 0$ is given by Eq. (7.1-42), with $z_3 = 0$. It can be seen that these reflection coefficients vary between zero and unity, depending upon the value of the parameters $k_1$, $k_2$, $k_0$, and $a_1 - a_0$.[†]

(ii) *Bistatic Scattering Amplitudes*

$$S_1(\theta) = \frac{k_0 a}{2} R_\perp \left( \frac{\theta'}{2} \right), \qquad S_2(\theta) = \frac{k_0 a}{2} R_\parallel \left( \frac{\theta'}{2} \right) \qquad (3.4\text{-}20)$$

where $R_\perp(\theta'/2)$ and $R_\parallel(\theta'/2)$ have the same definition as above.

(iii) *Forward Scattering Cross Section.* Again, the high-frequency forward scattering consists predominantly of the field diffracted at the shadow boundary of the sphere, and is independent of the sphere composition. The forward scattering cross section is then given in Eq. (3.2-34), and the bistatic cross section very near the forward direction (diffraction pattern) is given by Eq. (3.2-37).

3.4.1.2.6. *High-Frequency Limit—Scattering From the Outermost Surface* $(k_0\delta > 1.15/m_{l1})$. In this region, the outer layer is so thick in terms of wavelength that rays penetrating the sphere are sufficiently attenuated in this layer that they are negligible by the time they reach the second layer or core. Hence, all rays which enter the outer layer never leave the sphere and the sphere appears homogeneous of only one material, that of the outer layer. Thus, it may be treated in the same manner as in the previous section.

(i) *Backscattering Cross Section*

$$\sigma(0) = \pi a^2 \mid R(0)\mid^2 \qquad (3.4\text{-}21)$$

(ii) *Bistatic Scattering Amplitudes*

$$S_1(\theta') = \frac{k_0 a}{2} R_\perp \left( \frac{\theta'}{2} \right), \qquad S_2(\theta) = \frac{k_0 a}{2} R_\parallel \left( \frac{\theta'}{2} \right) \qquad (3.4\text{-}22)$$

where $R_\perp(\theta')$ and $R_\parallel(\theta')$ are the plane-wave reflection coefficients for a semi-infinite homogeneous medium given by Eqs. (7.1-9) and (7.1-10).

(iii) *Forward Scattering Cross Section.* The results here are identical to those above under Section 3.4.1.2.5, Part (iii).

---

[†] Since the high-frequency scattering cross sections for the coated sphere vary with the reflection coefficients for plane waves incident upon a two-layer infinite plane medium, one can control the high-frequency scattering properties of the sphere by varying the parameters of the coating. For a discussion of the maximization and minimization of $R_\parallel(\theta')$ and $R_\perp(\theta')$ in this manner, see Chapter 8.

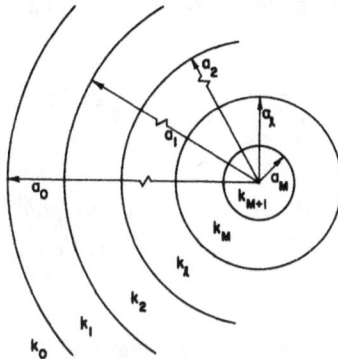

Fig. 3-15. Scattering geometry for an $M$-layer radially stratified sphere.

### 3.4.2. $M$–Concentric-Layered Sphere

This situation is the most general sphere problem for which all previous results are merely special cases (i.e., $M = 0$ and $a_1 = a_2 = \cdots = 0$ for the homogeneous zero-layer conducting and dielectric sphere, while $M = 1$ and $a_1 \neq 0$, but $a_2 = \cdots = 0$ for the one-layer sphere). Each of the layers is homogeneous and has concentric boundaries, as shown in Figure 3-15. This problem is of interest not only because it is a situation which can exist in practice, but also because it can be used to model radially inhomogeneous spheres; in the latter case, the more layers used (i.e., the greater the value $M$), the better is the model.[†]

#### 3.4.2.1. *Exact Solution*

The modal surface impedances, $Z_n(k_0 a_0)$, and admittances, $Y_n(k_0 a_0)$, to be used in the Mie coefficients of Eqs. (3.4-1) and (3.4-2) can be derived in a systematic manner using an iteration technique familiar in transmission-line theory. This iteration process is especially suited for machine computation. In general, the machine time required to determine the set of $Z_n(k_0 a_0)$ for $M$ layers is approximately $2M$ times as great as the time required for one layer, i.e., a homogeneous dielectric sphere; this is determined from the number of extra Bessel functions involved.

Define the surface impedance at the interface at radius $a_l$ as $Z_n^{(l)}$. First of all, $Z_n^{(M)}$ at the boundary between the core and the first layer is determined independently, and then from $Z_n^{(M)}$, the impedance $Z_n^{(M-1)}$ at the second interface from the center is determined. This process continues until $Z_n^{(0)}$ at the outer surface at $a_0$ is found; $Z_n(k_0 a_0)$ is then equal to $-i(1/\eta_0) Z_n^{(0)}$. The same

---

[†] See references 24 and 28 for discussions and application of this $M$-layered sphere model to the solution of scattering from radially inhomogeneous spheres.

process is used to determine $Y_n(k_0a_0) = -i\eta_0 Y_n^{(0)}$, starting with $Y_n^{(M)}$.[24]†
Define

$$P_n(x) = \frac{xj_n(x)}{[xj_n(x)]'} \quad \text{and} \quad Q_n(x) = \frac{xh_n^{(1)}(x)}{[xh_n^{(1)}(x)]'} \quad (3.4\text{-}23)$$

Also, define

$$U_n^{(l)} = \frac{[k_0m_{l+1}a_{l+1}j_n(k_0m_{l+1}a_{l+1})]'}{[k_0m_{l+1}a_l j_n(k_0m_{l+1}a_l)]'} \frac{[k_0m_{l+1}a_l h_n^{(1)}(k_0m_{l+1}a_l)]'}{[k_0m_{l+1}a_{l+1}h_n^{(1)}(k_0m_{l+1}a_{l+1})]'} \quad (3.4\text{-}24)$$

and

$$V_n^{(l)} = \frac{[k_0m_{l+1}a_{l+1}j_n(k_0m_{l+1}a_{l+1})]}{[k_0m_{l+1}a_l j_n(k_0m_{l+1}a_l)]} \frac{[k_0m_{l+1}a_l h_n^{(1)}(k_0m_{l+1}a_l)]}{[k_0m_{l+1}a_{l+1}h_n^{(1)}(k_0m_{l+1}a_{l+1})]} \quad (3.4\text{-}25)$$

where $k_l = k_0m_l$ . Then,

$$Z_n^{(M)} = \eta_{M+1}P_n(k_0m_{M+1}a_M) \quad \text{and} \quad Y_n^{(M)} = \frac{1}{\eta_{M+1}} P_n(k_0m_{M+1}a_M) \quad (3.4\text{-}26)$$

At the interface at radius $a_l$ , the impedance and admittance are

$$Z_n^{(l)} = \eta_{l+1}P_n(k_0m_{l+1}a_l) \left[1 - V_n^{(l)} \frac{1 - Z_n^{(l+1)}/\eta_{l+1}P_n(k_0m_{l+1}a_{l+1})}{1 - Z_n^{(l+1)}/\eta_{l+1}Q_n(k_0m_{l+1}a_{l+1})}\right]$$

$$\times \left[1 - U_n^{(l)} \frac{1 - \eta_{l+1}P_n(k_0m_{l+1}a_{l+1})/Z_n^{(l+1)}}{1 - \eta_{l+1}Q_n(k_0m_{l+1}a_{l+1})/Z_n^{(l+1)}}\right]^{-1} \quad (3.4\text{-}27)$$

$$Y_n^{(l)} = \frac{P_n(k_0m_{l+1}a_l)}{\eta_{l+1}} \left[1 - V_n^{(l)} \frac{1 - \eta_{l+1}Y_n^{(l+1)}/P_n(k_0m_{l+1}a_{l+1})}{1 - \eta_{l+1}Y_n^{(l+1)}/Q_n(k_0m_{l+1}a_{l+1})}\right]$$

$$\times \left[1 - U_n^{(l)} \frac{1 - P_n(k_0m_{l+1}a_{l+1})/\eta_{l+1}Y_n^{(l+1)}}{1 - Q_n(k_0m_{l+1}a_{l+1})/\eta_{l+1}Y_n^{(l+1)}}\right]^{-1} \quad (3.4\text{-}28)$$

where $l$ begins at $M - 1$ and ends at $l = 0$. Then,

$$Z_n(k_0a_0) = -i\frac{1}{\eta_0} Z_n^{(0)} \quad \text{and} \quad Y_n(k_0a_0) = -i\eta_0 Y_n^{(0)} \quad (3.4\text{-}29)$$

Hence, $Z_n(k_0a_0)$ and $Y_n(k_0a_0)$ for use in the Mie coefficients have been found.‡

---

† The results of Wait (reference 24), as used here, have been corrected where there were mistakes in sign.
‡ These modal surface impedances and admittances reduce to the proper values for $M = 0$ (homogeneous zero layer sphere) and for $M = 1$ (single layer or coated homogeneous spherical core).

### 3.4.2.2. *Approximate Solutions*

There are two regions where approximations may be obtained for the multilayer sphere.

3.4.2.2.1. *Low-Frequency Scattering* ($k_0 a_0 < 0.2$, $k_1 a_0 < 0.2$, $k_2 a_1 < 0.2$, etc.). In the low-frequency limit, an iteration technique is available for determining $Z_1(k_0 a_0)$ and $Y_1(k_0 a_0)$ for use in Eqs. (3.4-12) through (3.4-15). This iteration procedure is derivable from that of the preceding section by using small argument expansions for the Bessel functions, or it may be formulated more easily by the following method.

At the center core of radius $a_M$, one determines $f_M$ and $g_M$ as follows: (a) if the core is perfectly conducting, $f_M = 0$ and $g_M = \infty$; (b) if the core is a homogeneous material, then $f_M = (\mu_{M+1}/2) \cdot (a_M/a_0)$ and $g_M = \epsilon_{M+1}/2)(a_M/a_0)$. Then one proceeds to determine $f_{M-1}$, $g_{M-1}$, $f_{M-2}$, $g_{M-2}$,...., $f_l$, $g_l$,...., $f_1$, $g_1$, $f_0$, $g_0$, according to the following iteration technique:

$$K_l = \frac{(a_l/a_0)^3 \, [2f_l - \mu_l(a_l/a_0)]}{f_l + \mu_l(a_l/a_0)} \qquad (3.4\text{-}30)$$

and

$$f_{l-1} = \frac{(a_{l-1}/a_0)^3 + K_l}{2(a_{l-1}/a_0)^3 - K_l} \, [\mu_l(a_{l-1}/a_0)] \qquad (3.4\text{-}31)$$

The $g_l$, $g_{l-1}$, etc., functions are generated in exactly the same manner, but the $\mu_l$ are now replaced everywhere by $\epsilon_l$.

Then, the modal impedance and admittance at the outer surface are given by

$$\frac{iZ_1(k_0 a_0)}{k_0 a_0} = f_0, \qquad \frac{iY_1(k_0 a_0)}{k_0 a_0} = g_0 \qquad (3.4\text{-}32)$$

These may then be substituted into Eqs. (3.4-12) through (3.4-15).

If one of the parameters, $k_{l+1} a_l$, is not small, i.e., if that layer has a large refractive index, then the above method is not accurate. If $k_l a_{l+1}$ is much greater than unity due to a large permittivity or conductivity, then one can treat that layer as perfectly conducting, starting the iteration at $M = l$, and using the condition $f_l = 0$, $g_l = \infty$. If, on the other hand, $k_{l+1} a_l$ is neither very large nor very small, then one must use the exact method [i.e., Eqs. (3.4-23) to (3.4-29)] to find $Z_n(k_0 a_0)$ and $Y_n(k_0 a_0)$.

3.4.2.2.2. *High-Frequency Scattering* ($k_0 a_0 > 20$). In general, there is no simple approximate technique available unless rays which enter various layers are damped out sufficiently so that they may be neglected. The most obvious and trivial of these cases is when the rays which penetrate the first and second layers may be neglected.

(i) *External Scattering Only* $[k_0(a_0 - a_1) > 1.15/m_{I1}]$. In this case, rays which enter the outermost layers are sufficiently attenuated that they are neglected, and the equations of Section 3.4.1.2.6 apply.

(ii) *Scattering From the Second Layer* $[k_0(a_2 - a_1) > 1.15/m_{I2}]$. In .this this case, rays penetrate the outer surface and are reflected from the next interface, but those which penetrate that interface are attenuated and neglected. Then the equations of Section 3.4.1.2.5 apply.

### 3.4.3. Radially Inhomogeneous Spheres

Here, the wave number or refractive index of the sphere medium varies with radius $\rho$, i.e.,

$$k_1 = k_1(\rho) = \omega \sqrt{\mu_0 \epsilon_0} \, m_1(\rho) = k_0 m(\rho), \qquad \text{where} \quad \rho \leqslant a_0 \quad (3.4\text{-}33)$$

In the most general case, both $\epsilon_1$ and $\mu_1$ can be functions of $\rho$, and each may also be complex (i.e., involve loss). Thus,

$$m_1(\rho) = \sqrt{\epsilon_{r1}(\rho) \, \mu_{r1}(\rho)} \qquad (3.4\text{-}34)$$

#### 3.4.3.1. *Exact Solution*

Tai[29,30] and Garbacz[23,31] have shown that the Mie coefficients for the exact solution have the form of Eqs. (3.4-1) and (3.4-2), where now the normalized modal impedances, $Z_n(k_0 a_0)$, and admittances, $Y_n(k_0 a_0)$, are solutions of a Riccati differential equation. Let

$$z_n(k_0 a_0) = i Z_n(k_0 a_0), \qquad y_n(k_0 a_0) = i Y_n(k_0 a_0) \qquad (3.4\text{-}35)$$

and $x = k_0 \rho$; then $z_n$ and $y_n$ are solutions of the equations,

$$\frac{dz_n(x)}{dx} = \mu_{r1}(x) + \left[ \epsilon_{r1}(x) - \frac{n(n+1)}{x^2 \mu_{r1}(x)} \right] z_n^2(x) \qquad (3.4\text{-}36)$$

and

$$\frac{dy_n(x)}{dx} = \epsilon_{r1}(x) + \left[ \mu_{r1}(x) - \frac{n(n+1)}{x^2 \epsilon_{r1}(x)} \right] y_n^2(x) \qquad (3.4\text{-}37)$$

Although Eqs. (3.4-36) and (3.4-37) have been solved explicitly for certain distributions of $\epsilon_{r1}(x)$ and $\mu_{r1}(x)$, such as the spherical Luneberg lens,[29,30] in general, a solution is only obtainable numerically. Such a machine solution of the differential equation may in many cases be fast and desirable.

#### 3.4.3.2. *Approximate Solutions—Low-Frequency Region* $(k_0 a_0 < 0.5)$

In this region, only the lowest-order Mie coefficients, $A_1$ and $B_1$, are important. They reduce to the forms given in Eqs. (3.4-10) and (3.4-11). In this case, only $Z_1(k_0 a_0)$ and $Y_1(k_0 a_0)$ need be determined.

For certain of the more well-known radially inhomogeneous spheres, answers have already been obtained.[32]

### 3.4.3.2.1. *Luneberg Lens*

$$\mu_{r1}(\xi) = 1$$

$$\epsilon_{r1}(\xi) = \begin{cases} 2 - \xi^2 & \text{for} \quad 0 \leqslant \xi \leqslant 1 \\ 1 & \text{for} \quad \xi \geqslant 1 \end{cases}$$

where $\xi = r/a_0$.

For the Luneberg lens, whose permittivity distribution is given above, the first-order modal surface impedances and admittances are

$$\frac{iZ_1(k_0a_0)}{k_0a_0} = 0.50, \qquad \frac{iY_1(k_0a_0)}{k_0a_0} = 0.6911 \tag{3.4-38}$$

Using these results, the first Mie coefficients become

$$A_1 = 0, \qquad B_1 = 0.22494 \frac{(k_0a_0)^3}{2} \tag{3.4-39}$$

Hence, the low-frequency back and forward scattering cross sections of the Luneberg lens are

$$\sigma(0) = \pi a_0^2 (k_0a_0)^4 \cdot 0.05060 = \sigma(\pi) \tag{3.4-40}$$

### 3.4.3.2.2. *Eaton–Lippman Lens*

$$\mu_{r1}(\xi) = 1$$

$$\epsilon_{r1}(\xi) = \begin{cases} \dfrac{2}{\xi} - 1 & \text{for} \quad 0 \leqslant \xi \leqslant 1 \\ 1 & \text{for} \quad \xi \geqslant 1 \end{cases}$$

where $\xi = r/a_0$.

The solutions for the first-order modal impedances and admittances of the Eaton lens, whose permittivity and permeability distributions are given above, are

$$\frac{iZ_1(k_0a_0)}{k_0a_0} = 0.50, \qquad \frac{iY_1(k_0a_0)}{k_0a_0} = 0.8807 \tag{3.4-41}$$

Then the first two Mie coefficients become

$$A_1 = 0, \qquad B_1 = 0.40485 \frac{(k_0a_0)^3}{2} \tag{3.4-42}$$

and, in this case, the low-frequency backscattering and forward scattering cross sections are

$$\sigma(0) = \pi a_0^2 (k_0a_0)^4 \, 0.16390 = \sigma(\pi) \tag{3.4-43}$$

### 3.4.3.3. *Approximate Solutions—High-Frequency Region* ($k_0 a_0 > 20$)

When the radius is large compared to wavelength, optical methods may be used to determine the scattering properties of a radially inhomogeneous sphere. These optical methods may be broken into two general phenomenological categories: (a) treatment of reflection and refraction of waves at the sphere surface due to a radial discontinuity in refractive index, and (b) behavior of the waves within the sphere excluding the interactions with the surface.

The former category is solely responsible for the high frequency scattering behavior of the homogeneous dielectric sphere as discussed in Section 3.3. The methods are identical here. The analysis is straightforward, but the bookkeeping is laborious. References 17, 19, 20, and 21 provide the tools for such an analysis. Multiple internal and external reflections, jumps in phase as the internal rays pass through caustic lines and points, dispersion upon each reflection due to the curvature of the reflecting surface, and the necessity of applying the stationary phase principle in the cases of more complicated wavefronts, all enter into the analysis. When, however, the refractive index suffers no jump discontinuity at the sphere surface (i.e., is unity just inside the sphere), none of the phenomena in this category take place, and this part of the problem may be ignored.

The analysis of the behavior of the waves within the sphere due to the radial variation of refractive index will be discussed here briefly because of the relatively simple techniques available in the case of a sphere. This radial variation in composition produces the bending of the rays within the sphere and the shift in phase along the various ray paths. These techniques are useful in both the design and analysis of such spheres as lenses.

3.4.3.3.1. *Determination of the Ray Paths.* The ray paths within a radially inhomogeneous sphere can be described (due to axial symmetry) in polar coordinates by $\rho$ and $\theta$, as shown in Figure 3-16. The paths are thus

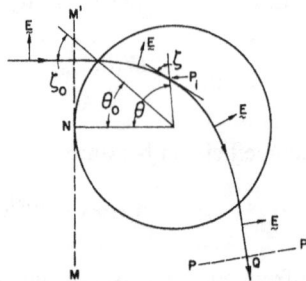

Fig. 3-16.  Tracing of the ray path through a radially inhomogeneous sphere.

identical with those of a radially inhomogeneous cylinder. The equation of
the ray at point $P_i$, which enters the sphere at point $a_0$, $\theta_0$, is then given by
solving[33]

$$\theta = \theta_0 + a_0 m_0 \sin \zeta_0 \int_{a_0}^{\rho} \frac{d\rho}{\rho \sqrt{m_1^2(\rho) \rho^2 - m_0^2 a_0^2 \sin^2 \zeta_0}} \qquad (3.4\text{-}44)$$

where $m_0 = m_1(a_0) =$ refractive index just inside sphere and $\zeta_0 =$ initial external angle between ray path and radius, $\rho$, ($\zeta_0 = \theta_0$, where $m_0 = 1$, as with
the Luneberg lens). If the refractive index at the inside of the sphere surface
is not unity (i.e., there is a discontinuity at the surface), then by Snell's law,
$\sin \zeta_0 = (1/m_0) \sin \theta_0$. It is assumed that $m_1(\rho)$, the refractive index of the
sphere medium, is known.

3.4.3.3.2. *Polarization Along the Ray Paths.* According to geometrical
optics, a monochromatic wave in the high-frequency limit is always polarized
in a plane perpendicular to its direction of propagation, so long as the radius
of curvature of the ray paths through an isotropic inhomogeneous medium is
much greater than wavelength. Thus, the ray entering the sphere, which has
its $E$ field polarized to the left as shown in Figure 3-16, will have its $E$-field
vector bent along the ray so that it is always perpendicular to it while still
remaining within the plane of the page.

However, the ray which enters the sphere of Figure 3-16 with its $E$ field
polarized perpendicular to the page, say out of the page, travels through the
sphere along an identical ray path and leaves with its $E$ field still perpendicular
to and out of the page.

In summary, one must take into account the polarization changes along
all possible rays emerging in a given direction to determine the polarization
and intensity of the scattered field.

3.4.3.3.3. *Phase Shift Along a Ray Path.* The phase of a wave traveling
through an inhomogeneous medium does not vary linearly with distance
along the ray path as it does in a homogeneous medium. The phase shift over
a small increment of distance, $ds$, along the phase path at radius, $\rho$, is
$d\phi = k_0 m_1(\rho) \, ds$. Hence the total phase shift at point $P_i$ on the ray path of
Figure 3-16, using the plane at the forward cap $M'NM$ as the reference, is
given by

$$\phi_{P_i} = k_0 a_0 (1 - \cos \theta_0) - k_0 \int_{a_0}^{\rho} \frac{m_1^2(\rho) \rho \, d\rho}{\sqrt{m_1^2(\rho) \rho^2 - m_0^2 a_0^2 \sin^2 \zeta_0}} \qquad (3.4\text{-}45)$$

If for a given inhomogeneous sphere the rays do not emerge parallel or
with a constant phase and amplitude distribution in a desired direction, then
one can choose a convenient "aperture" surface, normal to the desired
direction, and integrate the scattered field at the surface (both phase and
amplitude) over the aperture surface to obtain the scattered field and cross

section. At high frequencies, application of the stationary phase approximation should simplify the integration. These methods should prove useful in computing the phase and amplitude of the rays across the aperture surface selected.

3.4.3.3.4. *Luneberg Lens.* The Luneberg lens has the relative permeability and permittivity distributions as a function of radius given in Section 3.4.3.2.1. The lens has the desirable property that it focuses parallel rays (i.e., a plane wave) entering from one side onto a point on the opposite side of the sphere, as shown in Figure 3-17. If the back side of the lens is coated with a reflective surface, the rays will reemerge in the same pattern as they entered but in the reverse direction. In other words, a plane wave entering the sphere from the left is reflected back to the left, just as if it had fallen upon a flat plate or corner reflector.

The equation of the ray path through the lens for this refractive-index variation, as found from Eq. (3.4-44) is

$$1 - \frac{a_0^2}{\rho^2} \sin^2 \theta_0 = \cos \theta_0 \cos(2\theta - \theta_0) \qquad (3.4\text{-}46)$$

The radar cross section of the Luneberg lens with a reflective coating is the same as a flat plate of the same cross-sectional area, or

$$\sigma(0) = \pi k_0^2 a_0^4 \qquad (3.4\text{-}47)$$

In practice, the backscatter pattern of the Luneberg reflector is not aspect independent because the half-coated geometry destroys its 360° symmetry. Measured aspect patterns for actual Luneberg reflectors are given in Figure 8-52, and further details on their design and construction are found in Section 8.2.2.2.

3.4.3.3.5. *Eaton–Lippman Lens.* The Eaton–Lippman lens has the relative permeability and permittivity distributions as a function of radius given in Section 3.4.3.2.2. This distribution evolved in an attempt to redirect a plane wave back in the direction from which it came without a reflective

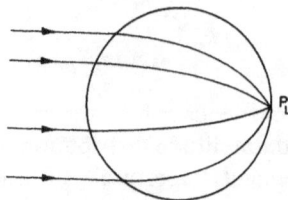

Fig. 3-17.   High-frequency ray paths in a Luneberg lens.

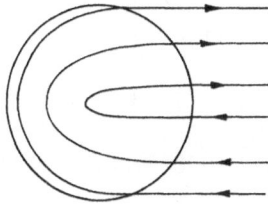

Fig. 3-18.   High-frequency ray paths in an Eaton–Lippman lens.

coating. The rays through such a lens are as shown in Figure 3-18. This lens, in contrast to the Luneberg reflector, is completely symmetrical and, hence, aspect insensitive. Unfortunately, the spherical version of the Eaton–Lippman lens does not operate as it was intended because of polarization cancellation in the backscatter direction. As discussed in Section 3.4.3.3.2, polarization information along the ray paths must be taken into account. In this case, the rays in the $E$ plane emerge with their polarization vector reversed, while those in the $H$ plane are unchanged; hence, there is a complete cancellation. Theoretically, therefore, the spherical Eaton–Lippman lens has zero back-scatter cross section. Such is not true of the cylindrical Eaton–Lippman lens discussed in Section 4.1.8.3; the cylindrical device does work as described, but only for incidence normal to the cylinder axis.

The equation for the ray path through this lens, as determined from Eq. (3.4-44), is

$$1 - \frac{a_0}{\rho} \sin^2 \theta_0 = \cos \theta_0 \cos(\theta - 2\theta_0) \qquad (3.4\text{-}48)$$

## 3.5.   TRANSIENT SCATTERING FROM A SPHERE

### 3.5.1.   Impulse Response

The impulse response, as discussed in Section 2.2.2.4, allows the scattered field from a sphere, excited by any transient waveform, to be obtained by convolution. In addition, an attractive feature of the impulse response is that it is merely the Fourier transform of the scattered field due to an incident harmonic plane wave as a function of frequency; this latter scattered field has been the subject of investigation of the first four sections of this chapter, and the results and analysis employed there can be simply extended to find the impulse response.

Define the impulse response at point $P$ as the electric field vector, $\mathbf{I}(P, t)$, that would be measured at time $t$ and point $P$, when a plane $x$-polarized impulse electric field vector of the form $\mathbf{E}^i = E_i \delta(t - t') \hat{\mathbf{x}}$ is incident upon the sphere from the $+z$ direction. $\mathbf{I}(P, t)$ has a Fourier transform given by

$$\mathbf{I}(P, \omega) = \int_{-\infty}^{\infty} \mathbf{I}(P, t) e^{-j\omega t} dt \qquad (3.5\text{-}1)$$

Then, $\mathbf{I}(P, \omega)$ is simply related to the scattered electric field, $\mathbf{E}^s(P, \omega)$, by the relationship

$$\mathbf{I}(P, \omega) = \frac{\mathbf{E}^s(P, \omega)}{|E^i|} \qquad (3.5\text{-}2)$$

Although several persons[34,35] have attempted to obtain results for $\mathbf{I}(P, t)$ by using the above equations along with the exact Mie solution for $\mathbf{E}^s(P, \omega)/|E^i|$, the results will not be presented here due to their complexity and the difficulty of readily interpreting them. Rather, curves will be shown for the scattering response to a very short incident pulse which approaches an impulse function in the limit. These results were obtained from References 36 and 37.

In the above references, Kennaugh has approximated an impulse function by a train of short cosine pulses as shown in Figure 3-19. If the spatial pulse width, $c\tau$, is much smaller than the sphere radius, a train of impulse functions is approximated quite well by cosine pulses. The spacing between the pulses is kept wide enough so that the scattering response due to one of the pulses is assumed to have died out before the next pulse excites the sphere. Then the input waveform is expanded into a Fourier series at multiples of the basic frequency, $\omega_0 = 2\pi/T$, where $T$ is the pulse spacing. Each of these monochromatic waves, which are terms of the Fourier series, strikes the sphere and produces a component of the impulse response given by

$$\mathbf{I}_n(P, t) = \text{Re}\left[A_n(n\omega_0) \frac{\mathbf{E}^s(P, n\omega_0)}{|E^i|} e^{-in\omega_0 t}\right] \qquad (3.5\text{-}3)$$

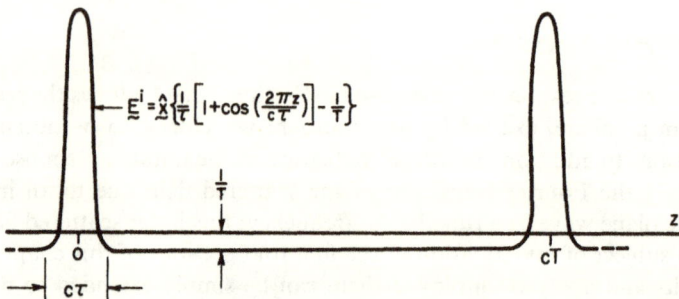

Fig. 3-19. Diagram showing the approximation to an impulse used by Kennaugh to obtain transient impulse response of sphere [after Kennaugh and Moffatt[36]].

where $A_n(n\omega_0)$ is the amplitude of the $n$th term of the Fourier series representing the input pulse train waveform, and $\mathbf{E}^s(P, n\omega_0)/|E^i|$ is the Mie series solution for the complex scattered field from the sphere at the single frequency $n\omega_0$. The total impulse response approximation is obtained by summing $\mathbf{I}_n(P, t)$ over all $n$ from zero to some upper limit $N$. Obviously, the larger the number, $N$, of terms used, and the smaller the $c\tau/a$ (the pulse length of each pulse relative to the sphere radius), the better is the approximation.

Figure 3-20 shows the impulse response approximation for the far-field backscattered waveform from a perfectly conducting sphere computed according to the above procedure.[36] The impulse approximation has the form $\mathbf{I}(P, t) = \hat{x}(c/2r) F_I(t)$, where $F_I(t)$ is plotted along an abscissa normalized by $2a$, the diameter of the sphere. The number of harmonics used is $N = 475$, and the spatial pulse width is $c\tau = a/19$. Plotted along with this is the physical-optics impulse response approximation obtained from Eq. 2.2-110 of Chapter 2. Both approximations predict an initial impulse in the scattered field corresponding to the reflection of the incident impulse from the forward cap. However, physical optics fails to predict to creeping wave components, the first of which occurs at time $\tau = a(2 + \pi)/c$, after the pulse strikes the sphere at $t = 0$. It is assumed that the spike representing the first creeping wave will also approach an impulse function, if enough terms and a short enough pulse length are used. Higher-order creeping wave components are also present, but are smaller and not shown in Figure 3-20. As one would expect, physical optics, although correct for small values of $t$ (corresponding to the higher frequencies), is not correct for larger values of $t$ (corresponding to the lower

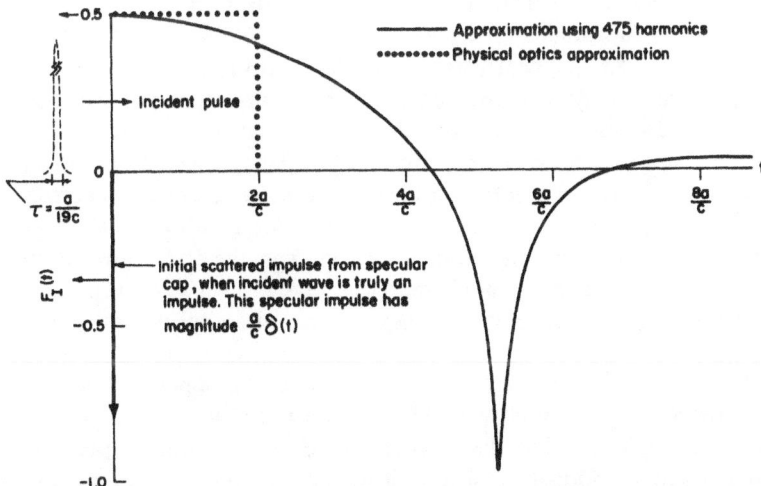

Fig. 3-20. Backscattering impulse response of a perfectly conducting sphere to an incident impulsive plane wave [after Kennaugh and Moffatt[36]].

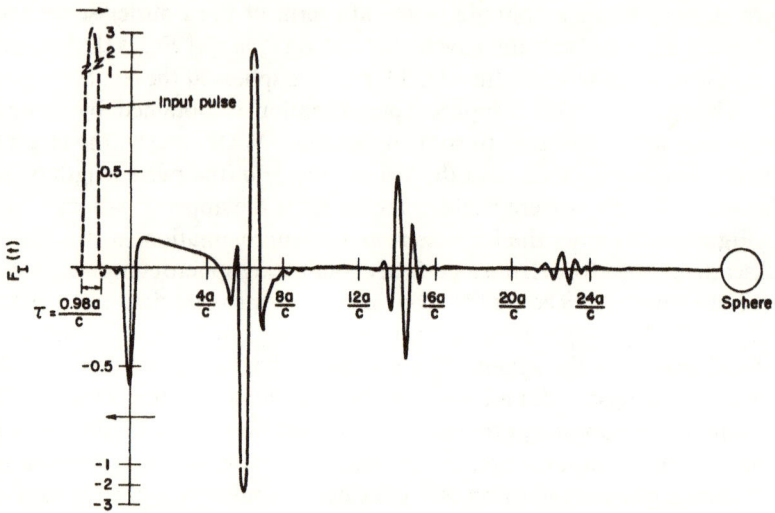

Fig. 3-21. Impulse response of a dielectric homogeneous sphere in the back-scattering direction to an incident impulsive plane wave (sphere material has refractive index $m = 1.486$) [after Kennaugh[37]].

frequencies). However, Figure 3-20 is plotted to scale and may be used to find the backscattering response to other incident transient signals by convolution, so long as these signals have no significant frequencies higher than $\omega = 2\pi/\tau = 38\pi c/a$.

Figure 3-21 shows the impulse response approximation for the far field backscattered waveform from a dielectric sphere of $m_1 = \sqrt{\epsilon_{r1}} = 1.486$.[37] The number of harmonics used here is only $N = 64$, and the spatial pulse width is $c\tau = 0.98a$, i.e., nearly as wide as the radius of the sphere. Although this incident pulse may be considered somewhat wide for an impulse approximation, the scattered wave clearly exhibits the behavior one would expect. The initial impulse represents the external specularly reflected ray, while the succeeding doublets represent the multiple internally reflected rays emerging in the backscattering direction. It is also evident that the chief response comes not from the externally reflected ray, but from the first of the internally reflected rays (the ordinate scale is representative of field strength).

Finally, Figure 3-22 shows the impulse response approximation for the far-field bistatic $E$-plane and $H$-plane scattered waves from a perfectly conducting sphere at increments of $30°$ around the sphere.[36] Again the number of pulses used was $N = 64$, and the pulse width is $c\tau = 0.784a$, about $\frac{3}{4}$ of the sphere radius. The higher peaks, representing the specular and creeping wave components, would approach impulse functions as the incident pulse width approached zero. The sphere at the center is shown with its size in the same scale as the waveforms in space.

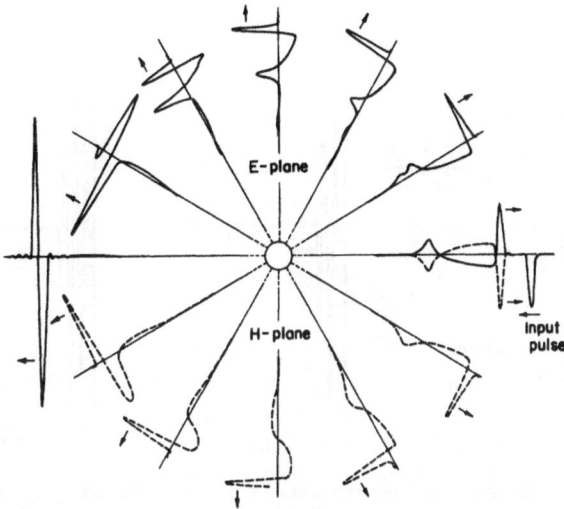

Fig. 3-22. Impulse response of a perfectly conducting sphere in various bistatic directions to an incident impulsive plane wave in both $E$ and $H$ planes [after Kennaugh and Moffatt[36]].

### 3.5.2. Backscattered Field for Pulse-Modulated Sine Wave Incident Field

The case where the transmitted or incident field is made up of a train of pulses of sine waves at a higher carrier frequency, $\omega_c$, is probably the rule rather than the exception with present-day radar systems. In most cases where the spatial pulse width, $cT_p$, is much larger than both the radius of the sphere, $a$, and the wavelength of the carrier frequency, $2\pi c/\omega_c$, the sine wave within the pulse can be considered monochromatic, having existed for all time. In this case, the scattered field and cross section are obtained by the methods of the preceding sections. The scattered field, and the power thus obtained, also appears as pulses of the same length. Hence, the transient nature of the incident field does not complicate the analysis.

However, when the spatial pulse width, $cT_p$, is of comparable size with the sphere radius, or smaller, then one must treat the incident field as a transient signal. The scattered field, $E^s(P, t)$, can be obtained from the impulse response, $I(P, t)$, of the scattered field at that same point, $P$, by a simple convolution of $I(P, t)$ with the incident time signal. In the case of backscattering, $I(P, t)$ has already been computed for two cases as shown in Figures 3-19 and 3-20.

Convolution is, of course, not the only method of obtaining such a transient response. If the impulse response of a given sphere in a given

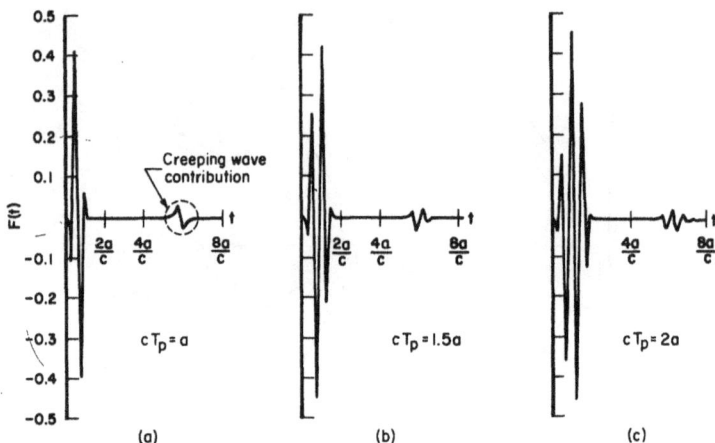

Fig. 3-23. Backscatter time response of a perfectly conducting sphere to an incident pulse on a sine wave carrier whose spatial length is comparable to the sphere size. Carrier wavelength is $\lambda_0 = a/24\pi$ and the pulse length varies between $1.0a$ and $2.0a$. Pulse envelope shape is sine-squared [after Kennaugh and Moffatt[36]].

direction has not already been determined, one can form the product of the Fourier transform of the incident signal and $\bar{\mathbf{I}}(P, \omega)$, as determined from the Mie series. The Fourier transform of this product then gives the time response of the scattered field to this incident signal.

Kennaugh and Moffatt[36] have used convolution with the impulse response to obtain the backscattered field from perfectly conducting spheres for various pulse widths, shapes, and carrier frequencies.

Figure 3-23 shows the normalized backscattered field function, $F(t)$, from a perfectly conducting sphere, where the incident wave is a planar $x$-polarized, sine-squared pulse at carrier frequency $\omega_c$.[†] It has the form

$$\mathbf{E}^i = \begin{cases} \hat{\mathbf{x}} \sin^2(\pi t/T_P) \sin \omega_c t & \text{for } 0 \leqslant t \leqslant T_P \\ 0 & \text{for } t < 0 \quad \text{and} \quad t > T_P \end{cases} \quad (3.5\text{-}4)$$

The wavelength used here is such that $\omega_c = 12c/a$, while the spatial pulse width is $cT_p = a$ for the curve (a), $cT_p = 1.5a$ for curve (b), and $cT_p = 2a$ for curve (c). The backscattered response at point $P$ is again assumed to begin at $t = 0$. The creeping wave contribution is clearly evident.

Figure 3-24 (from Reference 37) shows the response to two waveforms having somewhat lower carrier frequency and longer pulse length. The dotted curve is the response to an incident wave of sine-squared pulse form, as given

[†] The sine-squared radar pulse is quite realistic in practice when the pulse width is so narrow that it includes only a few cycles of the carrier.

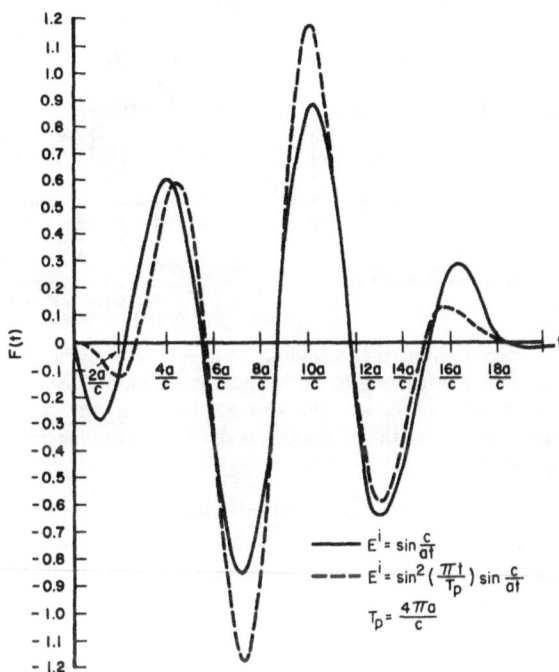

Fig. 3-24. Backscatter time response of a perfectly conducting sphere to an incident pulse on a sine-wave carrier whose spatial length is comparable to the sphere size. Carrier wavelength is $\lambda_0 = a/2\pi$ and the pulse length is $4\pi a$. The solid curve represents a square-pulse envelope shape, while the dashed shows a sine-squared pulse envelope shape, both of the same length, $4\pi a$ [from Kennaugh[37]].

by Eq. (3.5-4), while the solid curve is the response to an incident wave of a square pulse form, given by

$$\mathbf{E}^i = \begin{cases} \hat{\mathbf{x}} \sin \omega_c t & \text{for} \quad 0 \leqslant t \leqslant T_P \\ 0 & \text{for} \quad t < 0 \quad \text{and} \quad t > T_P \end{cases} \qquad (3.5\text{-}5)$$

The carrier frequency for both of these curves is $\omega_c = c/a$, and the spatial pulse length is $cT_p = 4\pi a$. The creeping wave contribution can no longer be distinguished from the specular portion because the pulse length is long enough that the incident wave is still striking the front cap when the creeping wave emerges into the backscattering direction.

Figure 3-25 shows similar backscattering time response curves to sine-squared pulses.[38] In both of these cases, the carrier frequency is $\omega_c = 2\pi c/a$, and spatial pulse width $cT_p = 4a$. Curve (a) is the return from a perfectly conducting sphere, while (b) is the return from a perfectly conducting sphere with a lossless dielectric coating of $\epsilon_r = 4$ and thickness $\delta = 0.1a$. In both

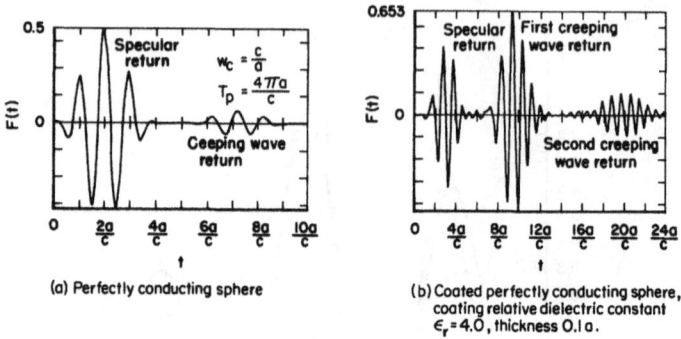

(a) Perfectly conducting sphere

(b) Coated perfectly conducting sphere, coating relative dielectric constant $\epsilon_r = 4.0$, thickness $0.1\,a$.

Fig. 3-25. Backscatter time response of a perfectly conducting sphere with and without a coating to an incident pulse on a sine-wave carrier whose spatial length is comparable to the sphere size. The carrier wavelength is $\lambda_0 = a/4\pi^2$ and the pulse shape is sine-squared with a length of $4a$. [after Rheinstein[38]].

of these cases, the pulse length is still small enough so that the creeping wave component can be distinguished; in fact, the dielectric coating, in the second case, enhances the creeping wave component so that it is larger than the specular component.

## 3.6. REFERENCES

1. Mie, G., Beiträge zur Optik trüber Medien, speziell kolloidaler Metallösungen, *Ann. Phys.* **25**:377 (1908).
2. Goodrich, R. F., Harrison, B. A., Kleinman, R. E., and Senior, T. B. A., Studies in Radar Cross Sections XLVII—Diffraction and Scattering by Regular Bodies—I: the Sphere, University of Michigan Radiation Laboratory (1961).
3. *Handbook of Mathematical Functions with Formulas, Graphs, and Mathematical Tables*, Abramowitz, M., and Stegun, I. A. (Eds.), U. S. Dept. of Commerce (1965).
4. Jahnke, E., Emde, F., and Losch, F., *Tables of Higher Functions*, McGraw-Hill, New York (1960).
5. Weil, H., Barach, M. L., and Kaplan, T. A., Studies in Radar Cross Sections—X-Scattering of Electromagnetic Waves by Spheres, University of Michigan Radiation Laboratory (1956).
6. Senior, T. B. A., and Goodrich, R. F., Scattering by a Sphere, *Proc. IEE* **111**:907 (1964).
7. Kouyoumjian, R. G., An Introduction to Geometrical Optics and Geometrical Theory of Diffraction. In: *Antenna and Scattering Theory: Recent Advances* (short course notes), The Ohio State University, 1966. The curves were obtained by Kouyoumjian and D. Voltmer.
8. King, R. W. P., and Wu, T. S., *The Scattering and Diffraction of Waves*, Harvard University Press (1959).
9. Volpert, A. R., and Potekhin, A. I., The Diffraction Field of an Ideally Conducting Sphere, *J. Exp. Teor. Phys.* **17**:803 (1947).
10. Van de Hulst, H. C., *Light Scattering by Small Particles*, John Wiley & Sons, New York (1957).
11. Aden, A. L., Scattering from Spheres with Sizes Comparable to the Wavelength, *J. Appl. Phys.* **22** (1951).
12. Aden, A. L., Scattering of Electromagnetic Waves From Spheres and Spherical Shells,

Geophysics Research Division, Air Force Cambridge Research Center, Geophysical Research Papers No. 15, 1952.

13. Atlas, D., Battan, L. J., Harper, W. G., Herman, B. M., Kerker, M., and Matievic, E., Back-Scatter by Dielectric Spheres, *IEEE Trans.* **AP-11**:68 (1963).

14. Herman, B. M., and Battan, L. J., Calculations of Mie Back-Scatter of Microwaves from Ice Spheres, *Institute of Atmosphere Physics*, University of Arizona, Scientific Report No. 12, 1959.

15. Kerker, M., and Matievic, E., Mie Scattering Functions for Refractive Index of 2.105, *J. Opt. Soc. Am.* **51**:87 (1961).

16. Stratton, J. A., *Electromagnetic Theory*, McGraw-Hill, New York (1941), p. 571.

17. Kouyoumjian, R. G., Peters, L., Jr., and Thomas, D. T., A Modified Geometrical Optics Method for Scattering by Dielectric Bodies, *IEEE Trans. Ant. Prop.* **AP-11**:690 (1963).

18. Atlas, D., and Glover, K. M., Backscatter by Dielectric Spheres with and Without Metal Caps. In Kerker, M., *Electromagnetic Scattering*, Pergamon Press, New York (1963), p. 231.

19. Thomas, D. T., Approximation for Backscatter From A Dielectric Sphere, The Antenna Laboratory, The Ohio State University, Report 1116-14, 1961.

20. Kawano, T., and Peters, L., Jr., An Extension of the Modified Geometrical Optics Methods for Radar Cross Sections of Dielectric Bodies, The Antenna Laboratory, The Ohio State University, Report 1116-38, January, 1964.

21. Kawano, T., and Peters, Jr., L., An Application of Modified Geometrical Optics Method for Bistatic Radar Cross Sections of Dielectric Bodies, The Antenna Laboratory, The Ohio State University, Report 1116-39, November, 1963.

22. Plonus, M. A., Inada, H., Closed Form Expression for the Mie Series for Large, Low Density, Dielectric Spheres, *Proc. IEEE* **53**:662 (1965).

23. Garbacz, R. J., Electromagnetic Scattering by Radially Inhomogeneous Spheres, The Antenna Laboratory, The Ohio State University, Report 1223-3, 1962.

24. Wait, J. R., Electromagnetic Scattering from Radially Inhomogeneous Spheres, *Appl. Sci. Res.* **10**:441 (1963).

25. Hiatt, R. E., Siegel, K. M., and Weil, H., The Ineffectiveness of Absorbing Coatings on Conducting Objects Illuminated by Long Wavelength Radar, *Proc. IRE* **48**:1636 (1960).

26. Rheinstein, J., Scattering of Electromagnetic Waves from Dielectric Coated Conducting Spheres, *IEEE Trans.* **AP-12**:334 (1964).

27. Herman, B. M., and Battan, L. J., Calculations of Mie Backscattering From Melting Ice Spheres, *J. Meteorology* **18**:468 (1961).

28. Mikulski, J. J., and Murphy, E. L., The Computation of Electromagnetic Scattering From Concentric Spherical Structures, *IEEE Trans.* **AP-11**:169 (1963).

29. Tai, C. T., The Electromagnetic Theory of the Spherical Luneberg Lens, The Antenna Laboratory, The Ohio State University, Report 667-17, August, 1956.

30. Tai, C. T., The Electromagnetic Theory of the Spherical Luneberg Lens, *Appl. Sci. Res.* **7**:113 (1959).

31. Garbacz, R. J., Electromagnetic Scattering From Radially Inhomogeneous Spheres, *Proc. IRE* **50**:1837 (1962).

32. Barrick, D. E., Low Frequency Solution to Electromagnetic Scatter from Radially Inhomogeneous Targets, to be published.

33. Luneberg, R. K., *Mathematical Theory of Optics*, University of California Press, California (1964).

34. Weston, V. H., Studies on Radar Cross Sections XXXIII—Exact Near Field and Far Field Solution for the Back-Scattering of a Pulse from a Perfectly Conducting Sphere, University of Michigan, Radiation Laboratory Report (1959).

35. Brown, W. P., Jr., A Theoretical Study of the Scattering of Electromagnetic Impulses by Finite Obstacles, California Institute of Technology, Antenna Laboratory Report No. 28, 1962.

36. Kennaugh, E. M., and Moffatt, D. L., Transient and Impulse Response Approximations, *Proc. IEEE* **53**:893 (1965).

37. Kennaugh, E. M., The Scattering of Transient Electromagnetic Waves by Finite Bodies, presented at URSI-IRE Fall Meeting, N. B. S., Boulder Colorado, December, 1960, and also The Antenna Laboratory, The Ohio State University, Final Engineering Report No. 1073-4, AF 19 (604)-6157 (1961).
38. Rheinstein, J., Scattering of Short Pulses of Electromagnetic Waves, *Proc. IEEE* **53**:1069 (1965).

# Chapter 4
# CYLINDERS
D. E. Barrick

## 4.1. INFINITELY LONG CIRCULAR CYLINDERS—NORMAL INCIDENCE

### 4.1.1. Form of Exact Series Solution

#### 4.1.1.1. *General Formulation*

Scattering from an infinitely long cylinder, when the incident field propagates normal to the cylinder axis (taken as the $z$ axis throughout this chapter, see Figure 4-1), is essentially a two-dimensional problem. This is evident because there is no variation in the incident field or in the cylinder properties in the axial or $z$ direction; hence, the scattered field will not vary in the $z$ direction. As a result, the problem is most easily solved in polar coordinates as shown in Figure 4-1. The cylinder surface is a natural coordinate surface in this system (i.e., $r =$ "$a$" $=$ constant), and boundary value techniques give an exact solution. There are obvious similarities between this problem and that of the sphere in the preceding chapter, and this can aid in obtaining an understanding of the scattering properties of infinite cylinders.

The incident wave is considered to be planar, and, in general, is polarized arbitrarily. However, it can be resolved into two components: (1) an $E$-field component parallel to the cylinder axis, and (2) an $E$-field component perpendicular to the same axis (or an $H$-field component parallel to the axis). Each case is then treated separately. The two cases will be distinguished by $\parallel$ or $\perp$, or $TM$ and $TE$, referring to (1) and (2) above, respectively.[†]

Since an electromagnetic field can be expressed entirely in terms of either the electric or the magnetic field, the problems can be solved for

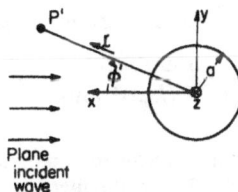

Fig. 4-1. Scattering geometry for an infinitely long circular cylinder with axis perpendicular to the page.

[†] For this section (4.1), the plane of incidence is considered to be the $x$-$z$ plane.

the two cases by: (a) using only the axial electric field for case (1), and (b) using only the axial magnetic field for case (2). In either case, both the incident and scattered fields possess only axial components. The boundary conditions at the cylinder can be expressed in terms of these axial components, and a scalar rather than a vector problem must be solved, which simplifies the solution considerably. (Such is not the case with a sphere, because the incident field strikes the sphere at different angles corresponding to different points on the sphere; hence, that problem retains a vector nature.)

Consider the incident wave to be resolvable into components of the form

$$\mathbf{E}^i = e^{-ik_0 x} \begin{cases} E_{\parallel}^i \hat{\mathbf{z}} \\ E_{\perp}^i \hat{\boldsymbol{\phi}} \end{cases} \tag{4.1-1}$$

where the field is incident from the left in Figure 4-1. The scattered field at point $P'$ can be expressed for each of the above two scalar field components as[†]

$$E_{\parallel}^s(P') = E_{\parallel}^i \sum_{n=0}^{\infty} (-i)^n \, \epsilon_n A_n H_n^{(1)}(k_0 r) \cos n\phi' \tag{4.1-2}$$

and

$$H_{\perp}^s(P') = H_{\perp}^i \sum_{n=0}^{\infty} (-i)^n \, \epsilon_n B_n H_n^{(1)}(k_0 r) \cos n\phi' \tag{4.1-3}$$

where

$$H_{\perp}^i = -\frac{1}{\eta} E_{\perp}^i \quad \text{and} \quad \epsilon_n = \begin{cases} 1 & \text{for } n = 0 \\ 2 & \text{for } n = 1, 2, 3, \dots \end{cases} \tag{4.1-4}$$

In the above equations, $H_n^{(1)}(k_0 r)$ is a cylindrical Hankel function of the first kind of order $n$ and argument $k_0 r$; it has the proper behavior for large $r$ corresponding to the time convention used here (see Reference 6 or 7 for properties and tables of this function). The constants $A_n$ and $B_n$ are functions only of the cylinder material and radius, and are given in the following sections for the various types of cylinders considered. It is again pointed out that in case (1), both the incident and scattered electric fields are polarized parallel to the cylinder axis; while in case (2), the incident and scattered magnetic fields are polarized parallel to this axis.

### 4.4.1.2. Far-Zone Simplification and Scattering Width

Simplification of the above series solution is possible when (i) $k_0 r \gg (k_0 a)^2$ and (ii) $k_0 r \gg 1$ (or $r \gg \lambda_0$). In this case, a large-argument

[†] The cylinder solution must be attributed to Lord Rayleigh, who appears to have been the first in the literature to treat the electromagnetic case (References 1 and 2). Many others since then have generalized and extended these results. More recent generalized treatments are found in References 3, 4, and 5.

expansion can be used for the Hankel function. In analogy with the case of the sphere [Eqs. (3.1-5) to (3.1-7)], the far-zone scattered fields are then written in terms of dimensionless complex scattering amplitude functions as follows:

$$E_\parallel{}^s(P') = E_\parallel{}^i \sqrt{\frac{2}{\pi}} \frac{e^{i(k_0 r - \pi/4)}}{\sqrt{k_0 r}} T_\parallel(\phi') \tag{4.1-5}$$

$$H_\perp{}^s(P') = H_\perp{}^i \sqrt{\frac{2}{\pi}} \frac{e^{i(k_0 r - \pi/4)}}{\sqrt{k_0 r}} T_\perp(\phi') \tag{4.1-6}$$

where

$$T_\parallel(\phi') = \sum_{n=0}^{\infty} (-1)^n \epsilon_n A_n \cos n\phi' \quad \text{and} \quad T_\perp(\phi') = \sum_{n=0}^{\infty} (-1)^n \epsilon_n B_n \cos n\phi' \tag{4.1-7}$$

In contrast with the exact Mie series for far-zone scattering from a sphere [Eq. (3.1-5)], the field from an infinite cylinder decreases as $1/\sqrt{r}$, rather than the $1/r$ dependence for objects of finite dimensions, such as spheres. Such must be the case in order for the total energy in the diverging scattered fields to be conserved.

According to Eq. (2.1-33), one can define an "echoing" or "scattering" width for an infinitely long cylinder in analogy with the "scattering cross section" for a finite object. (This scattering width is alternatively referred to as the scattering cross section per unit length). Then using Eq. (2.1-33), along with (4.1-5) through (4.1-7), one obtains

$$\sigma_\parallel{}^c(\phi') = \frac{4}{k_0} | T_\parallel(\phi')|^2 \quad \text{and} \quad \sigma_\perp{}^c(\phi') = \frac{4}{k_0} | T_\perp(\phi')|^2 \tag{4.1-8}$$

Two special scattering directions give further simplification:

(a) *Backscattering.* Here, $\phi' = 0$, and the above results simplify

$$T_\parallel(0) = \sum_{n=0}^{\infty} \epsilon_n (-1)^n A_n \quad \text{and} \quad T_\perp(0) = \sum_{n=0}^{\infty} \epsilon_n (-1)^n B_n \tag{4.1-9}$$

The backscattering widths become

$$\sigma_\parallel{}^c(0) = \frac{4}{k_0} \left| \sum_{n=0}^{\infty} \epsilon_n (-1)^n A_n \right|^2 \quad \text{and} \quad \sigma_\perp{}^c(0) = \frac{4}{k_0} \left| \sum_{n=0}^{\infty} \epsilon_n (-1)^n B_n \right|^2 \tag{4.1-10}$$

(b) *Forward Scattering.* Here, $\phi' = \pi$, and the scattering amplitudes simplify to

$$T_\parallel(\pi) = \sum_{n=0}^{\infty} \epsilon_n A_n \quad \text{and} \quad T_\perp(\pi) = \sum_{n=0}^{\infty} \epsilon_n B_n \tag{4.1-11}$$

The forward scattering widths then become

$$\sigma_{\parallel}{}^c(\pi) = \frac{4}{k_0} \left| \sum_{n=0}^{\infty} \epsilon_n A_n \right|^2 \quad \text{and} \quad \sigma_{\perp}{}^c(\pi) = \frac{4}{k_0} \left| \sum_{n=0}^{\infty} \epsilon_n B_n \right|^2 \quad (4.1\text{-}12)$$

### 4.1.1.3. *Use of Exact Solution for Infinitely Long Cylinder*

Scattering from an infinitely long circular cylinder at normal incidence appears at first glance to have little application for the radar engineer. Radar targets always have finite dimensions. Even if it were possible to conceive of a situation where the cylindrical target is very long, the incident wave from the transmitter would not be planar over the entire cylinder, since in the far zone this wave can be considered to come essentially from a point source. Nonetheless, the seemingly academic solution for the infinite cylinder does prove useful in the following ways:

(a) Scattering from the infinite circular cylinder is one of the very few electromagnetic scattering problems that can be solved exactly. Its solution, being scalar and two-dimensional in nature, is one of the simplest in form of all of the exact scattering problems thus far solved. Hence, one can gain needed insight into the general scattering process from it. Often a complicated problem must be analyzed in two dimensions before proceeding to the more complex three-dimensional case; for such situations, the infinite circular cylinder offers a convenient standard for comparison.

(b) The solutions for the infinite cylinder can and will be used later in the chapter to construct approximate solutions for cylinders of finite length.

(c) One can construct approximation techniques valid for other bodies from the exact solutions for the infinite cylinder. Creeping wave theory and the Fock theory (discussed in Sections 2.3.2 and 2.2.2.6, respectively) are examples of such techniques arising first from analysis of the cylinder problem.

(d) From the exact solution available for infinitely long cylinders, one can evaluate the effect of coatings and coating materials on the radar cross section. The exact solution for radially stratified or inhomogeneous cylinders is useful in evaluating such finite cylindrical focusing devices as Luneberg and Eaton lenses.

The exact scattered field solution, as formulated above (either in its exact form or the far-zone result), is a convergent infinite series. At very long and very short wavelengths compared to radius, for certain cases, approximations are obtainable from this solution. Most of the time, however, particularly in the medium- and short-wavelength regions, the series is difficult to simplify (e.g., the coefficients $A_n$ and $B_n$ are too complicated). In this case, machine summation of the series for specific parameters offers a practical means of obtaining answers. The Bessel and Hankel functions, appearing in the solution and also in the constants $A_n$ and $B_n$, can be generated by recurrence techniques as discussed in Section 3.1.3. The number of terms needed in the series varies as a function of $k_0a$ in the same manner as with the sphere of Section 3.1.3, and Figure 3-2 can also be used to estimate the number of terms required for the cylinder solution. Section 3.1.3 should be consulted for a more complete discussion of machine generation of the solution.

### 4.1.2. Perfectly Conducting Cylinder

#### 4.1.2.1. *Series Coefficients*

The coefficients $A_n$ and $B_n$ appearing in the exact solution for the perfectly conducting infinitely long cylinder at normal incidence are[3-5]

$$A_n = - \frac{J_n(k_0a)}{H_n^{(1)}(k_0a)} \tag{4.1-13}$$

and

$$B_n = - \frac{J_n'(k_0a)}{H_n^{(1)\prime}(k_0a)} \tag{4.1-14}$$

where the primes represent derivatives with respect to the expressed argument. $J_n(k_0a)$ is the usual cylindrical Bessel function of order $n$ and argument $k_0a$.[6,7] These coefficients are easily generated on a computer by recurrence techniques (see Section 3.2.1 for more discussion of numerical calculation of Bessel functions), or are found in the tables.

Curves of exact back, bistatic, and forward scattering widths have been computed by machine using these coefficients in Eqs. (4.1-8), (4.1-10), and (4.1-12). These curves are shown in Figures 4-2, 4-3, and 4-4. Figure 4-5 shows the same backscattering width but plotted to a different normalization, along with measured points, for the parallel polarization case (from Reference 5).

#### 4.1.2.2. *Surface Currents*

The surface current, $\mathbf{K}$ [defined in Eq. (2.2-34)], on any object is related to the total $\mathbf{H}$ field at the surface by $\mathbf{K} = \hat{\mathbf{n}} \times \mathbf{H}$. Having the series solutions (4.1-2) and (4.1-3) for the scattered fields along with the expressions for the incident fields, one can readily determine the surface currents.[4]

In the first case where the incident and scattered $E$ fields are parallel to the cylinder axis, the surface current is also directed along the axis and is

$$K_z = \frac{2E_\parallel{}^i}{\pi k_0 a \eta} \sum_{n=0}^{\infty} \frac{(-i)^n \, \epsilon_n \cos n\phi'}{H_n^{(1)}(k_0a)} \tag{4.1-15}$$

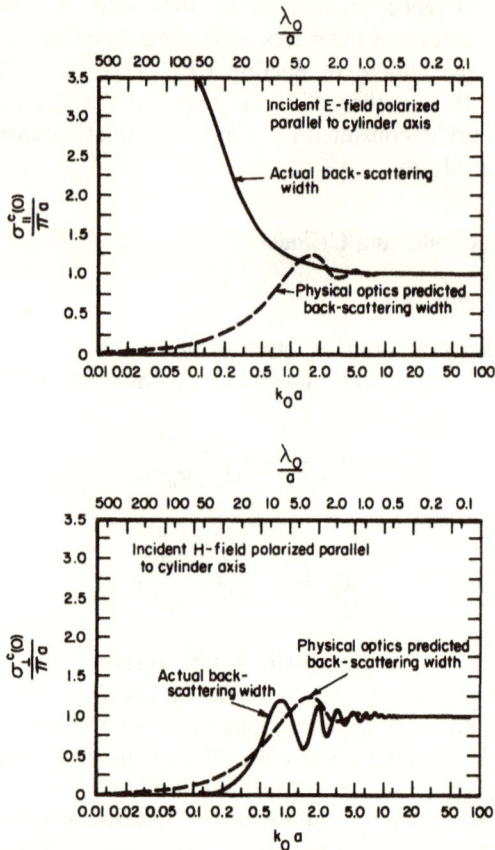

Fig. 4-2. Exact solution and the physical optics predicted normalized backscattering widths for an infinitely long, perfectly conducting circular cylinder —logarithmic $k_0a$ axis and linear ordinate.

In the second case where the incident and scattered $H$ fields are parallel to the cylinder axis, the surface current is circumferential, i.e., it flows around the cylinder, and is

$$K_\phi = \frac{-i2H_\perp^{\ i}}{\pi k_0 a} \sum_{n=0}^{\infty} \frac{(-i)^n \, \epsilon_n \cos n\phi'}{H_n^{(1)\prime}(k_0 a)} \qquad (4.1\text{-}16)$$

It is interesting to compare the actual surface currents on a cylinder, both in the illuminated and shadow regions, with the currents predicted by the physical-optics approximation (i.e., $\mathbf{K} = 2\hat{n} \times \mathbf{H}^i$ on the illuminated side and $\mathbf{K} = 0$ on the shadow side). This is done in Figure 4-6 at different points on the cylinder as a function of $k_0a$. It is easily seen how much

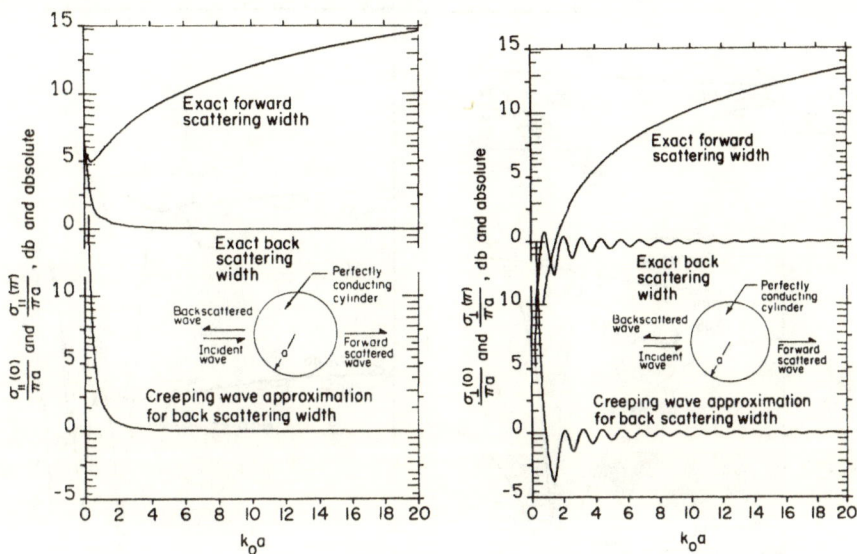

Fig. 4-3. Exact solution and the creeping wave approximation for the backscattering widths, and exact forward scattering widths for an infinitely long, perfectly conducting circular cylinder (linear $k_0a$ axis and logarithmic ordinate).

larger than unity $k_0a$ must be in order for the physical-optics predicted surface currents to approach the actual values.

4.1.2.3. *Approximations for the Low-Frequency Region* ($k_0a \ll 1$)

When $k_0a \ll 1$, the cylinder appears as a long, thin wire, and only the first term of the above series is nonvanishing. The Bessel functions in this term are represented by their small argument expansions. Hence, the far-zone scattered fields for each of the two polarization cases are[4]†

$$E_\parallel{}^s = E_\parallel{}^i \sqrt{\frac{\pi}{2}} \frac{e^{i(k_0r+\pi/4)}}{\sqrt{k_0r}[\ln(0.8905k_0a) - i\pi/2]} \qquad (4.1\text{-}17)$$

for the $E$ field parallel to the axis; and

$$H_\perp{}^s = H_\perp{}^i \sqrt{\frac{\pi}{2}} \frac{(k_0a)^2}{2} \frac{e^{i(k_0r-3\pi/4)}}{\sqrt{k_0r}} (1 + 2\cos\phi') \qquad (4.1\text{-}18)$$

for the $H$ field parallel to the axis.

From these expressions, the various low-frequency scattering widths and surface currents can be obtained.

† An oversight in Reference 4 in the argument of the logarithm has been corrected here.

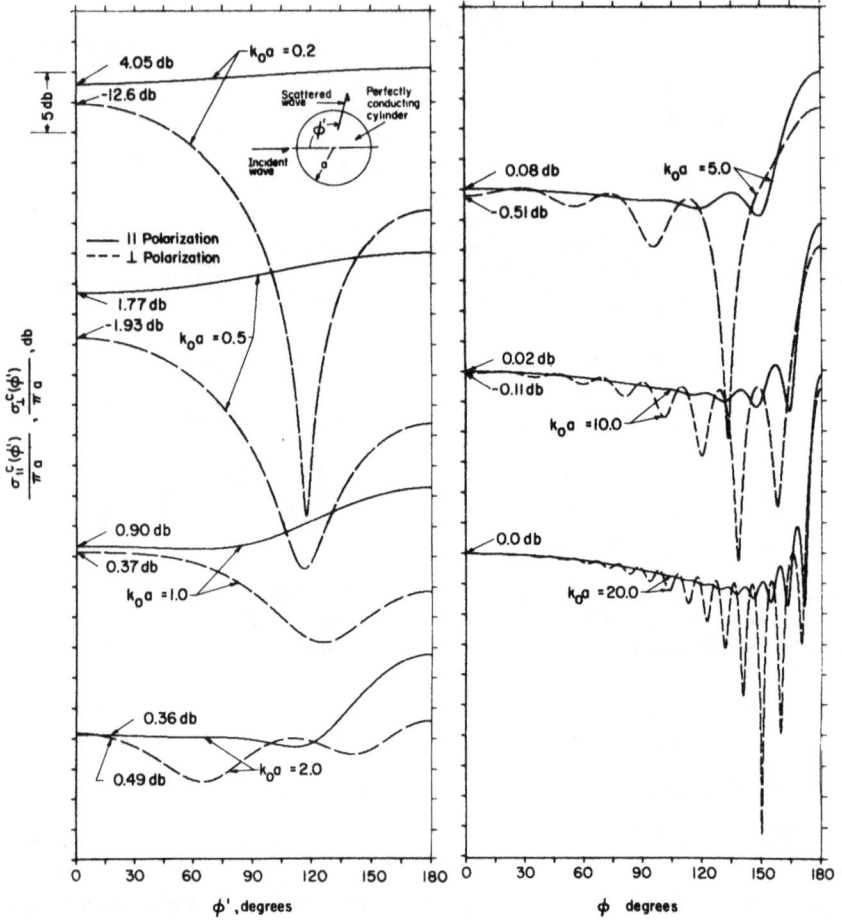

Fig. 4-4. Exact solution for the normalized bistatic scattering widths of an infinitely long, perfectly conducting circular cylinder for various values of $k_0a$.

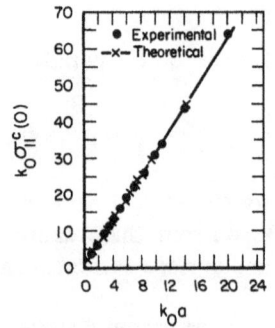

Fig. 4-5. Exact solution and measured points for the backscattering widths for an infinitely long, perfectly conducting circular cylinder with $E$ fields parallel to the axis (normalized as $\sigma k_0$) [from Tang[5]].

4.1.2.3.1. *Low-Frequency Scattering Widths* $(k_0 a \ll 1)$. For backscatter the scattering widths are

$$\sigma_\parallel{}^c(0) = \frac{\pi^2 a}{k_0 a[\ln^2(0.8905 k_0 a) + \pi^2/4]} \qquad (4.1\text{-}19)$$

for $E$ field parallel to the cylinder axis; and

$$\sigma_\perp{}^c(0) = \pi^2 a[\tfrac{9}{4}(k_0 a)^3] \qquad (4.1\text{-}20)$$

for $H$ field parallel to the cylinder axis. Equation (4.1-19) shows that the backscattered power goes slowly to zero as radius $a$ goes to zero, but for constant wire radius, the scattered power goes to infinity as frequency decreases to zero.[†] Equation (4.1-20) for $H$ field parallel to the axis, shows that the scattering width for this polarization behaves functionally more like the backscattering cross section for a small sphere. These behaviors are evident in Figure 4-2.

The bistatic scattering widths are

$$\sigma_\parallel{}^c(\phi') = \frac{\pi^2 a}{k_0 a[\ln^2(0.8905 k_0 a) + \pi^2/4]} \qquad (4.1\text{-}21)$$

for the $E$ field parallel to the cylinder axis; and

$$\sigma_\perp{}^c(\phi') = \pi^2 a[(k_0 a)^3 (1/2 + \cos \phi')^2] \qquad (4.1\text{-}22)$$

for the $H$ field parallel to the cylinder axis. The first equation, (4.1-21)' illustrates that when the electric field lies along the cylinder or wire axis' the scattered power radiates isotropically (i.e., is independent of scattering angle). The second equation gives a bistatic pattern identical in form to that for the $E$-phase pattern of a small sphere [see Eq. (3.2-22)].

The forward scattering widths are

$$\sigma_\parallel{}^c(\pi) = \frac{\pi^2 a}{k_0 a[\ln^2(0.8905 k_0 a) + \pi^2/4]} \qquad (4.1\text{-}23)$$

for the $E$ field parallel to the cylinder axis; and

$$\sigma_\perp{}^c(\pi) = \pi^2 a \left[\frac{(k_0 a)^3}{4}\right] \qquad (4.1\text{-}24)$$

for the $H$ field parallel to the cylinder axis. As with the sphere, this latter equation shows that scattered power in the backward direction is nine times that in the forward direction in the low-frequency limit.

[†] The scattered power cannot, in reality, increase without bound as the frequency goes to zero. As frequency decreases, a point is reached where $k_0 r$ is no longer much greater than unity, i.e., the observation point becomes part of the near field. Then Eq. (4.1-17) is no longer applicable since it applies only to the far zone, and Eq. (4.1-2) must be used instead.

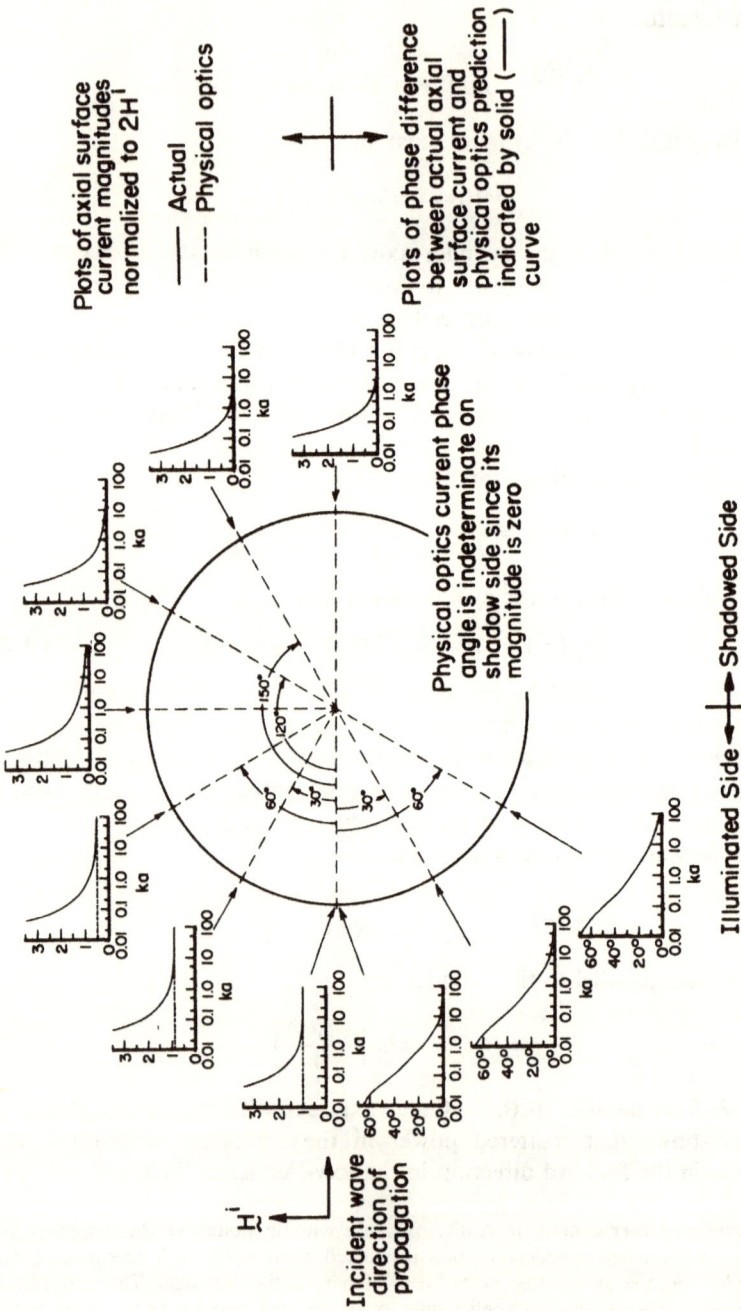

Fig. 4-6a.  Exact and physical-optics surface-current magnitudes and phase angles at different points around an infinitely long, perfectly conducting cylinder for the case of parallel polarization (*TM*).

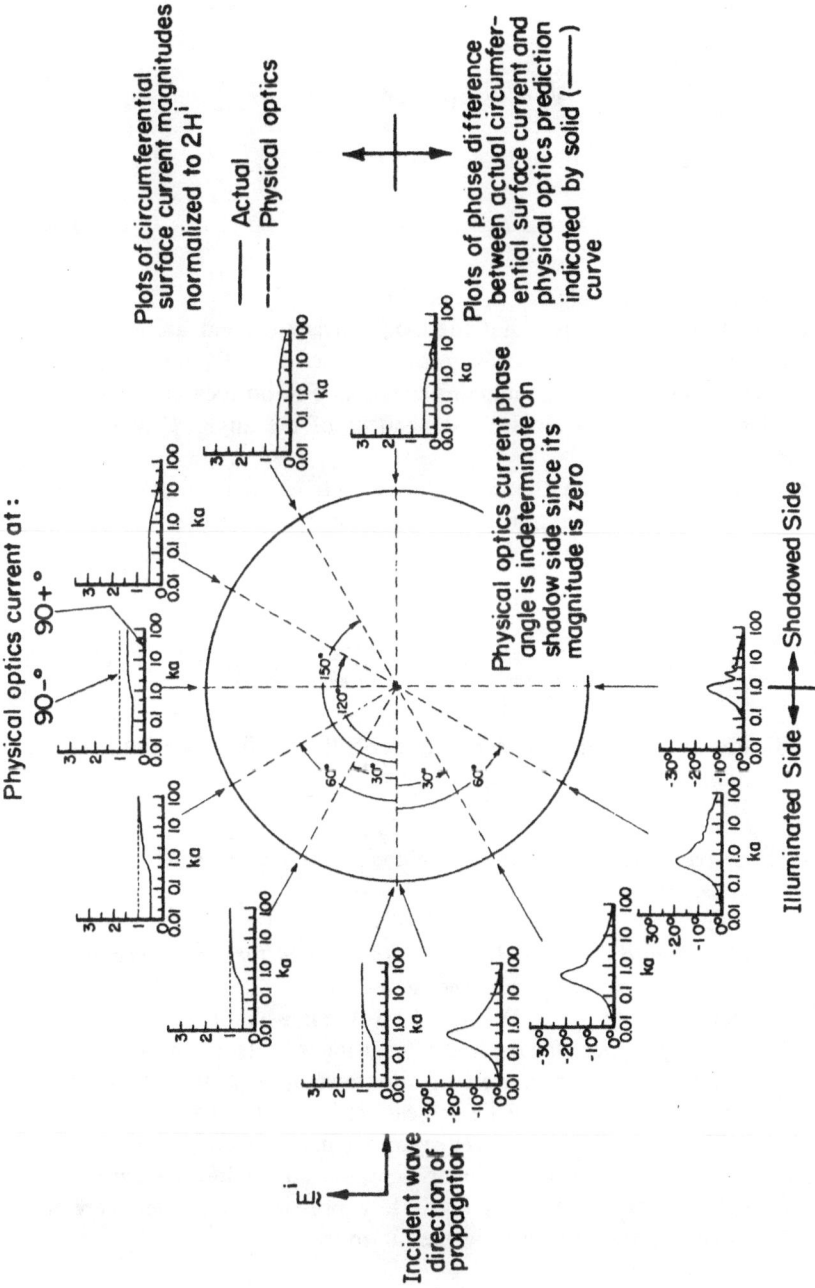

Fig. 4-6b. Same as Figure 4-6a, but for the case of perpendicular polarization (*TE*).

4.1.2.3.2. *Surface Currents* $(k_0 a \ll 1)$. The equations for the currents on the surface of the long thin cylinder (or wire) can also be simplified, giving

$$K_z = -i \sqrt{\frac{\epsilon_0}{\mu_0}} \frac{E_\parallel{}^i}{k_0 a[\ln(0.8905 k_0 a) - i\pi/2]} \qquad (4.1\text{-}25)$$

for the $E$ field parallel to the cylinder axis; and

$$K_\phi = -H_\perp{}^i \qquad (4.1\text{-}26)$$

for the $H$ field perpendicular to the cylinder axis.

The first equation shows that the axial surface current along the wire goes to infinity as $k_0 a$ approaches zero. For the second polarization, the circumferential surface current remains constant. For both cases, the current magnitudes and phase angles are independent of the angle $\phi'$ around the cylinder.

For the first case where the current is directed axially along the wire, it is more meaningful to compute the total current along the wire. This is done by multiplying $K_z$ by $a\, d\phi'$ and integrating from $\phi' = 0$ to $\phi' = 2\pi$. The result is

$$I_z = -i \sqrt{\frac{\epsilon_0}{\mu_0}} \frac{2\pi E_\parallel{}^i}{k_0[\ln(0.8905 k_0 a) - i\pi/2]} \qquad (4.1\text{-}27)$$

This indicates that the current in a long perfectly conducting wire lags the applied voltage or field strength by nearly 90°.

### 4.1.2.4. *Approximation for Upper-Middle and High-Frequency Region* $(k_0 a > 2)$

4.1.2.4.1. *Bistatic Creeping Wave Approximation for Resonance Region* $(0 \leqslant \phi' < \pi/2, \ k_0 a > 2)$. From the series of (4.1-2) and (4.1-3) along with the coefficients of (4.1-13) and (4.1-14), an asymptotic series valid in the limit of large $k_0 a$ can be derived after using a Watson transformation. Details and physical interpretation of this development for the cylinder are found in Section 2.3.2. The fields scattered into the backward region (i.e., $\phi' < \pi/2$) are broken into two components; an optical contribution (represented here by $E^{so}$), and a creeping wave contribution (represented here by $E^{sc}$). Only the first terms of those expressions are given here for each of the two polarization cases in the far zone:

$$E_\parallel{}^s = E_\parallel^{so} + E_\parallel^{sc} \qquad \text{and} \qquad H_\parallel{}^s = H_\parallel^{so} + H_\parallel^{sc} \qquad (4.1\text{-}28)$$

where

$$E_\parallel^{so} = -E_\parallel^i \sqrt{\frac{a\cos(\phi'/2)}{2r}} \exp\{ik_0[r - 2a\cos(\phi'/2)]\}$$

$$\times \left[1 + \frac{i}{2k_0a\cos(\phi'/2)}\left(\frac{1}{\cos^2(\phi'/2)} - \frac{3}{8}\right)\right.$$

$$\left. + \frac{1}{4(k_0a)^2\cos^2(\phi'/2)}\left(\frac{15}{128} - \frac{33}{8\cos^2(\phi'/2)} + \frac{5}{\cos^4(\phi'/2)}\right) + O[(k_0a)^{-3}]\right]$$

$$(4.1\text{-}29)$$

$$E_\parallel^{sc} = +E_\parallel^i \sqrt{\frac{a}{2r}} \exp[i(k_0r + \pi/12)](k_0a)^{-1/6}\,0.910721$$

$$\times \left[\exp\{ik_0a[\pi + \phi'][1 + (0.9358135 + i1.607129)(k_0a)^{-2/3}\right.$$

$$- (0.057397 - i0.0994145)(k_0a)^{-4/3} + O[(k_0a)^{-2}]\}$$

$$+ \exp\{ik_0a[\pi - \phi'][1 + (0.9358135 + i1.607129)(k_0a)^{-2/3}$$

$$\left. - (0.057397 - i0.0994145)(k_0a)^{-4/3} + O[(k_0a)^{-2}]\}\right] \qquad (4.1\text{-}30)$$

$$H_\perp^{so} = H_\perp^i \sqrt{\frac{a\cos(\phi'/2)}{2r}} \exp\{ik_0[r - 2a\cos(\phi'/2)]\}$$

$$\times \left[1 - \frac{i}{2k_0a\cos(\phi'/2)}\left(\frac{1}{\cos^2(\phi'/2)} + \frac{3}{8}\right)\right.$$

$$\left. + \frac{1}{4(k_0a)^2\cos(\phi'/2)}\left(\frac{15}{128} + \frac{33}{8\cos^2(\phi'/2)} - \frac{7}{\cos^4(\phi'/2)}\right) + O[(k_0a)^{-3}]\right]$$

$$(4.1\text{-}31)$$

$$H_\perp^{sc} = +H_\perp^i \sqrt{\frac{a}{2r}} \exp[i(k_0r + \pi/12)](k_0a)^{-1/6}\,1.531915$$

$$\times \left[\exp\{ik_0a[\pi + \phi'][1 + (0.404308 + i0.70028)(k_0a)^{-2/3}\right.$$

$$- (0.072732 - i0.1259755)(k_0a)^{-4/3} + O[(k_0a)^{-2}]\}$$

$$+ \exp\{ik_0a[\pi - \phi'][1 + (0.404308 + i0.70028)(k_0a)^{-2/3}$$

$$\left. - (0.072732 - i0.1257755)(k_0a)^{-4/3} + O[(k_0a)^{-2}]\}\right] \qquad (4.1\text{-}32)$$

4.1.2.4.2. *Backscattering Creeping-Wave Approximation for Resonance Region* ($\phi' = 0$, $k_0a > 2$). Equations (4.1-30) and (4.1-32) for the creeping wave components are valid only for $0 \leqslant \phi' < \pi/2$ (i.e., bistatic scattering into the backward half-space).

In the case of backscattering where $\phi' = 0$, the applicable equations are

$$E_{\parallel}^{so} = -E_{\parallel}^{i} \sqrt{\frac{a}{2r}} \exp[i(k_0 r - 2k_0 a)] \left[ 1 + \frac{i5}{16k_0 a} + \frac{127}{512(k_0 a)^2} + O[(k_0 a)^{-3}] \right]$$
$$(4.1\text{-}33)$$

$$E_{\parallel}^{sc} = +E_{\parallel}^{i} \sqrt{\frac{a}{2r}}$$

$$\times \exp\{i[k_0 r + k_0 a\pi + \pi/12 + 2.939945(k_0 a)^{1/3} - 0.180318(k_0 a)^{-1/3}]\}$$

$$\times [1.821442(k_0 a)^{-1/6} \exp[-5.048945(k_0 a)^{1/3} - 0.312320(k_0 a)^{-1/3}]]$$
$$(4.1\text{-}34)$$

$$H_{\perp}^{so} = H_{\perp}^{i} \sqrt{\frac{a}{2r}} \exp[i(k_0 r - 2k_0 a)] \left[ 1 - \frac{i11}{16k_0 a} - \frac{353}{512(k_0 a)^2} + O[(k_0 a)^{-3}] \right]$$
$$(4.1\text{-}35)$$

$$H_{\perp}^{sc} = +H_{\perp}^{i} \sqrt{\frac{a}{2r}}$$

$$\times \exp\{i[k_0 r + k_0 a\pi + \pi/12 + 1.2701695(k_0 a)^{1/3} - 0.2284945(k_0 a)^{-1/3}]\}$$

$$\times \{3.063830(k_0 a)^{-1/6} \exp[-2.200000(k_0 a)^{1/3} - 0.3957635(k_0 a)^{-1/3}]\}$$
$$(4.1\text{-}36)$$

These equations have been used to generate curves of backscattering width (Figure 4-3), and are plotted along with the exact solution. One can thereby readily determine the error involved in the use of the creeping wave theory at a given value of $k_0 a$.

4.1.2.4.3. *Bistatic Scattering Width in High-Frequency Limit* ($k_0 a > 20$). From Eqs. (4.1-28) through (4.1-32), one can see that as $k_0 a$ increases, the creeping wave contributions decrease, and only the first term of the specular or optical contribution remains. From this the bistatic scattering widths are

$$\sigma_{\parallel}^{c}(\phi') = \pi a \cos(\phi'/2) \quad \text{and} \quad \sigma_{\perp}^{c}(\phi') = \pi a \cos(\phi'/2) \quad (4.1\text{-}37)$$

for $|\pi - \phi'| > (k_0 a)^{-1/3}$.

4.1.2.4.4. *Backscattering Width in High-Frequency Limit* ($k_0 a > 20$). One can now use only the first term of Eqs. (4.1-33) and (4.1-35) in computing the backscattering widths. They are

$$\sigma_{\parallel}^{c}(0) = \pi a, \quad \sigma_{\perp}^{c}(0) = \pi a \quad (4.1\text{-}38)$$

4.1.2.4.5. *Forward Scattering Widths and Pattern in High-Frequency Limit* ($k_0 a > 20$). At and very near the forward direction, the scattered field is more commonly referred to as the diffracted field. It is the same in the high-frequency limit for all cylinders of width $2a$, regardless of the cylinder material, incident polarization, or cylinder cross section. Thus, the results given will apply to the diffracted field from any infinitely long cylinder at normal incidence.

Define $\alpha = \pi - \phi'$, then the diffracted field is given by (Reference 8, p. 465).

$$E^s = -E^i \sqrt{\frac{2k_0 a^2}{\pi r}}\, e^{i(k_0 r - \pi/4)} \left( \frac{\sin k_0 a\alpha}{k_0 a\alpha} \right) \qquad (4.1\text{-}39)$$

where the diffracted field has the same polarization state as the incident field. Thus, the scattering width at and near the forward direction becomes

$$\sigma^c(\phi') = \sigma^c(\pi - \alpha) = 4k_0 a^2 \left( \frac{\sin k_0 a\alpha}{k_0 a\alpha} \right)^2 \qquad (4.1\text{-}40)$$

and the forward scattering width ($\alpha = 0$) becomes

$$\sigma^c(\pi) = 4k_0 a^2 \qquad (4.1\text{-}41)$$

The familiar $(\sin \mu)/\mu$ diffraction pattern as discussed in Section 3.2.2.3 is displayed by Eq. (4.1-40).

### 4.1.3. Homogeneous Cylinder

#### 4.1.3.1. *Series Coefficients*

The coefficients $A_n$ and $B_n$, appearing in the exact solution for the infinitely long homogeneous cylinder of permeability $\mu_1$ and permittivity $\epsilon_1$ (both of which may be complex) at normal incidence, were first derived by Lord Rayleigh[1,2] and are

$$A_n = -\frac{(k_1/\mu_1)\, J_n(k_0 a)\, J_n'(k_1 a) - (k_0/\mu_0)\, J_n'(k_0 a)\, J_n(k_1 a)}{(k_1/\mu_1)\, H_n^{(1)}(k_0 a)\, J_n'(k_1 a) - (k_0/\mu_0)\, H_n'^{(1)}(k_0 a)\, J_n(k_1 a)} \qquad (4.1\text{-}42)$$

and

$$B_n = -\frac{(k_1/\epsilon_1)\, J_n(k_0 a)\, J_n'(k_1 a) - (k_0/\epsilon_0)\, J_n'(k_0 a)\, J_n(k_1 a)}{(k_1/\epsilon_1)\, H_n^{(1)}(k_0 a)\, J_n'(k_1 a) - (k_0/\epsilon_0)\, H_n'^{(1)}(k_0 a)\, J_n(k_1 a)} \qquad (4.1\text{-}43)$$

where

$$k_0^2 = \omega^2 \mu_0 \epsilon_0 \qquad \text{and} \qquad k_1^2 = \omega^2 \mu_1 \epsilon_1 \qquad (4.1\text{-}44)$$

The primes again represent derivatives with respect to the expressed arguments. These coefficients are only slightly more difficult to generate on a computer than were those for the perfectly conducting cylinder.

Curves of exact back, bistatic, and forward scattering widths have been computed using these coefficients in Eqs. (4.1-8), (4.1-10), and (4.1-12),

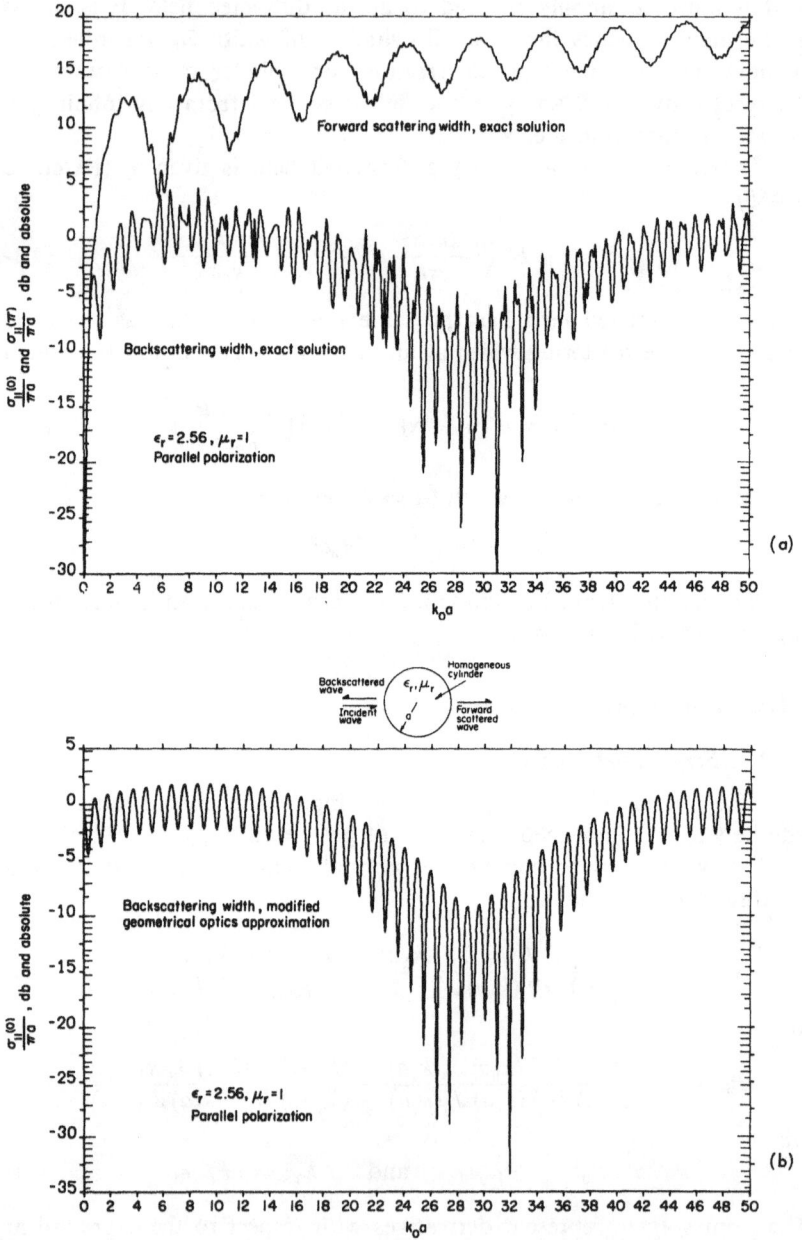

Fig. 4-7. Normalized scattering widths for an infinitely long, lossless homogeneous cylinder with $\epsilon_r = 2.56$, $\mu_r = 1$ for $E$ field parallel to cylinder axis ($TM$ mode): (a) exact solution for back and forward scattering widths and (b) modified geometrical optics approximation for backscattering width.

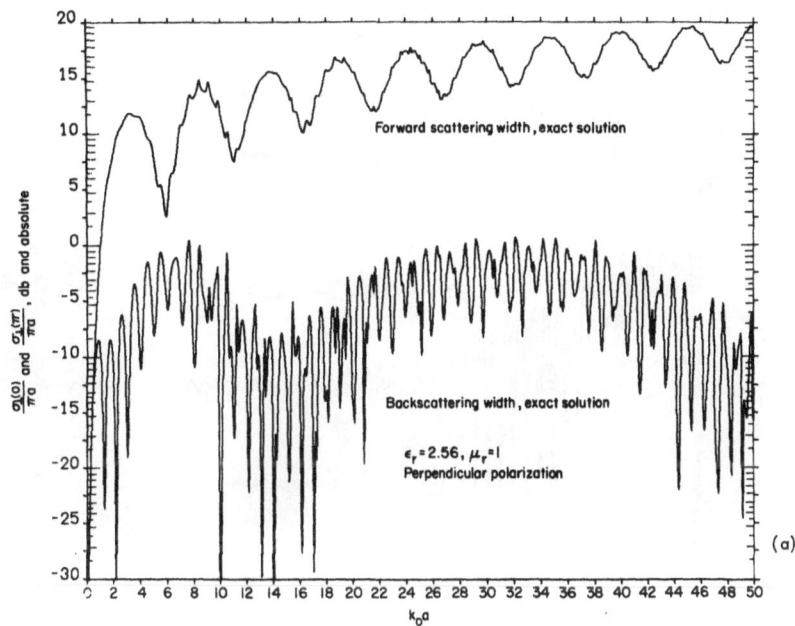

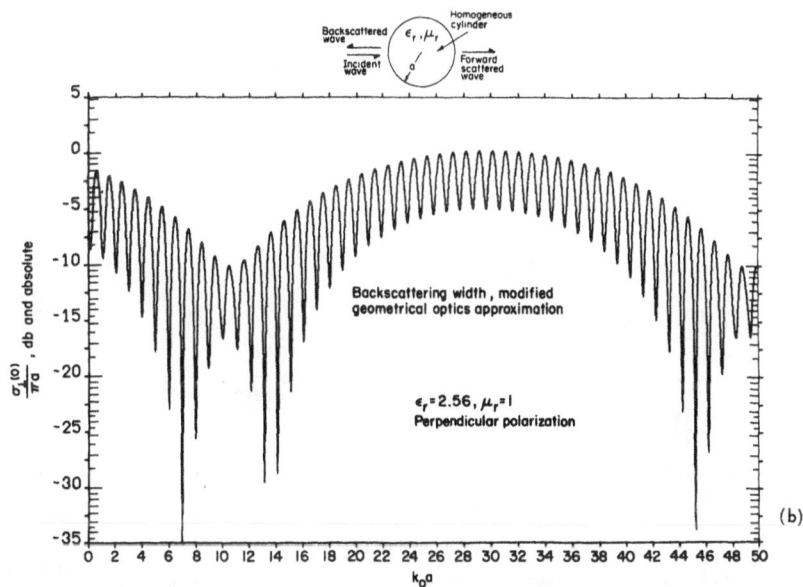

Fig. 4-8. Normalized scattering widths for an infinitely long, lossless homogeneous cylinder with $\epsilon_r = 2.56$, $\mu_r = 1$ for $E$ field perpendicular to cylinder axis (*TE* mode): (a) exact solution for back and forward scattering widths, and (b) modified geometrical optics approximation for backscattering width.

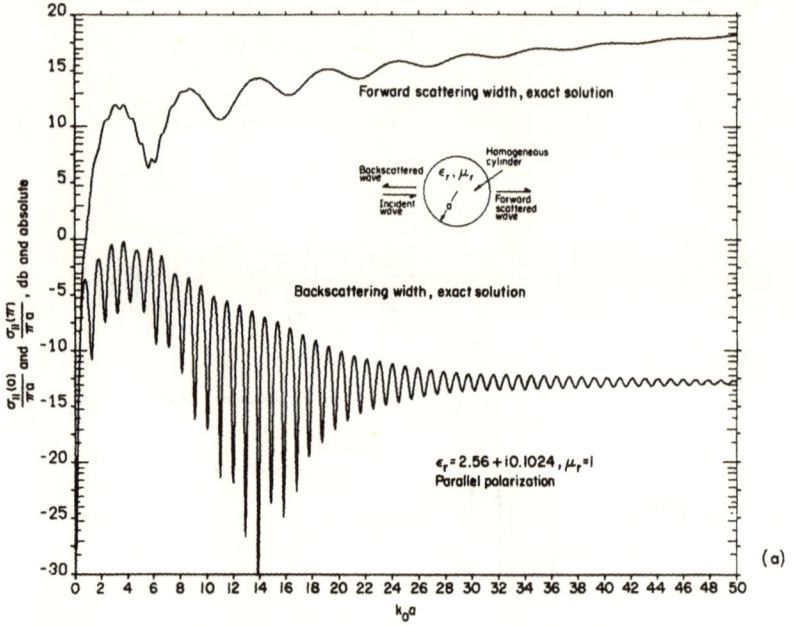

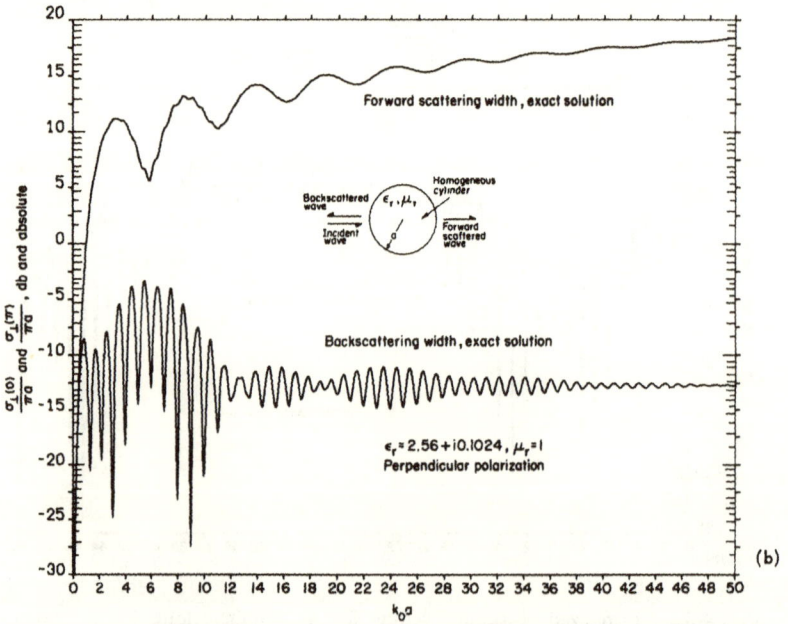

Fig. 4-9. Exact solution for normalized back and forward scattering widths of an infinitely long, slightly lossy homogeneous cylinder with $\epsilon_r = 2.56 + i0.1024$, $\mu_r = 1$: (a) parallel polarization (*TM*), and (b) perpendicular polarization (*TE*).

for a lossless dielectric cylinder of $\epsilon_r = 2.56$, and for a slightly lossy dielectric cylinder of $\epsilon_r = 2.56 + i0.1024$. They are shown in Figures 4-7 through 4-10.[†] It can be seen that the backscattering widths do not approach readily predictable values for large $k_0 a$ for the lossless cylinder, as they do for the perfectly conducting cylinder. Such was also the case for backscattering from the sphere.

### 4.1.3.2. Approximations for the Low-Frequency Region ($k_0 a \ll 1$, $k_1 a \ll 1$)

When $k_0 a \ll 1$, the cylinder appears as a long thin wire, and only the first two terms in the above series are nonvanishing. Representing the Bessel functions by their small argument expansions, the scattered far-zone fields become [from Eqs. (4.1-5)–(4.1-7)]

$$E_\parallel{}^s = E_\parallel{}^i \sqrt{\frac{2}{\pi}} \frac{e^{i(k_0 r - \pi/4)}}{\sqrt{k_0 r}} [A_0 - 2A_1 \cos \phi']$$

$$= E_\parallel{}^i \sqrt{\frac{\pi}{2}} \frac{e^{i(k_0 r + \pi/4)}}{\sqrt{k_0 r}} \frac{(k_0 a)^2}{2} \left[ \left( \frac{\epsilon_1}{\epsilon_0} - 1 \right) - 2 \left( \frac{\mu_1 - \mu_0}{\mu_1 + \mu_0} \right) \cos \phi' \right]$$

(4.1-45)

and

$$H_\perp{}^s = H_\perp{}^i \sqrt{\frac{2}{\pi}} \frac{e^{i(k_0 r - \pi/4)}}{\sqrt{k_0 r}} [B_0 - 2B_1 \cos \phi']$$

$$= H_\perp{}^i \sqrt{\frac{\pi}{2}} \frac{e^{i(k_0 r + \pi/4)}}{\sqrt{k_0 r}} \frac{(k_0 a)^2}{2} \left[ \left( \frac{\mu_1}{\mu_0} - 1 \right) - 2 \left( \frac{\epsilon_1 - \epsilon_0}{\epsilon_1 + \epsilon_0} \right) \cos \phi' \right]$$

(4.1-46)

4.1.3.2.1. *Bistatic Scattering Width* ($k_0 a \ll 1$, $k_1 a \ll 1$). From the above equations one can write the bistatic scattering widths at the angle $\phi'$ as

$$\sigma_\parallel{}^c(\phi') = \pi a \frac{\pi}{4} (k_0 a)^3 \left[ \left( \frac{\epsilon_1}{\epsilon_0} - 1 \right) - 2 \left( \frac{\mu_1 - \mu_0}{\mu_1 + \mu_0} \right) \cos \phi' \right]^2 \quad (4.1\text{-}47)$$

and

$$\sigma_\perp{}^c(\phi') = \pi a \frac{\pi}{4} (k_0 a)^3 \left[ \left( \frac{\mu_1}{\mu_0} - 1 \right) - 2 \left( \frac{\epsilon_1 - \epsilon_0}{\epsilon_1 + \epsilon_0} \right) \cos \phi' \right]^2 \quad (4.1\text{-}48)$$

From these equations it can be seen that for pure dielectric or lossy dielectric cylinders when $\mu_1 = \mu_0$, the scattered power is independent of $\phi'$ if the $E$ field is parallel to the cylinder axis, whereas it is proportional to $\cos^2 \phi'$ when the $E$ field is perpendicular to the axis.

---

[†] Figures 4-11 to 4-16 also give curves for the exact back, bistatic, and forward scattering widths for cylinders with $\epsilon_r = 1.5$, 1.05.

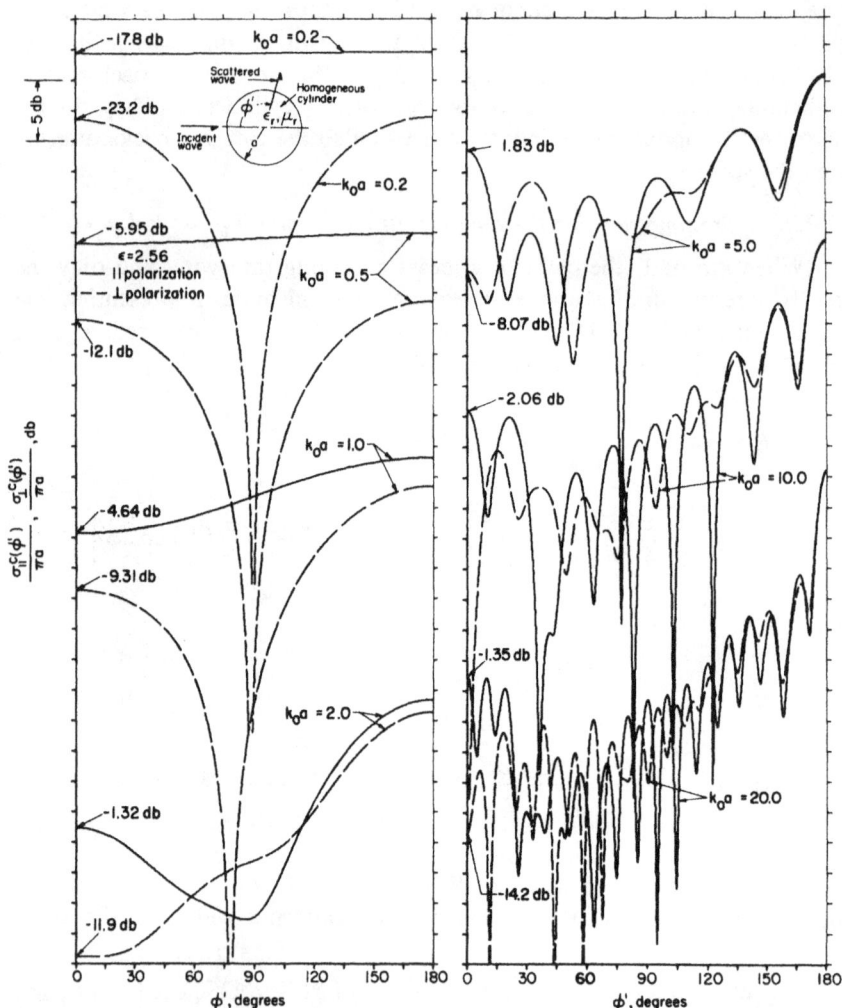

Fig. 4-10a. Exact solution for normalized bistatic scattering widths of an infinitely long, lossless homogeneous cylinder for various values of $k_0a$.

4.1.3.2.2. *Backscattering Width* ($k_0a \ll 1$, $k_1a \ll 1$). In this case, $\phi = 0$, and the above equations become

$$\sigma_\parallel{}^c(0) = \pi a \frac{\pi}{4} (k_0a)^3 \left[ \left( \frac{\epsilon_1}{\epsilon_0} - 1 \right) - 2 \left( \frac{\mu_1 - \mu_0}{\mu_1 + \mu_0} \right) \right]^2 \qquad (4.1\text{-}49)$$

$$\sigma_\perp{}^c(0) = \pi a \frac{\pi}{4} (k_0a)^3 \left[ \left( \frac{\mu_1}{\mu_0} - 1 \right) - 2 \left( \frac{\epsilon_1 - \epsilon_0}{\epsilon_1 + \epsilon_0} \right) \right]^2 \qquad (4.1\text{-}50)$$

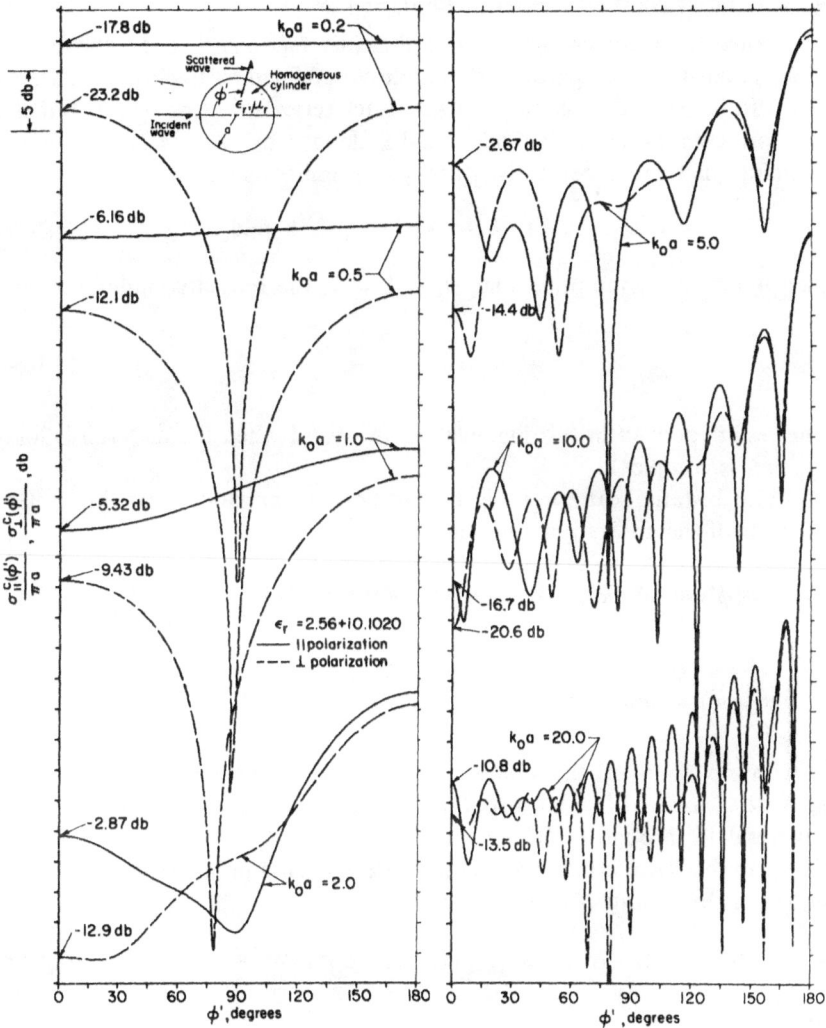

Fig. 4-10b. Same as Figure 4-10a, but for a lossy cylinder.

4.1.3.2.3. *Forward Scattering Width* $(k_0 a \ll 1, k_1 a \ll 1)$. In this direction, $\phi' = \pi$, resulting in

$$\sigma_{\parallel}{}^c(\pi) = \pi a \frac{\pi}{4} (k_0 a)^3 \left[ \left( \frac{\epsilon_1}{\epsilon_0} - 1 \right) + 2 \left( \frac{\mu_1 - \mu_0}{\mu_1 + \mu_0} \right) \right]^2 \qquad (4.1\text{-}51)$$

$$\sigma_{\perp}{}^c(\pi) = \pi a \frac{\pi}{4} (k_0 a)^3 \left[ \left( \frac{\mu_1}{\mu_0} - 1 \right) + 2 \left( \frac{\epsilon_1 - \epsilon_0}{\epsilon_1 + \epsilon_0} \right) \right]^2 \qquad (4.1\text{-}52)$$

4.1.3.3. *Rayleigh–Gans Region* $[2k_0a(m_1 - 1) < 1]$

When the cylinder material is sufficiently tenuous that the above condition is satisfied, then particularly simple closed form results for the scattered field can be derived directly from the exact series solution. The parameter $m_1$ represents the refractive index, and is defined by $m_1 = k_1/k_0 = \sqrt{\mu_r \epsilon_r}$. Defining the relative permittivity and permeability as follows,

$$\epsilon_r = 1 + \Delta_\epsilon \quad \text{and} \quad \mu_r = 1 + \Delta_\mu \tag{4.1-53}$$

then under the assumption that $m_1 - 1 \ll 1$, the refractive index becomes

$$m_1 \approx 1 + \frac{\Delta_\epsilon + \Delta_\mu}{2} \tag{4.1-54}$$

The results derived in this manner are also valid when the material is lossy (i.e., $m_1$ is complex).

The bistatic scattering widths for both polarizations can be written in terms of the above quantities as

$$\sigma_\parallel{}^c(\phi') = \pi^2 a(k_0a)^3 (\Delta_\epsilon - \Delta_\mu \cos \phi')^2 \left[ \frac{J_1[2k_0am_1 \cos (\phi'/2)]}{[2k_0am_1 \cos (\phi'/2)]} \right]^2 \tag{4.1-55}$$

and

$$\sigma_\perp{}^c(\phi') = \pi^2 a(k_0a)^3 (\Delta_\mu - \Delta_\epsilon \cos \phi')^2 \left[ \frac{J_1[2k_0am_1 \cos (\phi'/2)]}{[2k_0am_1 \cos (\phi'/2)]} \right]^2 \tag{4.1-56}$$

A comparison of the above equations with the exact solution when $\epsilon_r = 1.05$, $\mu_r = 1.0$, and also for $\epsilon_r = 1.5$ and $\mu_r = 1.0$, are shown in Figures 4-11 through 4-16.

The backscattering widths are directly obtainable from the above equations by setting $\phi' = 0$:

$$\sigma_\parallel{}^c(0) = \pi^2 a(k_0a)^3 (\Delta_\epsilon - \Delta_\mu)^2 \left[ \frac{J_1(2k_0am_1)}{2k_0am_1} \right]^2 \tag{4.1-57}$$

and

$$\sigma_\perp{}^c(0) = \pi^2 a(k_0a)^3 (\Delta_\mu - \Delta_\epsilon)^2 \left[ \frac{J_1(2k_0am_1)}{2k_0am_1} \right]^2 \tag{4.1-58}$$

The forward scattering widths are obtained by taking the limit of Eqs. (4.1-55) and (4.1-56) as $\phi' \to \pi$, or

$$\sigma_\parallel{}^c(\pi) = \pi^2 a(k_0a)^3 \left( \frac{\Delta_\epsilon + \Delta_\mu}{2} \right)^2 = \sigma_\perp{}^c(\pi) \tag{4.1-59}$$

The above backscatter and forward scattering widths are also compared with exact solutions in Figure 4-11.

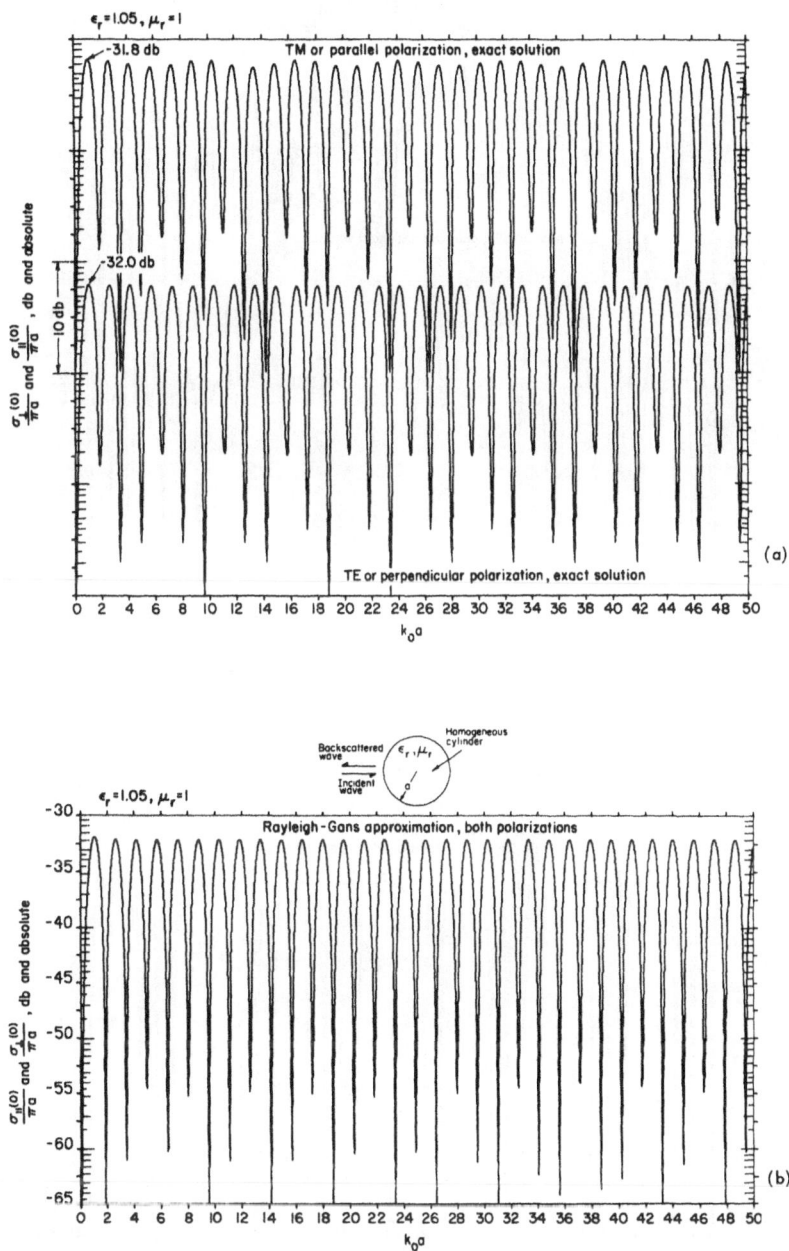

Fig. 4-11. Normalized backscattering widths of an infinitely long, tenuous homogeneous cylinder with $\epsilon_r = 1.05$, $\mu_r = 1$: (a) exact, and (b) Rayleigh–Gans approximation.

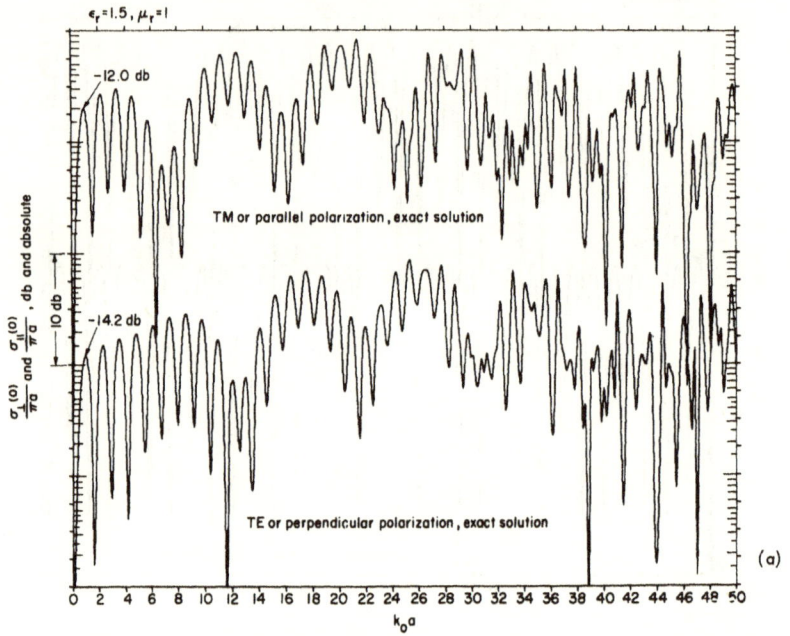

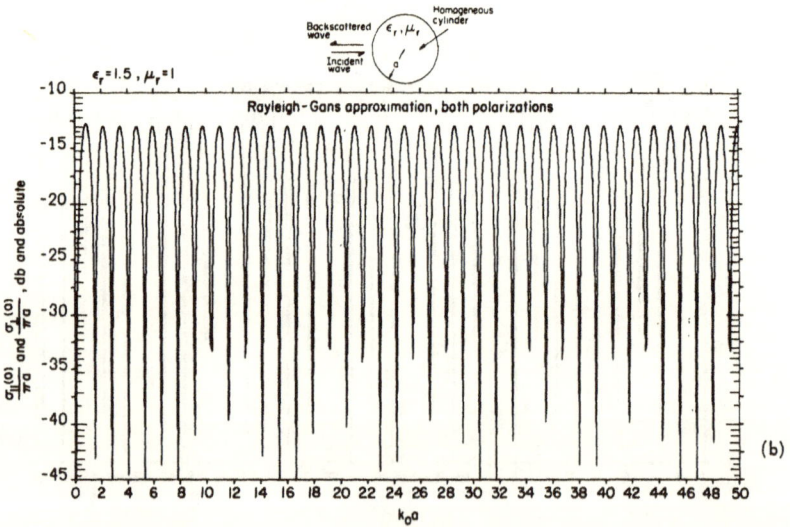

Fig. 4-12. Normalized backscattering widths of an infinitely long, less tenuous homogeneous cylinder with $\epsilon_r = 1.5$, $\mu_r = 1$: (a) exact, and (b) Rayleigh–Gans approximation.

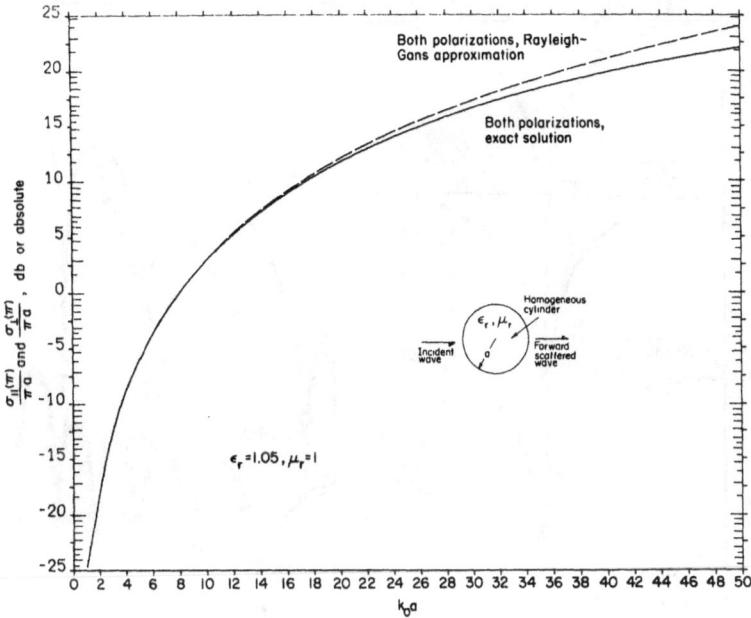

Fig. 4-13. Normalized forward scattering widths of an infinitely long, tenuous homogeneous cylinder with $\epsilon_r = 1.05$, $\mu_r = 1$.

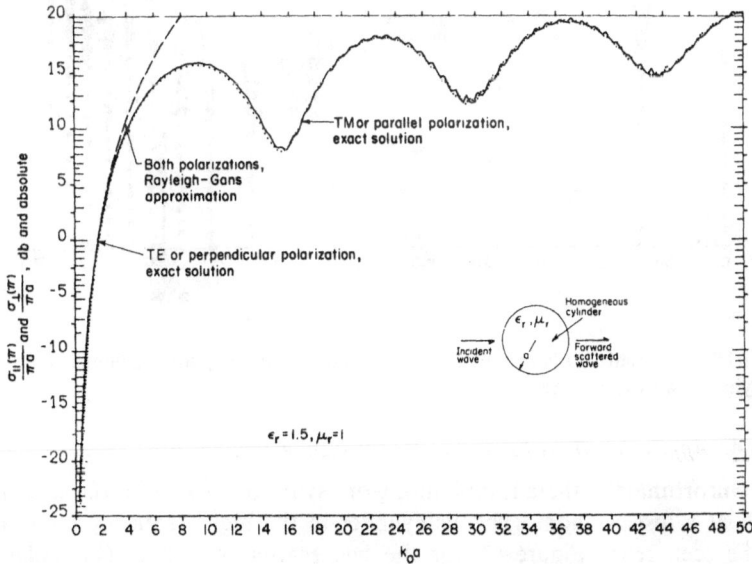

Fig. 4-14. Normalized forward scattering widths of an infinitely long, less tenuous homogeneous cylinder with $\epsilon_r = 1.5$, $\mu_r = 1$.

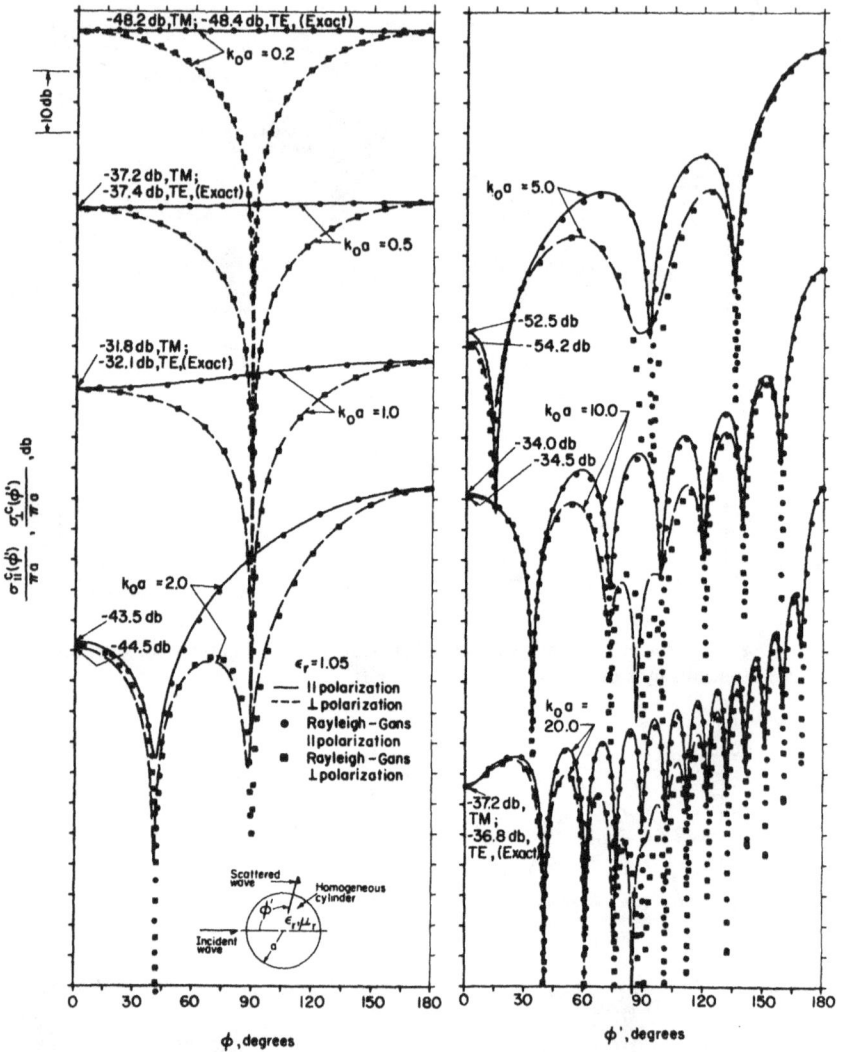

Fig. 4-15. Normalized bistatic scattering widths of an infinitely long, tenuous homogeneous cylinder with $\epsilon_r = 1.05$, $\mu_r = 1$.

### 4.1.3.4. Approximations for the High-Frequency Region ($k_0 a > 1$)

Unfortunately, there is no limiting or asymptotic value for the scattering
width of a lossless homogeneous cylinder in the high frequency limit. This
can be seen from Figure 4-8 for the backscattering width. This behavior
parallels that of the homogeneous sphere. Basically, there are many rays
which enter the cylinder, are reflected one or more times internally, and

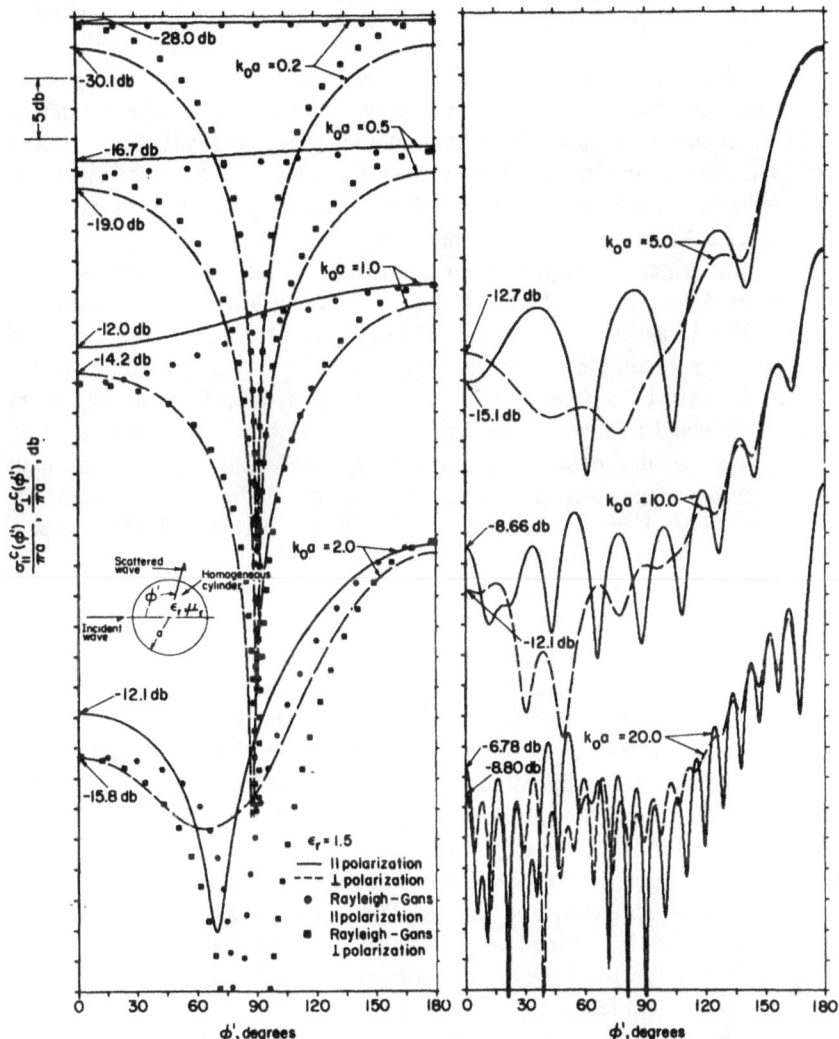

Fig. 4-16. Normalized bistatic scattering widths of an infinitely long, less tenuous, homogeneous cylinder with $\epsilon_r = 1.5$, $\mu_r = 1$.

then leave the cylinder in the desired scattering direction. As the frequency increases, these rays interfere with each other, giving rise to a random-appearing oscillatory behavior of the scattering width. This goes on indefinitely as the frequency continues to increase.

In a practical situation, however, there is some small loss (due to electric or magnetic conductivity) inside the medium, so that eventually as the cylinder radius becomes larger with respect to wavelength, all of the internal rays become attenuated as they propagate through the cylinder. Hence, in the limit none of the rays which enter the cylinder ever exit. In this limit, the only contribution to the scattered field comes from the specular externally reflected ray which never enters the cylinder. The scattering width then becomes identical to that of the perfectly conducting cylinder multiplied by a factor containing the Fresnel reflection coefficient for the medium. In this limit, the resulting scattering widths are given in Section 4.1.3.4.2.

A more rigorous approach to obtaining a high-frequency result for the scattered field is found in Reference 10. Here a Watson transformation is applied to the exact series solution. The scattered field can then be separated into terms interpretable as arising from creeping waves over the cylinder (similar in nature to those on the conducting cylinder), as arising from multiply reflected rays within the cylinder, as "rainbow" or stationary rays, and as the specular externally reflected ray. Unfortunately, even though the derivation is rigorous, practical numerical use of the results to compute the scattered field or scattering width is difficult because of the complex nature of the answer. In order to use the answer, one must evaluate the roots of a complex transcendental equation. Then the answer is expressed as a sum of terms evaluated at these roots, some of which contribute significantly in a given scattering direction, and most of which do not.

Because of such complications, a modified geometrical–optics approach has been formulated and used successfully for backscattering from the homogeneous cylinder by Thomas and Peters.[11,12] The justification for this approximation is based upon the rigorous formulation discussed in the preceding paragraph, but the approach is different. Instead of finding the general solution rigorously, one initially determines which rays will contribute the most in the backscattering direction for a given cylinder permeability, permittivity, and radius. Then one solves for these rays by geometrical optics (combined with physical optics) methods. These results compare favorably with those of the rigorous approach if all of the significant rays have been included.

The results of this method are presented only for the backscattering direction. The more general case of bistatic scattering can also be treated by this method, but will not be discussed due to lack of space. Note that these results are valid for both lossless and lossy cylinders.

4.1.3.4.1. *Modified Geometrical-Optics Approximation for Backscattering* ($k_0 a > 1$). Writing the parallel (or perpendicular) component of the backscattered field as a sum of all the significant ray contributions, then

$$E_\parallel^s = E_\parallel^i \sqrt{a/2r} \, e^{ik_0 r} [f_\parallel^{\,0} \, e^{i\psi_0} + f_\parallel^{a1} \, e^{i\psi_{a1}} + f_\parallel^{a2} \, e^{i\psi_{a2}} + \cdots$$

$$+ f_\parallel^{g1} \, e^{i\psi_{g1}} + f_\parallel^{g2} \, e^{i\psi_{g2}} + \cdots + f_\parallel^{s1} \, e^{i\psi_{s1}} + f_\parallel^{s2} \, e^{i\psi_{s2}} + \cdots] \quad (4.1\text{-}60)$$

A similar result holds for the other polarization, when $\|$ is replaced by $\perp$. The first term is the optical or specular component, the terms superscripted by "$a$" are axial rays, the terms superscripted by "$g$" are glory rays, and terms superscripted by "$s$" are stationary rays. Only rays which are reflected internally a small number of times will be considered here.

The backscattering width then becomes

$$\sigma_\|{}^c(0) = \pi a \, | f_\|{}^0 \, e^{i\psi_0} + f_\|^{a1} \, e^{i\psi_{a1}} + \cdots + f_\|^{g1} \, e^{i\psi_{g1}} + \cdots + f_\|^{s1} \, e^{i\psi_{s1}} + \cdots |^2$$

(4.1-61)

with an analogous result for $\sigma_\perp{}^c(0)$.

Expressions for the various ray amplitudes $f$, and phases $\psi$, are given below:

(a) *Specular Ray* (see Figure 4-17)

This ray is reflected from the front face and has the value[12]

$$f_\|{}^0 = R_\|^{ext}(0) \quad \text{and} \quad f_\perp{}^0 = R_\perp^{ext}(0) \quad (4.1-62)$$

(a) Specular Ray

(b) Axial Rays

(c) Glory Rays

(d) Stationary Rays

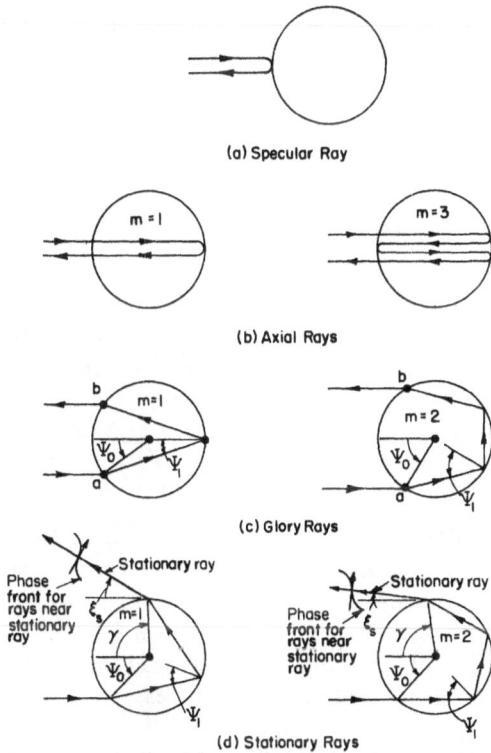

Fig. 4-17.   Diagram showing lowest-order rays used in modified geometrical-optics approximation.

with

$$\psi_0 = 0 \qquad (4.1\text{-}63)$$

i.e., the phase angles are referenced with respect to the front face. $R_{\parallel}^{ext}(\phi')$ and $R_{\perp}^{ext}(\phi')$ are the Fresnel reflection coefficients given in general by Eqs. (7.1-9), (7.1-10); and when $\phi' = 0$ (backscattering), they are both equal in magnitude, or

$$R_{\parallel}^{ext}(0) = -R_{\perp}^{ext}(0) = \frac{\sqrt{\mu_r} - \sqrt{\epsilon_r}}{\sqrt{\mu_r} + \sqrt{\epsilon_r}} \qquad (4.1\text{-}64)$$

(b) *Axial Rays* (see Figure 4-17)

Let $m$ be the number of internal reflections (for axial rays $m$ must be odd); then the contributions due to these rays are[†]

$$f_{\parallel/\perp}^{am} = [T_{\parallel/\perp}^{in}(0)(R_{\parallel/\perp}^{int}(0))^m \, T_{\parallel/\perp}^{out}(0)]\left[1 - \frac{m+1}{\sqrt{\epsilon_r\mu_r}}\right]^{-1/2} \qquad (4.1\text{-}65)$$

and

$$\psi_{am} = 2\sqrt{\mu_r\epsilon_r}\,(m+1)\,k_0 a - m\pi/2 \qquad (4.1\text{-}66)$$

where $T_{\parallel}(\phi')$ and $T_{\perp}(\phi')$ are the Fresnel transmission coefficients given in general by Eqs. (7.1-5) to (7.1-7). When $\phi' = 0$, they become

$$T_{\parallel}^{in}(0) = T_{\perp}^{in}(0) = \frac{2\sqrt{\mu_r}}{\sqrt{\mu_r} + \sqrt{\epsilon_r}} \qquad (4.1\text{-}67)$$

and

$$T_{\parallel}^{out}(0) = T_{\perp}^{out}(0) = \frac{2\sqrt{\epsilon_r}}{\sqrt{\mu_r} + \sqrt{\epsilon_r}} \qquad (4.1\text{-}68)$$

$T_{\parallel}^{in}$ corresponds to a plane wave traveling from free space into the medium with constitutive constants $\epsilon_r$, $\mu_r$, and $T_{\perp}^{out}$, a wave traveling out from the medium into free space. $R_{\parallel}^{int}$ and $R_{\perp}^{int}(0)$ are the reflection coefficients for the rays inside the cylinder and are given by

$$R_{\parallel}^{int}(0) = -R_{\perp}^{int}(0) = -\frac{\sqrt{\mu_r} - \sqrt{\epsilon_r}}{\sqrt{\mu_r} + \sqrt{\epsilon_r}} \qquad (4.1\text{-}69)$$

If the material is lossy, then $(m+1)\sqrt{\epsilon_r\mu_r}$ in the phase angle $\psi_{am}$ will have an imaginary part which will produce attenuation of the ray. Also, since the reflection coefficient appearing in (4.1-65) is always less than unity, its $m$th power is still smaller. Thus, in most practical cases, rays internally reflected many times (e.g., $m > 3$) are negligible in magnitude compared with those reflected only a few times.

---

[†] The first factor in braces represents the effects of interactions at the interface (i.e., transmission and reflection), while the second factor represents the effects of curvature of the interfaces on the egressing ray.

(c) *Glory Rays* (see Figure 4-17)

Again $m$ is the number of internal reflections. Note that there are always two emerging rays in the backscattering direction for a given $m$; one which enters at point $a$ (see Figure 4-17) and emerges at $b$, and the symmetrical ray (not shown) which enters at $b$ and leaves at $a$. They are always equal to each other in all cases; consequently, the results presented here will be multiplied by 2, and thus will include both rays

$$f^{gm}_{\parallel/\perp} = 2[T^{in}_{\parallel/\perp}(\Psi_0)(R^{int}_{\parallel/\perp}(\Psi_1))^m \, T^{out}_{\parallel/\perp}(\Psi_1)]$$

$$\left[\frac{\cos \Psi_0 \, \sqrt{\epsilon_r \mu_r - \sin^2 \Psi_0}}{\sqrt{\epsilon_r \mu_r - \sin^2 \Psi_0} - (m+1) \cos \Psi_0}\right]^{1/2} \tag{4.1-70}$$

and

$$\psi_{gm} = 2k_0 a(1 - \cos \Psi_0) + 2\sqrt{\epsilon_r \mu_r}\,(m+1)\,k_0 a \cos \Psi_1 - m\pi/2 \tag{4.1-71}$$

The Fresnel reflection and transmission coefficients appearing in the above equations can be obtained from Eqs. (7.1-4) to (7.1-7), and are

$$T^{in}_{\parallel}(\Psi) = \frac{2\sqrt{\mu_r/\epsilon_r}\,[1 - (\sin^2 \Psi)/\mu_r \epsilon_r]^{1/2}}{\cos \Psi + \sqrt{\mu_r/\epsilon_r}\,[1 - (\sin^2 \Psi)/\mu_r \epsilon_r]^{1/2}} \tag{4.1-72}$$

$$T^{in}_{\perp}(\Psi) = \frac{2 \cos \Psi}{\cos \Psi + \sqrt{\epsilon_r/\mu_r}\,[1 - (\sin^2 \Psi)/\mu_r \epsilon_r]^{1/2}} \tag{4.1-73}$$

$$T^{out}_{\parallel}(\Psi) = \frac{2\sqrt{\epsilon_r/\mu_r}\,(1 - \epsilon_r \mu_r \sin^2 \Psi)^{1/2}}{\cos \Psi + \sqrt{\epsilon_r/\mu_r}\,(1 - \epsilon_r \mu_r \sin^2 \Psi)^{1/2}} \tag{4.1-74}$$

$$T^{out}_{\perp}(\Psi) = \frac{2 \cos \Psi}{\cos \Psi + \sqrt{\mu_r/\epsilon_r}\,(1 - \mu_r \epsilon_r \sin^2 \Psi)^{1/2}} \tag{4.1-75}$$

$$R^{int}_{\parallel}(\Psi) = -\frac{\sqrt{\mu_r/\epsilon_r}\,\cos \Psi - [1 - (\sin^2 \Psi)/\mu_r \epsilon_r]^{1/2}}{\sqrt{\mu_r/\epsilon_r}\,\cos \Psi + [1 - (\sin^2 \Psi)/\mu_r \epsilon_r]^{1/2}} \tag{4.1-76}$$

$$R^{int}_{\perp}(\Psi) = \frac{\sqrt{\epsilon_r/\mu_r}\,\cos \Psi - (1 - \mu_r \epsilon_r \sin^2 \Psi)^{1/2}}{\sqrt{\epsilon_r/\mu_r}\,\cos \Psi + (1 - \mu_r \epsilon_r \sin^2 \Psi)^{1/2}} \tag{4.1-77}$$

The angles, $\Psi_0$ and $\Psi_1$, are the angle of incidence and angle of refraction of the ray striking the interface. They are determined from Snell's law, or

$$\sin \Psi_0 = \sqrt{\epsilon_r \mu_r}\,\sin \Psi_1 \tag{4.1-78}$$

and from one of the following equations obtained from geometrical considerations:

(i) $m = 1$ (one internal reflection)

$$2\Psi_0 - 4\Psi_1 = 0 \qquad (4.1\text{-}79)$$

(ii) $m = 2$ (two internal reflections)

$$\pi + 2\Psi_0 - 6\Psi_1 = 0 \qquad (4.1\text{-}80)$$

(iii) $m = 3$ (three internal reflections)

$$2\pi + 2\Psi_0 - 8\Psi_1 = 0 \quad \text{or} \quad 2\pi \qquad (4.1\text{-}81)$$

Using Eq. (4.1-78) and one of the following three equations applicable to the case, the angles $\Psi_0$ and $\Psi_1$ can be determined for a given refractive index $\sqrt{\epsilon_r \mu_r}$. However, a solution may not exist for all values of refractive index. For example, if $m = 1$, a solution exists only for $1 \leqslant \sqrt{\epsilon_r \mu_r} \leqslant 2$; this means, physically, that this ray ($m = 1$) does not exist if $\sqrt{\epsilon_r \mu_r}$ does not lie within this range.

More simply, $\Psi_0$ and $\Psi_1$ may be determined using Figure 4-18. One merely finds the intersection between the curve corresponding to the given

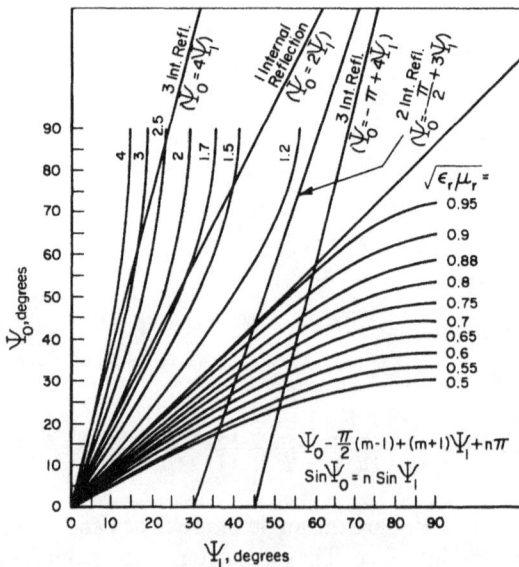

Fig. 4-18. Diagram for determination of internal angles at which glory rays can exist [after Lee et al.[12]].

value of $\sqrt{\epsilon_r \mu_r}$, and the straight line corresponding to the number of internal reflections, $m$. If no intersection exists, then that ray does not exist.

As before, $\sqrt{\epsilon_r \mu_r}$ may be complex; in this case, an imaginary term in Eq. (4.1-71) contributes to the attenuation of the higher order rays because of the factor $\sqrt{\epsilon_r \mu_r} (m + 1)$. Consider the example where $\epsilon_1 = 2.56$, $\mu_r = 1$. In this case, only the single internally reflected ray ($m = 1$) exists (of the three considered above), and the angles $\Psi_0$ and $\Psi_1$ are (from Figure 4-18) $\Psi_0 = 73.6°$ and $\Psi_1 = 36.8°$.

(d) *Stationary (or Rainbow) Rays* (see Figure 4-17)

As the incidence angle, $\theta_0$, of an entering ray varies, the emergence angle $\zeta$ also varies. It can happen, however, that the rate of change of $\zeta$ with $\theta_0$ falls to zero (i.e., $d\zeta/d\theta_0$) at some value of $\zeta$ which will be called $\zeta_s$. The ray emerging at this angle $\zeta_s$ is called a stationary ray. If the value of $\zeta_s$ is relatively small so that the stationary ray emerges near the backscattering direction, then it can contribute to the backscatter width. (Even though the ray does not emerge exactly in the backscattering direction). This is due to the fact that the phase front of the rays near the stationary ray passes through a stationary phase point (so that $d\psi_{sm}/d\zeta_s = 0$) which is cubic in nature. Thus, when one integrates emerging rays over a surface through the stationary phase point, a contribution at angles close to $\zeta_s$ is obtained. A more detailed discussion of stationary rays may be found in Reference 13.

The stationary ray contribution to backscattering is[†]

$$f^{sn}_{\parallel/\perp} = 2(k_0 a)^{1/6} \frac{h^{-1/3}}{\sqrt{2\pi}} [T^{in}_{\parallel/\perp}(\Psi_0)(R^{int}_{\parallel/\perp}(\Psi_1))^m T^{out}_{\parallel/\perp}(\Psi_1)]$$

$$\times 2^{3/2} | Ai(-z)| \left[-1 + 2(m + 1) \frac{\cos \Psi_0}{\sqrt{\epsilon_r \mu_r} \cos \Psi_1}\right]^{-1/2} \quad (4.1\text{-}82)$$

$$\psi_{sm} = k_0 a(1 - \cos \Psi_0) + 2 \sqrt{\epsilon_r \mu_r} (m + 1) k_0 a \cos \Psi_1$$
$$+ k_0 a(1 - \cos \gamma) - m\pi/2 \quad (4.1\text{-}83)$$

where the angle $\gamma$ is shown in Figure 4-17 and given below. The reflection and transmission coefficients are the same as those for the glory rays discussed previously. The quantity $Ai(-z)$ is the Airy function of argument $-z$, and its definition, properties, and tables can be found in Section 10.4 of Reference 6. The remaining parameters appearing in Eqs. (4.1-82) and (4.1-83) for the stationary ray are

$$z = \frac{(k_0 a)^{2/3}}{h^{1/3}} \zeta_s \quad (4.1\text{-}84)$$

---

[†] As with the glory ray, there are actually two rays which contribute; one which enters in the lower semicircle and exits from the upper as shown in Figure 4-17, and the symmetrical component which enters in the upper and exits from the lower semicircle. The two contributions are identical, and both are included here by the factor 2.

($\zeta_s$ in radians), where

$$h = \frac{m(m+2) \tan \Psi_0}{(m+1)^2 \cos^2 \Psi_0} \tag{4.1-85}$$

The quantities, $\Psi_0$ and $\Psi_1$, for the stationary ray are determined from

$$\cos \Psi_0 = \sqrt{\frac{\epsilon_r \mu_r - 1}{m(m+2)}} \quad \text{and} \quad \cos \Psi_1 = (m+1) \sqrt{\frac{\epsilon_r \mu_r - 1}{m(m+2) \epsilon_r \mu_r}}$$

$$\tag{4.1-86}$$

and the angles $\gamma$ and $\zeta_s$ are

$$\gamma = 2n\pi - \Psi_0 - (m+1)(\pi - 2\Psi_1) \tag{4.1-87}$$

(where $n = 1$ for $m = 1$ or 2, and $n = 1$ or 2 for $m = 3$) with

$$\zeta_s = \gamma - \psi_0 \tag{4.1-88}$$

A stationary ray for a given $m$ will not exist if the cosines defined in Eq. (4.1-86) are greater than unity or imaginary (e.g., if $\epsilon_r \mu_r < 1$). Even if the stationary ray does exist, it is significant only if $\zeta_s$ is less than approximately 20°.

As an example, consider a dielectric cylinder with $\epsilon_r = 2.56$. Then, using the above equations, one can show that only the singly-reflected ray ($m = 1$) produces a stationary ray near the backscattering direction. For this ray, $\Psi_0 = 44°$, $\Psi_1 = 25.7°$, and $\zeta_s = 15.0°$. Then $h = 1.39$ and $z = 0.232(k_0 a)^{2/3}$. It should be noted that for $k_0 a \gg 1$, $\text{Ai}(-z)$ is proportional to $(k_0 a)^{-1/6}$; this makes the magnitude of $f^{sm}$ independent of $k_0 a$ in the high-frequency limit.

Curves of the backscattering width for both polarizations with $\epsilon_r = 2.56$ are shown in Figures 4-7, and 4-8. In obtaining these curves, the following rays were included: (a) the specular ray, (b) the first two axial rays (for $m = 1$ and $m = 3$), (c) the glory ray for $m = 1$, and (d) the stationary ray contribution for $m = 1$. The actual magnitude and phase for each of these rays is given in Reference 14.

At this time, results employing this approach are available only for backscattering. However, the method can also be used for bistatic and forward scattering.[11,12]

4.1.3.4.2. *Scattering Widths in the High-Frequency Limit With Attenuation* [$\text{Im}(k_1 a) > 2$]. In the event that the cylinder medium has sufficient attenuation [i.e., $\text{Im}(k_1 a) > 2$], then the bistatic scattering widths are

$$\sigma_{\parallel}°(\phi') = \pi a \, | \, R_{\parallel}(\phi'/2)|^2 \cos^2(\phi'/2) \tag{4.1-89}$$

and

$$\sigma_{\perp}°(\phi') = \pi a \, | \, R_{\perp}(\phi'/2)|^2 \cos^2(\phi'/2) \tag{4.1-90}$$

where $R_\parallel(\alpha)$ and $R_\perp(\alpha)$ are the Fresnel reflection coefficients defined and evaluated in Section 7.1, Eqs. (7.1-9), and (7.1-10).

For backscatter, where $\phi' = 0$, Eqs. (4.1-89) and (4.1-90) reduce to

$$\sigma_\parallel{}^c(0) = \sigma_\perp{}^c(0) = \pi a \left| \frac{\sqrt{\mu_1} - \sqrt{\epsilon_1}}{\sqrt{\mu_1} + \sqrt{\epsilon_1}} \right|^2 \qquad (4.1\text{-}91)$$

and this is valid for $\text{Im}(k_1 a) > 5$.

For forward scatter when $k_0 a > 20$, as with the perfectly conducting cylinder, diffraction rather than reflection and refraction accounts for the significant portion of the forward scattered field. Thus, regardless of the cylinder material, the results are identical with those of Section 4.1.2.4.5 and are not repeated here. This behavior is illustrated in Figures 4-7 to 4-9.

### 4.1.4. Coated Perfectly Conducting Cylinder

#### 4.1.4.1. *Series Coefficients*

Let a plane wave be incident upon a coated perfectly conducting cylinder of radius $a_1$ as illustrated in Figure 4-19. The coating has a thickness $a_0 - a_1$ and is homogeneous with permittivity $\epsilon_1$ and permeability $\mu_1$, both of which may be complex if the coating has electric or magnetic losses. Both incident polarization cases are considered.[5,15] The series coefficients, $A_n$ and $B_n$, for each polarization are

$$A_n = - \frac{J_n(k_0 a_0) - i Z_n J_n'(k_0 a_0)}{H_n^{(1)}(k_0 a_0) - i Z_n H_n'^{(1)}(k_0 a_0)} \qquad (4.1\text{-}92)$$

and

$$B_n = - \frac{J_n(k_0 a_0) - i Y_n J_n'(k_0 a_0)}{H_n^{(1)}(k_0 a_0) - i Y_n H_n'^{(1)}(k_0 a_0)} \qquad (4.1\text{-}93)$$

These coefficients, when written in the above form, are general and can apply to any homogeneous, radially inhomogeneous, or stratified cylinder. They are similar in form to those for the series solution for a sphere [Eqs. (3.4-1) and (3.4-2)], where the constants $Z_n$ and $Y_n$ were called the modal surface impedances and admittances. The same significance and designations will be retained here.

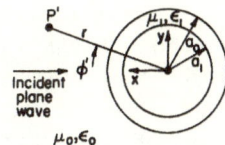

Fig. 4-19. Scattering geometry for an infinitely long, coated, perfectly conducting cylinder with axis perpendicular to the page.

For the coated cylinder, these impedances and admittances become

$$iZ_n = \frac{k_0\mu_1}{k_1\mu_0}\left[\frac{J_n(k_1a_0)\,H_n^{(1)}(k_1a_1) - H_n^{(1)}(k_1a_0)\,J_n(k_1a_1)}{J_n'(k_1a_0)\,H_n^{(1)}(k_1a_1) - H_n'^{(1)}(k_1a_0)\,J_n(k_1a_1)}\right]$$

$$= \frac{k_0\mu_1}{k_1\mu_0}\left[\frac{J_n(k_1a_0)\,N_n(k_1a_1) - N_n(k_1a_0)\,J_n(k_1a_1)}{J_n'(k_1a_0)\,N_n(k_1a_1) - N_n'(k_1a_0)\,J_n(k_1a_1)}\right] \qquad (4.1\text{-}94)$$

$$iY_n = \frac{k_0\epsilon_1}{k_1\epsilon_0}\left[\frac{J_n(k_1a_0)\,H_n'^{(1)}(k_1a_1) - H_n^{(1)}(k_1a_0)\,J_n'(k_1a_1)}{J_n'(k_1a_0)\,H_n'^{(1)}(k_1a_1) - H_n'^{(1)}(k_1a_0)\,J_n'(k_1a_1)}\right]$$

$$= \frac{k_0\epsilon_1}{k_1\epsilon_0}\left[\frac{J_n(k_1a_0)\,N_n'(k_1a_1) - N_n(k_1a_0)\,J_n'(k_1a_1)}{J_n'(k_1a_0)\,N_n'(k_1a_1) - N_n'(k_1a_0)\,J_n'(k_1a_1)}\right] \qquad (4.1\text{-}95)$$

Here $N_n(x)$ is the cylindrical Neumann function (or Bessel function of the second kind), and $H_n^{(1)}(x) = J_n(x) + iN_n(x)$ is the Hankel function of the first kind. Primes indicate derivatives of the functions with respect to the arguments.

Numerical machine solutions for the scattered fields and scattering widths using these coefficients require the computation of three sets of Bessel functions (with arguments $k_0a_0$, $k_1a_0$, and $k_1a_1$), as contrasted with two for the homogeneous cylinder (with arguments $k_0a$ and $k_1a$), and one set for the perfectly conducting cylinder (with argument $k_0a$). The machine time required for solution increases only in the amount of time required to compute the additional Bessel functions, which is generally not the major contribution to total machine time.

Curves of the exact backscattering widths (taken from Reference 5 and normalized as $\sigma k_0$) are shown in Figure 4-20 for various coating thicknesses along with some measured points. The $E$-field polarization is parallel to the cylinder axis. The center cylinder or core thickness in each figure is held constant (along with frequency), while the coating thickness is varied. It is seen that certain coating thicknesses can minimize or maximize the backscattered power at given frequencies. These results should be compared with that for a perfectly conducting uncoated cylinder of varying radius as illustrated in Figure 4-5.

Plots of backscattering width (normalized as $\sigma^0/\pi a_0$) are shown as a function of $k_0a_0$ for fixed cylinder configurations [$(a_1/a_0)$ constant] for both polarization states in Figures 4-21 through 4-23. It can be seen that only when the coating material is lossy does the backscattered power eventually decrease as the frequency increases.

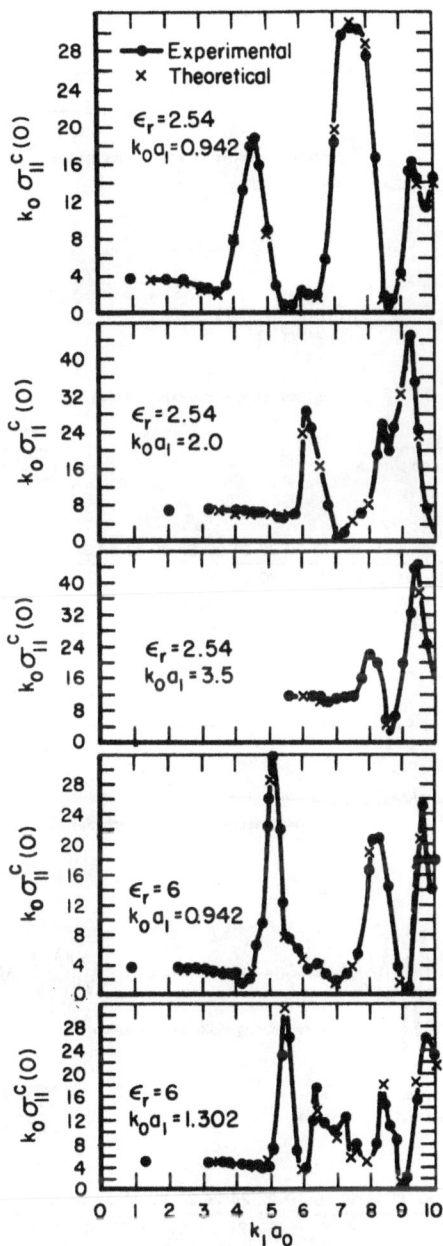

Fig. 4-20. Exact solution and measured points for backscattering widths for an infinitely long, coated, perfectly conducting cylinder as coating thickness varies (normalized as $\sigma k_0$) [after Tang[5]].

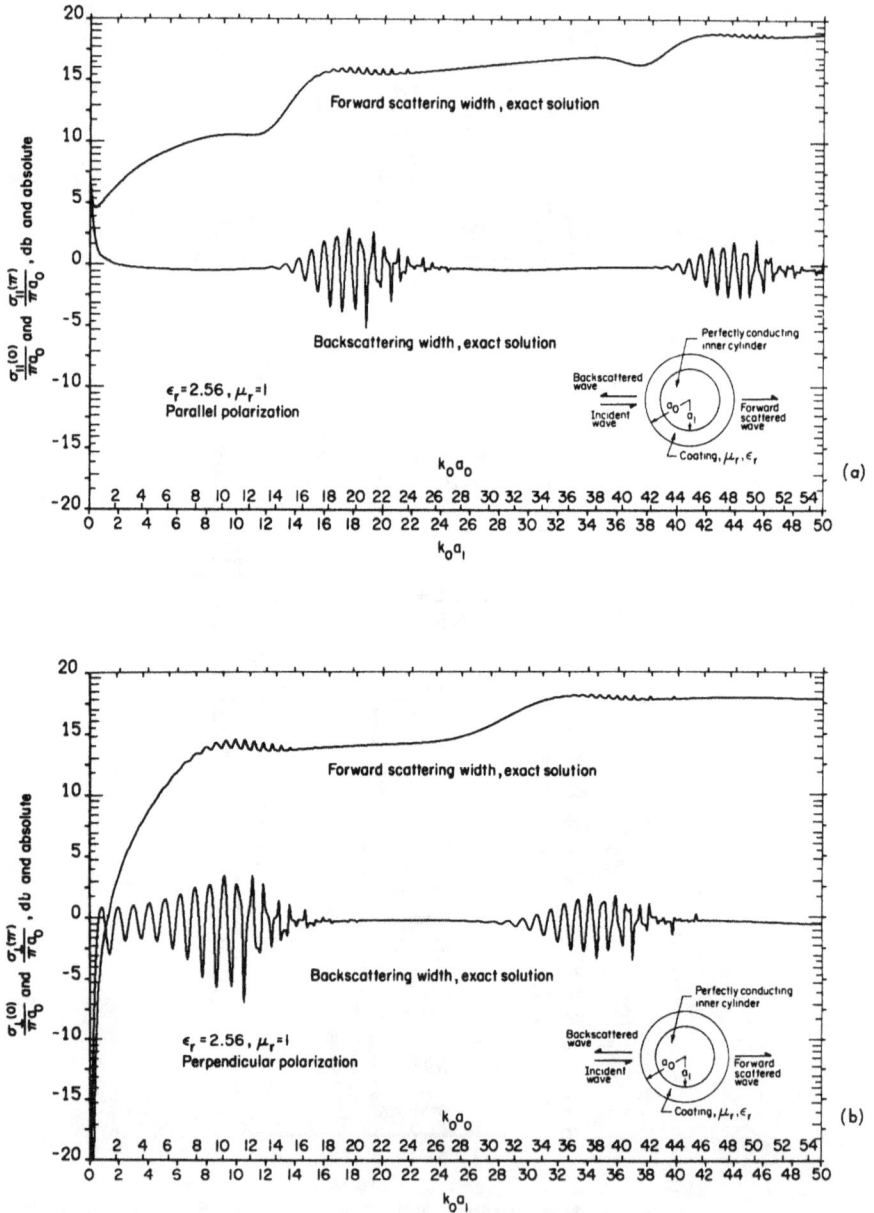

Fig. 4-21. Exact solution for back and forward scattering widths at parallel polarization of an infinitely long, coated, perfectly conducting cylinder: (a) lossless coating, and (b) lossy coating.

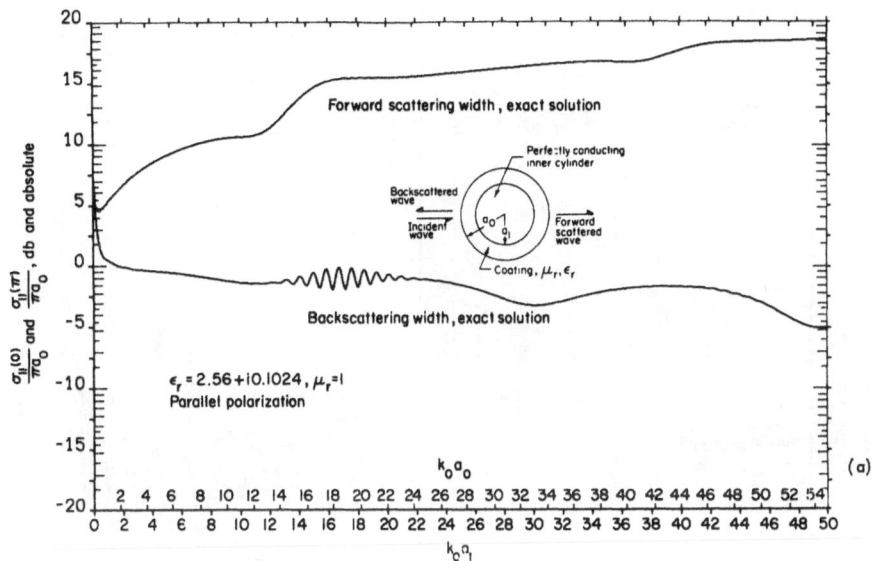

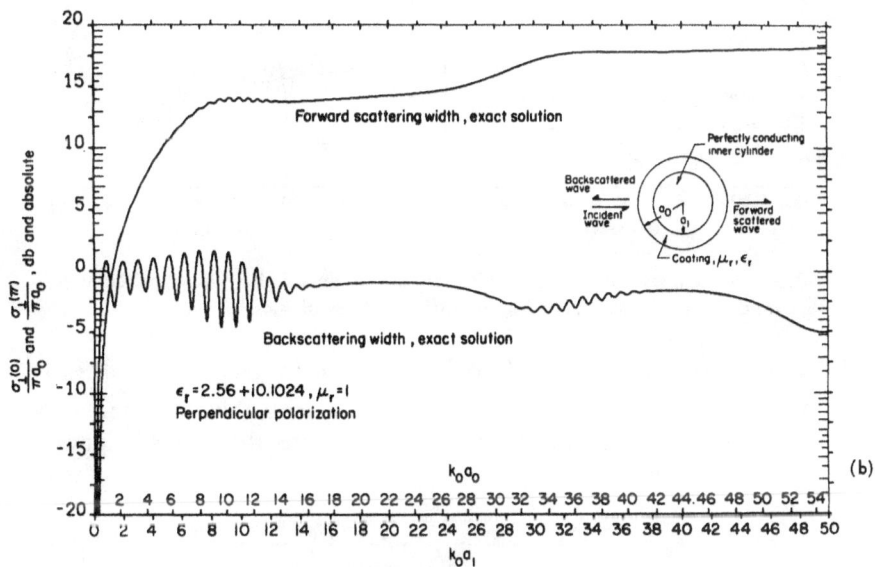

Fig. 4-22. Exact solution for back and forward scattering widths at perpendicular polarization of an infinitely long, coated, perfectly conducting cylinder: (a) lossless coating, and (b) lossy coating.

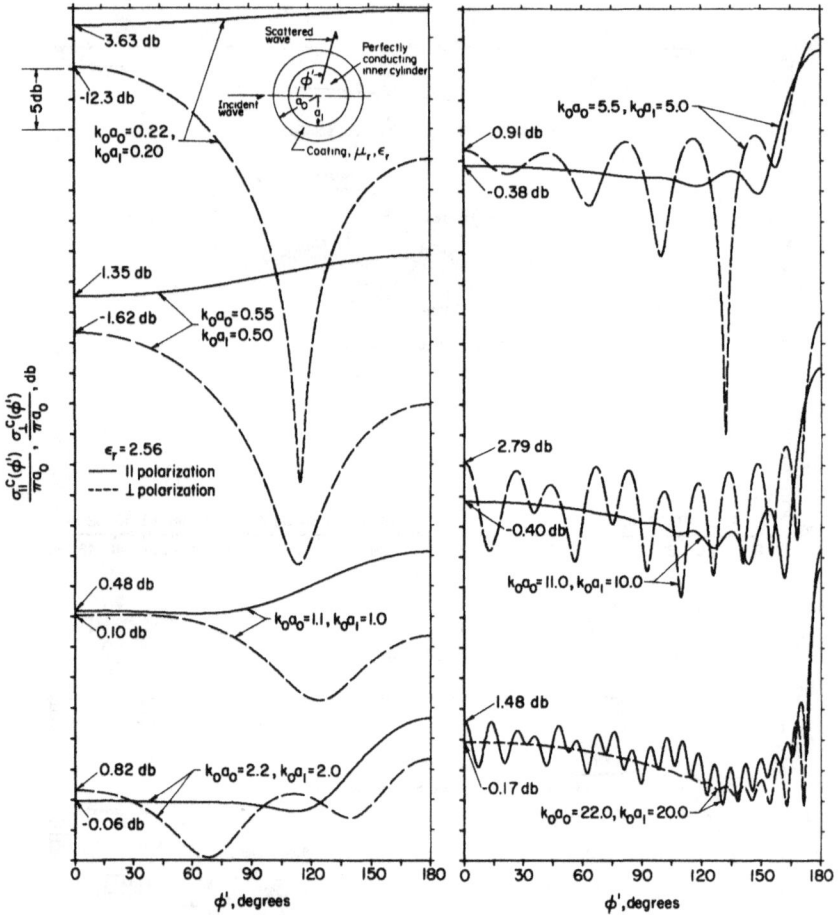

Fig. 4-23a. Normalized bistatic scattering widths of an infinitely long, perfectly conducting cylinder with a lossless coating for various values of $k_0 a_0$.

### 4.1.4.2. Approximations for the Low-Frequency Region ($k_0 a_0 \ll 1$, $k_1 a_0 \ll 1$)

When $k_0 a_0 \ll 1$, the cylinder appears as a long thin wire and only the first terms in the series solutions contribute to the scattered power. Then the first coefficients simplify to

$$A_0 = -i\,\frac{\pi}{2}\,\frac{1 + i\frac{1}{2}Z_0 k_0 a_0}{\dfrac{iZ_0}{k_0 a_0} - [\ln(0.8905 k_0 a_0) - i\pi/2]}$$

$$A_1 = i\,\frac{\pi}{4}\,(k_0 a_0)^2\,\frac{iZ_1/k_0 a_0 - 1}{iZ_1/k_0 a_0 + 1} \tag{4.1-96}$$

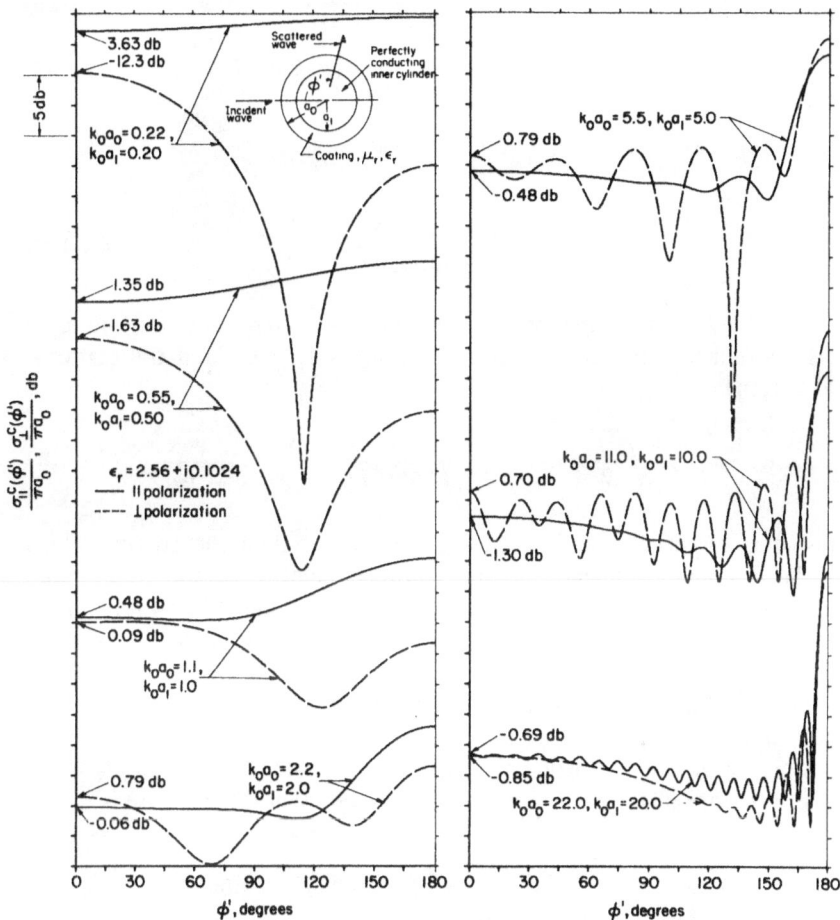

Fig. 4-23b.  Same as Figure 4-23a, but for a lossy coating.

and

$$B_0 = -i\,\frac{\pi}{2}\,\frac{1 + i\tfrac{1}{2}Y_0 k_0 a_0}{iY_0/k_0 a_0 - [\ln(0.8905 k_0 a_0) - i\pi/2]}$$

(4.1-97)

$$B_1 = i\,\frac{\pi}{4}\,(k_0 a_0)^2\,\frac{iY_1/k_0 a_0 - 1}{iY_1/k_0 a_0 + 1}$$

When in addition $k_1 a_0 \ll 1$, the surface impedances and admittances become

$$iZ_0 = \frac{(\mu_1/\mu_0)\ln(a_1/a_0)}{1 + \tfrac{1}{2}(k_1 a_0)^2 \ln(0.60653 a_1/a_0)}\,(k_0 a_0) \approx (\mu_1/\mu_0)\ln(a_1/a_0)\,k_0 a_0 \quad (4.1\text{-}98)$$

for $a_1 \neq 0$, and

$$iY_0 = -2(\mu_0/\mu_1) \frac{1}{1 - (a_1/a_0)^2} (k_0 a_0)^{-1} \qquad (4.1\text{-}99)$$

$$iZ_1 = (\mu_1/\mu_0) \frac{1 - (a_1/a_0)^2}{1 + (a_1/a_0)^2} (k_0 a_0)$$

$$iY_1 = (\epsilon_1/\epsilon_0) \frac{1 + (a_1/a_0)^2}{1 - (a_1/a_0)^2} (k_0 a_0) \qquad (4.1\text{-}100)$$

For the $E$ field parallel to the axis, only the $A_0$ term is significant since $A_1$ is of order $(k_0 a_0)^2$ greater than $A_0$ (so long as $a_1 \neq 0$), and the scattered field becomes

$$E_{\parallel}{}^s = E_{\parallel}{}^i \sqrt{\frac{\pi}{2}} \frac{e^{i(k_0 r + \pi/4)}}{\sqrt{k_0 r}} \frac{1}{\ln[0.89 k_0 a_0 (a_1/a_0)^{\mu_1/\mu_0}] - i\pi/2} \qquad (4.1\text{-}101)$$

Thus, it is essentially the conducting core rather than the coating which has the dominant influence on the scattered field, since the result more nearly resembles that for the perfect conducting cylinder [Eq. (4.1-17)] than that for the homogeneous cylinder [Eq. (4.1-45)]. This is to be expected because the $E$ field induces axial surface currents on the wire. These surface currents are far more significant in producing a scattered field than the dipole moments of the coating material.

For the $E$ field perpendicular to the axis, both the $B_0$ and $B_1$ terms of the series are of the same order. The scattered field for this case is

$$H_{\perp}{}^s = H_{\perp}{}^i \sqrt{\frac{\pi}{2}} \frac{e^{i(k_0 r + \pi/4)}}{\sqrt{k_0 r}} \frac{(k_0 a_0)^2}{2} \left[ \frac{\mu_1}{\mu_0} (1 - (a_1/a_0)^2) - 1 \right.$$

$$\left. - 2 \frac{\epsilon_1[1 + (a_1/a_0)^2] - \epsilon_0[1 - (a_1/a_0)^2]}{\epsilon_1[1 + (a_1/a_0)^2] + \epsilon_0[1 - (a_1/a_0)^2]} \cos \phi' \right] \qquad (4.1\text{-}102)$$

This result is similar to those for both the perfectly conducting cylinder and the homogeneous cylinder. It shows that the cylinder scatters poorly for this polarization direction (evidenced by the factor $(k_0 a_0)^2$) as compared with the axial direction.

4.1.4.2.1. *Bistatic Scattering Width* $(k_0 a_0 \ll 1, k_1 a_0 \ll 1)$. From the above equations the bistatic scattering widths become

$$\sigma_{\parallel}{}^c(\phi') = \sigma_{\parallel}{}^c(0) = \sigma_{\parallel}{}^c(\pi) = \frac{\pi^2 a_0}{(k_0 a_0)[\ln^2(0.89 k_0 a_0 (a_1/a_0)^{\mu_1/\mu_0}) + \pi^2/4]}$$

$$(4.1\text{-}103)$$

and

$$\sigma_\perp^c(\phi') = \pi a_0 \frac{\pi}{4} (k_0 a_0)^3 \left\{ \frac{\mu_1}{\mu_0} \left[ 1 - \left( \frac{a_1}{a_0} \right)^2 \right] - 1 \right.$$

$$\left. - 2 \frac{\epsilon_1 (1 + (a_1/a_0)^2) - \epsilon_0 (1 - (a_1/a_0)^2)}{\epsilon_1 (1 + (a_1/a_0)^2) + \epsilon_0 (1 - (a_1/a_0)^2)} \cos \phi' \right\}^2 \qquad (4.1\text{-}104)$$

Thus, one sees that for polarization parallel to the axis, the scattered power is radiated isotropically and increases as frequency decreases (for an infinitely long wire). For the other polarization, a nonisotropic scattering pattern is exhibited, but the scattered power level is much below that for the axial polarization.

4.1.4.2.2. *Backscattering Width* ($k_0 a_0 \ll 1$, $k_1 a_0 \ll 1$). In this case, $\sigma_\parallel^c(0)$ is given by Eq. (4.1-103), and $\sigma_\perp^c(0)$ is obtained from (4.1-104) by setting $\phi'$ equal to zero, or

$$\sigma_\perp^c(0) = \pi a_0 \frac{\pi}{4} (k_0 a_0)^3 \left\{ \frac{\mu_1}{\mu_0} \left[ 1 - \left( \frac{a_1}{a_0} \right)^2 \right] - 1 \right.$$

$$\left. - 2 \frac{\epsilon_1 (1 + (a_1/a_0)^2) - \epsilon_0 (1 - (a_1/a_0)^2)}{\epsilon_1 (1 + (a_1/a_0)^2) + \epsilon_0 (1 - (a_1/a_0)^2)} \right\}^2 \qquad (4.1\text{-}105)$$

It can be seen from this equation that by proper adjustment of the coating thickness and material parameters (i.e., $a_0 - a_1$, $\mu_1$, and $\epsilon_1$), it is possible either to enhance or reduce to zero the backscattered power.

4.1.4.2.3. *Forward Scattering Width* ($k_0 a_0 \ll 1$, $k_1 a_0 \ll 1$). Again, $\sigma_\parallel^c(\pi)$ is given by Eq. (4.1-103), and $\sigma_\perp^c(\pi)$ is obtained from (4.1-104) by setting $\phi' = \pi$, or

$$\sigma_\perp^c(\pi) = \pi a_0 \frac{\pi}{4} (k_0 a_0)^3 \left\{ \frac{\mu_1}{\mu_0} \left[ 1 - \left( \frac{a_1}{a_0} \right)^2 \right] - 1 \right.$$

$$\left. + 2 \frac{\epsilon_1 (1 + (a_1/a_0)^2) - \epsilon_0 (1 - (a_1/a_0)^2)}{\epsilon_1 (1 + (a_1/a_0)^2) + \epsilon_0 (1 - (a_1/a_0)^2)} \right\}^2 \qquad (4.1\text{-}106)$$

4.1.4.3. *Approximations for the High-Frequency Region* ($k_0 a_0 > 1$)

As with the homogeneous cylinder, there is no limiting or asymptotic value for the scattering width for large $k_0 a_0$ unless attenuation of the internal waves is sufficiently high. However, as can be seen from Figure 4-14, the scattering widths seem relatively smooth except for a few variations widely spaced. These variations occur when a ray trapped inside the coating can emerge in the scattering direction of interest so as to interfere constructively or destructively with the existing scattered field. The more or less stable value of the normalized scattering width is proportional to $| R_\parallel(\phi'/2)|^2$ and $| R_\perp(\phi'/2)|^2$, where these are the reflection coefficients for a plane wave of

*TM* or *TE* polarization incident upon a slab of material $\mu_1$, $\epsilon_1$ and thickness $a_0 - a_1$, backed by a perfect conductor (these are obtained from Eq. (7.1-42) by letting $z_3 \rightarrow \infty$).

A rigorous approach to obtaining a high-frequency result for the scattered field can be found in References 15 and 16. The exact series solution is transformed to a contour integral using a Watson transformation. Then the contour is changed and the result is rearranged to give several groups of terms. In the high-frequency limit, these groups become apparent as the optical contribution, the creeping wave contribution, trapped modes, and other rays penetrating the coating. Only a few of these rays contribute significantly in any scattering direction. The contribution of each ray is determined either by solving a complicated transcendental equation or by using a stationary phase approach. In either case, the general solution is not easily adapted for practical numerical use. The rigorous approach does yield two significant results: (1) It proves the validity of a geometrical-optics approach. Such an approach (see Section 4.1.3.4.1. and Reference 16) represents the scattered field as a sum of the various rays which can contribute significantly in a given direction. But rather than solve the exact equations, one determines at the outset which rays will exist and contribute significantly, and then proceeds to compute them based upon geometric principles. (2) The rigorous approach shows that the dominant contribution can come from "trapped rays," rays which enter the coating and circumnavigate the cylinder inside the coating, bouncing between the inner and outer coating interfaces.

For lack of space and lack of a concise formulation, this method will not be considered here. The above references should be consulted for more details concerning this approach.

### 4.1.4.4. *Scattering Widths in the High-Frequency Limit for a Lossy Coating*

In the event that the coating parameters $\epsilon$, $\mu$ are not purely real, the bistatic scattering widths are

$$\sigma_\parallel^c(\phi') = \pi a_0 \mid R_\parallel(\phi'/2)\mid^2 \cos^2(\phi'/2) \qquad (4.1\text{-}107)$$

and

$$\sigma_\perp^c(\phi') = \pi a_0 \mid R_\perp(\phi'/2)\mid^2 \cos^2(\phi'/2) \qquad (4.1\text{-}108)$$

The reflection coefficients, $R_\parallel$, $R_\perp$, in the above equations are to be interpreted as follows:

(i) $(\mathrm{Im}[k_1(a_0 - a_1)] < 5$, $\mathrm{Im}[k_1 a_0] > 2)$. In this case, rays which penetrate the coating reflect several times between the perfectly conducting core and the coating outer surface, and then exit in the direction $\phi'$. However, the attenuation is great enough so that trapped and creeping waves which must circumnavigate the cylinder at least once are too weak to contribute.

In this case, the cylinder appears like an infinite conducting plane coated with a homogeneous slab of thickness, $\tau = a_0 - a_1$. The reflection coefficients, $R_\parallel(\phi'/2)$ and $R_\perp(\phi'/2)$, are then those appropriate for a homogeneous planar slab, and can be obtained from Eq. (7.1-42) by letting $z_3 \to \infty$.

(ii) $(\text{Im}[k_1(a_0 - a_1)] > 5)$. In this case, rays which penetrate the coating are attenuated and never exit from the coating. Hence, the core is not even visible from the outside, and the cylinder looks like it is made entirely of lossy homogeneous material. The reflection coefficients, $R_\parallel(\phi'/2)$ and $R_\perp(\phi'/2)$, are merely the Fresnel coefficients for the homogeneous coating material as used in Eqs. (4.1-89) and (4.1-90), and given by Eqs. (7.1-9) and (7.1-10).

For backscatter the scattering widths are

$$\sigma_\parallel{}^c(0) = \pi a_0 \mid R_\parallel(0)\mid^2 \qquad (4.1\text{-}109)$$

and

$$\sigma_\perp{}^c(0) = \pi a_0 \mid R_\perp(0)\mid^2 \qquad (4.1\text{-}110)$$

The reflection coefficients appearing here (for normal incidence) have the same meaning and interpretation as for the bistatic case.

### 4.1.5. Sleeve Over a Perfectly Conducting Cylinder (Fig. 4-24)

#### 4.1.5.1. *Series Coefficients*

For the cylinder configuration of Figure 4-24, the series solution coefficients can be written in the form given by Eqs. (4.1-92) and (4.1-93), but now the modal impedances $Z_n$ and $Y_n$ take the form:

$$iZ_n \equiv z_{n,0}^{\text{int}} = \frac{k_0\mu_1}{k_1\mu_0} \frac{J_n(k_1a_0) - z_{n,1}^{\text{ext}}H_n(k_1a_0)}{J_n'(k_1a_0) - z_{n,1}^{\text{ext}}H_n'(k_1a_0)} \qquad (4.1\text{-}111)$$

$$iY_n \equiv y_{n,0}^{\text{int}} = \frac{k_0\epsilon_1}{k_1\epsilon_0} \frac{J_n(k_1a_0) - y_{n,1}^{\text{ext}}H_n(k_1a_0)}{J_n'(k_1a_0) - y_{n,1}^{\text{ext}}H_n'(k_1a_0)} \qquad (4.1\text{-}112)$$

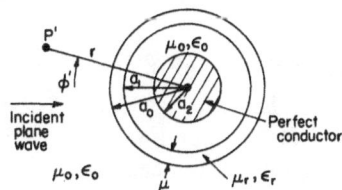

Fig. 4-24. Scattering geometry for an infinitely long, perfectly conducting circular cylinder with a shell spaced away from it.

where

$$z_{n,1}^{\text{ext}} = \frac{J_n(k_1a_1) - z_{n,1}^{\text{int}}J_n'(k_1a_1)}{H_n(k_1a_1) - z_{n,1}^{\text{int}}H_n'(k_1a_1)} \tag{4.1-113}$$

$$y_{n,1}^{\text{ext}} = \frac{J_n(k_1a_1) - y_{n,1}^{\text{int}}J_n'(k_1a_1)}{H_n(k_1a_1) - y_{n,1}^{\text{int}}H_n'(k_1a_1)} \tag{4.1-114}$$

and

$$z_{n,1}^{\text{int}} = \frac{k_1\mu_0}{k_0\mu_1} \frac{J_n(k_0a_1) - z_{n,2}^{\text{ext}}H_n(k_0a_1)}{J_n'(k_0a_1) - z_{n,2}^{\text{ext}}H_n'(k_0a_1)} \tag{4.1-115}$$

$$y_{n,1}^{\text{int}} = \frac{k_1\epsilon_0}{k_0\epsilon_1} \frac{J_n(k_0a_1) - y_{n,2}^{\text{ext}}H_n(k_0a_1)}{J_n'(k_0a_1) - y_{n,2}^{\text{ext}}H_n'(k_0a_1)} \tag{4.1-116}$$

with

$$z_{n,2}^{\text{ext}} = \frac{J_n(k_0a_2)}{H_n(k_0a_2)} \quad \text{and} \quad y_{n,2}^{\text{ext}} = \frac{J_n'(k_0a_2)}{H_n'(k_0a_2)} \tag{4.1-117}$$

An iteration pattern in the determination of the series coefficients is apparent. One begins at the innermost interface (at radius $a_2$) and iterates outward by determining modal impedances (or admittances) internal and external to each interface. At the innermost interface, these modal impedances, $z_{n,2}^{\text{ext}}$, and admittances, $y_{n,2}^{\text{ext}}$, are merely the scattering coefficients, $A_n$ and $B_n$, for a perfectly conducting cylinder [(Eq. (4.1-13) and (4.1-14)]. This iteration procedure is readily amenable to computer determination of the exact scattered field or scattering width from such a cylindrical structure.

Figures 4-25 and 4-26 show the effect of varying the spacing, $a_2 - a_1$, on the back, and forward, scattering widths for both polarization directions. The inner cylinder radius, $a_2$, the shell thickness, $a_0 - a_1$, the frequency, and the shell's electrical properties are considered fixed, while the spacing, $a_2 - a_1$, is varied over several wavelengths ($k_0a_2 = 10$, $k_0(a_0 - a_1) = 2$). In Figure 4-25, the shell material is taken to be a lossless dielectric with $\epsilon_r = 2.56$; while in Figure 4-26, the shell material is a typical ferrite with $\epsilon_r = 13.0$ and $\mu_r = 3.5$.

### 4.1.5.2. Approximations for the Low-Frequency Region ($k_0a_0 \ll 1$, $k_1a_0 \ll 1$)

When $k_0a_0 \ll 1$, the first two series coefficients simplify as given by Eqs. (4.1-96) and (4.1-97). When $k_1a_0 \ll 1$ also, the modal impedances and admittances can all be simplified by using small-argument expansions

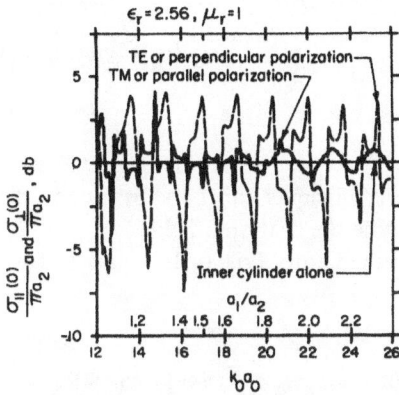

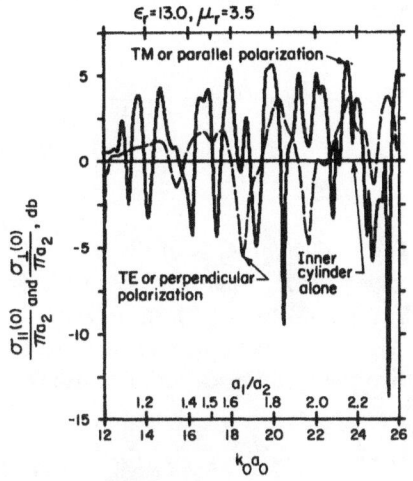

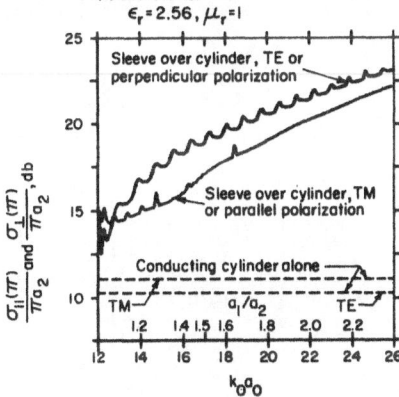

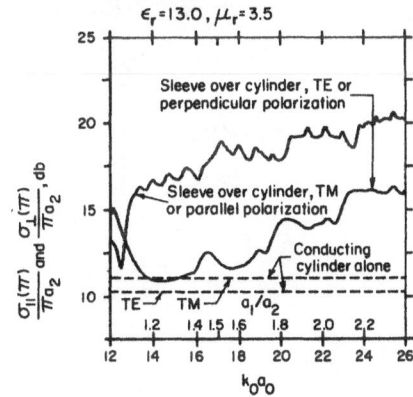

Fig. 4-25. Exact solutions for normalized scattering widths for an infinitely long, perfectly conducting circular cylinder with a shell of constant thickness, $(a_0 - a_1)/a_2 = 0.2$, as a function of shell spacing.

Fig. 4-26. Exact solutions for normalized scattering widths for an infinitely long, perfectly conducting circular cylinder with a shell of constant thickness, $(a_0 - a_1)/a_2 = 0.2$, as a function of shell spacing.

for the Bessel functions. Then the series coefficients which are significant in the scattering process are

$$A_0 = i(\pi/2)\frac{1}{\ln[0.89k_0a_2(a_0/a_1)^{1-\mu_1/\mu_0}] - i\pi/2} \qquad (4.1\text{-}118)$$

$$B_0 = -i(\pi/4)(k_0a_0)^2\,[1 + (a_2/a_0)^2 - (a_1/a_0)^2 - (\mu_1/\mu_0)(1 - (a_1/a_0)^2)] \qquad (4.1\text{-}119)$$

and

$$B_1 = -i(\pi/4)(k_0a_0)^2 \frac{\left(\begin{array}{l}\{1 + (a_1/a_0)^2 - (\epsilon_1/\epsilon_0)[1 - (a_1/a_0)^2]\}[1 - (a_2/a_1)^2] \\ - \{1 + (a_1/a_0)^2 - (\epsilon_0/\epsilon_1)[1 - (a_1/a_0)^2]\}[1 + (a_2/a_1)^2]\end{array}\right)}{\left(\begin{array}{l}\{1 + (a_1/a_0)^2 + (\epsilon_1/\epsilon_0)[1 - (a_1/a_0)^2]\}[1 - (a_2/a_1)^2] \\ + \{1 + (a_1/a_0)^2 + (\epsilon_0/\epsilon_1)[1 - (a_1/a_0)^2]\}[1 + (a_2/a_1)^2]\end{array}\right)}$$

$$(4.1\text{-}120)$$

The coefficient $A_1$ is of the order of $(k_0a_0)^2$, so that it may be neglected with respect to $A_0$ when $k_0a_0$ is small. These coefficients can be substituted into Eqs. (4.1-5) to (4.1-7), in order to determine the scattered fields.

If the above coefficients are substituted into Eqs. (4.1-7) and (4.1-8), then the bistatic scattering widths are

$$\sigma_{\parallel}^{c}(\phi') = \sigma_{\parallel}^{c}(0) = \sigma_{\parallel}^{c}(\pi) = \frac{\pi^2 a_0}{(k_0a_0)[\ln^2[0.89k_0a_2(a_0/a_1)^{1-\mu_1/\mu_0}] + \pi^2/4]}$$

$$(4.1\text{-}121)$$

and

$$\sigma_{\perp}^{c}(\phi') = \pi a_0(\pi/4)(k_0a_0)^3 \left[ [1 + (a_2/a_1)^2 - (a_1/a_0)^2 - (\mu_1/\mu_0)(1 - (a_1/a_0)^2)] \right.$$
$$\left. - 2\frac{\left(\begin{array}{l}\{1 + (a_1/a_0)^2 - (\epsilon_1/\epsilon_0)[1 - (a_1/a_0)^2]\}[1 - (a_2/a_1)^2] \\ - \{1 + (a_1/a_0)^2 - (\epsilon_0/\epsilon_1)[1 - (a_1/a_0)^2]\}[1 + (a_2/a_1)^2]\end{array}\right)}{\left(\begin{array}{l}\{1 + (a_1/a_0)^2 + (\epsilon_1/\epsilon_0)[1 - (a_1/a_0)^2]\}[1 - (a_2/a_1)^2] \\ + \{1 + (a_1/a_0)^2 + (\epsilon_0/\epsilon_1)[1 - (a_1/a_0)^2]\}[1 + (a_2/a_1)^2]\end{array}\right)} \cos \phi' \right]^2$$

$$(4.1\text{-}122)$$

From these equations it is apparent that the perfectly conducting core dominates when the polarization is parallel to the axis; the shell is nearly invisible. The scattered power for this polarization is radiated isotropically, independent of scattering angle. Power scattered for the other polarization is considerably smaller due to the factor $(k_0a_0)^3$, and it is sensitive to the scattering angle.

### 4.1.5.3. Approximations for the High-Frequency Region $(k_0a_0 > 1)$

As with the coated cylinder, there are no simple results available. The description of the scattering process given there is also applicable here, except that the situation is more complicated. Trapped waves can exist not only in the shell material, but also in the free space region between the shell and the core. Reference 17 examines an approximate expansion and its general validity.

When the shell material is lossy, attenuation of the waves trapped inside the shell takes place. Waves trapped between the shell and core are not attenuated, however, and may contribute to the scattered field. Only when $\text{Im}[k_1(a_0 - a_1)] > 2$ do waves which enter the shell become attenuated sufficiently that they can never exit. In this case, the entire structure from the outside looks like a homogeneous cylinder with $\epsilon_1$, $\mu_1$. Then the back and bistatic scattering widths are given by Eqs. (4.1-89) to (4.1-91).

The forward scattering cross section in the high-frequency limit is given in Section 4.1.2.4.5; the $a$ used there is replaced by $a_0$, the outermost radius.

### 4.1.6. Hollow Cylindrical Shell (Fig. 4-27)

#### 4.1.6.1. Series Coefficients

The series solution coefficients for the hollow cylindrical shell of Figure 4-27 can be written in the form of Eqs. (4.1-92) and (4.1-93). The modal impedances appearing in these coefficients are readily obtained from those of the preceding section [Eqs. (4.1-111) to (4.1-117)] by letting $a_2 \to 0$. In this case, $z_{n,0}^{int}$, $y_{n,0}^{int}$, $z_{n,1}^{ext}$, $y_{n,1}^{ext}$ remain the same as given by (4.1-111)–(4.114). However, $z_{n,2}^{ext}$ and $y_{n,2}^{ext}$ approach zero, making

$$z_{n,1}^{int} = \frac{k_1\mu_0}{k_0\mu_1}\frac{J_n(k_0a_1)}{J_n'(k_0a_1)} \quad \text{and} \quad y_{n,1}^{int} = \frac{k_1\epsilon_0}{k_0\epsilon_1}\frac{J_n(k_0a_1)}{J_n'(k_0a_1)} \quad (4.1\text{-}123)$$

Figures 4-28 to 4-30 show the exact back, forward, and bistatic scattering cross sections for a dielectric cylindrical shell with $(a_0 - a_1)/a_0 = 0.1$. In the first figure, the shell has a dielectric constant $\epsilon_1 = 2.56$. In the second figure some small loss is added so that $\epsilon_1 = 2.56 + i0.1024$ (loss tangent $= 0.04$). The independent variable in the figures is frequency, or $k_0$.

#### 4.1.6.2. Approximations for the Low-Frequency Region ($k_0a_0 \ll 1$, $k_1a_0 \ll 1$)

When $k_0a_0 \ll 1$, the first two series coefficients simplify as given by Eqs. (4.1-96) and (4.1-97). When $k_1a_0 \ll 1$, the modal impedances and admittances can also be simplified by using small-argument expansions for

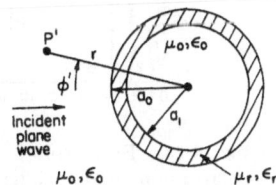

Fig. 4-27. Scattering geometry for an infinitely long, hollow cylindrical shell.

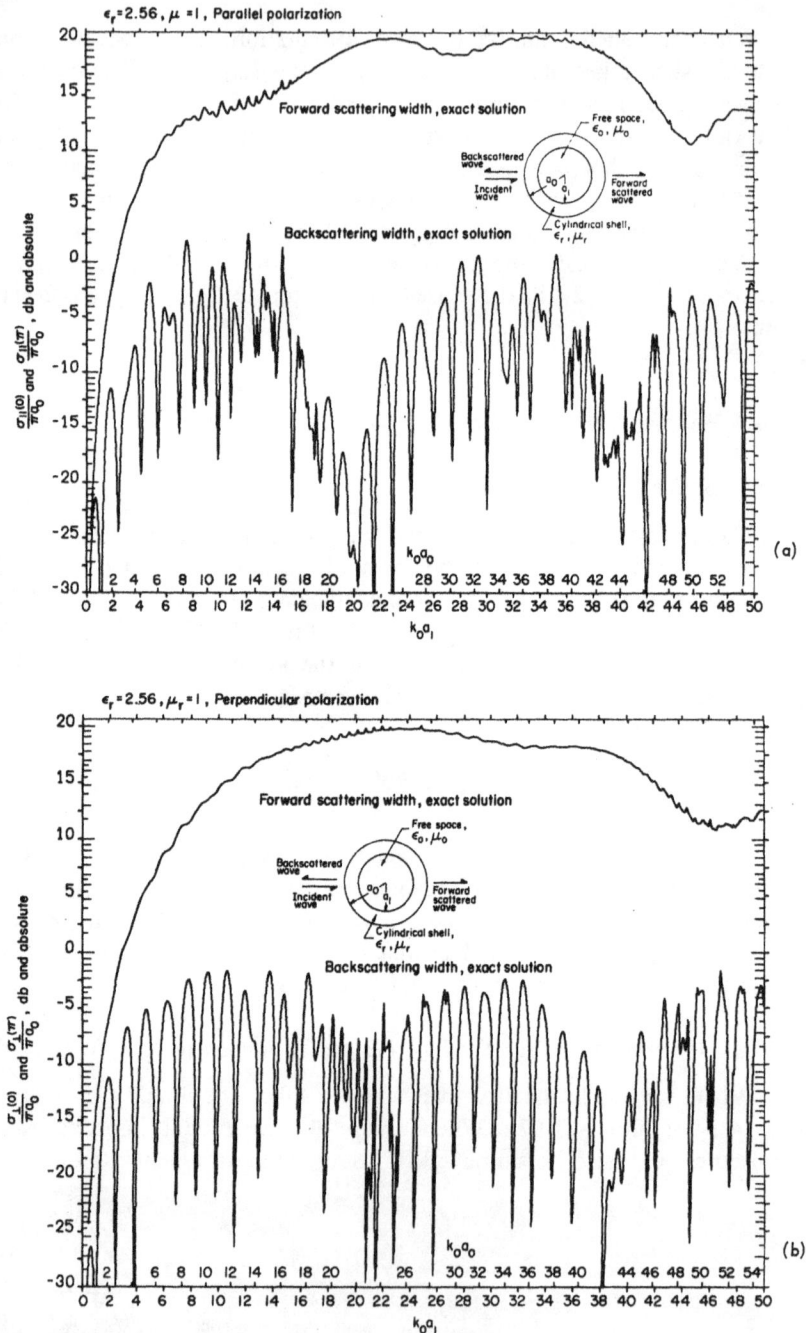

Fig. 4-28. Exact solutions for back and forward scattering widths at parallel polarization for an infinitely long, hollow cylindrical shell of thickness $(a_0 - a_1)/a_1 = 0.1$: (a) lossless shell, and (b) lossy shell.

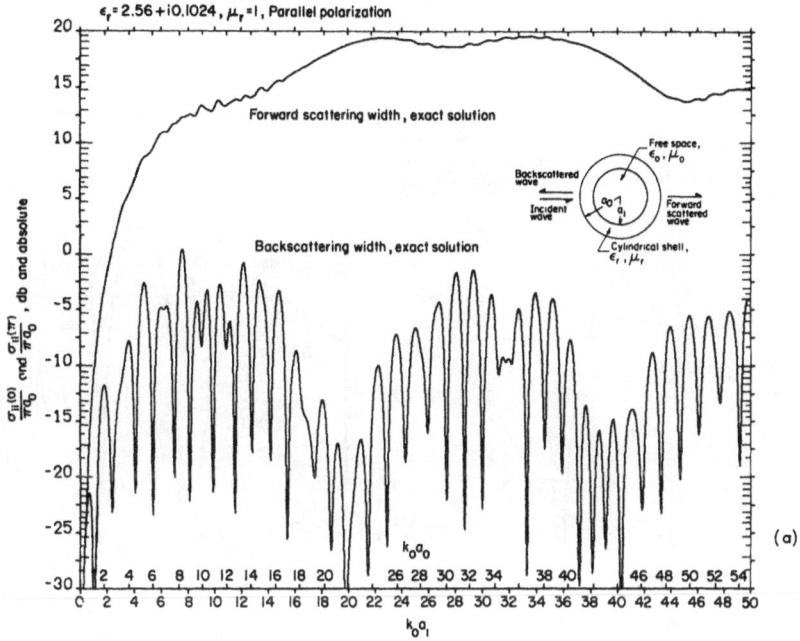

(a)

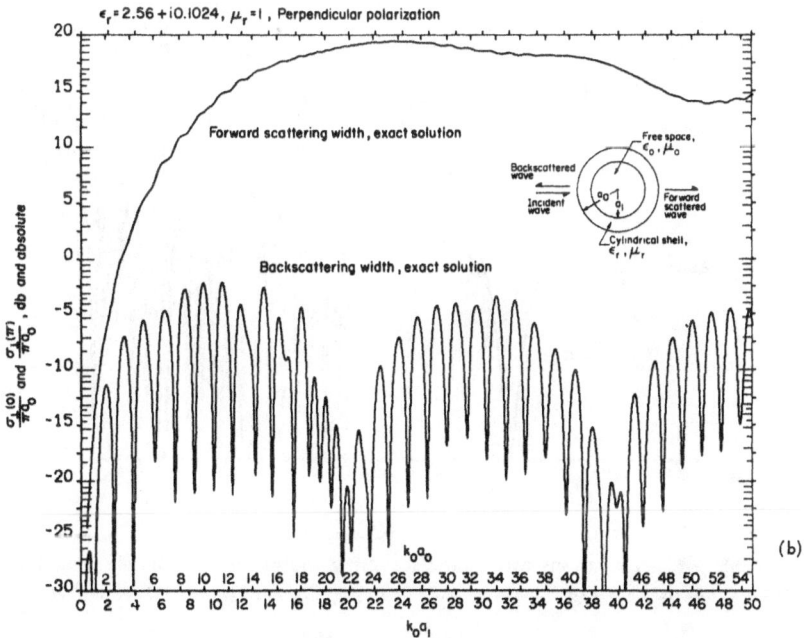

(b)

Fig. 4-29. Exact solutions for back and forward scattering widths at perpendicular polarization for an infinitely long, hollow cylindrical shell of thickness $(a_0 - a_1)/a_1 = 0.1$: (a) lossless shell, and (b) lossy shell.

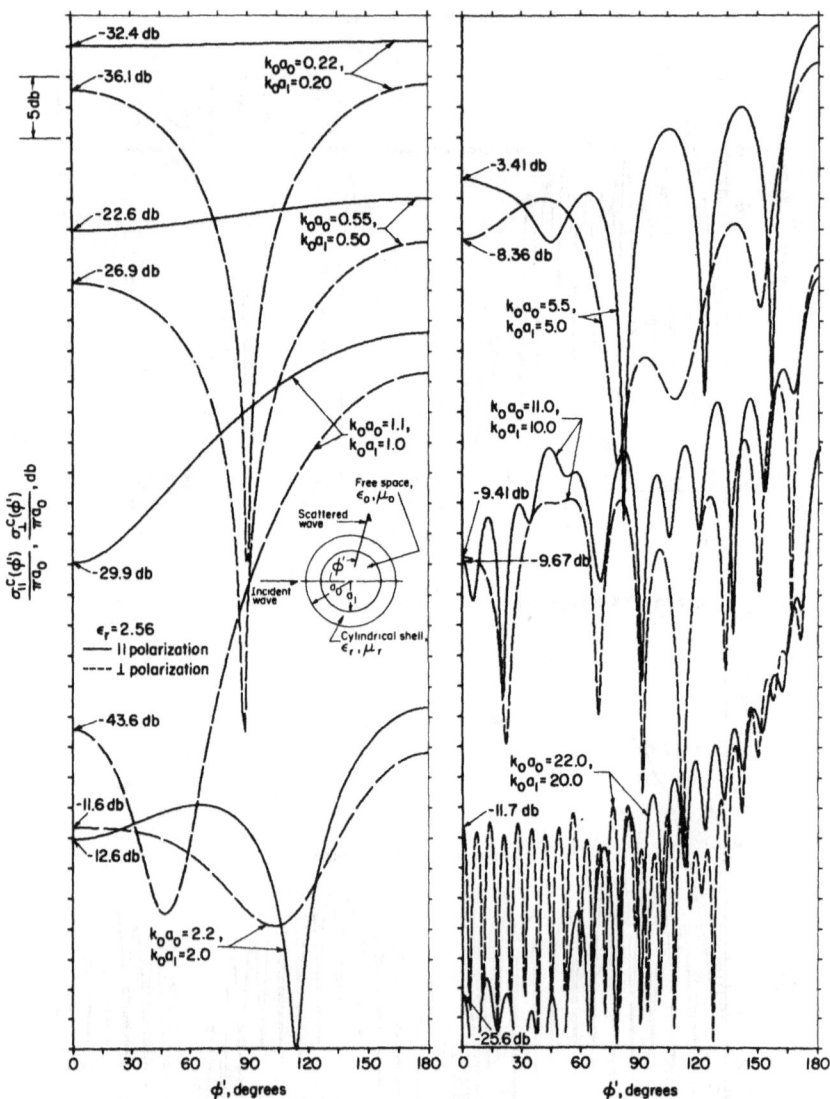

Fig. 4-30a.   Exact solutions for bistatic scattering widths for an infinitely long, hollow, lossless cylindrical shell of thickness $(a_0 - a_1)/a_1 = 0.1$, for various values of $k_0a_0$.

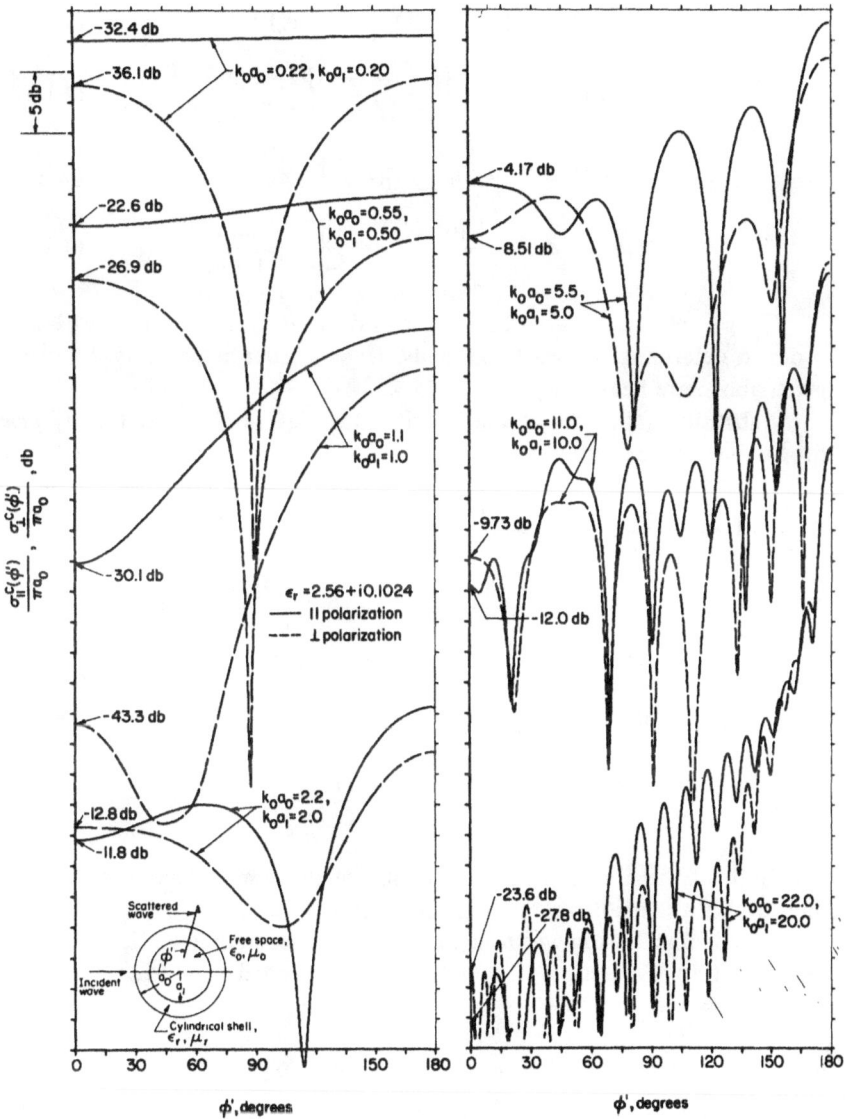

Fig. 4-30b.   Same as Figure 4-30a, but for a lossy shell.

the Bessel functions. Then the first two series coefficients for each polarization are significant and are given below:

$$A_0 = i(\pi/4)(k_0a_0)^2 \, (\epsilon_1/\epsilon_0 - 1)[1 - (a_1/a_0)^2] \tag{4.1-124}$$

$$A_1 = i(\pi/4)(k_0a_0)^2 \, \frac{[(\mu_1/\mu_0)^2 - 1][1 - (a_1/a_0)^2]}{(\mu_1/\mu_0 + 1)^2 - (a_1/a_0)^2 \, (\mu_1/\mu_0 - 1)^2} \tag{4.1-125}$$

$$B_0 = i(\pi/4)(k_0a_0)^2 \, (\mu_1/\mu_0 - 1)[1 - (a_1/a_0)^2] \tag{4.1-126}$$

$$B_1 = i(\pi/4)(k_0a_0)^2 \, \frac{[(\epsilon_1/\epsilon_0)^2 - 1][1 - (a_1/a_0)^2]}{(\epsilon_1/\epsilon_0 + 1)^2 - (a_1/a_0)^2 \, (\epsilon_1/\epsilon_0 - 1)^2} \tag{4.1-127}$$

These coefficients are then substituted into Eqs. (4.1-5) to (4.1-7) in order to determine the scattered fields. Higher-order coefficients than those given above are negligible.

Substituting the above coefficients into Eqs. (4.1-7) and (4.1-8) gives for the bistatic scattering widths:

$$\sigma_{\parallel}{}^c(\phi') = \pi a_0(\pi/4)(k_0a_0)^3 \left[ (\epsilon_1/\epsilon_0 - 1)[1 - (a_1/a_0)^2] \right.$$

$$\left. - 2 \frac{[(\mu_1/\mu_0)^2 - 1][1 - (a_1/a_0)^2]}{(\mu_1/\mu_0 + 1)^2 - (a_1/a_0)^2 \, (\mu_1/\mu_0 - 1)^2} \cos \phi' \right]^2 \tag{4.1-128}$$

and

$$\sigma_{\perp}{}^c(\phi') = \pi a_0(\pi/4)(k_0a_0)^3 \left[ (\mu_1/\mu_0 - 1)[1 - (a_1/a_0)^2] \right.$$

$$\left. - 2 \frac{[(\epsilon_1/\epsilon_0)^2 - 1][1 - (a_1/a_0)^2]}{(\epsilon_1/\epsilon_0 + 1)^2 - (a_1/a_0)^2 \, (\epsilon_1/\epsilon_0 - 1)^2} \cos \phi' \right]^2 \tag{4.1-129}$$

Both of the above cross sections are of the same order of magnitude, and exhibit a nonisotropic pattern in general.

The backscattering widths are obtained immediately from Eqs. (4.1-128) and (4.1-129) by setting $\phi' = 0$, and the forward scattering widths by setting $\phi' = \pi$.

### 4.1.6.3. Approximations for the High-Frequency Region $(k_0a_0 > 1)$

As with the coated cylinder, there are no simple results available. The description of the scattering process given there is also applicable here. However, in this situation, trapped waves can exist not only in the dielectric shell but also inside the hollow cavity.

When the shell material is lossy, attenuation of the trapped waves inside the shell takes place; for shell diameter sufficiently large, these waves may be neglected. However, the waves trapped inside the hollow cavity are not attenuated and can only be neglected when the shell is so thick and lossy that waves from the outside cannot penetrate the shell (i.e., $Im[k_1(a_0 - a_1)] > 2$). In that case, only the externally-reflected wave contributes to the scattered field, and the back and bistatic scattering widths are given by Eqs. (4.1-89) to (4.1-91).

The forward scattering width in the high-frequency limit is given in Section 4.1.2.4.5; the $a$ used there is replaced by $a_0$, the outermost radius.

### 4.1.7. $M$-Layer Stratified Cylinder

#### 4.1.7.1. Series Coefficients

All previously considered cylinders in this chapter can be considered special cases of the general $M$-layered cylinder shown in Figure 4-31. They were treated individually first because of the tendency of the general formulation to obscure particular details in each case.

The scattering series solution coefficients can be obtained most directly and logically from an iteration process starting at the innermost interface (at radius $a_M$). The pattern should be apparent from Section 4.1.5; a modal impedance (or admittance) is computed just inside and outside the $i$th layer (at radius $a_i$), i.e., $z_{n,i}^{int}$ and $z_{n,i}^{ext}$. These quantities depend upon those at the $(i + 1)$th layer.

At the innermost interface, two situations may exist: The inner cylinder may be perfectly conducting, or it may be homogeneous with finite constitutive constants. The model impedances at the outside of this innermost cylinder are given below for each case:

(a) *Innermost Cylinder Is Perfectly Conducting.*

$$z_{n,M}^{ext} = \frac{J_n(k_M a_M)}{H_n(k_M a_M)} \quad \text{and} \quad y_{n,M}^{ext} = \frac{J_n'(k_M a_M)}{H_n'(k_M a_M)} \qquad (4.1\text{-}130)$$

where $k_M = k_0 \sqrt{\epsilon_M \mu_M}$ is the wave number in the layer surrounding the inner core.

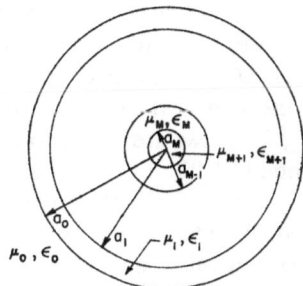

Fig. 4-31. Scattering geometry for an infinitely long $M$-layered stratified circular cylinder.

(b) *Innermost Cylinder Is Homogeneous* (with $\epsilon_{M+1}$, $\mu_{M+1}$, and $k_{M+1} = k_0 \sqrt{\epsilon_{M+1}\mu_{M+1}}$).

$$z_{n,M}^{\text{ext}} = \frac{J_n(k_M a_M) - \xi_{M+1}[J_n(k_{M+1}a_M)/J_n'(k_{M+1}a_M)] J_n'(k_M a_M)}{H_n(k_M a_M) - \xi_{M+1}[J_n(k_{M+1}a_M)/J_n'(k_{M+1}a_M)] H_n'(k_M a_M)} \quad (4.1\text{-}131)$$

and

$$y_{n,M}^{\text{ext}} = \frac{J_n(k_M a_M) - (1/\xi_{M+1})[J_n(k_{M+1}a_M)/J_n'(k_{M+1}a_M)] J_n'(k_M a_M)}{H_n(k_M a_M) - (1/\xi_{M+1})[J_n(k_{M+1}a_M)/J_n'(k_{M+1}a_M)] H_n'(k_M a_M)} \quad (4.1\text{-}132)$$

where in general

$$\xi_i = \sqrt{\frac{\mu_i \epsilon_{i-1}}{\epsilon_i \mu_{i-1}}} \quad (4.1\text{-}133)$$

From the above sets of equations for $z_{n,M}^{\text{ext}}$ and $y_{n,M}^{\text{ext}}$ at the outside of the innermost layer, one proceeds to find $z_{n,M-1}^{\text{int}}$, $y_{n,M-1}^{\text{int}}$ at the inside of the next layer at $a_{M-1}$ as follows:

$$z_{n,M-1}^{\text{int}} = \xi_M \frac{J_n(k_M a_{M-1}) - z_{n,M}^{\text{ext}} H_n(k_M a_{M-1})}{J_n'(k_M a_{M-1}) - z_{n,M}^{\text{ext}} H_n'(k_M a_{M-1})} \quad (4.1\text{-}134)$$

and

$$y_{n,M-1}^{\text{int}} = (1/\xi_M) \frac{J_n(k_M a_{M-1}) - y_{n,M}^{\text{ext}} H_n(k_M a_{M-1})}{J_n'(k_M a_{M-1}) - y_{n,M}^{\text{ext}} H_n'(k_M a_{M-1})} \quad (4.1\text{-}135)$$

From these, the impedances external to the $M-1$ interface are computed:

$$z_{n,M-1}^{\text{ext}} = \frac{J_n(k_{M-1}a_{M-1}) - z_{n,M-1}^{\text{int}} J_n'(k_{M-1}a_{M-1})}{H_n(k_{M-1}a_{M-1}) - z_{n,M-1}^{\text{int}} H_n'(k_{M-1}a_{M-1})} \quad (4.1\text{-}136)$$

and

$$y_{n,M-1}^{\text{ext}} = \frac{J_n(k_{M-1}a_{M-1}) - y_{n,M-1}^{\text{int}} J_n'(k_{M-1}a_{M-1})}{H_n(k_{M-1}a_{M-1}) - y_{n,M-1}^{\text{int}} H_n'(k_{M-1}a_{M-1})} \quad (4.1\text{-}137)$$

The iteration is repeated at the $M-2$ interface employing (4.1-134) and (4.1-135), where $M$ is now replaced by $M-1$ to arrive at $z_{n,M-2}^{\text{int}}$ and $y_{n,M-2}^{\text{int}}$. These, then, are substituted into (4.1-136) and (4.1-137) (again replacing $M$ by $M-1$) to calculate $z_{n,M-2}^{\text{ext}}$ and $y_{n,M-2}^{\text{ext}}$. This process continues until one arrives at the impedances just inside of the outer interface, i.e., $z_{n,0}^{\text{int}}$ and $y_{n,0}^{\text{int}}$. These are equal to $iZ_n$ and $iY_n$, respectively, for use in Eqs. (4.1-92) and (4.1-93) for the series coefficients $A_n$ and $B_n$.

In general, the time required on a computer to calculate these series coefficients, $A_n$ and $B_n$, varies in direct proportion to the number of layers, $M$; one must calculate Bessel functions and their derivatives at each interface.

Often a radially inhomogeneous cylinder is modeled by such a stratified structure; this model obviously becomes more accurate as one increases the number of layers, $M$, in the model.

### 4.1.7.2. Approximations for the Low-Frequency Region ($k_0 a_0 \ll 1$, $k_0 a_1 \ll 1$, $k_1 a_i \ll 1$, $k_i a_2 \ll 1$, etc.)

In the low-frequency limit, only the first terms of the series solution are significant. Thus, one must find $A_n$ and $B_n$ only for $n = 0$ and 1. These are obtained by iteration after an expansion of the Bessel functions in their power series. The cases of the innermost cylinder perfectly conducting and a homogeneous material are considered separately.

### 4.1.7.2.1. Innermost Cylinder Is Perfectly Conducting.

In this situation the inner wire or conducting cylinder has a strong scattering effect when the $E$ field is polarized parallel to the axis. In this case, only $A_0$ need be used since $A_1$ is of order $(k_0 a_0)^2$ compared to $A_0$. $A_0$ may be obtained by iteration, and the result is[†]

$$A_0 = i\pi/2 \frac{1}{\ln[0.89 k_0 a_0 (a_1/a_0)^{\mu_1/\mu_0} (a_2/a_1)^{\mu_2/\mu_0} (a_3/a_2)^{\mu_3/\mu_0} \cdots (a_M/a_{M-1})^{\mu_M/\mu_0}] - i\pi/2}$$

(4.1-138)

For the $H$ field polarized parallel to the axis, both coefficients $B_0$ and $B_1$ are of the same order, and must be included. Define

$$p_M = \frac{\mu_M}{\mu_{M-1}} [1 - (a_M/a_{M-1})^2]$$

(4.1-139)

then form

$$p_{M-1} = \frac{\mu_{M-1}}{\mu_{M-2}} \left[1 - \left(\frac{a_{M-1}}{a_{M-2}}\right)^2 p_M\right]$$

(4.1-140)

$$p_{M-2} = \frac{\mu_{M-2}}{\mu_{M-3}} \left[1 - \left(\frac{a_{M-2}}{a_{M-3}}\right)^2 p_{M-1}\right]$$

(4.1-141)

until by iteration $p_1$ is obtained. Then the coefficient $B_0$ is given by

$$B_0 = -i(\pi/4)(k_0 a_0)^2 [1 - p_1]$$

(4.1-142)

In order to obtain $B_1$, define

$$q_M = \frac{\epsilon_M}{\epsilon_{M-1}} \frac{1 + (a_M/a_{M-1})^2}{1 - (a_M/a_{M-1})^2}$$

(4.1-143)

---

[†] This equation is not valid in the limit as $a_M \to 0$; i.e., when the radius of the perfectly-conducting core approaches zero. Higher-order terms which were neglected here must be included in that case. That case, however, becomes identical to the homogeneous inner core, which will be treated next.

then form

$$q_{M-1} = \frac{\epsilon_{M-1}}{\epsilon_{M-2}} \frac{1 + (a_{M-1}/a_{M-2})^2 (q_M - 1)/(q_M + 1)}{1 - (a_{M-1}/a_{M-2})^2 (q_M - 1)/(q_M + 1)} \qquad (4.1\text{-}144)$$

$$q_{M-2} = \frac{\epsilon_{M-2}}{\epsilon_{M-3}} \frac{1 + (a_{M-2}/a_{M-3})^2 (q_{M-1} - 1)/(q_{M-1} + 1)}{1 - (a_{M-2}/a_{M-3})^2 (q_{M-1} - 1)/(q_{M-1} + 1)} \qquad (4.1\text{-}145)$$

until by iteration $q_1$ is obtained. Then the coefficient $B_1$ is given by

$$B_1 = i(\pi/4)(k_0 a_0)^2 (q_1 - 1)/(q_1 + 1) \qquad (4.1\text{-}146)$$

Using the above results, the bistatic scattering widths become

$$\sigma_{\parallel}{}^c(\phi') = \sigma_{\parallel}{}^c(0) = \sigma_{\parallel}{}^c(\pi)$$

$$= \frac{\pi^2 a_0}{k_0 a_0 [\ln^2(0.89 k_0 a_0 (a_1/a_0)^{\mu_1/\mu_0} (a_2/a_1)^{\mu_2/\mu_0} \cdots (a_M/a_{M-1})^{\mu_M/\mu_0}) + \pi^2/4]}$$

$$(4.1\text{-}147)$$

and

$$\sigma_{\perp}{}^c(\phi') = \pi a_0 (\pi/4)(k_0 a_0)^3 \left[ (p_1 - 1) - 2\frac{q_1 - 1}{q_1 + 1} \cos \phi' \right]^2 \qquad (4.1\text{-}148)$$

The scattering is isotropic when the $E$ field is parallel to the axis as illustrated by the first equation. When the $H$ field is parallel to the axis, the magnitude is much smaller, due to the factor $(k_0 a_0)^3$, and a definite nonisotropic pattern is exhibited.

The back and forward scattering widths for the parallel $E$ field are given by (4.1-147), while that for the parallel $H$ field backscatter width is obtained from (4.1-148) by setting $\phi' = 0$, and the forward scatter width by setting $\phi' = \pi$.

4.1.7.2.2. *Innermost Cylinder Is Homogeneous.* In this situation, the first two terms, $A_0$ and $A_1$ (as well as $B_0$ and $B_1$ for the other polarization), are of the same order of magnitude. These coefficients are obtained by iteration and the results are given below.

Define

$$r_M{}^A = \frac{\epsilon_M}{\epsilon_{M-1}} \left[ 1 - \left(\frac{a_M}{a_{M-1}}\right)^2 \left(1 - \frac{\epsilon_{M+1}}{\epsilon_M}\right) \right] \qquad (4.1\text{-}149)$$

then form

$$r_{M-1}^A = \frac{\epsilon_{M-1}}{\epsilon_{M-2}} \left[ 1 - \left(\frac{a_{M-1}}{a_{M-2}}\right)^2 r_M{}^A \right] \qquad (4.1\text{-}150)$$

$$r_{M-2}^A = \frac{\epsilon_{M-2}}{\epsilon_{M-3}} \left[ 1 - \left(\frac{a_{M-2}}{a_{M-3}}\right)^2 r_{M-1}^A \right] \qquad (4.1\text{-}151)$$

until by iteration $r_1^A$ is obtained. Then the coefficient $A_0$ is given by

$$A_0 = -i(\pi/4)(k_0 a_0)^2 [1 - r_1^A] \qquad (4.1\text{-}152)$$

To obtain $A_1$, define

$$s_M^A = \frac{\mu_M}{\mu_{M-1}} \frac{1 - (a_M/a_{M-1})^2 (1 - \mu_{M+1}/\mu_M)/(1 + \mu_{M+1}/\mu_M)}{1 + (a_M/a_{M-1})^2 (1 - \mu_{M+1}/\mu_M)/(1 + \mu_{M+1}/\mu_M)} \qquad (4.1\text{-}153)$$

then form

$$s_{M-1}^A = \frac{\mu_{M-1}}{\mu_{M-2}} \frac{1 - (a_{M-1}/a_{M-2})^2 s_M^A}{1 + (a_{M-1}/a_{M-2})^2 s_M^A} \qquad (4.1\text{-}154)$$

$$s_{M-2}^A = \frac{\mu_{M-2}}{\mu_{M-3}} \frac{1 - (a_{M-2}/a_{M-3})^2 s_{M-1}^A}{1 + (a_{M-2}/a_{M-3})^2 s_{M-1}^A} \qquad (4.1\text{-}155)$$

until by iteration $s_1^A$ is obtained. Then the coefficient $A_1$ is given by

$$A_1 = -i(\pi/4)(k_0 a_0)^2 \frac{1 - s_1^A}{1 + s_1^A} \qquad (4.1\text{-}156)$$

The coefficients, $B_0$ and $B_1$, are obtained in identically the same manner as $A_0$ and $A_1$, by using Eqs. (4.1-149)–(4.1-156), except that $\epsilon$ everywhere is replaced by $\mu$, and $\mu$ by $\epsilon$ to generate $r_1^B$ and $s_1^B$.

Using the above results, the bistatic scattering widths become

$$\sigma^c(\phi') = \pi a_0 (\pi/4)(k_0 a_0)^3 \left[ (r_1^A - 1) - 2\frac{s_1^A - 1}{s_1^A + 1} \cos \phi' \right]^2 \qquad (4.1\text{-}157)$$

$$\sigma^c(\phi') = \pi a_0 (\pi/4)(k_0 a_0)^3 \left[ (r_1^B - 1) - 2\frac{s_1^B - 1}{s_1^B + 1} \cos \phi' \right]^2 \qquad (4.1\text{-}158)$$

In this case, both scattering widths are generally of the same order, $(k_0 a_0)^3$, and the patterns exhibited are nonisotropic.

The backscattering widths are obtained directly from (4.1-157) and (4.1-158) by substituting $\phi' = 0$, and the forward scattering widths by substituting $\phi' = \pi$.

### 4.1.7.3. *Approximations for the High-Frequency Region* $(k_0 a_0 > 1)$

As was the case for the simple coated cylinder, there are no simple results available in the high frequency region. Trapped waves, in general, can exist in all of the layers. It is only when one or more of the layers are lossy, so that waves cannot penetrate or be trapped, that simplification is possible. In particular, when the outermost layer is sufficiently lossy $(\text{Im}[k_1(a_0 - a_1)] > 2)$, then Eqs. (4.1-89) to (4.1-91) may be used to find the scattering widths.

The forward scattering width in the high frequency limit is given in Section 4.1.2.4.5; the $a$ used there is replaced by $a_0$, the outermost radius.

### 4.1.8. Radially Inhomogeneous Cylinder

#### 4.1.8.1. *Exact Series Solution*

In this situation, the permittivity and/or permeability inside the cylinder can vary with radial distance from the axis as illustrated in Figure 4-32. Then the radial dependence of the fields inside the cylinder no longer satisfies the Bessel differential equation as is the case when the material is homogeneous. Instead of solving the new second-order differential equations (which depend upon $\epsilon_r$ and $\mu_r$ and their derivatives), one can convert directly to equations for the modal impedances, $iZ_n$ and $iY_n$, found in Eqs. (4.1-92) and (4.1-93). Hence the series solutions have the same form as before, but the modal surface impedances in the series solution coefficients now satisfy Riccati nonlinear differential equations. Define

$$iZ_n = \beta z_n(\beta), \qquad iY_n = \beta y_n(\beta) \qquad (4.1\text{-}159)$$

with $\beta = k_0 a$; then $z_n(\beta)$ and $y_n(\beta)$ satisfy the following differential equations inside the cylinder material

$$z_n'(w) = \frac{\mu_r}{w} + \left(\epsilon_r w - \frac{n^2}{\mu_r w}\right) z_n{}^2(w) \qquad (4.1\text{-}160)$$

and

$$y_n'(w) = \frac{\epsilon_r}{w} + \left(\mu_r w - \frac{n^2}{\epsilon_r w}\right) y_n{}^2(w) \qquad (4.1\text{-}161)$$

where $w = k_0 r$. In general, $\mu_r$ and $\epsilon_r$ are known functions of $w$ inside the cylinder. One would, in most cases, have to integrate the above equations numerically.

#### 4.1.8.2. *Low-Frequency Approximations* ($k_0 a < 0.2$)

When frequency is low enough, such that the approximation $k_0 a \ll 1$ is valid, then the above results can be simplified since the modal impedances are needed only for $n = 0, 1$.

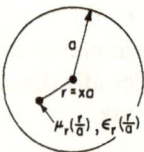

Fig. 4-32. Scattering geometry for an infinitely long, radially inhomogeneous circular cylinder.

Results for the cylindrical Luneberg lens and the Eaton–Lippman lens in the low frequency limit are as follows:

(a) *Cylindrical Luneberg Lens* ($k_0a < 0.2$): $\epsilon_r = 2 - u^2$; $\mu_r = 1$ *for* $0 \leqslant u \leqslant 1$, *where* $u = r/a$.

The series solution coefficients become

$$
\begin{aligned}
A_0 &= i(\pi/4)(k_0a)^2/2, & A_1 &= 0 \\
B_0 &= 0, & B_1 &= i(\pi/4)[1.8992(k_0a_0)^2]
\end{aligned}
\tag{4.1-162}
$$

and the scattering widths are

$$
\sigma_{\parallel}{}^c(\phi') = (\pi^2a/16)(k_0a)^3
\tag{4.1-163}
$$

$$
\sigma_{\perp}{}^c(\phi') = 0.03607\pi^2a(k_0a)^3 \cos^2 \phi'
\tag{4.1-164}
$$

The back and forward scattering widths are obtained from these equations by setting $\phi' = 0$ and $\phi' = \pi$, respectively.

(b) *Cylindrical Eaton–Lippman Lens* ($k_0a < 0.2$): $\epsilon_r = (2/u) - 1$; $\mu_r = 1$ *for* $0 < u \leqslant 1$, *where* $u = r/a$.

The series solution coefficients become

$$
\begin{aligned}
A_0 &= i(\pi/4)(k_0a)^2, & A_1 &= 0 \\
B_0 &= 0, & B_1 &= i(\pi/4)\,[0.43616(k_0a)^2]
\end{aligned}
\tag{4.1-165}
$$

and the scattering widths are

$$
\sigma_{\parallel}{}^c(\phi') = (\pi^2a/4)(k_0a)^3
\tag{4.1-166}
$$

$$
\sigma_{\perp}{}^c(\phi') = 0.19024\pi^2a(k_0a)^3 \cos^2 \phi'
\tag{4.1-167}
$$

The back and forward scattering widths are obtained from these equations by setting $\phi' = 0$ and $\phi' = \pi$, respectively.

### 4.1.8.3. *High-Frequency Approximations* ($k_0a > 10$)

For the above range of $k_0a$, the only reasonable alternative to the exact series solution is geometrical optics or a ray-tracing approximation. The analysis using ray tracing, in this case, is identical to that for the sphere as given in Section 3.4.3.3. Even the equations given there apply here also. One must keep in mind, however, that the rays entering the sphere form a surface which is rotationally symmetric about the direction of propagation, while the rays entering the cylinder form cylindrical surfaces along the z axis.

Examples of radially inhomogeneous cylindrical structures or lenses found in practice are the Luneberg lens and the Eaton–Lippman lens. These lenses perform ideally according to ray-tracing techniques only in the high-frequency limit.

The cylindrical Luneberg lens focuses a plane wave striking it onto a line on the back side (see Figure 3-17). If a reflecting surface is placed in this position, the wave reemerges as a plane wave in the backscattering direction. This Luneberg reflector behaves like a corner reflector, and the reflecting area is $2aL$, where $L$ is the length of the cylinder.

The cylindrical Eaton–Lippman lens performs the same as the Luneberg reflector above, except that no reflecting surface is needed on the backside of the lens. All rays entering this lens bend around and reemerge parallel and in phase in the backscattering direction (see Figure 3-18). Hence, the lens "reflects" rays equally well from all sides (as long as they are normal to the cylinder axis), while the Luneberg reflector or a corner reflector is not symmetric with respect to the $z$ axis in this performance. It should be noted that the cylindrical Eaton reflector described here reflects like a corner reflector or flat plate, whereas the spherical Eaton lens does not work at all due to a polarization cancellation.

When a plane wave strikes the cylindrical Luneberg reflector or cylindrical Eaton–Lippman lens, propagating normal to the cylinder axis, the cylinders reflect identical to the infinite strip of width $2a$ at normal incidence. In the high-frequency limit, all of the incident energy is back-scattered, and the backscattering width, in this case, is

$$\sigma^o(0) = \sigma^c(0) = 4k_0 a^2 \qquad (4.1\text{-}168)$$

which is also the forward scatter width. As mentioned above, the Luneberg reflector must be oriented so that the incident wave passes into the lens instead of striking the reflecting surface first.

## 4.2. INFINITELY LONG CIRCULAR CYLINDERS—OBLIQUE INCIDENCE

### 4.2.1. Form of Exact Series Solution

#### 4.2.1.1. *General Formulation*

Let a plane wave whose propagation direction lies in the $x$-$z$ plane be incident upon a cylinder whose axis coincides with the $z$ axis (Figure 4-33).

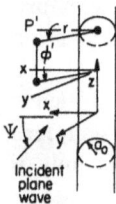

Fig. 4-33. Scattering geometry for an infinitely long, circular cylinder at oblique incidence.

The direction of propagation of the incident wave makes an angle, $\Psi$, with the $x$-$y$ plane (normal to the cylinder axis). The scattered field at point $P'$ is desired, where $P'$ lies in a plane intersecting the $z$ axis which makes an angle $\phi'$ with respect to the $x$ axis. (For the special case, where $\Psi = 0$, i.e., for normal incidence, the problem reduces to that considered in Section 4.1 and illustrated in Figure 4-1).

The scattered field from this infinitely long cylinder propagates in a direction which makes an angle, $\Psi$, with respect to the $x$-$y$ plane. Consequently, backscattering can only occur when the incident wave strikes the cylinder at normal incidence. In general, the scattered field exists for all $\phi'$. Thus, the propagation vectors of the scattered field from the cylinder form a cone in the forward direction which has a half-angle of $\pi/2 - \Psi$ with respect to the $+z$ axis.

Define the $TM$ wave as that component of the incident field with its electric field vector in the $x - z$ plane (and with its magnetic field vector perpendicular to the $x - z$ plane). Then the $TE$ wave is the orthogonal component of the incident field with its electric field vector perpendicular to the $x - z$ plane.[†] Thus, the $z$-components of the incident $E$ and $H$ fields are[18]

$$E_z^{TM} = E_0^{TM} \cos \Psi \exp[ik_0(z \sin \Psi - x \cos \Psi)] \tag{4.2-1}$$

and

$$H_z^{TE} = H_0^{TE} \cos \Psi \exp[ik_0(z \sin \Psi - x \cos \Psi)] \tag{4.2-2}$$

where

$$H_0^{TE} = -\sqrt{\frac{\epsilon_0}{\mu_0}} E_{0y}^{TE} \tag{4.2-3}$$

The scattered fields for the $TM$ and $TE$ components of the incident field are

(a) *TM Incident*

$$E_z^{sTM} = E_0^{TM} \cos \Psi \, e^{ik_0 z \sin \Psi} \sum_{n=-\infty}^{\infty} (-i)^n C_n^{TM} H_n^{(1)}(k_0 r \cos \Psi) \, e^{in\phi'} \tag{4.2-4}$$

$$E_\phi^{sTM} = E_0^{TM} \cos \Psi \, e^{ik_0 z \sin \Psi} \sum_{n=-\infty}^{\infty} \left[ \frac{-n \sin \Psi}{k_0 r \cos^2 \Psi} C_n^{TM} H_n^{(1)}(k_0 r \cos \Psi) \right.$$

$$\left. - \frac{i}{\cos \Psi} \sqrt{\frac{\mu_0}{\epsilon_0}} \, \bar{C}_n H_n^{(1)\prime}(k_0 r \cos \Psi) \right] (-i)^n \, e^{i\phi'} \tag{4.2-5}$$

---

[†] In terms of the $\|$, $\perp$ notation introduced in Chapter 2, a $TM$ wave corresponds to $\|$ polarization, and $TE$ to $\perp$ polarization, with the plane of incidence being the $x$-$z$ plane.

$$H_z^{sTM} = \sqrt{\frac{\epsilon_0}{\mu_0}} \, E_0^{TM} \cos \Psi \, e^{ik_0 z \sin \Psi} \sum_{n=-\infty}^{\infty} (-i)^n \, \bar{C}_n H_n^{(1)}(k_0 r \cos \Psi) \, e^{in\phi'}$$

$$(4.2\text{-}6)$$

$$H_\phi^{sTM} = \sqrt{\frac{\epsilon_0}{\mu_0}} \, E_0^{TM} \cos \Psi \, e^{ik_0 z \sin \Psi} \sum_{n=-\infty}^{\infty} \left[ \frac{-n \sin \Psi}{k_0 r \cos^2 \Psi} \, \bar{C}_n H_n^{(1)}(k_0 r \cos \Psi) \right.$$

$$\left. + \frac{i}{\cos \Psi} \, C_n^{TM} H_n^{(1)\prime}(k_0 r \cos \Psi) \right] (-i)^n \, e^{in\phi'} \qquad (4.2\text{-}7)$$

(b) *TE Incident*

$$H_z^{sTE} = H_0^{TE} \cos \Psi \, e^{ik_0 z \sin \Psi} \sum_{n=-\infty}^{\infty} (-i)^n \, C_n^{TE} H_n^{(1)}(kr \cos \Psi) \, e^{in\phi'} \qquad (4.2\text{-}8)$$

$$H_\phi^{sTE} = H_0^{TE} \cos \Psi \, e^{ik_0 z \sin \Psi} \sum_{n=-\infty}^{\infty} \left[ \frac{-n \sin \Psi}{k_0 r \cos^2 \Psi_0} \, C_n^{TE} H_n^{(1)}(k_0 r \cos \Psi) \right.$$

$$\left. - \frac{i}{\cos \Psi} \sqrt{\frac{\epsilon_0}{\mu_0}} \, \bar{C}_n H_n^{(1)\prime}(k_0 r \cos \Psi) \right] (-i)^n \, e^{in\phi'} \qquad (4.2\text{-}9)$$

$$E_z^{sTE} = - \sqrt{\frac{\mu_0}{\epsilon_0}} \, H_0^{TE} \cos \Psi \, e^{ik_0 z \sin \Psi} \sum_{n=-\infty}^{\infty} (-i)^n \, \bar{C}_n H_n^{(1)}(k_0 r \cos \Psi) \, e^{in\phi'}$$

$$(4.2\text{-}10)$$

$$E_\phi^{sTM} = - \sqrt{\frac{\mu_0}{\epsilon_0}} \, H_0^{TE} \cos \Psi \, e^{ik_0 z \sin \Psi} \sum_{n=-\infty}^{\infty} \left[ \frac{-n \sin \Psi}{k_0 r \cos^2 \Psi} \, \bar{C}_n H_n^{(1)}(k_0 r \cos \Psi) \right.$$

$$\left. + \frac{i}{\cos \Psi} \, C_n^{TE} H_n^{(1)\prime}(k_0 r \cos \Psi) \right] (-i)^n \, e^{in\phi'} \qquad (4.2\text{-}11)$$

In the above expressions, $r$ is the radial distance from the observation point to the cylinder axis. The constants, $C_n^{TM}$, $C_n^{TE}$, and $\bar{C}_n$, are functions of the outer cylinder radius, $a_0$, and the cylinder material; they will be given later for three different types of cylinders.

There are also scattered field components in the $r$ direction. They can be determined simply using Maxwell's equations and the above results.

The above expressions are exact and apply both in the near field and far field of the cylinder. However, in the far field, the above results will be simplified as given below.

### 4.2.1.2. *Far-Zone Simplification and Scattering Width*

When $k_0 r \cos \Psi \gg 1$ and $k_0 r \gg (k_0 a_0)^2 \cos \Psi$, the field point of interest is said to be in the far zone. Under these restrictions, the scattered field solutions simplify to

(a) *TM Incident*

$$E_z^{sTM} = \sqrt{\frac{2}{\pi}} \frac{E_0^{TM} \sqrt{\cos \Psi} \exp[i(k_0 z \sin \Psi + k_0 r \cos \Psi - \pi/4)]}{\sqrt{k_0 r}}$$

$$\times \sum_{n=-\infty}^{\infty} (-1)^n C_n^{TM} e^{in\phi'} \qquad (4.2\text{-}12)$$

$$E_\phi^{sTM} = \sqrt{\frac{2}{\pi}} \frac{E_0^{TM} \exp[i(k_0 z \sin \Psi + k_0 r \cos \Psi - \pi/4)]}{\sqrt{k_0 r} \cos \Psi}$$

$$\times \sum_{n=-\infty}^{\infty} (-1)^n \bar{C}_n e^{in\phi'} \qquad (4.2\text{-}13)$$

$$H_z^{sTM} = \sqrt{\frac{2}{\pi}} \frac{\epsilon_0}{\mu_0} \frac{E_0^{TM} \sqrt{\cos \Psi} \exp[i(k_0 z \sin \Psi + k_0 r \cos \Psi - \pi/4)]}{\sqrt{k_0 r}}$$

$$\times \sum_{n=-\infty}^{\infty} (-1)^n \bar{C}_n e^{in\phi'} \qquad (4.2\text{-}14)$$

$$H_\phi^{sTM} = -\sqrt{\frac{2}{\pi}} \frac{\epsilon_0}{\mu_0} \frac{E_0^{TM} \exp[i(k_0 z \sin \Psi + k_0 r \cos \Psi - \pi/4)]}{\sqrt{k_0 r} \cos \Psi}$$

$$\times \sum_{n=-\infty}^{\infty} (-1)^n C_n^{TM} e^{in\phi'} \qquad (4.2\text{-}15)$$

(b) *TE Incident*

$$H_z^{sTE} = \sqrt{\frac{2}{\pi}} \frac{H_0^{TE} \sqrt{\cos \Psi} \exp[i(k_0 z \sin \Psi + k_0 r \cos \Psi - \pi/4)]}{\sqrt{k_0 r}}$$

$$\times \sum_{n=-\infty}^{\infty} (-1)^n C_n^{TE} e^{in\phi'} \qquad (4.2\text{-}16)$$

$$H_\phi^{sTE} = \sqrt{\frac{2}{\pi}} \frac{H_0^{TE} \exp[i(k_0 z \sin \Psi + k_0 r \cos \Psi - \pi/4)]}{\sqrt{k_0 r} \cos \Psi}$$

$$\times \sum_{n=-\infty}^{\infty} (-1)^n \bar{C}_n e^{in\phi'} \qquad (4.2\text{-}17)$$

$$E_z^{sTE} = -\sqrt{\frac{2}{\pi}\frac{\mu_0}{\epsilon_0}}\, \frac{H_0^{TE}\sqrt{\cos\Psi}\,\exp[i(k_0 z \sin\Psi + k_0 r \cos\Psi - \pi/4)]}{\sqrt{k_0 r}}$$

$$\times \sum_{n=-\infty}^{\infty} (-1)^n \bar{C}_n\, e^{in\phi'} \qquad (4.2\text{-}18)$$

$$E_\phi^{sTE} = \sqrt{\frac{2}{\pi}\frac{\mu_0}{\epsilon_0}}\, \frac{H_0^{TE}\exp[i(k_0 z \sin\Psi + k_0 r \cos\Psi - \pi/4)]}{\sqrt{k_0 r \cos\Psi}}$$

$$\times \sum_{n=-\infty}^{\infty} (-1)^n C_n^{TE}\, e^{in\phi'} \qquad (4.2\text{-}19)$$

An important fact is evident from the above equations for the scattered far field: the scattered field, in general, has a component perpendicular to the plane containing the $z$ axis and observation point, $P'$. Thus, for example, even though the incident $E$-field vector for $TM$ polarization lies entirely in the $x - z$ plane (i.e., there is no $E_\phi^{iTM}$ component in the incident field), there will be an $E_\phi^{sTM}$ component in the scattered field, as well as the $E$-field component in the scattering plane. This is due to the existence of the constants $\bar{C}_n$. In the special case of normal incidence ($\Psi = 0$), $\bar{C}_n = 0$, and this is no longer true. Also, it will be seen that for the perfectly conducting cylinder, $\bar{C}_n = 0$ for any incidence angle; while for the homogeneous and coated conducting cylinders, $\bar{C}_n \neq 0$. Hence, the perfectly conducting cylinder does not "depolarize" in the manner described above at oblique incidence, while the homogeneous and coated cylinders do depolarize (i.e., cross polarization occurs, or in terms of the scattering matrix formulation of Section 2.1.4, the nondiagonal terms $a_{12}$, $a_{21}$ are nonzero.)

In the far zone, there is also a component of the scattered field in the $r$ direction; this exists because the $E$ and $H$ fields must be perpendicular to the direction of propagation. These radial components for both modes are related to the $z$ components as follows:

$$E_r^{sTM} = -E_z^{sTM}\tan\Psi, \qquad H_r^{sTM} = -H_z^{sTM}\tan\Psi$$

$$E_r^{sTE} = -E_z^{sTE}\tan\Psi, \qquad H_r^{sTE} = -H_z^{sTE}\tan\Psi \qquad (4.2\text{-}20)$$

In this section, two scattering widths will be defined. One will relate the scattered $E$-field intensity (for the $TM$ component) in the scattering plane to the incident $E$-field intensity in the incidence plane ($x - z$ plane); this will be called the polarized contribution. The other will relate the scattered $E$-field intensity (for the $TM$ component), perpendicular to the scattering plane ($E_\phi^{sTM}$ in this case), to the incident $E$-field intensity in the incidence plane; this is the depolarized contribution. The same applies to the $H$ fields for the $TE$ component.

4.2.1.2.1. *General Bistatic Scattering Widths.*[†] The scattering width for oblique incidence is defined as the ratio of scattered to incident field intensities times $2\pi S_0$, where $S_0 =$ distance along the scattered field propagation path from the cylinder to point $P'$. By this definition, $S_0 = r/\cos \Psi$. Hence, for the polarized and depolarized contributions discussed above for each mode, these scattering widths become:

(a) *TM Incident*

(i) Polarized Component.

$$\sigma_p^{cTM}(\phi') = \frac{4}{k_0 \cos^2 \Psi} \left| \sum_{n=-\infty}^{\infty} (-1)^n C_n^{TM} e^{in\phi'} \right|^2 \qquad (4.2\text{-}21)$$

(ii) Depolarized Component.

$$\sigma_d^{cTM}(\phi') = \frac{4}{k_0 \cos^2 \Psi} \frac{\mu_0}{\epsilon_0} \left| \sum_{n=-\infty}^{\infty} (-1)^n \bar{C}_n e^{in\phi} \right|^2 \qquad (4.2\text{-}22)$$

(b) *TE Incident*

(i) Polarized Component.

$$\sigma_p^{cTE}(\phi') = \frac{4}{k_0 \cos^2 \Psi} \left| \sum_{n=-\infty}^{\infty} (-1)^n C_n^{TE} e^{in\phi'} \right|^2 \qquad (4.2\text{-}23)$$

(ii) Depolarized Component.

$$\sigma_d^{cTE}(\phi') = \frac{4}{k_0 \cos^2 \Psi} \frac{\epsilon_0}{\mu_0} \left| \sum_{n=-\infty}^{\infty} (-1)^n \bar{C}_n e^{in\phi'} \right|^2 \qquad (4.2\text{-}24)$$

4.2.1.2.2. *Incidence Plane Scattering Width* ($\phi' = 0$). This is the specularly scattered field in the plane of incidence, and the scattering widths are

(a) *TM Incident*

(i) Polarized Component.

$$\sigma_p^{cTM}(0) = \frac{4}{k_0 \cos^2 \Psi} \left| \sum_{n=-\infty}^{\infty} (-1)^n C_n^{TM} \right|^2 \qquad (4.2\text{-}25)$$

(ii) Depolarized Component.

$$\sigma_d^{cTM}(0) = \frac{4}{k_0 \cos^2 \Psi} \frac{\mu_0}{\epsilon_0} \left| \sum_{n=-\infty}^{\infty} (-1)^n \bar{C}_n \right|^2 \qquad (4.2\text{-}26)$$

[†] Note again that there is no backscattered power and, consequently, no backscattering width. All scattered power propagates in a cone which makes an angle of $\pi/2 - \Psi$ with respect to the cylinder axis.

(b) *TE Incident*

(i) Polarized Component.

$$\sigma_p^{cTE}(0) = \frac{4}{k_0 \cos^2 \Psi} \left| \sum_{n=-\infty}^{\infty} (-1)^n C_n^{TE} \right|^2 \tag{4.2-27}$$

(ii) Depolarized Component.

$$\sigma_d^{cTE}(0) = \frac{4}{k_0 \cos^2 \Psi} \frac{\epsilon_0}{\mu_0} \left| \sum_{n=-\infty}^{\infty} (-1)^n \bar{C}_n \right|^2 \tag{4.2-28}$$

4.2.1.2.3. *Incidence Plane Scattering Width* ($\phi' = \pi$). This is the forward scattered field in the plane of incidence, and the scattering widths are

(a) *TM Incident*

(i) Polarized Component.

$$\sigma_p^{cTM}(\pi) = \frac{4}{k_0 \cos^2 \Psi} \left| \sum_{n=-\infty}^{\infty} C_n^{TM} \right|^2 \tag{4.2-29}$$

(ii) Depolarized Component.

$$\sigma_d^{cTM}(\pi) = \frac{4}{k_0 \cos^2 \Psi} \frac{\mu_0}{\epsilon_0} \left| \sum_{n=-\infty}^{\infty} \bar{C}_n \right|^2 \tag{4.2-30}$$

(b) *TE Incident*

(i) Polarized Component.

$$\sigma_p^{cTE}(\pi) = \frac{4}{k_0 \cos^2 \Psi} \left| \sum_{n=-\infty}^{\infty} C_n^{TE} \right|^2 \tag{4.2-31}$$

(ii) Depolarized Component.

$$\sigma_d^{cTE}(\pi) = \frac{4}{k_0 \cos^2 \Psi} \frac{\epsilon_0}{\mu_0} \left| \sum_{n=-\infty}^{\infty} \bar{C}_n \right|^2 \tag{4.2-32}$$

## 4.2.2. Perfectly Conducting Cylinder

The coefficients, $C_n^{TM}$, $C_n^{TE}$, and $\bar{C}_n$, appearing in the exact solution for the scattered field at oblique incidence when the cylinder is perfectly conducting are[18,19]

$$C_n^{TM} = - \frac{J_n(k_0 a_0 \cos \Psi)}{H_n^{(1)}(k_0 a_0 \cos \Psi)} = C_{-n}^{TM} \tag{4.2-33}$$

and

$$C_n^{TE} = - \frac{J_0'(k_0 a_0 \cos \Psi)}{H_n^{(1)\prime}(k_0 a_0 \cos \Psi)} = C_{-n}^{TE} \tag{4.2-34}$$

with

$$\bar{C}_n = 0 = \bar{C}_{-n} \tag{4.2-35}$$

The primes in (4.2-34) indicate differentiation of the Bessel functions with respect to the argument shown in parentheses. These coefficients are readily generated on a computer by recurrence techniques (see Section 3.2.1). When substituted into Eqs. (4.2-12) to (4.2-32), they provide the scattered fields and various scattering widths for the cylinder.

It should be noted from comparison of Eqs. (4.2-21) and (4.2-23) with $\sigma^c$ at normal incidence that the only difference is that the argument $k_0 a$ at normal incidence is replaced everywhere by $k_0 a_0 \cos \Psi$. Hence, Figures 4-2 to 4-4 for normal incidence also apply here for oblique incidence with the following scaling changes:[†]

(1) The quantity $k_0 a$ should be replaced by $k_0 a_0 \cos \Psi$. This occurs on the abscissas for $\sigma^c(0)$, and $\sigma^c(\pi)$, and on the labels of the curves for $\sigma^c(\phi')$.
(2) The ordinate axis should be changed from $\sigma/\pi a$ to $\sigma(\cos \Psi)/\pi a_0$. It should be noted that the subscripts $\|$ and $\perp$ used there are synonymous with $TM$ and $TE$, respectively.

Similarly, all the scattering width equations of Section 4.1.2 are valid for the oblique incidence case if $a$ is replaced by $a_0 \cos \Psi$, and $\sigma^c$ for normal incidence is divided by $\cos^2 \Psi$.

### 4.2.3. Homogeneous Cylinder

#### 4.2.3.1. Series Coefficients

Let the cylinder material have relative permeability, $\mu_r$, and permittivity, $\epsilon_r$ (in general these may be complex). Then the exact series solution coefficients are[9,18,19]

$$C_n^{TM} = - \frac{V_n P_n - q_n^2 J_n(x_0) H_n^{(1)}(x_0) J_n^2(x_1)}{P_n N_n - [q_n H_n^{(1)}(x_0) J_n(x_1)]^2} \tag{4.2-36}$$

$$C_n^{TE} = - \frac{M_n N_n - q_n^2 J_n(x_0) H_n^{(1)}(x_0) J_n^2(x_1)}{P_n N_n - [q_n H_n^{(1)}(x_0) J_n(x_1)]^2} \tag{4.2-37}$$

$$\bar{C}_n = i \frac{2}{\pi x_0} \left[ \frac{s_0 q_n J_n^2(x_1)}{P_n N_n - [q_n H_n^{(1)}(x_0) J_n(x_1)]^2} \right] \tag{4.2-38}$$

where

$$x_0 = k_0 a_0 \cos \Psi \tag{4.2-39}$$

$$x_1 = k_0 a_0 \sqrt{\epsilon_r \mu_r - \sin^2 \Psi} \tag{4.2-40}$$

$$q_n = \frac{n \sin \Psi}{k_0 a_0} \left( \frac{1}{\epsilon_r \mu_r - \sin^2 \Psi} - \frac{1}{\cos^2 \Psi} \right) \tag{4.2-41}$$

$$V_n = s_1 J_n(x_0) J_n'(x_1) - s_0 J_n'(x_0) J_n(x_1) \tag{4.2-42}$$

[†] This simple scaling technique is only applicable for perfectly conducting cylinders.

$$P_n = r_1 H_n^{(1)}(x_0) J_n'(x_1) - s_0 H_n^{(1)\prime}(x_0) J_n(x_1) \tag{4.2-43}$$

$$N_n = s_1 H_n^{(1)}(x_0) J_n'(x_1) - s_0 H_n^{(1)\prime}(x_0) J_n(x_1) \tag{4.2-44}$$

$$M_n = r_1 J_n(x_0) J_n'(x_1) - s_0 J_n'(x_0) J_n(x_1) \tag{4.2-45}$$

and

$$s_0 = \frac{1}{\cos \Psi}, \quad s_1 = \frac{\epsilon_r}{\sqrt{\epsilon_r \mu_r - \sin^2 \Psi}}, \quad r_1 = \frac{\mu_r}{\sqrt{\epsilon_r \mu_r - \sin^2 \Psi}} \tag{4.2-46}$$

One notes immediately that, unlike the perfectly conducting cylinder, the homogeneous-cylinder series coefficients are much more complex. The significant difference about the nature of the scattered fields, in this case, is the presence of both *TM* and *TE* modes in the scattered field, even though the incident field may have been either strictly *TM* or *TE*. Stated another way, there is a $z$ component for the scattered $H$ field, even though there was no $z$ component in the incident $H$ field (i.e., *TM* incident field). This "depolarizing" effect is evidenced by the presence of the nonzero, $\bar{C}_n$, series coefficients. These coefficients are zero only at normal incidence ($\Psi = 0$), or for perfectly conducting or reflecting cylinder material (where $\mu_r$ or $\epsilon_r$ becomes infinite).

Symmetry relations among the series coefficients yield the following properties:

$$C_n^{TM} = C_{-n}^{TM}, \quad C_n^{TE} = C_{-n}^{TE}, \quad \bar{C}_n = -\bar{C}_{-n}, \quad \bar{C}_0 = 0 \tag{4.2-47}$$

From these properties, it can be shown that the far-zone scattered field in the plane of incidence (where $\phi' = 0$ or $\pi$) is not depolarized; this means that $E_\phi^{sTM}$, $H_z^{sTM}$, $H_\phi^{sTE}$, and $E_z^{sTE}$ [see Eqs. (4.2-12) to (4.2-19)] are all zero for $\phi' = 0$ or $\pi$. At other angles, $\phi'$, such is not the case.

From these series solution coefficients, it is not difficult to find the scattered field at any angle of incidence, $\psi$, and any scattering angle, $\phi'$, on a computer. Results[9] generated in this manner are shown in Figures 4-34 and 4-35. These figures show the quantities, $\sigma_p^{\circ TM}(\phi')(\cos^2 \Psi)/\pi a_0$ and $\sigma_p^{\circ TE}(\phi')(\cos^2 \Psi)/\pi a_0$ versus $k_0 a_0$, for $\phi' = 0$, $\Psi = 45°$ and $84°$, and $\epsilon_r = 4.0$ and $2.56$ ($\mu_r = 1$).[†] (Similar results at normal incidence, $\Psi = 0$, have already been shown for $\epsilon_r = 2.56$ in Figure 4-7). These figures show the behavior of the far-zone scattered power for scattering in the plane of incidence for both *TM* and *TE* incident waves. The region shown is the resonance region ($0.2\pi < k_0 a_0 < 15\pi$). Results valid in the low-frequency limit will be given in the next section.

---

[†] The scattering widths, $\sigma_p^{TM}(\phi')$ and $\sigma_p^{TE}(\phi')$, are defined formally in Eqs. (4.2-21) and (4.2-23). $\sigma_d^{TM}(\phi')$ and $\sigma_d^{TE}(\phi')$ are zero here for $\phi' = 0$ (or $\phi' = \pi$).

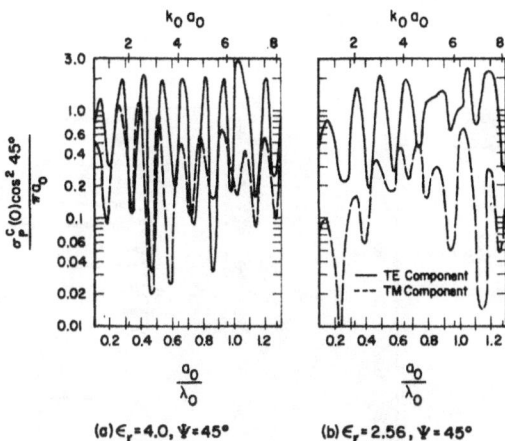

Fig. 4-34. Exact solutions for scattering widths for an infinitely long, homogeneous cylinder for oblique incidence and scattering in the plane of incidence.

### 4.2.3.2. Approximations for the Low-Frequency Limit $(k_0 a_0 \sqrt{\epsilon_r \mu_r - \sin^2 \Psi} < 0.5)$

When the cylinder radius or frequency is such that the above inequality is valid, then the series solution coefficients, $C_n^{TM}$, $C_n^{TE}$, and $C_n$, can be simplified by using small argument expansions for the Bessel functions.[20] Only the first two coefficients (for $n = 0$ and 1) are significant in this limit.

4.2.3.2.1. Scattered Fields $(k_0 a_0 \sqrt{\epsilon_r \mu_r - \sin^2 \Psi} < 0.5)$. The $z$ and $\phi$ components of the far-zone scattered field for each mode, as obtained from Eqs. (4.2-12) to (4.2-19), then become:

(a) TM Incident Field

$$E_z^{sTM} = \sqrt{\frac{\pi}{2}} \frac{E_0^{TM} \sqrt{\cos \Psi} \exp[i(k_0 z \sin \Psi + k_0 r \cos \Psi + \pi/4)]}{\sqrt{k_0 r}}$$

$$\times \frac{(k_0 a_0)^2}{2} \left[ (\epsilon_r - 1) \cos^2 \Psi \right.$$

$$\left. - 2 \frac{(\epsilon_r + 1)(\mu_r - 1) + \sin^2 \Psi(\epsilon_r - 1)(\mu_r + 1)}{(\epsilon_r + 1)(\mu_r + 1)} \cos \phi' \right] \quad (4.2\text{-}48)$$

$$E_\phi^{sTM} = \sqrt{2\pi} \frac{E_0^{TM} \exp[i(k_0 z \sin \Psi + k_0 r \cos \Psi - \pi/4)]}{\sqrt{k_0 r \cos \Psi}} (k_0 a_0)^2$$

$$\times \frac{(\epsilon_r \mu_r - 1)}{(\epsilon_r + 1)(\mu_r + 1)} \sin \Psi \sin \phi' \quad (4.2\text{-}49)$$

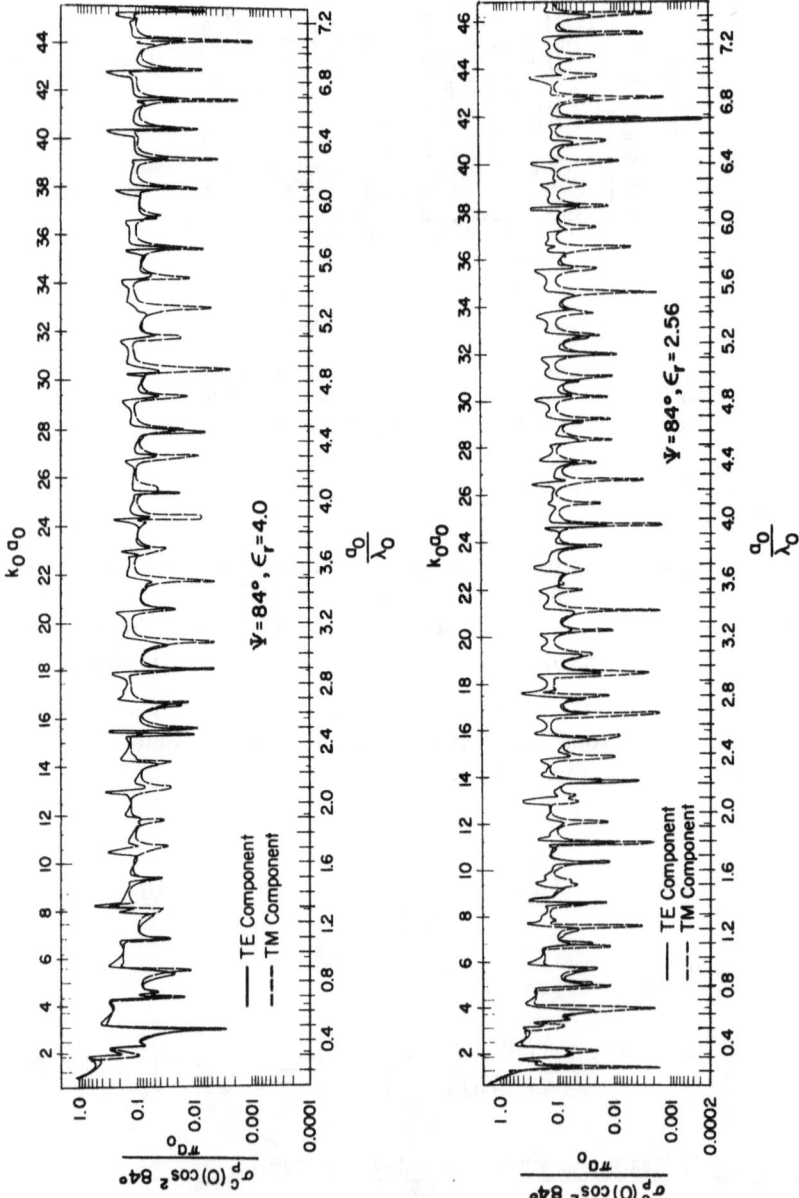

Fig. 4-35. Exact solutions for scattering widths for an infinitely long, homogeneous cylinder for oblique incidence and scattering in the plane of incidence.

$$H_z^{sTM} = \sqrt{2\pi}\sqrt{\frac{\epsilon_0}{\mu_0}}\frac{E_0^{TM}\sqrt{\cos\Psi}\exp[i(k_0 z\sin\Psi + k_0 r\cos\Psi - \pi/4)]}{\sqrt{k_0 r}}$$

$$\times (k_0 a_0)^2 \frac{(\epsilon_r\mu_r - 1)}{(\epsilon_r + 1)(\mu_r + 1)}\sin\Psi\sin\phi' \tag{4.2-50}$$

$$H_\phi^{sTM} = -\sqrt{\frac{\pi}{2}}\sqrt{\frac{\epsilon_0}{\mu_0}}\frac{E_0^{TM}\exp[i(k_0 z\sin\Psi + k_0 r\cos\Psi + \pi/4)]}{\sqrt{k_0 r}\cos\Psi}\frac{(k_0 a_0)^2}{2}$$

$$\times \Bigg[(\epsilon_r - 1)\cos^2\Psi$$

$$-2\frac{(\epsilon_r + 1)(\mu_r - 1) + \sin^2\Psi(\epsilon_r - 1)(\mu_r + 1)}{(\epsilon_r + 1)(\mu_r + 1)}\cos\phi'\Bigg] \tag{4.2-51}$$

(b) *TE Incident Field*

$$H^{sTE} = \sqrt{\frac{\pi}{2}}\frac{H_0^{TE}\sqrt{\cos\Psi}\exp[i(k_0 z\sin\Psi + k_0 r\cos\Psi + \pi/4)]}{\sqrt{k_0 r}}\frac{(k_0 a_0)^2}{2}$$

$$\times \Bigg[(\mu_r - 1)\cos^2\Psi$$

$$-2\frac{(\epsilon_r - 1)(\mu_r + 1) + \sin^2\Psi(\epsilon_r + 1)(\mu_r - 1)}{(\epsilon_r + 1)(\mu_r + 1)}\cos\phi'\Bigg] \tag{4.2-52}$$

$$H_\phi^{sTE} = \sqrt{2\pi}\frac{H_0^{TE}\exp[i(k_0 z\sin\Psi + k_0 r\cos\Psi - \pi/4)]}{\sqrt{k_0 r}\cos\Psi}(k_0 a_0)^2$$

$$\times \frac{(\epsilon_r\mu_r - 1)}{(\epsilon_r + 1)(\mu_r + 1)}\sin\Psi\sin\phi' \tag{4.2-53}$$

$$E_z^{sTE} = -\sqrt{2\pi}\sqrt{\frac{\mu_0}{\epsilon_0}}\frac{H_0^{TE}\sqrt{\cos\Psi}\exp[i(k_0 z\sin\Psi + k_0 r\cos\Psi - \pi/4)]}{\sqrt{k_0 r}}$$

$$\times (k_0 a_0)^2 \frac{(\epsilon_r\mu_r - 1)}{(\epsilon_r + 1)(\mu_r + 1)}\sin\Psi\sin\phi' \tag{4.2-54}$$

$$E_\phi^{sTE} = \sqrt{\frac{\pi}{2}}\sqrt{\frac{\mu_0}{\epsilon_0}}\frac{H_0^{TE}\exp[i(k_0 z\sin\Psi + k_0 r\cos\Psi + \pi/4)]}{\sqrt{k_0 r}\cos\Psi}\frac{(k_0 a_0)^2}{2}$$

$$\times \Bigg[(\mu_r - 1)\cos^2\Psi$$

$$-2\frac{(\epsilon_r - 1)(\mu_r + 1) + \sin^2\Psi(\epsilon_r + 1)(\mu_r - 1)}{(\epsilon_r + 1)(\mu_r + 1)}\cos\phi'\Bigg] \tag{4.2-55}$$

4.2.3.2.2. *Bistatic Scattering Width* $(k_0 a_0\sqrt{\epsilon_r\mu_r - \sin^2\Psi} < 0.5)$. Remembering that the scattered field propagates only at an angle, $\pi/2 - \Psi$, with respect to the $+z$ axis, one can obtain low-frequency scattering widths in this direction for arbitrary polar angle $\phi'$ from Eqs. (4.2-21) to (4.2-25).

(a) *TM Incident*

$$\sigma_p^{cTM}(\phi') = \frac{\pi^2 a_0}{4 \cos^2 \Psi} (k_0 a_0)^3 \left[ (\epsilon_r - 1) \cos^2 \Psi \right.$$

$$\left. - 2 \frac{(\epsilon_r + 1)(\mu_r - 1) + \sin^2 \Psi (\epsilon_r - 1)(\mu_r + 1)}{(\epsilon_r + 1)(\mu_r + 1)} \cos \phi' \right]^2 \quad (4.2\text{-}56)$$

$$\sigma_d^{cTM}(\phi') = \sigma_d^{cTE}(\phi') = \frac{4\pi^2 a_0}{\cos^2 \Psi} (k_0 a_0)^3 \left[ \frac{\epsilon_r \mu_r - 1}{(\epsilon_r + 1)(\mu_r + 1)} \sin \Psi \sin \phi' \right]^2$$

$$(4.2\text{-}57)$$

(b) *TE Incident*

$$\sigma_p^{cTE}(\phi') = \frac{\pi^2 a_0}{4 \cos^2 \Psi} (k_0 a_0)^3 \left[ (\mu_r - 1) \cos^2 \Psi \right.$$

$$\left. - 2 \frac{(\epsilon_r - 1)(\mu_r + 1) + \sin^2 \Psi (\epsilon_r + 1)(\mu_r - 1)}{(\epsilon_r + 1)(\mu_r + 1)} \cos \phi' \right]^2 \quad (4.2\text{-}58)$$

The depolarized scattering width in this case is identical to that for *TM* polarization, and is given by Eq. (4.2-57). These results also apply to the special cases of $\phi' = 0$ and $\phi' = \pi$ (scattering in the plane of incidence).

4.2.3.3. *Approximations for the High-Frequency Limit with Attenuation* $[\mathrm{Im}(k_0 a_0 \sqrt{\mu_r \epsilon_r}) > 5]$

If the cylinder material is not lossy, there will be no high-frequency limit for the scattering width. Rays (or energy) entering the cylinder are not absorbed; rather, they are reflected internally many times, and at each internal reflection, some of the energy leaves the cylinder and contributes to the scattered field. At certain combinations of scattering angle, frequency, and cylinder radius, the scattered field may be very large due to the in-phase addition of various emerging, internally-reflected rays. The behavior of the scattered field in this case is similar to that for the homogeneous cylinder at normal incidence (Section 4.1.3.4) and for the homogeneous sphere (Section 3.3.2.2). The only promising approximate method of attacking this problem for oblique incidence is the modified geometrical-optics approach of Section 4.1.3.4; this method apparently has not yet been applied to the cylinder at oblique incidence.

In most cases, the cylinder material has some loss, however small. Generally, this loss term in the permittivity and/or permeability increases with frequency, and the path length in wavelengths through the cylinder medium increases with frequency so that eventually the scattered field approaches a high-frequency limit. Thus in the high-frequency limit, all rays which enter the cylinder are attenuated sufficiently in passing through the material that none of them effectively reemerge from the cylinder to contribute to the scattered field. When $\mathrm{Im}(k_0 a_0 \sqrt{\epsilon_r \mu_r}) > 5$, this condition occurs. Then the only contribution to the scattered field comes from the external specularly-reflected ray which never enters the cylinder. The field externally reflected at a given angle is proportional to the Fresnel reflection coefficients. Knowing the location of the specular point on the cylinder (from a knowledge of the incidence direction, $\Psi$, and the desired bistatic angle, $\phi'$), one can determine the angle of incidence. Also, it is necessary to resolve the incident $E$ field into components in and perpendicular to the plane of incidence.[†] This is due to the existence of two different Fresnel reflection coefficients for each of these components. Let the incident field vector be given in general by $\hat{\tau}$. The angle of incidence, $\gamma$, is then defined to be the angle between the incident wave propagation direction and the normal to the surface at the specular point. For given $\Psi$ and $\phi'$, it is determined from

$$\cos \gamma = \cos \Psi \cos(\phi'/2) \qquad (4.2\text{-}59)$$

(a) *Scattering for Component in the Plane of Incidence*

Let

$$\hat{\zeta} = \hat{x} \sin^2 \Psi \cos(\phi'/2) + \hat{y} \sin(\phi'/2) + \hat{z} \sin \Psi \cos \Psi \cos(\phi'/2) \qquad (4.2\text{-}60)$$

be a unit vector in the plane of incidence, and perpendicular to the direction of propagation of the incident wave. The component of the scattered field in the plane of incidence is then given by

$$E^s = \hat{\zeta} \cdot \hat{\tau} E_0 \sqrt{\frac{a_0 \cos(\phi'/2)}{2S_0 \cos \Psi}} \, R_{\parallel}(\gamma) \, e^{ik_0 S_0} \qquad (4.2\text{-}61)$$

where $S_0$ represents the distance along the scattering direction from the cylinder to the far-zone field point of interest. The Fresnel coefficient is given by Eq. (7.1-10).

The scattering width for this component of the scattered field is

$$\sigma^c(\phi') = \pi a_0 \frac{\cos(\phi'/2)}{\cos \Psi} \, | \, \hat{\zeta} \cdot \hat{\tau} \, R_{\parallel}(\gamma)|^2 \qquad (4.2\text{-}62)$$

(b) *Scattering for Component Perpendicular to Plane of Incidence*

Let

$$\hat{\eta} = -\hat{x} \sin \Psi \sin(\phi'/2) + \hat{y} \sin \Psi \cos(\phi'/2) - \hat{z} \cos \Psi \sin(\phi'/2) \qquad (4.2\text{-}63)$$

---

[†] The plane of incidence here is taken to mean the plane including the incident field propagation direction and the normal to the cylinder surface at the specular point. This meaning of incidence plane is different from the previous definition used to define *TM* and *TE*; the incidence plane defined previously included the incident field propagation direction and the cylinder axis.

be the unit vector perpendicular to the plane of incidence. The component of the scattered field perpendicular to the plane of incidence is then

$$E^s = \hat{\eta} \cdot \hat{t} \, E_0 \sqrt{\frac{a_0 \cos(\phi/2)}{2S_0 \cos \Psi}} \, R_\perp(\gamma) \, e^{ik_0 S_0} \qquad (4.2\text{-}64)$$

The scattering width for this component of the scattered field is then

$$\sigma^c(\phi') = \pi a_0 \frac{\cos(\phi'/2)}{\cos \Psi} | \, \hat{\eta} \cdot \hat{t} \, R_\perp(\gamma)|^2 \qquad (4.2\text{-}65)$$

### 4.2.4. Coated Perfectly Conducting Cylinder

The geometrical configuration for a coated, perfectly conducting cylinder is illustrated in Figure 4-33. The actual cylinder is made up of an inner perfectly conducting cylinder of radius $a_1$, covered with an outer homogeneous layer of radius, $a_0$ (or thickness $a_0 - a_1$), and relative permittivity and permeability, $\epsilon_r$ and $\mu_r$ (cylinder construction is shown in Figure 4-13).

#### 4.2.4.1. Series Coefficients

The exact series solution coefficients for this cylinder are[9]

$$C_n^{TM} = - \frac{V_n P_n - q_n^2 J_n(x_0) \, H_n^{(1)}(x_0)}{P_n N_n - [q_n H_n^{(1)}(x_0)]^2} \qquad (4.2\text{-}66)$$

$$C_n^{TE} = - \frac{M_n N_n - q_n^2 J_n(x_0) \, H_n^{(1)}(x_0)}{P_n N_n - [q_n H_n^{(1)}(x_0)]^2} \qquad (4.2\text{-}67)$$

$$\bar{C}_n = i \, \frac{2}{\pi x_0} \left[ \frac{s_0 q_n}{P_n N_n - [q_n H_n^{(1)}(x_0)]^2} \right] \qquad (4.2\text{-}68)$$

where $x_0$, $x_1$, $q_n$, $s_0$, $s_1$, and $r_1$ are the same as defined in Eqs. (4.2-39) to (4.2-46), while $V_n$, $P_n$, $N_n$, and $M_n$ are

$$V_n = J_n(x_0) - Z_n J_n'(x_0) \qquad (4.2\text{-}69)$$

$$P_n = H_n^{(1)}(x_0) - Y_n H_n^{(1)\prime}(x_0) \qquad (4.2\text{-}70)$$

$$N_n = H_n^{(1)}(x_0) - Z_n H_n^{(1)\prime}(x_0) \qquad (4.2\text{-}71)$$

$$M_n = J_n(x_0) - Y_n J_n'(x_0) \qquad (4.2\text{-}72)$$

with

$$Z_n = \frac{s_0}{s_1} \frac{J_n(x_1) H_n^{(1)}(x_2) - H_n^{(1)}(x_1) J_n(x_2)}{J_n'(x_1) H_n^{(1)}(x_2) - H_n^{(1)'}(x_1) J_n(x_2)}$$

$$= \frac{s_0}{s_1} \frac{J_n(x_1) N_n(x_2) - N_n(x_1) J_n(x_2)}{J_n'(x_1) N_n(x_2) - N_n'(x_1) J_n(x_2)} \qquad (4.2\text{-}73)$$

$$Y_n = \frac{s_0}{r_1} \frac{J_n(x_1) H_n^{(1)'}(x_2) - H_n^{(1)}(x_1) J_n'(x_2)}{J_n'(x_1) H_n^{(1)'}(x_2) - H_n^{(1)'}(x_1) J_n'(x_2)}$$

$$= \frac{s_0}{r_1} \frac{J_n(x_1) N_n'(x_2) - N_n(x_1) J_n'(x_2)}{J_n'(x_1) N_n'(x_2) - N_n'(x_1) J_n'(x_2)} \qquad (4.2\text{-}74)$$

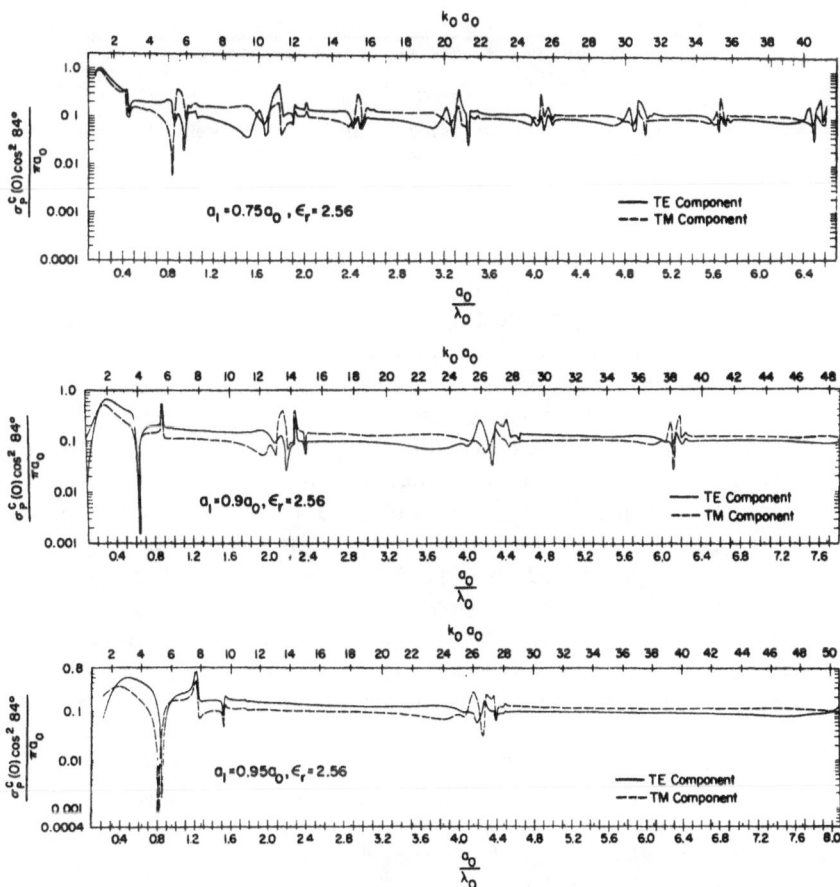

Fig. 4-36. Exact solutions for scattering widths for an infinitely long, coated, perfectly conducting cylinder for oblique incidence and scattering in the plane of incidence.

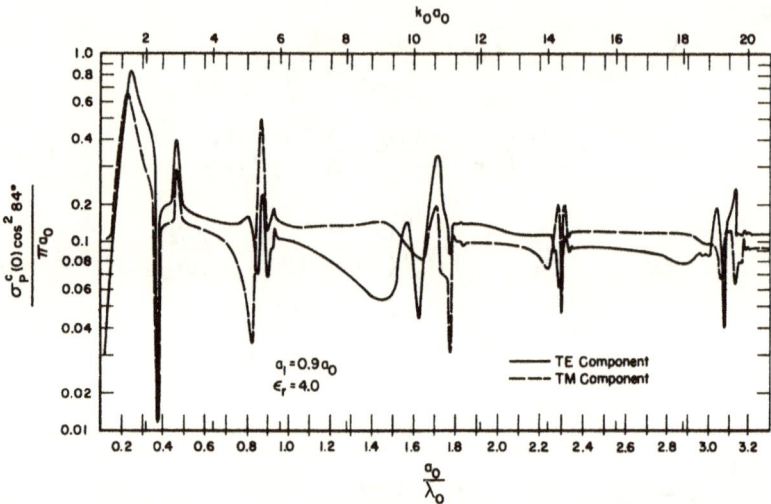

Fig. 4-37. Exact solutions for scattering widths for an infinitely long, coated, perfectly conducting cylinder for oblique incidence and scattering in the plane of incidence.

and

$$x_2 = k_0 a_1 \sqrt{\epsilon_r \mu_r - \sin^2 \Psi} \qquad (4.2\text{-}75)$$

The coated perfectly conducting cylinder, in general, has a depolarized component in the scattered field (i.e., there are both TM and TE components in the scattered field, even though the incident field may be strictly TM or TE). The same symmetry relations among the scattering series coefficients [Eq. (4.2-47)] are true here as for the homogeneous cylinder. The "depolarized" component in the scattered field becomes zero at normal incidence ($\Psi = 0$) or for scattering in the plane of incidence ($\phi' = 0$ or $\pi$).

This exact series solution can be summed on a computer to give the scattering widths as defined by Eqs. (4.2-21) to (4.2-24). The results of such a process for two dielectric constants ($\epsilon_r = 4.0$ and 2.56, where $\mu_r = 1$) are shown in Figures 4-36 to 4-38 (from Reference 9). The angle of incidence in Figures 4-36 and 4-37 is $\Psi = 84°$. Three ratios of $a_1/a_0$ were chosen: 0.90, 0.75, and 0.95. The curves present the quantities $\sigma_p^{cTM}(\phi')(\cos^2 \Psi)/\pi a_0$ and $\sigma_p^{cTE}(\phi')(\cos^2 \Psi)/\pi a_0$ versus $k_0 a_0$ for $\phi' = 0$ (in the plane of incidence). In Figure 4-38, curves of $\sigma_p^{cTM}(0)(\cos^2 \Psi)/\pi a_0$ and $\sigma_p^{cTE}(0)(\cos^2 \Psi)/\pi a_0$ for $\Psi$, varying from 0 to 84°, are shown for fixed values of $k_0 a_0$, where $a_1/a_0 = 0.90$.

### 4.2.4.2. Approximations for the Low-Frequency Limit

$$(k_0 a_0 \sqrt{\epsilon_r \mu_r - \sin^2 \Psi} < 0.5)$$

When the cylinder radius or the frequency is such that the above inequality is valid, then the series solution coefficients, $C_n^{TM}$, $C_n^{TE}$, and $\bar{C}_n$,

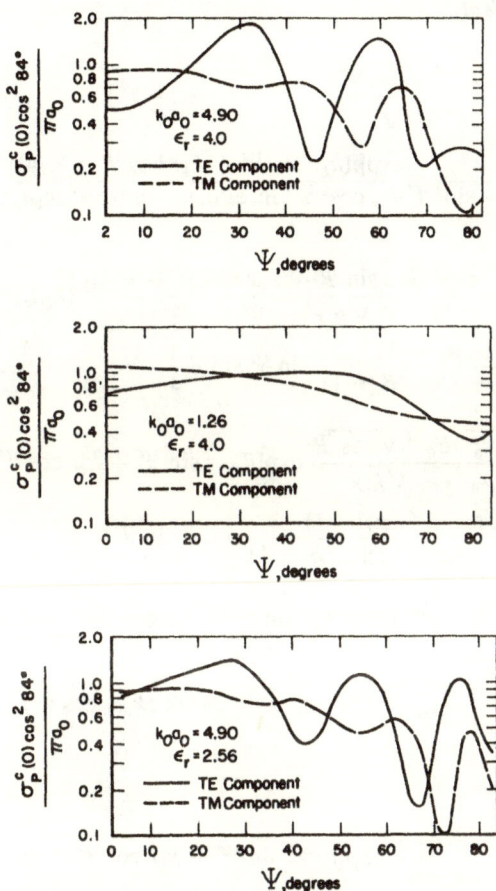

Fig. 4-38. Exact solutions for scattering widths for an infinitely long, coated, perfectly conducting cylinder for oblique incidence and scattering in the plane of incidence.

can be simplified by using small-argument expansions for the Bessel functions. For the $TM$ incident mode, only $C_0^{TM}$ is of importance; while for the $TE$ incident mode, both $C_0^{TE}$ and $C_1^{TE}$ are important. For both cases, $\bar{C}_1$ is nonvanishing.

4.2.4.2.1. *Scattered Fields* $(k_0 a_0 \sqrt{\epsilon_r \mu_r - \sin^2 \Psi} < 0.5)$. The $z$ and $\phi$ components of the far-zone scattered field for each mode, as obtained from Eqs. (4.2-12) to (4.2-19), become:

(a) *TM Incident*

$$E_z^{sTM} = -\sqrt{\frac{\pi}{2}} \frac{E_0^{TM} \sqrt{\cos \Psi}}{\sqrt{k_0 r}}$$

$$\times \frac{\exp[i(k_0 z \sin \Psi \cdot + k_0 r \cos \Psi + \pi/4)]}{[(\epsilon_r \mu_r - \sin^2 \Psi)/\epsilon_r \cos^2 \Psi] \ln(a_0/a_1) - [\ln(0.8950 k_0 a_0 \cos \Psi) - i\pi/2]}$$

$$(4.2\text{-}76)$$

$$E_\phi^{sTM} = \sqrt{2\pi} \frac{E_0^{TM} \exp[i(k_0 z \sin \Psi + k_0 r \cos \Psi - \pi/4)]}{\sqrt{k_0 r \cos \Psi}} (k_0 a_0)^2$$

$$\times \left[ \frac{(\epsilon_r \mu_r - 1)}{(\epsilon_r B + 1)(\mu_r/B + 1)} \sin \Psi \sin \phi' \right] \qquad (4.2\text{-}77)$$

$$H_z^{sTM} = \sqrt{2\pi} \sqrt{\frac{\epsilon_0}{\mu_0}} \frac{E_0^{TM} \sqrt{\cos \Psi}}{\sqrt{k_0 r}} \exp[i(k_0 z \sin \Psi + k_0 r \cos \Psi - \pi/4)]$$

$$\times (k_0 a_0)^2 \left[ \frac{(\epsilon_r \mu_r - 1)}{(\epsilon_r B + 1)(\mu_r/B + 1)} \sin \Psi \sin \phi' \right] \qquad (4.2\text{-}78)$$

$$H_\phi^{sTM} = +\sqrt{\frac{\pi}{2}} \sqrt{\frac{\epsilon_0}{\mu_0}} \frac{E_0^{TM} \exp[i(k_0 z \sin \Psi + k_0 r \cos \Psi + \pi/4)]}{\sqrt{k_0 r \cos \Psi}}$$

$$\times \left[ \left( \frac{\epsilon_r \mu_r - \sin^2 \Psi}{\epsilon_r \cos^2 \Psi} \right) \ln a_0/a_1 - [\ln(0.8905 k_0 a_0 \cos \Psi) - i\pi/2] \right]^{-1}$$

$$(4.2\text{-}79)$$

(b) *TE Incident*

$$H_z^{sTE} = \sqrt{\frac{\pi}{2}} \frac{H_0^{TE} \sqrt{\cos \Psi}}{\sqrt{k_0 r}} \exp[i(k_0 z \sin \Psi + k_0 r \cos \Psi + \pi/4)] \left[ \frac{(k_0 a_0)^2}{2} \right]$$

$$\times \left[ \left( \mu_r \left[ 1 - \left( \frac{a_1}{a_0} \right)^2 \right] - 1 \right) \cos^2 \Psi \right.$$

$$\left. - 2 \frac{(\epsilon_r B - 1)(\mu_r/B + 1) + \sin^2 \Psi(\epsilon_r B + 1)(\mu_r/B - 1)}{(\epsilon_r B + 1)(\mu_r/B + 1)} \cos \phi' \right]$$

$$(4.2\text{-}80)$$

$$H_\phi^{sTE} = \sqrt{2\pi} \frac{H_0^{TE} \exp[i(k_0 z \sin \Psi + k_0 r \cos \Psi - \pi/4)]}{\sqrt{k_0 r \cos \Psi}} (k_0 a_0)^2$$

$$\times \left[ \frac{\epsilon_r \mu_r - 1}{(\epsilon_r B + 1)(\mu_r/B + 1)} \sin \Psi \sin \phi' \right] \qquad (4.2\text{-}81)$$

$$E_z^{sTE} = -\sqrt{2\pi} \sqrt{\frac{\mu_0}{\epsilon_0}} \frac{H_0^{TE} \sqrt{\cos \Psi}}{\sqrt{k_0 r}} \exp[i(k_0 z \sin \Psi + k_0 r \cos \Psi - \pi/4)]$$

$$\times (k_0 a_0)^2 \left[ \frac{\epsilon_r \mu_r - 1}{(\epsilon_r B + 1)(\mu_r/B + 1)} \sin \Psi \sin \phi' \right] \qquad (4.2\text{-}82)$$

$$E_\phi^{sTE} = \sqrt{\frac{\pi}{2}\frac{\mu_0}{\epsilon_0}}\frac{H_0^{TE}\exp[i(k_0 z\sin\Psi + k_0 r\cos\Psi + \pi/4)]}{\sqrt{k_0 r\cos\Psi}}\frac{(k_0 a_0)^2}{2}$$

$$\times\left[\left[\mu_r\left(1-\frac{a_1^2}{a_0^2}\right)-1\right]\cos^2\Psi\right.$$

$$\left. - 2\frac{(\epsilon_r B-1)(\mu_r/B+1)+\sin^2\Psi(\epsilon_r B+1)(\mu_r/B-1)}{(\epsilon_r B+1)(\mu_r/B+1)}\cos\phi'\right]$$

$$\tag{4.2-83}$$

where

$$B = \frac{1+(a_1/a_0)^2}{1-(a_1/a_0)^2} \tag{4.2-84}$$

It should be noted that Eqs. (4.2-76) and (4.2-79) are not valid in the limit as $a_1/a_0 \to 0$ (i.e., the central conductor thickness approaches zero), because there are terms omitted in $(k_0 a_0)^2$ which are significant in that case. All of the other equations here are valid in both limits, as $a_1/a_0 \to 0$ or $a_1/a_0 \to 1$. Again, as at normal incidence, one sees that when there is an incident electric field component parallel to the cylinder axis, the central conducting core causes the structure to resemble the perfectly conducting cylinder case much more nearly than the homogeneous cylinder case; the scattered $E$ field in the axial direction is, in general, larger by a factor $(k_0 a_0)^{-2}$.

4.2.4.2.2. *Bistatic Scattering Width* $(k_0 a_0\sqrt{\epsilon_r\mu_r - \sin^2\Psi} < 0.5)$.
Remembering that the scattered field propagates only in a direction which makes an angle $\pi/2 - \Psi$ with respect to the $+z$ axis, one obtains the following low-frequency scattering widths for arbitrary polar angle $\phi'$ about the cylinder:

(a) *Incident TM Component*

$$\sigma_p^{cTM}(\phi') = \frac{\pi^2 a_0}{\cos\Psi(k_0 a_0\cos\Psi)}\left[\left[\left(\frac{\epsilon_r\mu_r-\sin^2\Psi}{\epsilon_r\cos^2\Psi}\right)\ln\frac{a_0}{a_1}\right.\right.$$

$$\left.\left. - \ln(0.8905 k_0 a_0\cos\Psi)\right]^2 + \pi^2/4\right]^{-1} \tag{4.2-85}$$

$$\sigma_d^{cTM}(\phi') = \sigma_d^{cTE}(\phi') = \frac{4\pi^2 a_0}{\cos^2\Psi}(k_0 a_0)^3\left[\frac{\epsilon_r\mu_r-1}{(\epsilon_r B+1)(\mu_r/B+1)}\sin\Psi\sin\phi'\right]^2 \tag{4.2-86}$$

(b) *Incident TE Component*

$$\sigma_p^{cTE}(\phi') = \frac{\pi^2 a_0}{4\cos^2\Psi}(k_0 a_0)^3\left[\left[\mu_r\left(1-\frac{a_1^2}{a_0^2}\right)-1\right]\cos^2\Psi\right.$$

$$\left. - 2\frac{(\epsilon_r B-1)(\mu_r/B+1)+\sin^2\Psi(\epsilon_r B+1)(\mu_r/B-1)}{(\epsilon_r B+1)(\mu_r/B+1)}\cos\phi'\right]^2 \tag{4.2-87}$$

$\sigma_d^{cTE}(\phi')$ is given by Eq. (4.2-86). These results also apply in the special cases of $\phi' = 0$ or $\pi$ (scattering in the plane of incidence).

4.2.4.3. *Approximations for the High-Frequency Limit with Attenuation* $[\text{Im}(k_0 a_0 \sqrt{\mu_r \epsilon_r}) > 5]$

As in the case of the homogeneous cylinder, there is in general no high-frequency limit if the coating material is not lossy. In that case, rays which enter the coating material can bounce between the inner core and the outer surface many times. These waves propagate in the coating layer and are said to be trapped. At each reflection against the outer interface, energy emerges and is added to the scattered fields. Thus at various frequencies, with given coating material properties, scattering angles, and cylinder radii, these field components can add constructively or destructively with the externally reflected field, producing variations in the scattered field intensity. However, one notes from Figures 4-36 and 4-37 (as well as Figures 4-21 and 4-22) that the scattering widths seem relatively smooth, except for a few widely spaced variations.

In most practical situations, however, the coating material has some loss. Thus as the frequency or cylinder size increases, all trapped rays are sufficiently attenuated so that they cannot completely circumnavigate the cylinder within the coating and emerge with any significant power. When $\text{Im}(k_0 a_0 \sqrt{\mu_1 \epsilon_1}) > 5$, this condition holds. In this case the scattered fields and scattering widths can be found in the same manner as for the homogeneous cylinder of Section 4.2.3.3, except that instead of using the reflection coefficients $R_\parallel(\gamma)$ and $R_\perp(\gamma)$ for a semi-infinite homogeneous medium, one uses the reflection coefficients $R_\parallel(\gamma)$ and $R_\perp(\gamma)$ for a layer of material of $\mu_1$, $\epsilon_1$, with thickness $a_0 - a_1$, and backed by a perfect conductor. These latter reflection coefficients can be obtained from Eq. (7.1-42) by taking the limit as $z_3 \to 0$.

## 4.3. CIRCULAR CYLINDERS OF FINITE LENGTH

For most of this section, unless otherwise noted, the scattering geometry and symbols as illustrated in Figure 4-39 will be used. The plane of incidence will be defined here as the plane with the incident wave propagation direction and the cylinder axis ($z$ axis here), and is shown in Figure 4-39 as the $x - z$ plane. The plane of scattering will be defined as the plane with the scattering propagation direction and the cylinder axis, and is shown in Figure 4-39 as rotated by azimuth angle $\phi'$ from the $x$ axis.

The incident and scattered $E$ field can always be resolved into components in and perpendicular to the incidence and scattering planes. The former situation ($E$ field in the plane) defines a $TM$ wave (or $\parallel$ polarization), while the latter defines a $TE$ wave (or $\perp$ polarization). Let the true incident and scattered $E$-field polarization directions of interest be rotated clockwise about the directions of propagation by angles $\gamma_i$ and $\gamma_s$ from the incidence

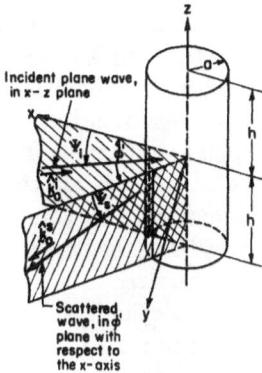

Fig. 4-39. Scattering geometry for a circular cylinder of finite length.

and scattering planes when looking in the directions of propagation. $\gamma_i = 0$ and $\gamma_s = 0$ then correspond to *TM* incident and scattered fields.

In general, a scattered field from a cylinder of finite length propagates in all directions, so that one can consider the scattered field at arbitrary angle $\Psi_s$. Such is not true for an infinitely long cylinder, as treated previously. In that case, a scattered field propagates only in the direction $\Psi_s = -\Psi_i$, where $\phi'$ is still arbitrary. Thus for cylinders whose length is very much greater than the radius, one should expect to find that the scattered field at and near $\Psi_s = -\Psi_i$ is much larger than the scattered field in other directions. This should help, in general, in identifying the scattering and aspect angles which will give the strongest return.

### 4.3.1. Cylinder Radius Small ($k_0a < 1$)

When the cylinder radius is small, it makes essentially no difference how the cylinder ends are terminated as far as the scattered fields are concerned. The cylinder ends may be flat, rounded, or pointed, and the results are the same.

#### 4.3.1.1. *Cylinder Length Small (Rayleigh or Low-Frequency Region)* ($h \ll \lambda$)

In this case, all of the cylinder dimensions are very small compared to wavelength. In this low-frequency region, the target features which affect the scattering properties are its volume and the frequency. The shape of the object, which plays the dominant role in the high-frequency region, has little effect upon the scattering characteristics. Hence, one can model the cylinder by a spheroid of equal volume in order to approximate the scattering cross section, since there is no exact solution presently available for the scattered field from a finite cylinder. Exact solutions are available for scattering from spheroids in the low-frequency limit (see Section 5.1). The error incurred in such an approximation for the scattering cross section should be less than 10%, which is negligible, in general, compared with radar cross-section measurement accuracies (10% corresponds to 0.4 db).

An exception to the target volume dependence of the scattered field occurs when the target surface is perfectly conducting, and when one of the dimensions of the object shrinks to zero (or becomes very small). For the cylinder, the case where $h \to 0$ corresponds to a disc, whereas the case where $a \to 0$ corresponds to a fine wire. In both situations, the volume approaches zero, but for certain incident polarization directions, the scattered field is not zero. This case in each limit will be treated separately.

Only cylinders with perfectly conducting surfaces and cylinders made of homogeneous materials will be treated here. This is due to lack of results for coated or otherwise inhomogeneous cylinders.

### 4.3.1.1.1. *Perfectly Conducting Cylinders*

4.3.1.1.1.1. *"Prolate Cylinder"* $(h > a)$. When the cylinder length is greater than its diameter, it can be modeled by a prolate spheroid with little error. Equations (5.1-56) to (5.1-59) give the low-frequency fields scattered from a prolate spheroid with semi-axes $a$ and $b$ $(a > b)$, where the $x$ axis is the axis of symmetry. The coordinate system there is convertible to that of Figure 4-39 by making the following substitutions in Eqs. (5.1-56) to (5.1-59):

$$\cos \theta \to 0$$
$$\sin \theta \to 1$$
$$\sin \phi \to -\cos \Psi_i \tag{4.3-1}$$
$$\cos \phi \to -\sin \Psi_i$$

and

$$\cos \theta' \to -\sin \phi' \cos \Psi_s$$
$$\sin \theta' \to \sqrt{1 - \sin^2 \phi' \cos^2 \Psi_s}$$
$$\cos \phi' \to \frac{-\sin \Psi_s}{\sqrt{1 - \sin^2 \phi' \cos^2 \Psi_s}} \tag{4.3-2}$$
$$\sin \phi' \to \frac{-\cos \phi' \cos \Psi_s}{\sqrt{1 - \sin^2 \phi' \cos^2 \Psi_s}}$$

In general, the scattered field from both the finite cylinder and the spheroid in the low frequency limit are proportional to their volumes. Only in the wire and disc limits $(a/h \to 0, \infty)$ and for perfectly-conducting material, does the scattered field no longer depend upon the volume. In the wire limit, the scattered field for a prolate spheroid is easily derived from Eqs. (5.1-56) to (5.1-59), (5.1-65), and (5.1-66). The scattered field for the cylinder in this limit has been derived by other methods.[21] Thus, using the constraint that the spheroid and cylinder volumes must be identical when the cylinder is as wide as it is long, and the constraint that the scattered fields for the two must also be identical in the wire limit, one can arrive at a set of dimensions for the equivalent spheroid in terms of the cylinder dimensions on both limits. Then an "educated guess" for the fit between these two limits is made (similar to that of Reference 22). Thus the dimensions for the equivalent prolate spheroid model for the finite cylinder in the low-frequency limit are

$$a_e = h\left[\left(\frac{3}{2}\right)^{1/3} + \left[1 - \left(\frac{3}{2}\right)^{1/3}\right]e^{-4a/h}\right] \tag{4.3-3}$$

$$b_e = a\left[\left(\frac{3}{2}\right)^{1/3} + \left[\frac{1}{2} - \left(\frac{3}{2}\right)^{1/3}\right]e^{-4a/h}\right] \tag{4.3-4}$$

where $a_e$ and $b_e$ are the semi-axes of the equivalent spheroid to be used in Eqs. (5.1-56) to (5.1-59) in place of $a$, $b$. It is believed that if the cylinder is modeled in this manner, an error of no more than 10% (0.4 db) will be incurred for the cross section.

In particular, the backscattering cross sections employing (5.1-56) to (5.1-59) ($\Psi_i = \Psi_s$) are written here for the $E$-field components in, and perpendicular to the plane of incidence:

$$\sigma^A_{\|,\|}(\Psi) = \frac{16\pi}{9}k_0^4\left[\frac{\sin^2\Psi}{I_{b_e}} + \frac{\cos^2\Psi}{I_{a_e}} + \frac{1}{\dfrac{2}{a_eb_e^2} - I_{b_e}}\right]^2 \tag{4.3-5}$$

$$\sigma^B_{\perp,\perp}(\Psi) = \frac{16\pi}{9}k_0^4\left[\frac{1}{I_{b_e}} + \frac{\sin^2\Psi}{\dfrac{2}{a_eb_e^2} - I_{b_e}} + \frac{\cos^2\Psi}{\dfrac{2}{a_eb_e^2} - I_{a_e}}\right]^2 \tag{4.3-6}$$

$$\sigma^B_{\|\perp}(\Psi) = \sigma^B_{\perp\|}(\Psi) = 0 \tag{4.3-7}$$

where $I_{a_e}$ and $I_{b_e}$ are given in Eqs. (5.1-65) and (5.1-66) and Figure 5-2, and 5-3, using $a_e$ and $b_e$, as given above. The parallel and perpendicular subscript designations refer to the $E$-field polarization direction, with respect to the plane of incidence ($x - z$ plane of Figure 4-39).

4.3.1.1.1.2. *Wire Limit of Cylinder* ($h \gg a$). When the radius of the cylinder is much less than the length but the length is still considerably smaller than the wavelength, the scattered field from the actual cylinder can be found by several approximations (i.e., variational procedures, antenna theory, etc.; see reference 21). The equivalent prolate spheroid results in this limit are also valid, and they are made to conform to the cylinder results in this limit by choosing the dimensions of the spheroid model properly, as shown in the last section. In this limit, the $I_{a_e}$ and $I_{b_e}$ for the spheroid become

$$I_{a_e} \approx \frac{2}{h^3}\left[\ln\frac{4h}{a} - 1\right], \qquad I_{b_e} \approx \frac{4}{ha^2} \xrightarrow[a\to 0]{} \infty \tag{4.3-8}$$

Then the scattered fields and scattering cross sections become [from Eqs. (5.1-56) to (5.1-59)]

$$E^s_{\|\|} \approx \frac{1}{3} k_0^2 \frac{e^{ik_0 r}}{r} E^i_{\|} \frac{h^3 \cos \Psi_i \cos \Psi_s}{\ln \dfrac{4h}{a} - 1} ; \qquad E^s_{\|\perp} = E_\perp{}^s = E^s_{\perp\perp} \to 0 \qquad (4.3\text{-}9)$$

and

$$\sigma_{\|\|}(\Psi_i, \Psi_s, \phi') = \frac{4\pi}{9} k_0^4 h^6 \frac{\cos^2 \Psi_i \cos^2 \Psi_s}{\left[\ln \dfrac{4h}{a} - 1\right]^2} ; \qquad \sigma_{\|\perp} = \sigma_{\perp\|} = \sigma_{\perp\perp} \to 0$$
$$(4.3\text{-}10)$$

These results show that the only significant scattered power from a short wire arises from an incident $E$-field component along the $z$ axis (in the plane of incidence), and this scattered field itself is polarized with its $E$ field entirely in the scattering plane. This is the expected behavior of a short dipole.

If the incident and scattered $E$ fields of interest do not lie entirely in the plane of incidence and scattering, but are rotated about their respective propagation directions by angles $\gamma_i$ and $\gamma_s$ from these planes, the bistatic cross section [Eq. (4.3-10)] will have $\cos^2 \gamma_i \cos^2 \gamma_s$ as a factor.

It often is desirable to know the average backscattering cross section from a group of randomly oriented short dipoles. The average backscattering cross section from a single dipole over all possible random orientations is obtained from Eq. (4.3-10) by setting $\Psi_i = \Psi_s$, introducing the factor $\cos^4 \gamma$ mentioned in the preceding paragraph to account for polarization effects, and then averaging over all $\Psi$ and $\gamma$. (Here it is assumed that the desired scattering and incidence polarization directions coincide as they normally will in monostatic radar applications. If they do not, or if an average bistatic cross section is needed, this can be obtained from Eq. (4.3-10) by multiplying by the factor $\cos \gamma_i{}^2 \cos \gamma_s{}^2$, and averaging over all dipole orientations (Reference 23 presents such results). The average dipole scattering cross section when the incidence and scattered polarization directions coincide is[21]

$$\bar{\sigma}^B = \frac{4\pi}{45} k_0^4 h^6 \frac{1}{[\ln(4h/a) - 1]^2} \qquad (4.3\text{-}11)$$

4.3.1.1.1.3. *"Oblate Cylinder"* $(h \leqslant a)$. When the cylinder length is shorter than its diameter, it can be modeled by an oblate spheroid with little error. The dimensions of the equivalent spheroid are chosen so that the volume of the spheroid in the limit $h = a$ is the same as that of the cylinder (here assumed to be flat on the ends). In the disc limit of the oblate spheroid, the scattered field is no longer proportional to volume since the volume vanishes. In this limit, the cylinder is assumed to scatter in the same manner as the spheroid having the same radius. Using these constraints, the semi-axes $a_e$ and $c_e$ $(a_e > c_e)$ of the equivalent spheroid in terms of the cylinder dimensions are

$$a_e = a\left[\left(\frac{3}{2}\right)^{1/3} + \left[1 - \left(\frac{3}{2}\right)^{1/3}\right] e^{-4h/a}\right] \qquad (4.3\text{-}12)$$

$$c_e = \left(\frac{3}{2}\right)^{1/3} h \qquad (4.3\text{-}13)$$

As in Section 4.3.1.1.1.1, a smooth transition of the dimensions between the two limits is postulated. The error involved in estimating the cross section in this manner is believed to be less than 10 % (0.4 db).

The dimensions of the equivalent spheroid, as given above, may now be used in Eqs. (5.1-83) to (5.1-86) for the scattered fields from an oblate spheroid in the low-frequency limit. In order to conform to the coordinates of Figure 4-39, one must let $\theta$, $\theta'$, $\phi$, and $\phi'$ in those equations be defined in terms of the angles of Figure 4-39 as follows: $\theta \to \Psi_i + \pi/2$, $\theta' \to \Psi_s + \pi/2$, $\phi \to 0$, $\phi' \to \phi'$. The quantities $I_a$ and $I_c$, found in those equations, are defined in Eqs. (5.1-91) and (5.1-92), and plotted in Figures 5-2 and 5-4.

In particular, the backscattering cross sections $(\Psi_i = \Psi_s)$ for the E-field components in and perpendicular to the plane of incidence are

$$\sigma^B_{\parallel\parallel}(\Psi) = \frac{16\pi}{9} k_0^4 \left[\frac{\sin^2\Psi}{I_{a_e}} + \frac{\cos^2\Psi}{I_{c_e}} + \frac{1}{2/a_e^2 c_e - I_{a_e}}\right] \qquad (4.3\text{-}14)$$

$$\sigma^B_{\perp\perp}(\Psi) = \frac{16\pi}{9} k_0^4 \left[\frac{1}{I_{a_e}} + \frac{\sin^2\Psi}{2/a_e^2 c_e - I_{a_e}} + \frac{\cos^2\Psi}{2/a_e^2 c_e - I_{c_e}}\right] \qquad (4.3\text{-}15)$$

$$\sigma^B_{\parallel\perp}(\Psi) = \sigma^B_{\perp\parallel}(\Psi) = 0 \qquad (4.3\text{-}16)$$

where $I_{a_e}$ and $I_{c_e}$ are given in terms of $a_e$ and $c_e$ by Eqs. (5.1-91) and (5.1-92).

4.3.1.1.1.4. *Disc Limit of Cylinder* $(h \ll a)$. When the length of the cylinder, $h$, approaches zero, it is assumed that the cylinder scatters in the same manner as a disc of the same radius. In turn, the scattering behavior of the disc in the low-frequency limit is the same as that of the oblate spheroid in the limit $c_e \to 0$.[22] The dimensions of the equivalent spheroid in Eqs. (4.3-12) and (4.3-13) have been chosen such that the behavior in this limit for the spheroid is correct. However, it is illuminating to give the results in the limit separately in terms of the actual cylinder dimensions. The quantities $I_{a_e}$ and $I_{c_e}$ of Eqs. (5.1-91) and (5.1-92) become

$$I_{a_e} \approx \frac{1}{a^3}\frac{\pi}{2}, \qquad I_{c_e} \approx \frac{2}{a^2 h} \xrightarrow[h\to 0]{} \infty \qquad (4.3\text{-}17)$$

Thus the scattered fields from the cylinder in the disc limit become

$$E^s_{\parallel\parallel} \approx \frac{4}{3\pi} k_0^2 \frac{e^{ik_0 r}}{r} E^i_{\parallel} a^3 \sin \Psi_i \sin \Psi_s \cos \phi' \qquad (4.3\text{-}18)$$

$$E^s_{\perp\parallel} \approx \frac{4}{3\pi} k_0^2 \frac{e^{ik_0 r}}{r} E^i_{\parallel} a^3 \sin \Psi_i \sin \phi' \qquad (4.3\text{-}19)$$

$$E^s_{\parallel\perp} \approx \frac{4}{3\pi} k_0^2 \frac{e^{ik_0 r}}{r} E^i_{\perp} a^3 \sin \Psi_s \sin \phi' \qquad (4.3\text{-}20)$$

$$E^s_{\perp\perp} \approx \frac{4}{3\pi} k_0^2 \frac{e^{ik_0 r}}{r} E^i_{\perp} a^3 \left(\cos \phi' + \frac{1}{2} \cos \Psi_i \cos \Psi_s\right) \qquad (4.3\text{-}21)$$

From these equations, the bistatic scattering cross sections in the disc limit are

$$\sigma_{\parallel\parallel}(\Psi_i, \Psi_s, \phi') = \frac{64}{9\pi} k_0^4 a^6 \sin^2 \Psi_i \sin^2 \Psi_s \cos^2 \phi' \qquad (4.3\text{-}22)$$

$$\sigma_{\perp\parallel}(\Psi_i, \Psi_s, \phi') = \frac{64}{9\pi} k_0^4 a^6 \sin^2 \Psi_i \sin^2 \phi' \qquad (4.3\text{-}23)$$

$$\sigma_{\parallel\perp}(\Psi_i, \Psi_s, \phi') = \frac{64}{9\pi} k_0^4 a^6 \sin^2 \Psi_s \sin^2 \phi' \qquad (4.3\text{-}24)$$

$$\sigma_{\perp\perp}(\Psi_i, \Psi_s, \phi') = \frac{64}{9\pi} k_0^4 a^6 \left[\cos \phi' + \frac{1}{2} \cos \Psi_i \cos \Psi_s\right]^2 \qquad (4.3\text{-}25)$$

The above equations simplify considerably for backscattering, where $\Psi_i = \Psi_s$ and $\phi' = 0$, or

$$\sigma^B_{\parallel\parallel}(\Psi) = \frac{64}{9\pi} k_0^4 a^6 \sin^4 \Psi$$

$$\sigma^B_{\parallel\perp}(\Psi) = \sigma^B_{\perp\parallel}(\Psi) = 0 \qquad (4.3\text{-}26)$$

$$\sigma^B_{\perp\perp}(\Psi) = \frac{64}{9\pi} k_0^4 a^6 \left[1 + \frac{1}{2} \cos^2 \Psi\right]^2$$

4.3.1.1.2. *Homogeneous Cylinders.* When the cylinder material is not perfectly conducting but is homogeneous with finite relative permittivity, $\epsilon_r$, and permeability, $\mu_r$ (which may be complex if the material is lossy), then the scattered field from both the finite cylinder and the spheroid in the low-frequency limit is everywhere proportional to the volume. This is true even in the wire limit ($h \gg a$) and the disc limit ($h \ll a$), in contrast to the perfectly conducting case. Thus one can expect, for example, that the field scattered from a very thin dielectric cylinder will approach zero as $a^2$ in the limit $a \to 0$, since the cylinder volume depends upon $a^2$ when length is constant. One can approximate the field scattered from a finite homogeneous cylinder in the Rayleigh limit by the field scattered from an equivalent spheroid with the same permeability, permittivity, and volume. The length to width ratio of the spheroid is kept the same as that for the cylinder.

When $h > a$ (i.e., "prolate" cylinder), the cylinder will be modeled by a prolate spheroid of semi-axes $a_e$ and $b_e$ ($a_e > b_e$). Using the constraints mentioned in the previous paragraph, these quantities are given in terms of the cylinder dimensions as

$$a_e = \left(\frac{3}{2}\right)^{1/3} h, \qquad b_e = \left(\frac{3}{2}\right)^{1/3} a \qquad (4.3\text{-}27)$$

Then the scattered fields are given by Eqs. (5.1-56) to (5.1-59), and $I_a$ and $I_b$ in these equations are defined in terms of $a_e$ and $b_e$ by Eqs. (5.1-65) and (5.1-66). The substitutions given by Eqs. (4.3-1) and (4.3-2) should be made in Eqs. (5.1-56) to (5.1-59), in order to conform to the coordinates of Figure 4-39. Bistatic and backscattering cross sections follow directly from these equations.

When $h < a$ (i.e., "oblate" cylinder), the cylinder will be modeled by an oblate spheroid of semi-axes $a_e$ and $c_e$ ($a_e > c_e$). Using the constraints mentioned in a previous paragraph, these quantities are given in terms of the cylinder dimensions as

$$a_e = \left(\frac{3}{2}\right)^{1/3} a, \qquad c_e = \left(\frac{3}{2}\right)^{1/3} h \qquad (4.3\text{-}28)$$

The scattered fields are given by Eqs. (5.1-83) to (5.1-86); $I_a$ and $I_c$ in these equations are defined in terms of $a_e$ and $c_e$ by Eqs. (5.1-91) and (5.1-92). The angles $\theta$, $\theta'$, $\phi$, and $\phi'$ there are related to $\Psi_i$, $\Psi_s$, and $\phi'$ of Figure 4-39 by $\theta \rightarrow \Psi_i + \pi/2$, $\theta' \rightarrow \Psi_s + \pi/2$, $\phi' \rightarrow \phi'$, $\phi = 0$. Bistatic and back-scattering cross sections follow directly from these equations.

It is believed that the error involved in these approximations is less than 10%.

### 4.3.1.2. Cylinder Length of Same Order as Wavelength (Resonance Region)

When the cylinder length is of the same order as wavelength but its diameter is very small, one has the situation familiar to antenna specialists, the dipole. The cylinder or "dipole" to be considered here is perfectly conducting, but the effects of finite wire conductivity are also discussed. From antenna theory, one recognizes that at certain lengths, which are nearly multiples of a half-wavelength, the dipole "is resonant." When this happens the radiated field in given directions becomes much stronger because of resonant currents. One would then be led to expect a similar increase in the scattered field at certain "dipole" lengths, such as $\lambda_0/2$. Such is indeed the case.

The thinness of the dipole permits one to assume that the total current along its surface is essentially concentrated along a line. Hence, the azimuthal pattern (variation of the scattered field with azimuth angle $\phi'$ for given $\Psi_i$ and $\Psi_s$) is constant for all $\phi'$. Also, only the component of the incident $E$ field along the cylinder axis can excite currents in the wire, and in turn, this current radiates a scattered $E$ field only in the scattering plane (i.e., the plane containing the cylinder axis and the scattered wave propagation direction). Thus, the polarization properties of the fields are easily accounted for. As might be expected, the currents induced in the wire and the resulting scattered fields are relatively insensitive to the wire radius for the perfectly-conducting wire so long as the wire is thin. Nevertheless, there must be some dependence upon wire radius because as the wire radius approaches zero in the limit, no current can be induced in the wire since there is no surface or volume over which the current can flow.

Many theories have been constructed to account for all these factors. In recent years several have been quite successful in exhibiting the proper behavior. The first thorough study devoted to this problem was probably that of Reference 21 which has been often cited by other workers. In this work, Van Vleck, Block, and Hamermesh employ two methods of analyzing the problem. The first has been referred to as the EMF method because it uses the principle of the conservation of energy as a constraint to equate the power due to the induced EMF in the dipole with the reradiated power due to the induced current. The authors of Reference 21 call it Method A. The second method, called Method B, uses an approximate or first-order solution to an integral equation. There is very little difference in the results from these two methods, but Method B seems to offer solutions which are somewhat more tractable and general. The authors of Reference 21 seem to favor this second method. Nonetheless, Reference 24 uses Method A to generate curves for the scattering cross sections.

In reference 25, Tai has set forth a third approach which employs variational methods. Later works based upon Tai's approach have considered wires of finite conductivity, instead of perfectly conducting wires,[26] and bistatic as well as backscattering.[27] Tai's approach seems to offer more variety in the choice of the form of the current and somewhat more generality. However, for the same degree of complexity, the solution does not appear to differ radically from other results; in some regions it is an improvement, while in others it is not as good.

Other approaches to the problem are based upon the assumption of an ideal sinusoidal current variation for $2h = n\lambda_0/2$,[23] or of currents predicted by transmission line theory.[28] Still another approach relates the bistatic scattered field to waves diffracted between the ends of the wire and predicted by geometrical diffraction theory.[29] All of these approaches have interesting points, and offer advantages for particular situations but disadvantages in others. If a thorough understanding of this problem and the available solutions is desired, all of these references should be consulted. This list of references is by no means exhaustive, but is representative of the approaches taken.

It should be emphasized that all of the known approaches are approximations. The exact solution to a seemingly simple problem such as this is unattainable at present. Most of these approaches become more accurate in the limits of $h/a \to \infty$ and $h/\lambda_0 > \frac{1}{4}$.

Merely for convenience, the results of Method B of Reference 21 will be given here. This method obtains the backscattering cross section of perfectly conducting wires at arbitrary aspect angles and polarizations. Also obtained is the backscatter cross section averaged over all possible dipole orientations.[†]

The expressions are accurate to within about 15% (0.6 db) at the resonance peaks, and are valid for broadside backscatter ($\Psi_i = \Psi_s = 0$). Results given here are limited to backscatter because of its relevance in most radar situations, and due to the complexity of bistatic solutions. Bistatic cross sections are presented in References 27 and 29. Bistatic cross sections, averaged over all possible orientations for half- and fullwave dipoles for linear and circular polarization states, are given in Reference 23. Solutions valid for wire material of finite conductivity are given in Reference 26.

*4.3.1.2.1. Backscattering Cross Section for Perfectly Conducting Wire* ($2h > \lambda/4$). Let the cylindrical wire be oriented along the $z$ axis, as shown in Figure 4-39, with its ends at $z = \pm h$. Since only backscattering is considered, $\Psi_i = \Psi_s = \Psi$. Assume for generality that the polarization direction of the incident $E$-field is rotated by an angle $\gamma_i$ about its direction of propagation from the plane of incidence, and that the desired scattered field polarization is rotated by angle $\gamma_s$ about its propagation direction from this same plane. (If the same antenna is used for transmitting and receiving, $\gamma_s = \gamma_i$.) The backscattering cross section for these two polarization states is[21]

$$\sigma^B_{\gamma_i \gamma_s}(\Psi) = \frac{16\pi}{k_0^2} \cos^2 \gamma_i \cos^2 \gamma_s [a_1^2(F'^2 + F''^2)$$

$$+ (a_2 + a_3)^2 (G'^2 + G''^2) \cos(k_0 h \sin \Psi)$$

$$+ (a_2 - a_3)^2 (H'^2 + H''^2) \sin^2(k_0 h \sin \Psi)$$

$$- 2(a_2^2 - a_3^2)(G'H' + G''H'') \sin(k_0 h \sin \Psi) \cos(k_0 h \sin \Psi)$$

$$+ 2a_1(a_2 + a_3)(F'G' + F''G'') \cos(k_0 h \sin \Psi)$$

$$- 2a_1(a_2 - a_3)(F'H' + F''H'') \sin(k_0 h \sin \Psi)] \qquad (4.3\text{-}29)$$

---

[†] This situation would be encountered with randomly oriented dipole chaff particles. The West–Ford orbiting dipole belt is another example of randomly oriented dipoles.

The parameters in Eq. (4.3-29) are defined as

$$a_1 = \frac{\sin(2k_0h \sin \Psi)}{2 \sin \Psi}, \quad a_2 = \frac{\sin[k_0h(1 - \sin \Psi)]}{(1 - \sin \Psi)}, \quad a_3 = \frac{\sin[k_0h(1 + \sin \Psi)]}{(1 + \sin \Psi)}$$

$$(4.3\text{-}30)$$

$$F' = \frac{\Omega}{\Omega^2 + \pi^2}, \quad F'' = \frac{\pi}{\Omega^2 + \pi^2}, \quad \text{with} \quad \Omega = -2 \ln(0.8905 k_0 a)$$

$$(4.3\text{-}31)$$

$$G' = \frac{1/2\Lambda(k_0h)}{\Lambda^2(k_0h) + \Xi^2(k_0h)} - \frac{\pi G''}{2\Omega}, \quad G'' = \frac{1/2\Xi(k_0h)}{\Lambda^2(k_0h) + \Xi^2(k_0h)} \qquad (4.3\text{-}32)$$

$$H' = \frac{1/2\Lambda(k_0h - \pi/2)}{\Lambda^2(k_0h - \pi/2) + \Xi^2(k_0h - \pi/2)} - \frac{\pi H''}{2\Omega},$$

$$(4.3\text{-}33)$$

$$H'' = \frac{1/2\Xi(k_0h - \pi/2)}{\Lambda^2(k_0h - \pi/2) + \Xi^2(k_0h - \pi/2)}$$

where

$$\Lambda(k_0h) = -(\Omega - \Delta) \cos k_0h + (\pi/4) \sin k_0h \qquad (4.3\text{-}34)$$

$$\Xi(k_0h) = 1/2(\ln 7.12 k_0 a) \sin k_0h - (\pi/4) \cos k_0h \qquad (4.3\text{-}35)$$

and

$$\Delta = -1/2 \ln \left( \frac{k_0h}{4.11} \right) \qquad (4.3\text{-}36)$$

The above expressions are valid for $2h > \lambda_0/4$; at the first resonance ($2h$ near $\lambda_0/2$), the backscattering cross section is in error by only 6% for a wire with $2h/a = 900$ as compared with a more exact expression. At larger values of $h$, the error is still smaller.

Although the above equations are complicated, they exhibit the expected behavior for the backscattering cross section. They also show that the maximum cross sections do not occur at exactly the odd resonance lengths, $2h = n\lambda_0/2$, but at a slightly smaller length than these values, i.e., at $2h = 0.95 \lambda_0/2$, $1.94 \lambda_0/2$, $2.935 \lambda_0/2$, $3.93 \lambda_0/2$, $4.925 \lambda_0/2$, etc., for $2h/a = 900$. The values observed experimentally are $2h = 0.96 \lambda_0/2$, $1.88 \lambda_0/2$, $2.94 \lambda_0/2$, $3.90 \lambda_0/2$, $4.92 \lambda_0/2$, etc. This result for the backscattering cross section checks quite well with experimental observations at the resonance points.

A curve, generated from this cross section at normal incidence and optimum polarization ($\Psi = 0$, $\gamma_i = \gamma_s = 0$) for varying frequency ($2h/\lambda_0$ varies), is shown in Figure 4-40 (from Reference 21). From this figure, the agreement between theory and experiment appears quite good (never more than 9% error). It should be noted that the first resonance peak is considerably higher than the successive valley, but the difference between succeeding peaks and valleys becomes less. Thus long wires do not exhibit

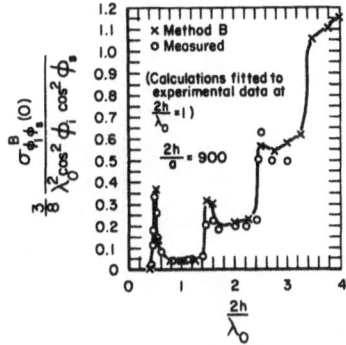

Fig. 4-40. Backscattering cross section of a thin wire oriented broadside to the incident wave as a function of wire-length to wavelength ratio [after Van Vleck et al.[(21)]].

a large peak–valley resonance effect in the backscattered power at broadside as the length to wavelength ratio varies.

Several curves of backscattering cross section at various wire length to wavelength ratios are plotted *vs.* aspect angle, $\Psi$, in Figure 4-41. These have been made using Eq. 4.3-29. One observes the appearance of maxima in the scattered power, at aspect angles other than broadside, as the length of the wire increases with respect to the wavelength.

*4.3.1.2.2. Backscattering Cross Section for Randomly-Oriented Perfectly Conducting Wire* $(2h > \lambda/4)$. If the wire can assume any orientation with respect to the transmitting and receiving antennas, then a more meaningful measure of the scattering properties is the backscattering cross section averaged over all possible, equally likely, orientations of the wire, $\bar{\sigma}^B_{\gamma_i\gamma_s}(\Psi)$. Formally, this average is defined as

$$\bar{\sigma}^B = \frac{1}{4\pi} \int_0^{2\pi} d\gamma \int_{-\pi/2}^{\pi/2} d\Psi \sigma^B_{\gamma_i\gamma_s}(\Psi)$$

where $\gamma_i = \gamma$ and $\gamma_s$ is related to $\gamma_i$ in some known manner.

Figure 4-42 shows the variation in the average cross section when $\gamma_i = \gamma_s$ versus $2h/\lambda_0$ for various length-to-radius ratios. As can be seen, a change in the radius has little effect upon the scattered power. Figure 4-43 shows this variation for $2h/a = 900$ over a much greater length to wavelength ratio. The percentage difference between resonance and antiresonance diminishes as the length of the wire (compared to wavelength) increases; for instance, the first peak is ten times as high as the first valley, but the ninth peak is only three times as high as the ninth valley. For infinitely long wires, this difference between peaks and valleys will disappear entirely. The behavior of the average backscattering cross section, as demonstrated by this curve, conforms with experimental observation quite well.

*4.3.1.2.3. Simplified Back and Bistatic Scattering Cross Sections for Half- and Full-Wave Dipole* $(2h = \lambda_0/2 \text{ and } \lambda_0)$. In the case of half- and

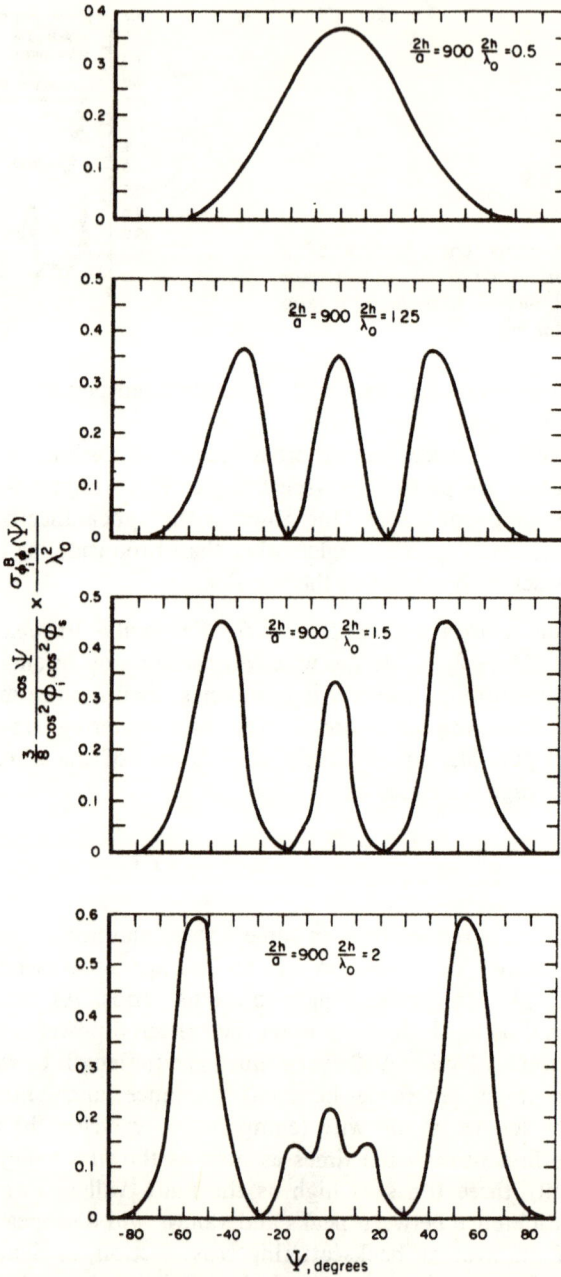

Fig. 4-41.   Backscattering cross section of a thin wire versus incidence angle for different wire lengths [after Van Vleck et al.[21]].

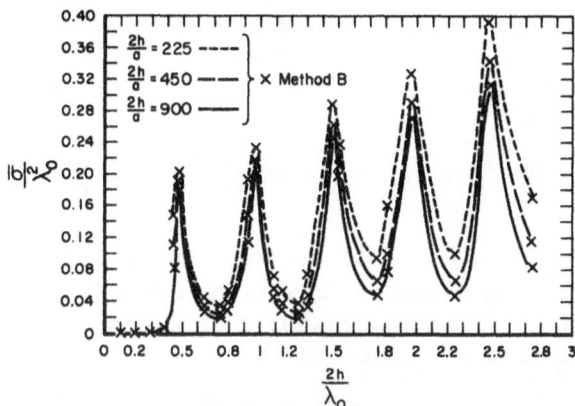

Fig. 4-42. Average backscattering cross section of a randomly-oriented thin wire as a function of wire-length to wavelength ratio for various length-to-radius values [after Van Vleck
et al.[(21)]].

full-wave dipoles, it is often desirable to have a simple expression for the
scattering cross section. The above equations are applicable in this case,
but they are so complicated that it is difficult to ascertain the cross section
easily. For these special cases, it is convenient to assume a sinusoidal current
distribution along the dipole with zero current at the ends.[(23)†] Antenna

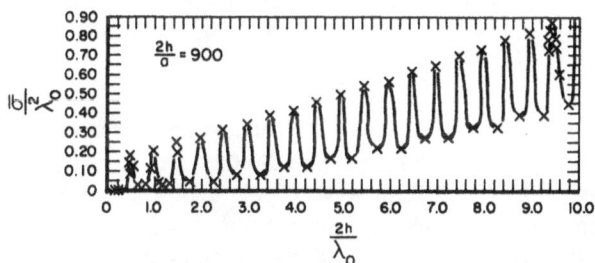

Fig. 4-43. Average backscattering cross section of a randomly-
oriented thin wire as a function of wire length-to-wavelength
ratio [after Van Vleck et al.[(21)]].

† One cannot extend this "assumed" sinusoidal current distribution to resonant dipole
lengths much beyond the full-wave resonance length. This assumption would indicate
that scattered power for a given incidence angle, $\Psi_i$, is symmetric about broadside, i.e.,
is the same in the specular and back directions ($\Psi_s = \pm \Psi_i$). Such is clearly not the
case; as the wire length becomes longer, more power is scattered into the specular direction than into the back direction. This means that the current distribution on
the wire, induced by the incident wave, is not sinusoidal and symmetrical about
the midpoint as assumed here. Only for half and full resonance lengths, and angles
$\Psi_i$ and $\Psi_s$ not too far from broadside, is this assumption valid. For plots of current
vs. incidence angle on a halfwave dipole, see reference 27.

theory is then used to predict the scattering cross sections. One assumes that the resonant lengths are exact multiples of a half wavelength. A simple expression for the bistatic and backscattering cross sections for the half-wave dipole is

$$\sigma_{\gamma_i\gamma_s}(\varPsi_i, \varPsi_s) = 0.86\lambda_0^2 \cos^2 \gamma_i \cos^2 \gamma_s \left[\frac{\cos(\pi/2 \sin \varPsi_i)}{\cos \varPsi_i}\right]^2 \left[\frac{\cos(\pi/2 \sin \varPsi_s)}{\cos \varPsi_s}\right]^2$$

(4.3-37)

According to this expression, the maximum cross section occurs for backscattering at broadside ($\varPsi_s = \varPsi_i = 0$) when the polarization directions of the transmitter and receiver lie in the plane of incidence ($\gamma_i = \gamma_s = 0$). This maximum backscatter cross section becomes

$$\sigma_{00}(0, 0) = 0.86\lambda_0^2$$

(4.3-38)

Comparing this with the more exact theory and curves of Section 4.3.1.2.1, at the first resonance the backscattering cross section has a value $\approx 0.99\lambda_0^2$. Thus, the above approximate expression is not greatly in error. One should notice that the approximate expressions of this section do not include any dependence upon wire radius; this is because it was assumed that the current on the perfectly conducting dipole when acting as an antenna is constant, independent of the radius. Such is obviously not true in practice, but is not greatly in error because the function $1/\ln k_0 a$ is extremely slowly varying. Only when $k_0 a$ is very small does this factor significantly affect the current.

The average backscattering cross section over all orientations when the antennas are polarized in the same direction is (from reference 23)

$$\bar{\sigma} \approx 0.155\lambda_0^2$$

(4.3-39)

When the polarization directions of receiver and transmitter are crossed, this average drops to $\frac{1}{3}$ of that in Eq. (4.3-39).

The gain of this ideal half-wave dipole acting as an antenna is 1.64, and maximum radiation intensity occurs at broadside.

For a full-wave dipole, a simplified expression is

$$\sigma_{\gamma_i\gamma_s}(\varPsi_i, \varPsi_s) = 0.93 \lambda_0^2 \cos_0^2 \gamma_i \cos^2 \gamma_s \left[\frac{\sin(\pi \sin \varPsi_i)}{\cos^2 \varPsi_i}\right]^2 \left[\frac{\sin(\pi \sin \varPsi_s)}{\cos^2 \varPsi_s}\right]^2$$

(4.3-40)

The maximum cross sections for this ideal full-wave dipole occur when $\varPsi_i = \pm\varPsi_s = \pm55°$, whereas zero backscattering cross section occurs at broadside, or $\varPsi_i = \varPsi_s = 0$. At the maxima, the cross section is

$$\sigma_{00}(\pm55°, \pm55°) = 0.93 \lambda_0^2$$

(4.3-41)

The average cross section over all possible dipole orientations, in this case, is[23]

$$\bar{\sigma} \approx 0.167 \lambda_0^2$$

(4.3-42)

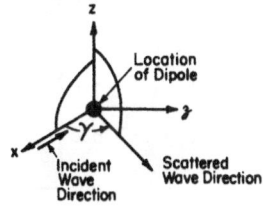

Fig. 4-44. Scattering geometry for bistatic scattering from a randomly oriented half- or full-wave dipole.

The gain of this ideal full-wave dipole acting as an antenna is 1.71, and the maximum radiation intensity occurs at $\pm 55°$ from broadside.

Average bistatic cross sections for half- and full-wave dipoles versus angle $\beta$, between incident and scattering direction, have been obtained in Reference 23. The coordinate system of Figure 4-44 is used, and the dipole at the origin may be oriented in any equally likely position. The scattering direction lies in the $x - y$ plane, and the scattered return, in this case, depends upon the polarization direction. Vertical polarization is taken here to mean that the $E$ field lies in the plane formed by the direction of propagation (incident or scattered field) and the $z$ axis. Then the average bistatic cross sections for the ideal half- and full-wave dipoles are shown in Figure 4-45 for the linear polarization states. The average bistatic cross sections for circular polarization states are also shown in Figure 4-45. For this latter case, it makes no difference whether the incident and/or receiving antennas are right or left circularly polarized. For such a random cloud of dipoles, the depolarized and polarized return using circular polarization is the same, and is another distinctive property of dipole scatterers.

4.3.1.2.4. *Effect of Finite Conductivity on Scattering Properties.* Up to now, only thin dipoles which were perfectly conducting have been considered. Most metals available for dipoles are less than perfectly conducting at certain frequencies. Consequently, the resistance per unit length along the dipole may not be negligible compared with the radiation impedance. This becomes more serious for very thin wires, since the resistance per unit length increases with decreasing wire cross section.

A rigorous analysis of the scattered field from wires with finite conductivity can be found in Reference 26. A less rigorous but more straightforward discussion is found in Reference 23. Rather than reproduce those results, it will simply be noted that for most typical dipole materials (copper, aluminum, iron, etc.), the assumption of perfect conductivity introduces no serious errors. This is well illustrated by the curves of Reference 26, where finite conductivity is considered.

Curves of the frequency (versus wire diameter), below which finite conductivity will affect the scattering properties of wires, are shown in Figure 4-46 for various wire materials (from Reference 24). So long as the

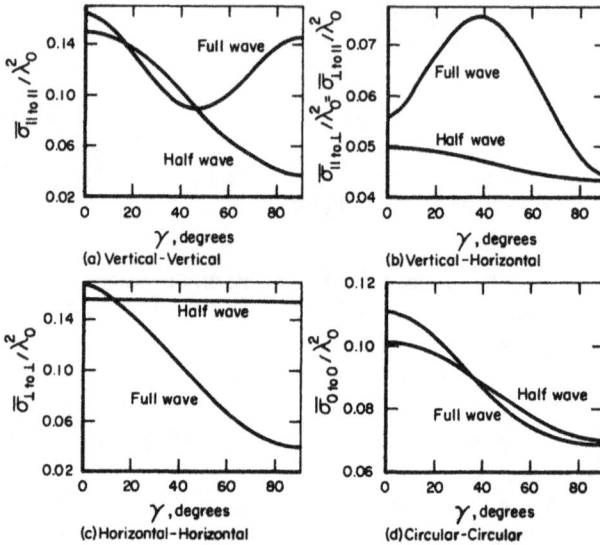

Fig. 4-45. Average bistatic scattering cross section for a randomly oriented half- or full-wave dipole as a function of bistatic angle [after Mack and Reiffen[23]].

frequency is above this value, the assumption of perfect conductivity is valid. Below it, the scattered field begins to drop in magnitude, and the methods of Reference 23 or 26 must be used to take this into account.

### 4.3.1.3. Cylinder Length Much Greater Than Wavelength ($h \gg \lambda$)

When the wire length increases beyond several wavelengths, the "resonance" effect of the scattered field at lengths which are multiples of a half wavelength disappears. As the wire length increases, the scattered field becomes more concentrated in the specular direction ($\Psi_s = -\Psi_i$ in Figure 4-39). In this limit, the results derived previously for the infinitely long cylinder can be used to find the scattering cross section from a long finite wire.

The method used here is that suggested by Van de Hulst.[30] The steps and approximations involved are outlined below.

(a) Very near the cylinder radially, the fields scattered by an incident plane wave are assumed to be the same as those scattered by an infinitely long cylinder of the same radius.

(b) There is assumed to be a zone where the scattered field is diverging as $1/\sqrt{r}$ (typical of two-dimensional scattering). This zone lies between the radial bounds $a^2/\lambda_0 \ll r \ll h^2/\lambda_0$. Then the approximation is made that the scattered field in this region is: (i) zero for $|z| > h$, and (ii) equal to that from an infinitely long cylinder for $|z| < h$.

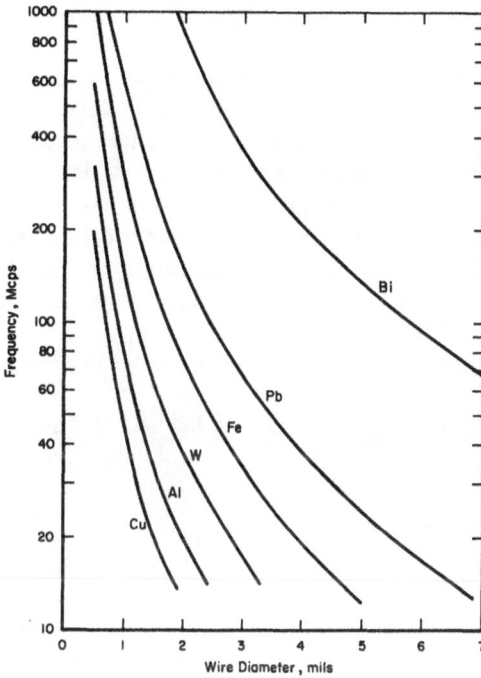

Fig. 4-46. Frequency versus wire diameter below which finite conductivity affects scattering properties of a wire for different wire materials [after Albini[24]].

(c) This field is then integrated over a plane $(x^1 - z^1)$ normal to the desired scattering direction, $\phi'$ (Figure 4-47). Huygen's principle is then employed to give the far-zone scattered field $(h^2/\lambda_0 > r)$.

(d) The integration in the $z^1$ direction is straightforward, since the fields varies only as $\exp[ik_0z^1 \sin \Psi_i]$ in that direction and is zero for $|z^1| > h$. The integration in the $x^1$ direction is performed using the stationary phase

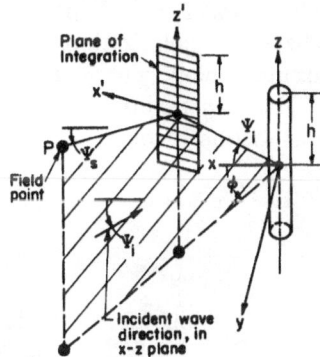

Fig. 4-47. Scattering geometry and plane of stationary-phase integration for scattering from a long cylinder.

method; the effective field over the plane of integration in the $x^1$ direction is thus assumed to be constant at its value at $x^1 = 0$. This is the field obtained from Section 4.2 at $\Psi_i = \Psi$ and polar angle $\phi'$. In this manner, the field at the far-zone point $P$ (lying in the $\phi'$-plane at angle $\Psi_s$ from the $x - z$ plane, positive in the same direction as $\Psi_i$) is determined.

The results of this method are valid for bistatic scattering in any direction. The approximations involved are similar to those of physical optics. They are germane in this case to the cylindrical nature of the scattered fields near the wire. Physical optics, itself, cannot be used here because the cylinder radius is too small compared to wavelength for the tangent-plane approximation to be valid.

The bistatic scattering cross section derived in this manner is

$$\sigma(\Psi_i, \Psi_s, \phi') = \frac{4k_0 h^2}{\pi} \cos^2 \Psi_s \sigma^c(\Psi_i, \phi') \left[ \frac{\sin[k_0(\sin \Psi_i + \sin \Psi_s) h]}{k_0(\sin \Psi_i + \sin \Psi_s) h} \right]^2 \tag{4.3-43}$$

where $\sigma^c(\Psi_i, \phi')$ is the scattering width for an infinitely long cylinder at oblique incidence, as given in Section 4.2. It may be for either $TM$ or $TE$ incident polarizations (indicated by superscripts in that section), and for the polarized or depolarized scattered polarization directions (indicated by subscripts $p$ and $d$ in that section). Equation (4.3-43) is valid for perfectly conducting, homogeneous, or coated cylinders, as well as for both $TM$ and $TE$ polarizations including any depolarized scattering component. Reasonably good results can be obtained with Eq. (4.3-43), even for cylinders as short as $2h \approx \lambda_0$.

One application of Eq. (4.3-43) of particular interest is a straight, long, thin, perfectly conducting wire of length $2h$ and diameter $2a$ ($a \ll \lambda$, $h > \lambda$). In this case, $\sigma^c(\Psi_i, \varphi')$ in Eq. (4.3-43) is determined by applying the rules of Section 4.2.2 to Eq. (4.1-21) for a thin, infinitely long, perfectly conducting cylinder at normal incidence. As shown there, only the $TM$ incident field contributes significantly to the scattering. In addition, the scattered field polarization is for all practical purposes $TM$, i.e., it lies in the $\varphi'$ plane. (The other polarizations contribute to the scattering widths relative to this main component with order $(k_0 a)^4$.)

For general bistatic scattering from such a long, thin wire, the scattering cross section becomes

$$\sigma(\Psi_i, \Psi_s, \varphi')$$

$$= 4\pi h^2 \frac{\cos^2 \Psi_s}{\cos^2 \Psi_i} \cos^2 \gamma_i \cos^2 \gamma_s \left[ \frac{1}{\ln^2(0.8905 k_0 a \cos \Psi_i) + \pi^2/4} \right]$$

$$\times \left[ \frac{\sin[k_0(\sin \Psi_i + \sin \Psi_s) h]}{k_0(\sin \Psi_i + \sin \Psi_s) h} \right]^2 \tag{4.3-44}$$

This expression is independent of $\varphi'$, as expected. The angles $\gamma_i$, $\gamma_s$ define the directions of the desired incident and scattered polarization states with respect to the $TM$ planes. The scattered field is strongest in the specular cone, where $\Psi_s = -\Psi_i$. It possesses a lobe-like structure away from this direction, as given by the last factor in square brackets.

For backscatter, $\Psi_i = \Psi_s = \Psi$, and the above equation reduces to

$$\sigma(\Psi) = \frac{2\pi h^2 \cos^2 \gamma_i \cos^2 \gamma_s}{[\ln^2(0.8905 k_0 a \cos \Psi) + \pi^2/4]} \left[ \frac{\sin[2k_0 h \sin \Psi]}{2k_0 h \sin \Psi} \right]^2 \quad (4.3\text{-}45)$$

### 4.3.2. Cylinder Radius of Same Order as Wavelength ($0.5 < k_0 a < 10$)

#### 4.3.2.1. *Length Considerably Less Than Radius* ($2h \ll a$)

When the cylinder length is considerably less than the radius, it is also small with respect to wavelength. The cylinder (with flat ends) resembles a disc. The radius of the cylinder is large enough that its end faces would exhibit resonance region scattering properties. However, the length is small enough so that the $TM$ "broadside" return ($\Psi_i = \Psi_s = 0$ in Figure 4-39) would be very small. The end faces are close enough together that there is negligible interference effect from waves multiply-diffracted by the end edges and interacting with each other. Consequently, the scattering properties of such a cylinder would be quite similar to those of a disc of the same radius. This applies to either perfectly conducting or homogeneous cylinder materials. The reader is referred to the results of Section 7.5 for the circular disc.

#### 4.3.2.2. *Length of Same Order as Radius* ($h \approx a$)

When both cylinder length and diameter are of the same order as wavelength, there appears to be no valid approximation for computing the scattered field. The body is too big in terms of wavelength to use the low-frequency or Rayleigh region results, and too small to use high-frequency approximations, such as geometrical or physical optics. The only resort for valid results seems to be measured data. Only for backscattering from a perfectly conducting cylinder does there appear to be any extensive measurements.

For the dimension region considered here, Carswell[31] has experimentally determined the backscattering cross section versus aspect angle from a finite cylinder. The actual dimensions of the cylinder used by Carswell have been normalized here to the wavelength so that the curves will be valid at any frequency or size, so long as the dimensions ($h$ and $a$) are both of the same order as wavelength. Figure 4-48 shows the $TE$ measured backscattering cross section, normalized to $\pi a^2$, for cylinders of various lengths but with the same radius ($k_0 a = 0.72$); it shows the cylinder as it

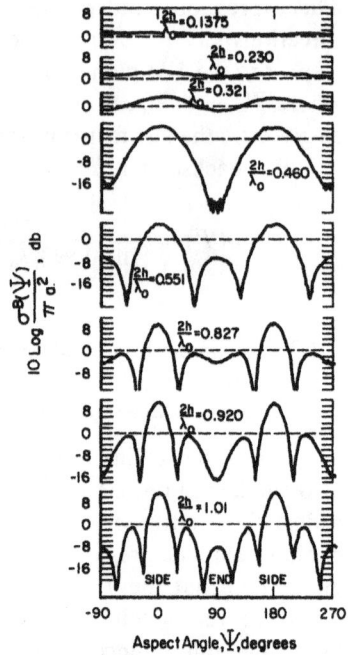

Fig. 4-48. Measured backscattering cross section for a perfectly conducting circular cylinder with flat ends as a function of aspect angle for *TE* polarization (*H* field parallel to cylinder axis) [after Carswell[31]].

is rotated in aspect angle from end-on through broadside and back to end-on again. (*TE* means the electric field polarization is perpendicular to the plane formed by the incident propagation direction and the cylinder axis.) Carswell states that the results for the other polarization, *TM*, are qualitatively similar and, consequently, does not show them.

One can see that at the shortest length, the backscattering cross section is independent of the aspect angle. This is to be expected, because all dimensions are such that the cylinder is in the Rayleigh region ($2a/\lambda_0 = 0.230$, $2h/\lambda_0 = 0.1375$). It will be recalled that in the Rayleigh region or at the low-frequency limit, the scattered field is proportional to volume and does not depend upon the shape or orientation of the target, so long as none of the length-width-height ratios are exceptionally large or small.

Figure 4-49 shows measured normalized backscattering cross sections for end-on and broadside aspects for two different diameters as a function of cylinder length. Both polarization modes are shown. The end-on cross section is not different for the two polarizations, but at broadside, there is considerable difference, as would be expected.

### 4.3.2.3. *Length Much Greater Than Diameter* ($h \gg a$)

In this situation, the length (compared to wavelength) is sufficiently long for a high-frequency approximation. However, the diameter is still

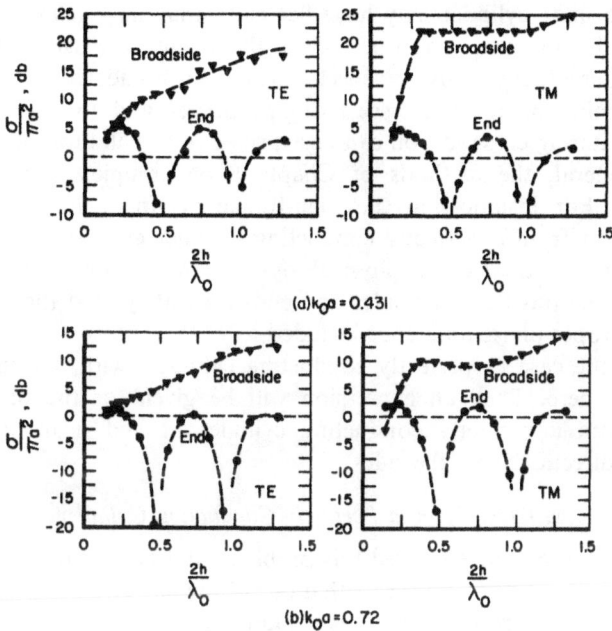

Fig. 4-49. Measured backscattering cross section for broadside and end-on aspects of a perfectly conducting circular cylinder with flat ends as a function of cylinder length [after Carswell[31]].

too small to apply any of the conventional high-frequency approximations, such as physical optics. The approximation discussed in Section 4.3.1.3 is particularly suited to this situation, since only the length need be long compared to wavelength in order to use that method. The value of $\sigma^c(\Psi_i, \phi')$ to be used in Eq. (4.3-43), however, must be obtained from the exact solution of Section 4.2.

The only region of error is at and near the end-on aspects; the approximation of Section 4.3.1.3 predicts zero cross section end-on, but in reality the finite end area, $\pi a^2$, is indeed visible and does scatter significantly when the cylinder radius is in the resonance region.

### 4.3.3. Cylinder Radius and Length Much Larger Than Wavelength $(k_0 a > 10, 2h > \lambda_0)$

Only the case where length is much larger than wavelength will be considered here. The other case, where length is less than wavelength, is close enough to the disc in its scattering behavior that it can be treated satisfactorily by the methods of Section 7.5.

When all dimensions of a cylinder are much larger than wavelength, the scattered field depends upon all of the features of the cylinder. For instance, a finite cylinder may have flat ends, may have hemispherical end caps, may be hollow,..., etc. For each of these cases, the backscattered field will be different, especially at aspects away from broadside. Thus, the type of end termination affects the scattering properties, and no single expression for the scattering cross section can be expected to be valid for all these cases.

In general, the methods of Chapter 8 on complex bodies must be employed. For instance, in the high-frequency limit, the backscattering cross section from a perfectly conducting cylinder capped with end hemispheres can be treated by physical optics. One computes the scattered field contributions from the sides and ends separately, and then adds them with the proper phase difference included.

Only the case of perfectly conducting cylinders with flat ends will be considered here. Then an expression will be given for the bistatic cross section from a perfectly conducting cylinder at and near the specular reflection direction from the sides.

### 4.3.3.1. Backscattering From a Perfectly Conducting Cylinder With Flat Ends

Many authors have treated this problem, and no attempt is made here to review the literature. It appears that one of the most thorough and general solutions to this problem has been obtained by Ufimtsev.[32,33] Basically, he uses physical optics and stationary phase to compute the fields scattered from the cylinder sides and flat ends. To this, he adds the contributions from the edges computed by a modification of the Sommerfeld–MacDonald technique. He corrects the discrepancy in the edge contributions near the end-on aspect by postulating that the diffracted far field from the edges came from an asymptotic Bessel function expansion with argument $k_0 a \cos \Psi_s$. Thus, near $\Psi_s \to \pi/2$ (end-on), the Bessel functions approach zero and the edge contributions become negligible compared to the physical-optics contribution from the end. When such an "interpolating" procedure is not done, the edge-diffracted fields near end-on aspect go to infinity (a situation which is clearly in error), and this requires a caustic correction to be made.

Since this cylinder is axially symmetric, there is no depolarization when $TM$ and $TE$ waves are incident upon the cylinder (i.e., an incident $TM$ wave, with its $H$ field perpendicular to the plane formed by the propagation direction and cylinder axis, produces only a $TM$ scattered field in any direction). Hence, only the two scattered fields need be given.

Ufimtsev derives expressions for bistatic scattering with the following restriction: $\Psi_i$ and $\Psi_s$ (Fig. 4-39) must be of the same sign. The reason for this is twofold: (i) If $\Psi_i$ and $\Psi_s$ are different in sign, then three different edges are visible when looking along $\hat{k}_i$ and $-\hat{k}_s$ (incidence and negative scattering propagation direction). Thus, not all of the illuminated edges contribute directly to the scattered field; one of the edges is hidden from the scattered field direction. The edge-diffraction theory of Ufimtsev does not take this more difficult situation into account. (ii) If $\Psi_i$ and $\Psi_s$ are of the same sign (e.g., positive), then the same end face at $z = -h$ is visible to the incidence and scattering directions (looking along $\hat{\mathbf{k}}_i$ and $-\hat{\mathbf{k}}_s$). However, when $\Psi_i > 0$ and $\Psi_s < 0$, then the flat end face at $z = -h$ is illuminated by the incident wave, but in the high-frequency limit cannot contribute to the scattered field, since only the face at $z = +h$ is visible from the scattered field direction. Thus Ufimtsev's results, valid for $\Psi_i$ and $\Psi_s > 0$, cannot be used directly for $\Psi_i > 0$ and $\Psi_s < 0$. If the reader desires the bistatic scattered fields for $\Psi_i$ and $\Psi_s > 0$, he is advised to consult either reference 32 or 33; because of this serious restriction on the bistatic direction, only the results for backscattering will be presented here.

For $\Psi_i = \Psi_s = \Psi$ ($0 \leqslant \Psi \leqslant \pi/2$) and $\phi' = 0$, the *TM* (*E* field lies in plane formed by $\hat{\mathbf{k}}_i$ and $\hat{\mathbf{z}}$) backscattering cross section is

$$\sigma^{TM}(\Psi) = \pi a^2 \mid F^{TM}(\Psi)\mid^2 \qquad (4.3\text{-}46)$$

where $F^{TM}(\Psi)$ is directly proportional to the *TM* backscattered field divided by the *TM* incident field strength (will be defined below).

For $\Psi_i = \Psi_s = \Psi$ ($0 \leqslant \Psi \leqslant \pi/2$) and $\phi' = 0$, the *TE* (*E* field lies perpendicular to plane formed by $\mathbf{k}_i$ and $\mathbf{z}$) backscattering cross section is

$$\sigma(\Psi^{TE}) = \pi a^2 \mid F^{TE}(\Psi)\mid^2 \qquad (4.3\text{-}47)$$

where $F^{TE}(\Psi)$ is directly proportional to the *TE* backscattered field divided by the *TE* incident field strength.

The parameters, $F^{TM}(\Psi)$ and $F^{TE}(\Psi)$, used above are

$$F^{TM}(\Psi) = [MJ_1(2k_0a \cos \Psi) + iNJ_2(2k_0a \cos \Psi)]\, e^{-i2k_0h \sin \Psi}$$
$$- g[J_1(2k_0a \cos \Psi) - iJ_2(2k_0a \cos \Psi)]\, e^{i2k_0h \sin \Psi} \qquad (4.3\text{-}48)$$

and

$$F^{TE}(\Psi) = [\overline{M}J_1(2k_0a \cos \Psi) + i\overline{N}J_2(2k_0a \cos \Psi)]\, e^{-i2k_0h \sin \Psi}$$
$$- f[J_1(2k_0a \cos \Psi) - iJ_2(2k_0a \cos \Psi)]\, e^{i2k_0h \sin \Psi} \qquad (4.3\text{-}49)$$

where $J_1(x)$ and $J_2(x)$ are the cylindrical Bessel functions of argument $x$ and orders 1 and 2. The functions $M, N, f$, and $g$ depend strictly upon the aspect angle $\Psi$, and are

$$\left. \begin{matrix} M \\ M \end{matrix} \right\} = \frac{1}{\sqrt{3}} \left[ \mp \frac{1}{\frac{1}{2} + \cos(4\Psi/3)} \pm \frac{1}{\frac{1}{3} + \cos[\frac{2}{3}(\pi + 2\Psi)]} \right] \qquad (4.3\text{-}50)$$

$$\left. \begin{matrix} N \\ N \end{matrix} \right\} = \frac{1}{\sqrt{3}} \left[ -4 \mp \frac{1}{\frac{1}{2} + \cos(4\Psi/3)} \mp \frac{1}{\frac{1}{2} + \cos[\frac{2}{3}(\pi + 2\Psi)]} \right] \qquad (4.3\text{-}51)$$

$$\left. \begin{matrix} g \\ f \end{matrix} \right\} = \frac{1}{\sqrt{3}} \left[ -2 \mp \frac{1}{\frac{1}{2} + \cos(4\Psi/3)} \right] \qquad (4.3\text{-}52)$$

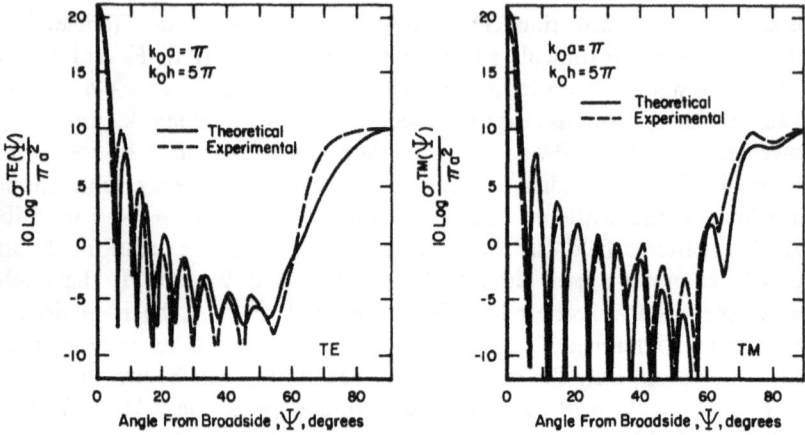

Fig. 4-50. Physical diffraction theory and experimental results for backscattering cross sections of a perfectly conducting circular cylinder with flat ends as a function of aspect angle [after Ufimtsev[33]].

These results are correct for any aspect from end-on ($\Psi = \pi/2$) to broadside ($\Psi = 0$) in the high-frequency limit. Surprisingly, they give remarkably good agreement even where $k_0 a$ and $k_0 h$ are not extremely large. Examples of $TM$ and $TE$ backscattering cross sections (divided by $\pi a^2$) are plotted in Figure 4-50, and compared with measurements for $k_0 a = \pi$ and $k_0 h = 5\pi$ (from Reference 33). The predicted and measured curves agree everywhere to within 2 db, and the lobes match up quite well. Figure 4-51 shows the same cross section (for either $TM$ or $TE$) predicted by physical optics alone. It is seen, in this case, that physical optics gives poor results for angles greater than 30° away from broadside ($\Psi > 30°$).

### 4.3.3.2. Scattering From Cylinder Near the Specular Direction

When the scattering direction is restricted to be near the specular direction (to within about 45° on either side of the specular direction), a simple physical-optics or high-frequency approximation is available. The specular direction is defined to have scattering angle $\Psi_s = -\Psi_i$ (Figure 4-39), but arbitrary $\phi'$. In this specular region, the strong return from the cylinder sides makes the contributions from the ends and edges negligible.

For backscattering, this region exists at and near the broadside aspect (out to about 25° away from broadside).

Physical optics, in this case, gives essentially the same result as derived and presented in Eq. (4.3-43), which is actually somewhat more general, since it applies to either $TM$ or $TE$ polarizations and arbitrary cylinder material, whereas physical optics is useful only for perfectly conducting cylinders.

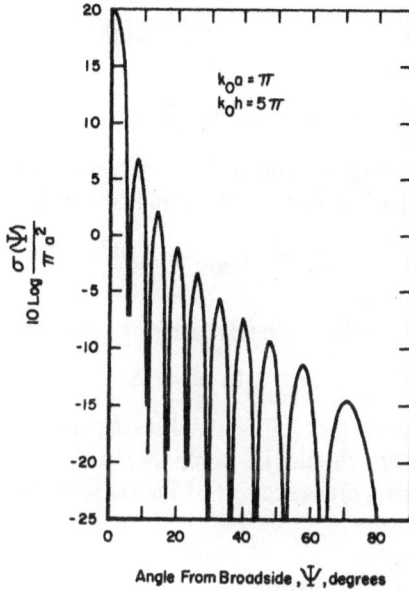

Fig. 4-51. Physical optics approximation for backscattering cross section of a perfectly conducting circular cylinder with flat ends as a function of aspect angle.

Equation (4.3-43) is then valid at and near the specular direction. One must know the value of $\sigma^c(\Psi_i, \phi')$, the scattering width of the infinitely long cylinder of the same material and radius. Unfortunately, in the high-frequency limit, only the perfectly conducting cylinder has a simple closed form expression for this scattering width. For other cylinder materials, this scattering width must be obtained from the exact series solutions or one of the approximations discussed in Sections 4.1 and 4.2.

For the perfectly conducting cylinder, however, the scattering width does exist in the high-frequency limit, and is the same for any incident polarization (i.e., the cylinder does not depolarize in this approximation). Using Eq. (4.1-37), and the technique discussed in Section 4.2.2 along with Eq. (4.3-43), gives for the bistatic cross section of a finite-length perfectly conducting cylinder near the specular direction,

$$\sigma(\Psi_i, \Psi_s, \phi') = 4k_0 a h^2 \frac{\cos^2 \Psi_s \cos \phi'/2}{\cos \Psi_i} \left[ \frac{\sin[k_0(\sin \Psi_i + \sin \Psi_s) h]}{k_0(\sin \Psi_i + \sin \Psi_s) h} \right]^2$$

(4.3-53)

In the above equation, $2h$ is the cylinder length and $a$ is the radius.

In the specular direction, $\Psi_s = -\Psi_i$, and the cross section and scattered power are maximum, or

$$\sigma(\Psi_i, -\Psi_i, \phi') = 4k_0ah^2 \cos \Psi_i \cos \phi'/2 \qquad (4.3\text{-}54)$$

In the backscattering direction, $\Psi_s = \Psi_i$ and $\phi' = 0$. In this case the backscattering cross section for incidence angles near broadside becomes

$$\sigma(\Psi_i, \Psi_i, 0) = 4k_0ah^2 \cos \Psi_i \left[ \frac{\sin(2k_0h \sin \Psi_i)}{2k_0h \sin \Psi_i} \right]^2 \qquad (4.3\text{-}55)$$

At broadside ($\Psi_i = 0$), the backscattering cross section is

$$\sigma(0, 0, 0) = 4k_0ah^2 \qquad (4.3\text{-}56)$$

Figure 4-51 shows a plot of the backscattering cross section predicted by physical optics; this should be compared with Figure 4-50, where the measured backscattering cross section of the same cylinder is shown.

## 4.4. ELLIPTICAL CYLINDERS

### 4.4.1. Infinitely Long, Perfectly Conducting Elliptical Cylinder at Normal Incidence

#### 4.4.1.1. *Series Solution*

Elliptic cylinders appear to be the only class of cylinders with finite cross section for which an exact scattering solution can be found (the circular cylinders considered previously are special cases of elliptic cylinders). The scattered field from an elliptic cylinder for an incident plane wave can be determined by boundary value techniques. The results will appear as an infinite series of eigenfunctions, just as for the circular cylinder.

Define the incident wave polarization directions as ‖ (*TM*, or *E* field parallel to cylinder axis) and ⊥ (*TE*, or *E* field perpendicular to cylinder axis). Then the scattered field will be of the same polarization as the incident field, and the problem reduces to a scalar one for each polarization. The incident wave strikes the cylinder at angle $\phi$ from its major axis (of half length $a$), while the scattered field is desired at point $P'$ which is at angle $\phi'$ with respect to the major axis as illustrated in Figure 4-52. The minor axis has half-length $b$. Interest shall be restricted to far-zone scattered fields only. This implies that $k_0r \gg [k_0(a + b)/2]^2$, and $k_0r \gg 1$.

The scattering problem as formulated above is, thus, reduced to a two-dimensional scalar problem. Elliptic cylinder coordinates are employed because the cylinder boundary is a natural coordinate surface in that system. Boundary conditions for the two polarizations are as follows: (i) *TM* case, total *E* field must vanish at cylinder surface; (ii) *TE* case,

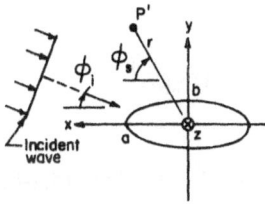

Fig. 4-52. Scattering geometry for an infinitely long, elliptic cylinder with incident field propagating normal to cylinder axis.

derivative of total $H$ field, with respect to surface normal, must vanish at the cylinder boundary.

Whereas the eigenfunctions for the circular cylindrical coordinate system are Bessel functions and trigonometric functions, for the elliptic cylinder coordinate system, the eigenfunctions are the ordinary Mathieu functions and the modified Mathieu functions. For the sake of brevity and simplicity, notations similar to those for circular cylinders will be used so that the similarities will be obvious. Detailed derivations of the solution can be found in References 34–37. Essentially, the development of References 34, 35 will be followed here; these two references employ the notation of Reference 38.

Denoting the incident fields as

$$\mathbf{E}^i = e^{-ik_0 r \cos(\phi'-\phi)} E_0^{TM} \hat{z} \qquad (4.4\text{-}1)$$

for the *TM* case, and

$$\mathbf{H}^i = e^{-ik_0 r \cos(\phi'-\phi)} H_0^{TE} \hat{z} \qquad (4.4\text{-}2)$$

for the *TE* case, these incident plane waves are expanded into a series of Mathieu functions. The scattered fields are then expressed in terms of a Mathieu function series with unknown coefficients. These coefficients are determined by the application of the boundary conditions given above. The resulting far-zone fields are [compare with Eqs. (4.1-5) to (4.1-7)]:

(i) *TM* Case

$$E^{sTM}(P') = E_0^{TM} \sqrt{\frac{2}{\pi}} \frac{\exp[i(k_0 r - \pi/4)]}{\sqrt{k_0 r}} T_{\|}(\phi', \phi) \qquad (4.4\text{-}3)$$

where

$$T_{\|}(\phi', \phi) = \left[ \sum_{n=0}^{\infty} (-1)^n Ae_n Se_n(h, \cos \phi') Se_n(h, \cos \phi)/Me_n(h) \right]$$

$$+ \left[ \sum_{n=1}^{\infty} (-1)^n Ao_n So_n(h, \cos \phi') So_n(h, \cos \phi)/Mo_n(h) \right] \qquad (4.4\text{-}4)$$

(ii) *TE* Case

$$H^{sTE}(P') = H_0^{TE} \sqrt{\frac{2}{\pi}} \frac{\exp[i(k_0 r - \pi/4)]}{\sqrt{k_0 r}} T_{\perp}(\phi', \phi) \qquad (4.4\text{-}5)$$

where

$$T_\perp(\phi', \phi) = \left[ \sum_{n=0}^{\infty} (-1)^n Be_n Se_n(h, \cos \phi') Se_n(h, -\cos \phi)/Me_n(h) \right]$$

$$+ \left[ \sum_{n=1}^{\infty} (-1)^n Bo_n So_n(h, \cos \phi') So_n(h, -\cos \phi)/Mo_n(h) \right] \quad (4.4\text{-}6)$$

In the above equations, the quantities, $Se_n(h, \cos \phi)$ and $So_n(h, \cos \phi)$, represent the angular Mathieu functions as defined in reference 38; and $Me_n(h)$ and $Mo_n(h)$ are normalization constants, which are equal to those of Reference 38 divided by $2\pi$. The angular Mathieu functions are analogous to $\cos n\phi$ and $\sin n\phi$ of the circular cylinder-scattered field solution. The constants $Ae_n$, $Ao_n$, $Be_n$, and $Bo_n$ depend upon the cylinder size, shape, and material only. The quantity $h$ is defined by

$$h = k_0 a \sqrt{1 - b^2/a^2} \quad (4.4\text{-}7)$$

For the perfectly conducting elliptic cylinder, the constants $A$, $B$ are given as follows:

$$Ae_n = \frac{Je_n(h, \cosh \zeta_0)}{He_n(h, \cosh \zeta_0)}, \qquad Ao_n = \frac{Jo_n(h, \cosh \zeta_0)}{Ho_n(h, \cosh \zeta_0)} \quad (4.4\text{-}8)$$

$$Be_n = \frac{(\partial/\partial\zeta_0) Je_n(h, \cosh \zeta_0)}{(\partial/\partial\zeta_0) He_n(h, \cosh \zeta_0)}, \qquad Bo_n = \frac{(\partial/\partial\zeta_0) Jo_n(h, \cosh \zeta_0)}{(\partial/\partial\zeta_0) Ho_n(h, \cosh \zeta_0)} \quad (4.4\text{-}9)$$

The quantities $Je_n(h, g)$, $He_n(h, g)$, $Jo_n(h, g)$, and $Ho_n(h, g)$ are the radial Mathieu functions, and are analogous to the Bessel and Hankel functions of the circular cylinder solution. The variable, $\zeta_0$, defines the surface of the cylinder in the elliptic coordinate system, just as $r = a$ defined the cylinder surface in the circular system. The quantity, $\cosh \zeta_0$, appearing in the above equations, is related to $a$ and $b$ of the ellipse as follows:

$$\cosh \zeta_0 = \frac{1}{\sqrt{1 - b^2/a^2}} \quad (4.4\text{-}10)$$

The above solutions for the scattered fields from an elliptic cylinder are of the same form and behavior of those for the circular cylinder, and one would expect to be able to sum the series solution numerically in the same manner. However, there are differences involved with considerably more inherent difficulty in the above solution as contrasted with the circular solution. Chief among these differences are the following:

(a) The solution for the elliptical cylinder contains two more parameters than that of the circular cylinder. While the parameters of the circular cylinder solution were $k_0 a$ and $\phi'$, the additional parameters here are $b/a$ (axial ratio) and $\phi$ (incidence angle). Hence, there are considerably more aspects and axial ratios to consider if one wishes to determine the complete scattering behavior of this cylinder.

(b) In the circular cylinder system, the variable, $r$, could represent a circular cylinder of any relative radius, $k_0 a$; the eigenvalues and angular eigenfunctions were independent of this cylinder radius, (i.e., the eigenvalues were all positive integers, $n$, and the eigenfunctions were $\cos n\phi$). In the elliptic coordinate system, however, the eigenvalues, angular eigenfunctions, and normalization constants, $[Me_n(h)$ and $Mo_n(h)]$, are all functions of the parameter $h$ ($h = k_0 a \sqrt{1 - /a^2 b^2}$). If one wishes to determine the back-scattered power from a cylinder of constant axial ratio, $b/a$, as a function of frequency (or $k_0 a$), then one must, essentially, recompute the eigenvalues, eigenfunctions, and normalization constants for each yalue of $h$. Another way of stating the same problem is that for a given axial ratio (or $1 - b^2/a^2$), but for varying size with respect to wavelength ($k_0 a$ varies), one must solve the problem for each $k_0 a$ in a different elliptical coordinate system because the interfocal distance in terms of wavelength must vary. This fact increases computation complexity considerably.

(c) The Bessel and trigonometric functions of the circular cylinder system have been studied in much greater detail than Mathieu functions. Consequently, relatively few of the useful mathematical relationships widely used for Bessel and trigonometric function generation and manipulation have counterparts for Mathieu functions. For example, no recurrence or addition formulas are known for Mathieu functions; for Bessel and trigonometric functions, recurrence formulas provide the most useful method for their numerical generation. Very few tables of any depth exist for Mathieu functions. For computational purposes, each Mathieu function [e.g., $Se_n(h, \cos \phi)$ or $Je_n(h, \cosh \zeta_0)$] is generated from a series of trigonometric or Bessel functions whose coefficients are determined for separate values of $h$. (See References 38, 39, and 40 for discussion of generation of Mathieu functions.)

Scattering widths are defined according to Eqs. (4.1-8) for the two polarizations by using $T_\parallel(\phi', \phi)$ and $T_\perp(\phi', \phi)$ in place of $T_\parallel(\phi')$ and $T_\perp(\phi')$ there. Backscattering takes place when $\phi' = \phi$, whereas forward scattering occurs when $\phi' = \pi + \phi$.

Normalized backscattering and forward scattering widths for given values of axial ratio, $b/a$, and aspect angles, $\phi$, have been generated on a computer and are shown in Figures 4-53 to 4-59. These widths are normalized so that the ordinates represent the quantities $\sigma_\parallel{}^c(\phi)/\pi[(a + b)/2]$ and $\sigma_\perp{}^c(\phi)/\pi[(a + b)/2]$. Thus, when $a/b = 1$ (circular cylinders), these quantities become the standard normalized widths employed previously in this chapter. When $b \to 0$ and $a = $ constant (i.e., the perfect conducting cylinder degenerates to a strip of width $2a$), then the above quantities are easily related to the scattering widths of Section 7.4.

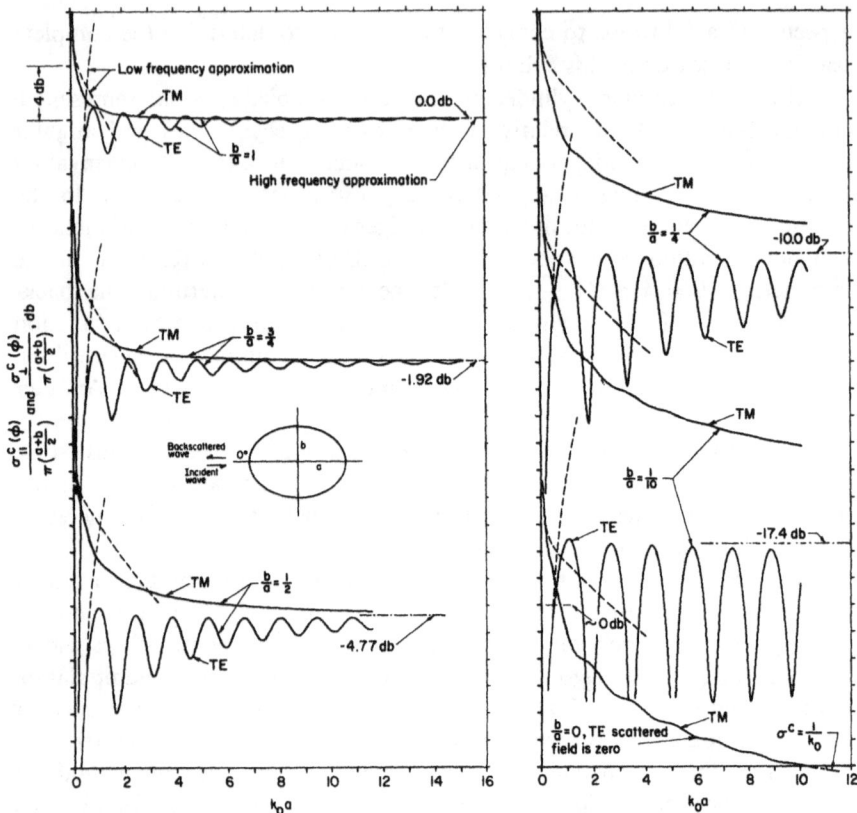

Fig. 4-53a.   Exact solutions for backscattering widths of an infinitely long, perfectly conducting elliptic cylinder as a function of size (or frequency) for various axial ratio values.

### 4.4.1.2. *Low-Frequency Approximations* ($k_0a < 0.5$, $k_0b < 0.5$)

In the low-frequency limit, approximations to the scattered field are obtained by expressing the coefficients of the above series solutions in their small argument expansions, and employing only the first few terms. This is done in Reference 34. Only the lowest-order terms in $k_0a$ are retained here. The resulting scattering amplitudes are

(i) *TM* Case

$$T_{\parallel}(\phi', \phi) = i\,\frac{\pi}{2}\,\frac{1}{\ln 0.8905k_0[(a+b)/2] - i\pi/2} \qquad (4.4\text{-}11)$$

(ii) *TE* Case

$$T_{\perp}(\phi', \phi) = -i\,\frac{\pi}{4}\,(k_0a)^2 \left[\frac{b}{a} + \left(1 + \frac{b}{a}\right)\left(\frac{b}{a}\cos\phi'\cos\phi + \sin\phi'\sin\phi\right)\right]$$
$$(4.4\text{-}12)$$

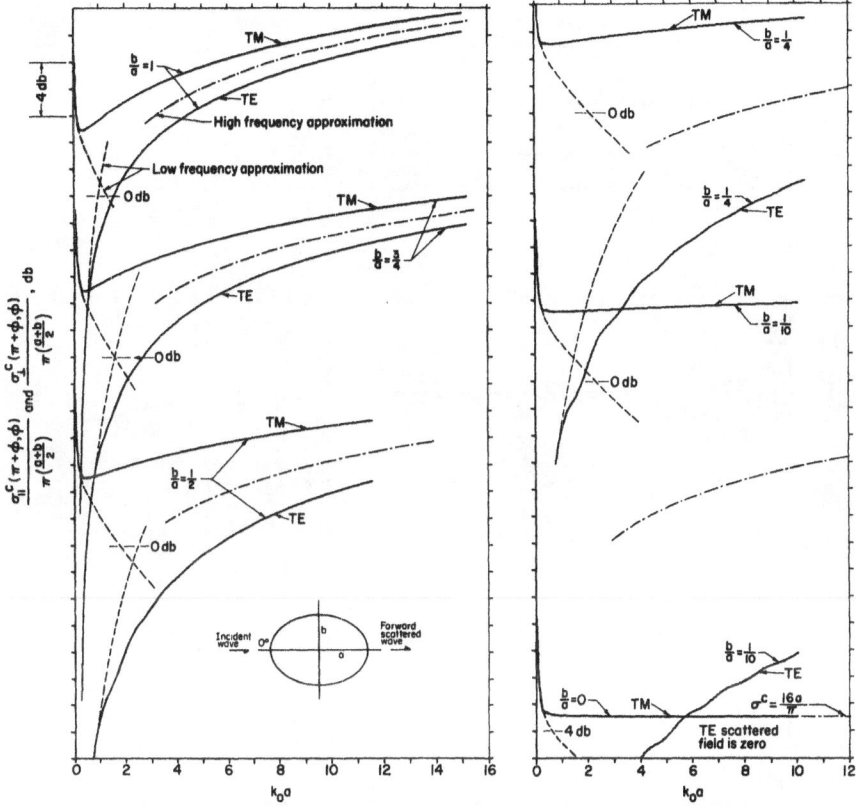

Fig. 4-53b. Exact solutions for forward scattering width of an infinitely long, perfectly conducting elliptical cylinder as a function of size (or frequency) for various axial ratio values.

The above results show that the *TM* scattered field in this limit is stronger than the *TE* field by a factor of the order of $(k_0a)^2$. This was also true for the circular cylinder. The *TM* scattered field is independent of the angles of incidence and scattering, i.e., the scattering pattern is isotropic.

Both of the above results are also valid in the strip limit ($b/a \to 0$) and for the circular cylinder limit ($b/a \to 1$).

The above two equations may be substituted into Eq. (4.1-8) to give the bistatic scattering widths, or

(i) *TM* Case

$$\sigma_\parallel{}^c(\phi', \phi) = \frac{\pi^2(a + b)/2}{[\sqrt{k_0(a + b)/2} \ \sqrt{\ln^2[0.8905k_0(a + b)/2] + \pi^2/4}]^2} \qquad (4.4\text{-}13)$$

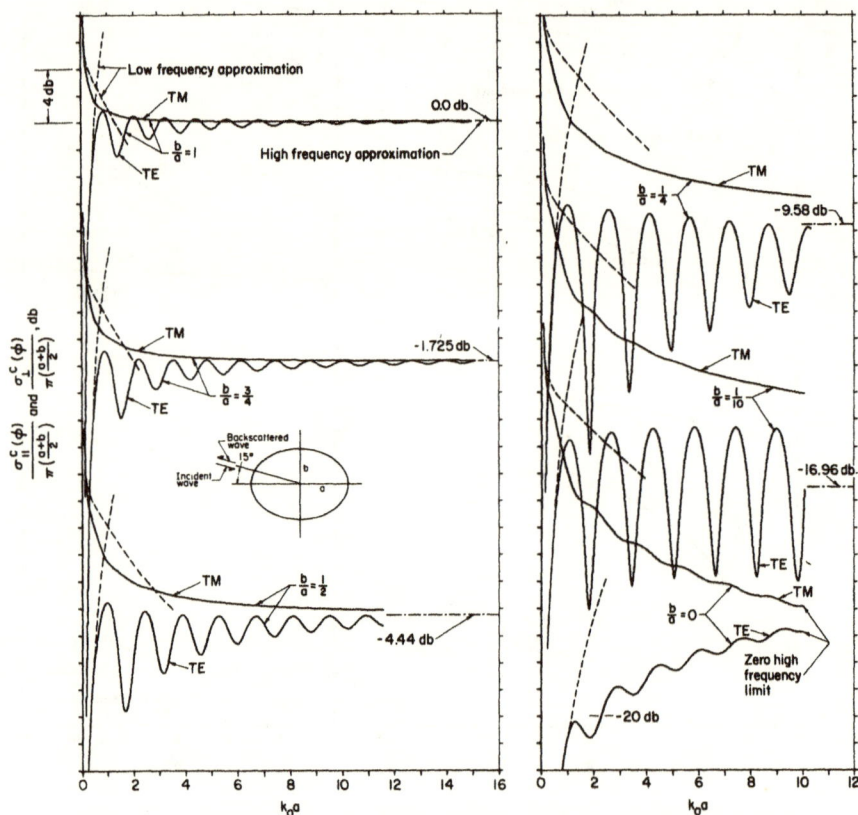

Fig. 4-54a. Exact solutions for backscattering widths of an infinitely long, perfectly conducting elliptic cylinder as a function of size (or frequency) for various axial ratio values.

(ii) *TE* Case

$$\sigma_\perp^c(\phi', \phi) = \frac{\pi^2 a}{4} (k_0 a)^3 \left[ \frac{b}{a} + \left(1 + \frac{b}{a}\right)\left(\frac{b}{a} \cos \phi' \cos \phi + \sin \phi' \sin \phi\right)\right]^2$$

(4.4-14)

The backscatter and forward scatter widths are obtained from Eqs. (4.4-13) and (4.4-14) by setting $\phi' = \phi$ and $\phi' = \pi + \phi$, respectively.

### 4.4.1.3. *High-Frequency Approximations* $(k_0 a > 5, k_0 b > 5)$

A solution for the scattered field from an elliptic cylinder, valid in the resonance and high-frequency regions (i.e., $k_0 a \gg 1$, $k_0 b \gg 1$), can be obtained in terms of specular and creeping wave contributions, just as was done for the circular cylinder. A transformation of the exact Mathieu

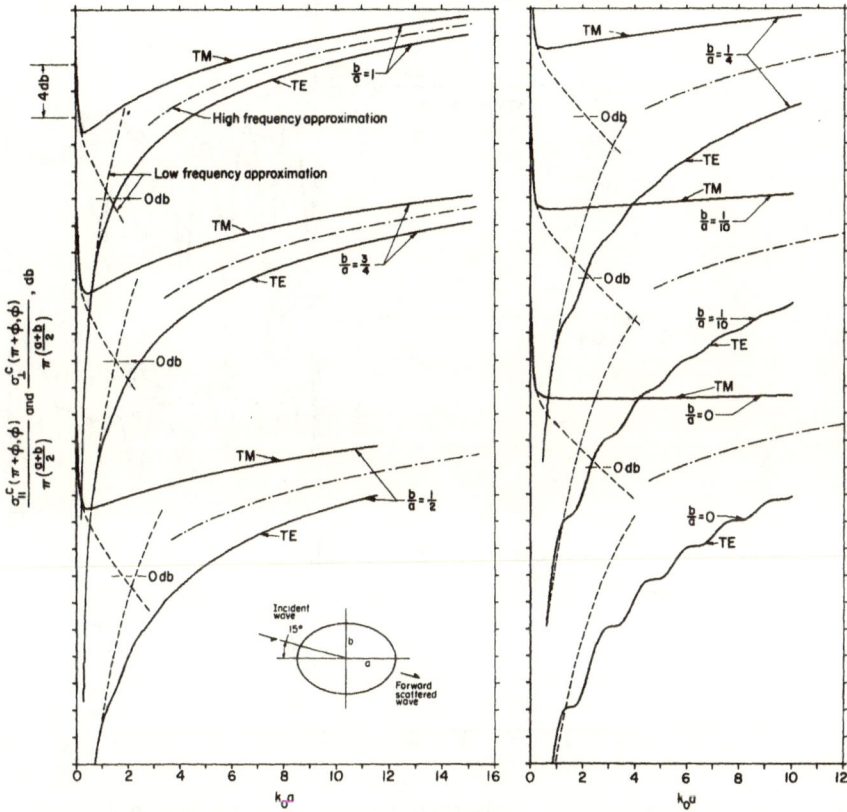

Fig. 4-54b. Exact solutions for forward scattering widths of an infinitely long, perfectly conducting elliptic cylinder as a function of size (or frequency) for various axial ratio values.

function series solution was used in Reference 41, while in Reference 42 the same problem was approached using a Green's function formalism.

In Section 2.2.2.5, the creeping wave contributions to the backscattered field from an elliptic perfectly conducting cylinder are formulated in terms of the Fock functions. The first asymptotic series term of the Fock function is equivalent to the creeping wave obtainable by formal expansion of the Mathieu function series, as was done in Reference 41. The expressions obtained in Section 2.2, i.e., Eqs. (2.2-230) and (2.2-231) can be used to give the creeping wave contributions to the backscattered field at any aspect angle, $\phi$. These will not be repeated here. They must be added to the specular contributions given below to obtain approximations valid in the resonance and high-frequency regions.

Creeping wave contributions to the general bistatic field are much

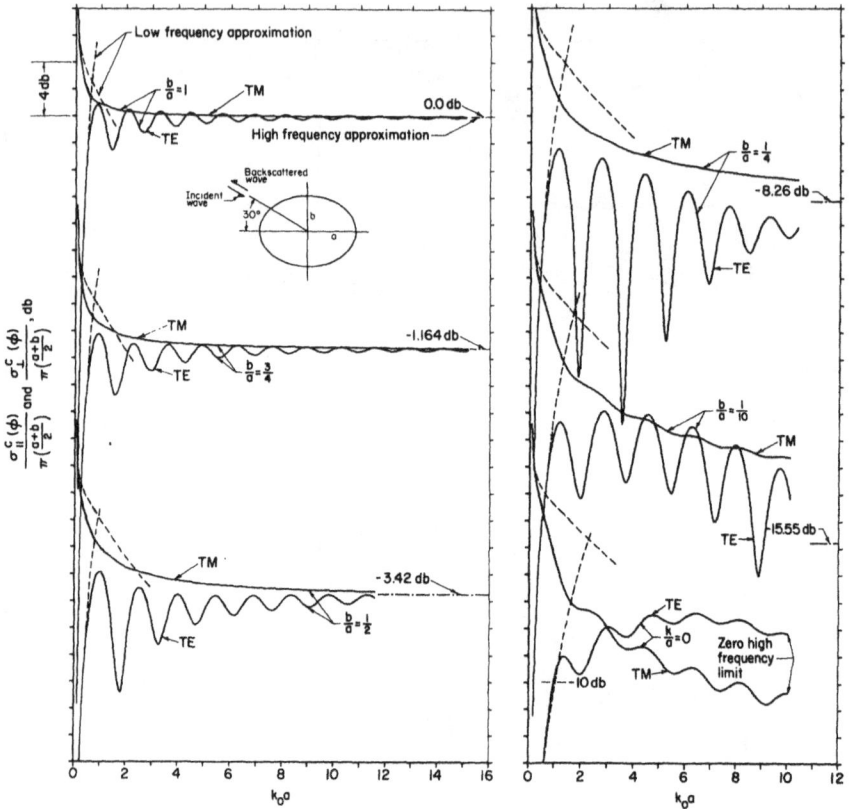

Fig. 4-55a. Exact solutions for backscattering widths of an infinitely long, perfectly conducting elliptic cylinder, as a function of size (or frequency) for various axial ratio values.

more complex; however, they can be obtained using the formalism given in Section 2.2.5.

Even though complex expressions are obtained, it is undoubtedly easier to determine the scattered fields in the resonance and high-frequency regions employing the Fock function method of Section 2.2.2.5, than it is to sum the exact Mathieu function series solution, even using a computer. This is especially true for $k_0a$ and $k_0b > 5$.

In the interest of providing a simple, closed-form answer for the scattered fields, only the optical or specular contribution is given here. It is easily derived from geometrical optics and is valid in the high-frequency limit. As can be ascertained from comparison of this answer with the exact solutions shown in Figures 4-53 to 4-59, the error incurred is at most approximately 15% in the vicinity of $k_0a \approx 5$ and $k_0b \approx 5$, and decreases rapidly for larger values.

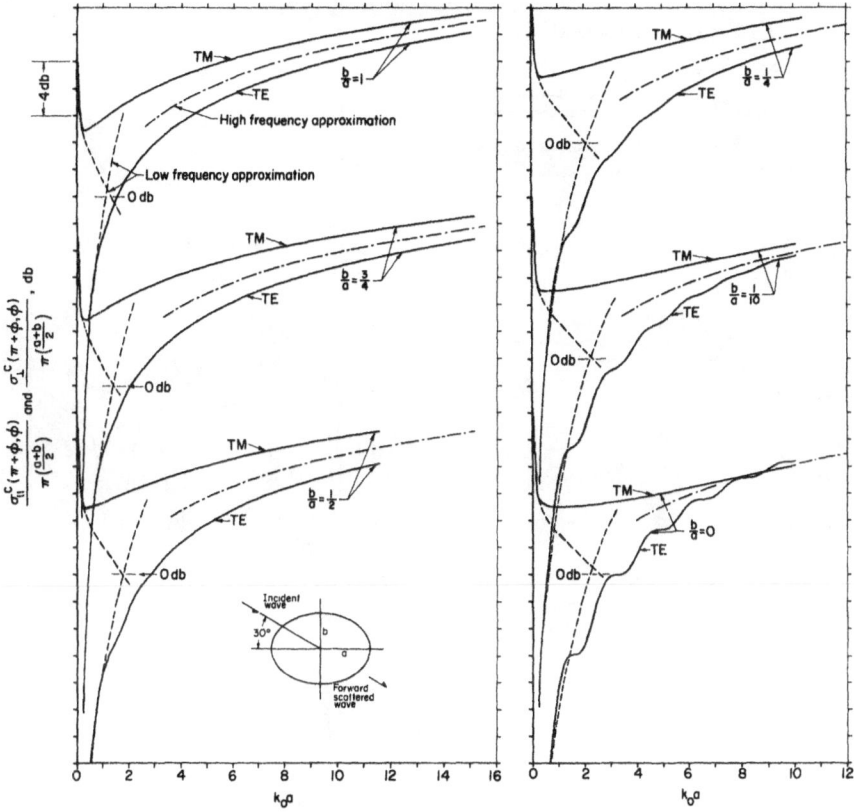

Fig. 4-55b. Exact solutions for forward scattering widths of an infinitely long, perfectly conducting elliptic cylinder as a function of size (or frequency) for various axial ratio values.

The scattering amplitudes in the high-frequency limit are

(i) *TM* Case

$$T_{\parallel}(\phi, \phi') = -\sqrt{\pi \cos\left(\frac{\phi' - \phi}{2}\right)} \frac{\sqrt{k_0\rho}}{2} \exp\left[-ik_0 af + i\frac{\pi}{4}\right] \quad (4.4\text{-}15)$$

(ii) *TE* Case

$$T_{\perp}(\phi, \phi') = -T_{\parallel}(\phi, \phi') \quad (4.4\text{-}16)$$

where $\rho$ is the radius of curvature of the specular point, i.e.,

$$\rho = \frac{a^2 b^2}{\{a^2 \cos^2[(\phi' + \phi)/2] + b^2 \sin^2[(\phi' + \phi)/2]\}^{3/2}} \quad (4.4\text{-}17)$$

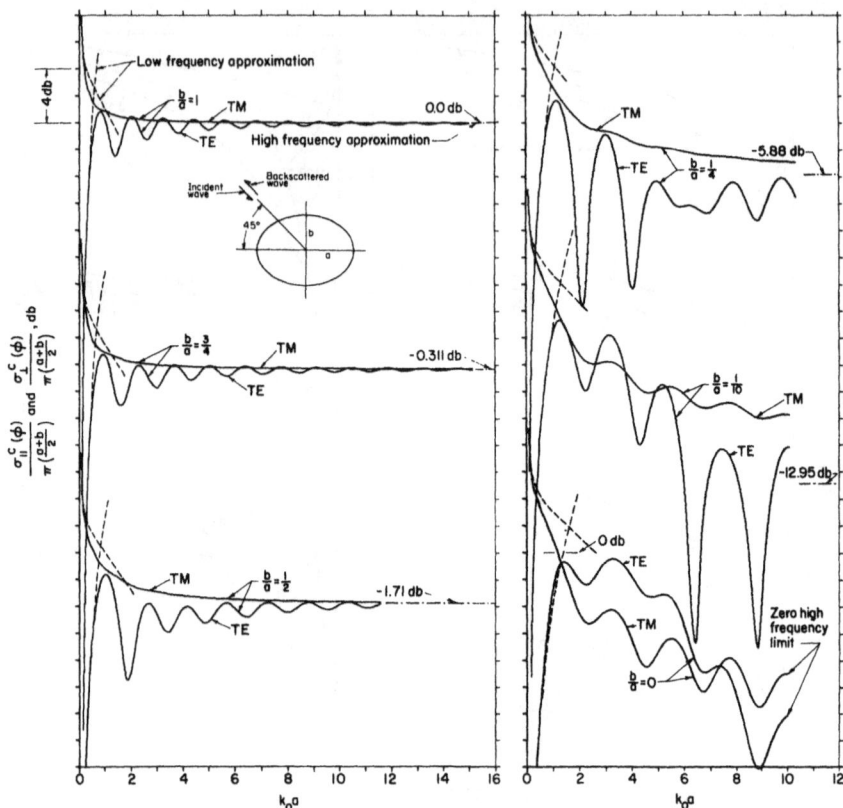

Fig. 4-56a. Exact solutions for backscattering widths of an infinitely long, perfectly conducting elliptic cylinder as a function of size (or frequency) for various axial ratio values.

and $f$ is the phase factor accounting for the path length difference, or

$$f = \frac{(\cos\phi + \cos\phi')\cos[(\phi + \phi')/2] + (b^2/a^2)(\sin\phi + \sin\phi')\sin[(\phi + \phi')/2]}{\{\cos^2[(\phi + \phi')/2] + (b^2/a^2)\sin^2[(\phi + \phi')/2]\}^{1/2}}$$

$$(4.4\text{-}18)$$

These results are valid in the high-frequency limit except in the vicinity of the forward direction, i.e., the condition

$$\pi - |\phi' - \phi| > (k_0^2 a^2 \sin^2\phi + k_0^2 b^2 \cos^2\phi)^{-1/6}$$

must hold.

4.4.1.3.1. *Bistatic Scattering Width.* The above results, when substituted into Eqs. (4.1-8), give the following scattering widths

$$\sigma_{\parallel}^c(\phi', \phi) = \sigma_{\perp}^c(\phi', \phi) = \frac{\pi a^2 b^2}{\{a^2 \cos^2[(\phi' + \phi)/2] + b^2 \sin^2[(\phi' + \phi)/2]\}^{3/2}}$$

$$(4.4\text{-}19)$$

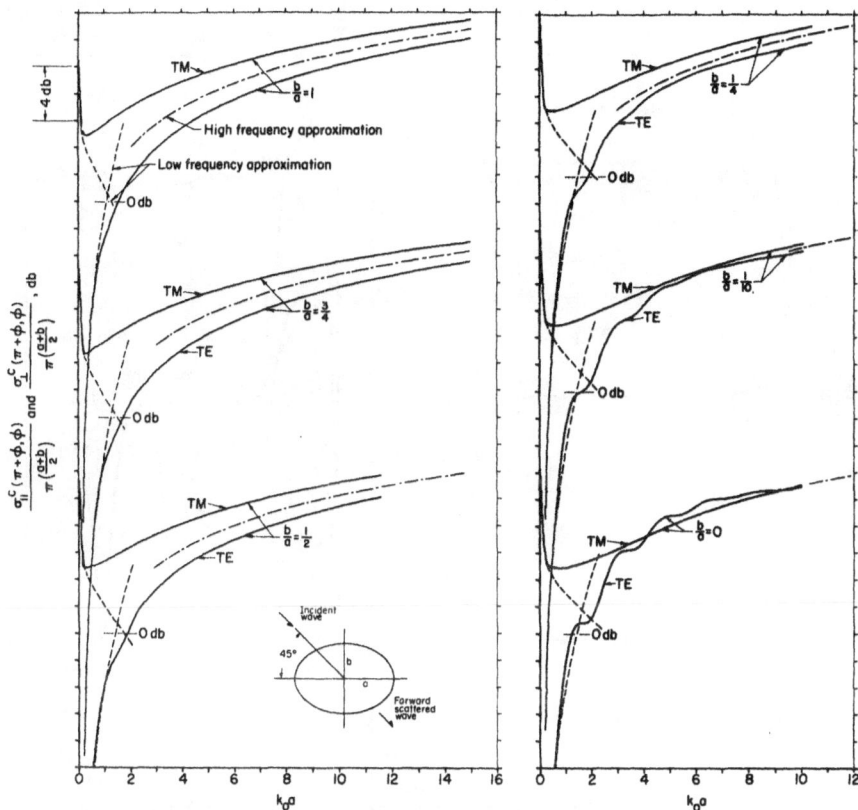

**Fig. 4-56b.** Exact solutions for forward scattering widths of an infinitely long, perfectly conducting elliptic cylinder as a function of size (or frequency) for various axial ratio values.

4.4.1.3.2. *Backscattering Width.* The above equation is immediately applicable to backscattering, where $\phi' = \phi$,

$$\sigma_{\parallel}^{c}(\phi, \phi) = \sigma_{\perp}^{c}(\phi, \phi) = \frac{\pi a^2 b^2}{[a^2 \cos^2 \phi + b^2 \sin^2 \phi]^{3/2}} \qquad (4.4\text{-}20)$$

4.4.1.3.3. *Forward Scattering Pattern and Width.* The high-frequency forward scattering pattern depends upon the width of the geometrical shadow immediately behind the cylinder; this width is $2c$, where

$$c = \sqrt{a^2 \sin^2 \phi + b^2 \cos^2 \phi} \qquad (4.4\text{-}21)$$

Define

$$\alpha = \pi + \phi - \phi' \qquad (4.4\text{-}22)$$

then the scattering amplitudes are

$$T_{\parallel}(\pi + \phi - \alpha, \phi) = T_{\perp}(\pi + \phi - \alpha, \phi) = -k_0 c \left( \frac{\sin(k_0 c \alpha)}{k_0 c \alpha} \right) \qquad (4.4\text{-}23)$$

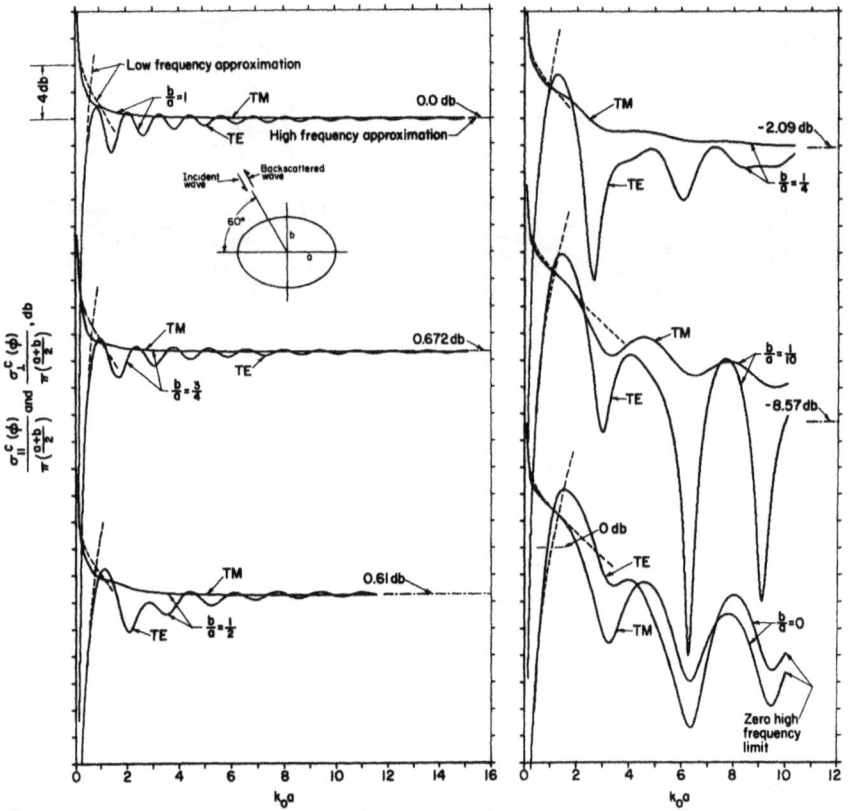

Fig. 4-57a. Exact solutions for backscattering widths of an infinitely long, perfectly conducting elliptic cylinder as a function of size (or frequency) for various axial ratio values.

Hence, the scattering widths near the forward direction are

$$\sigma_{\parallel}^c(\pi + \phi - \alpha, \phi) = \sigma_{\perp}^c(\pi + \phi - \alpha, \phi) = 4k_0c^2 \left(\frac{\sin k_0 c\alpha}{k_0 c\alpha}\right)^2 \quad (4.4\text{-}24)$$

and the scattering width in the exact forward direction ($\alpha = 0$) is

$$\sigma_{\parallel}(\pi + \phi, \phi) = \sigma_{\perp}^c(\pi + \phi, \phi) = 4k_0c^2 = 4k_0(a^2 \sin^2 \phi + b^2 \cos^2 \phi) \quad (4.4\text{-}25)$$

All of the results in this section are valid only near the forward direction, i.e., for

$$|\alpha| < (k_0^2a^2 \sin \phi + k_0^2b^2 \cos^2 \phi)^{-1/6}$$

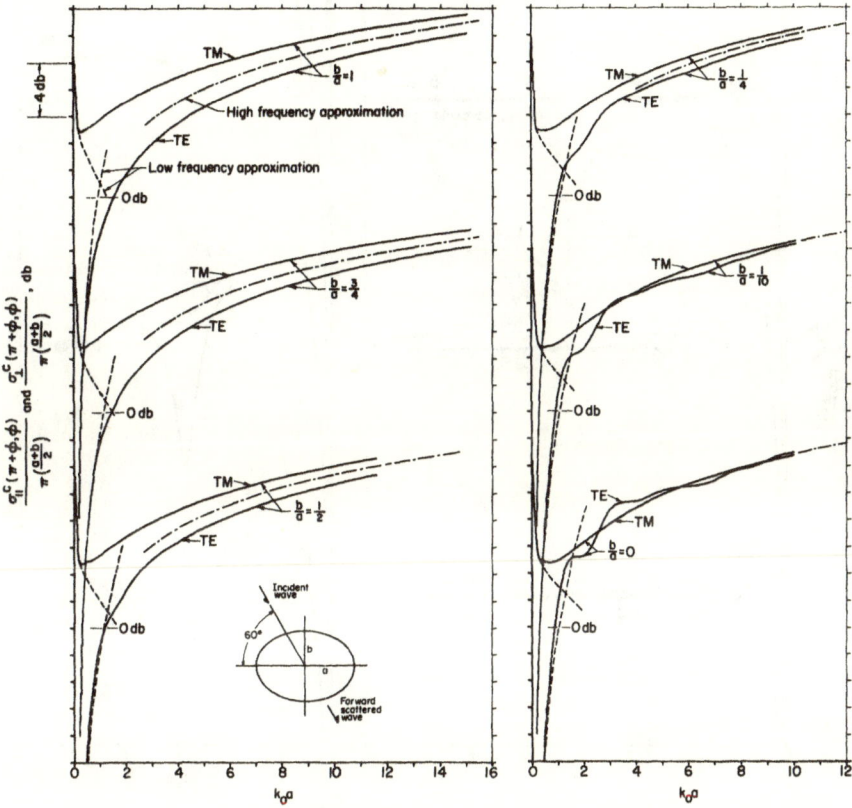

Fig. 4-57b. Exact solutions for forward scattering widths of an infinitely long, perfectly conducting elliptic cylinder as a function of size (or frequency) for various axial ratio values.

### 4.4.2. Infinitely Long Homogeneous Elliptical Cylinder at Normal Incidence ($\mu_1$, $\varepsilon_1$ inside cylinder)

#### 4.4.2.1. *Series Solution*

The formal series solutions for the scattered fields from a homogeneous elliptical cylinder for both polarizations, when the incident wave propagates normal to the cylinder axis, have the same form as Eqs. (4.4-3) to (4.4-6) for the perfectly conducting cylinder. One must find the series coefficients, $Ae_n$, $Ao_n$, $Be_n$, and $Bo_n$, by applying the boundary conditions that the tangential $E$ and $H$ fields must be continuous across the boundary. However, the problem for the homogeneous cylinder is not as simply solved as that for the conducting cylinder of the preceding section, and explicit general

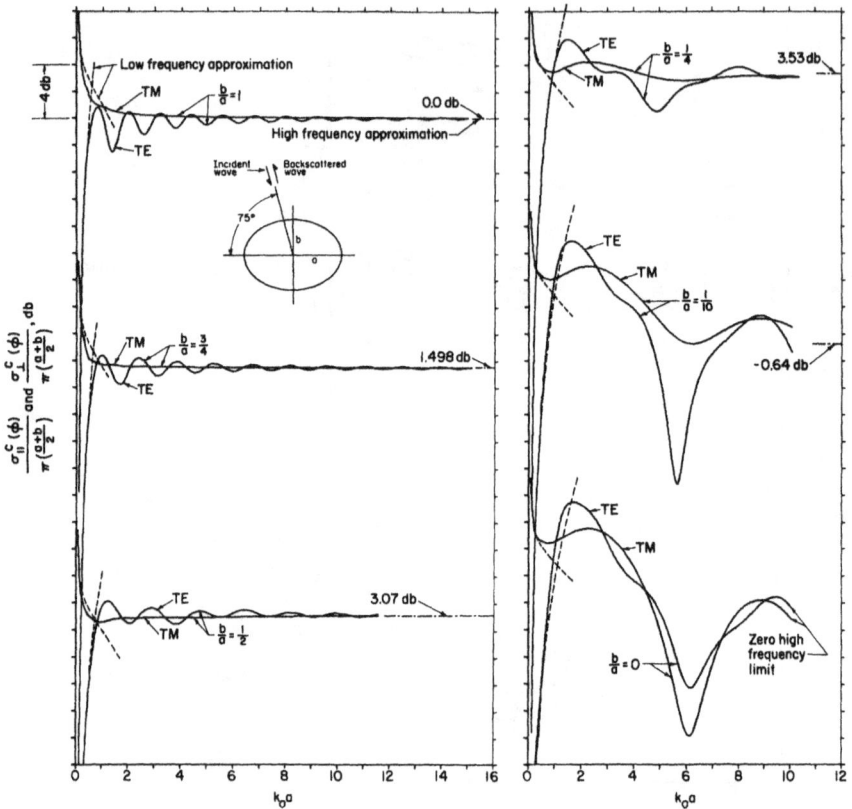

Fig. 4-58a. Exact solutions for backscattering widths of an infinitely long, perfectly conducting elliptic cylinder as a function of size (or frequency) for various axial ratio values.

expressions for these coefficients cannot be found. The reason for the added complexity is the following.

Fields outside the cylinder are expressed in series whose terms involve the Mathieu functions; $Me_n(h_0)$, $Se_n(h_0, \cos \eta_s)$, $Se_n(h_0, -\cos \eta_i)$, $He_n(h_0, \cosh \zeta)$, $Je_n(h_0, \cosh \zeta)$ (as well as the odd functions). Fields inside the cylinder are expressed in series where the Mathieu function family is now $Me_n(h_1)$, $Se_n(h_1, \cos \eta_s)$, $Se_n(h_1, -\cos \eta_i)$, $Je_n(h_1, \cosh \zeta)$. Here, $h_0$ is as defined in Eq. (4.4-7), while $h_1$ is the same as $h_0$ but with wave number $k_0$ replaced by $k_1$ ($k_1 = \sqrt{\epsilon_r \mu_r} k_0$). Consequently, the angular Mathieu functions vary with the permittivity and permeability of the media in which the fields exist. (In the circular cylinder problem, the angular eigenfunctions were $\cos n\phi$ and $\sin n\phi$, which were independent of the permeability and permittivity of the media.) Hence, the fields, when substituted into the

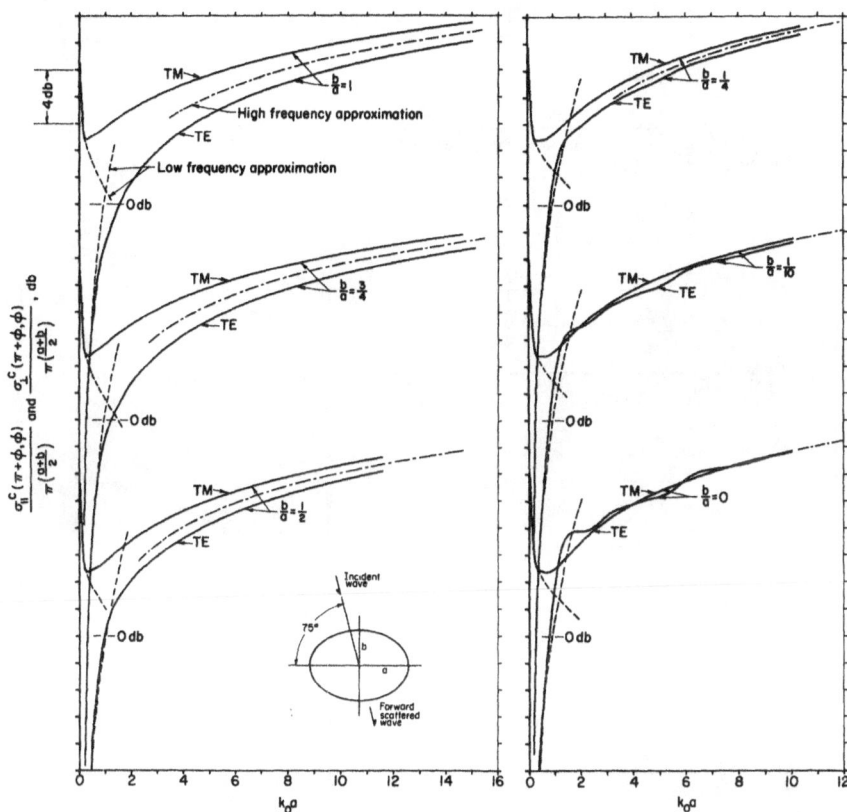

Fig. 4-58b. Exact solutions for forward scattering widths of an infinitely long, perfectly conducting elliptic cylinder as a function of size (or frequency) for various axial ratio values.

boundary conditions, produce equations containing both families of angular Mathieu functions. The coefficients $Ae_n$, $Ao_n$, $Be_n$, and $Bo_n$ cannot be determined simply by equating the coefficients of the angular functions to zero for each series term, as was done in all previous cases. In general, the coefficients can be determined by truncating the series properly, multiplying by $Se_n(h_1, \cos \eta)$ and $So_n(h_1, \cos \eta)$, and employing the orthogonality property of these functions to integrate over all $\eta$. Then one ends up with a set of linear $2N + 1$ equations in $2N + 1$ coefficients ($Ae_n$ and $Ao_n$ for $0 \leqslant n \leqslant N$). These are then solved, usually numerically, for the coefficients.

The problem is formulated in more detail in References 35, 36 and 37. Since solution is laborious, only a few backscattering widths have been computed up to now. In Reference 37, several scattering widths for constant

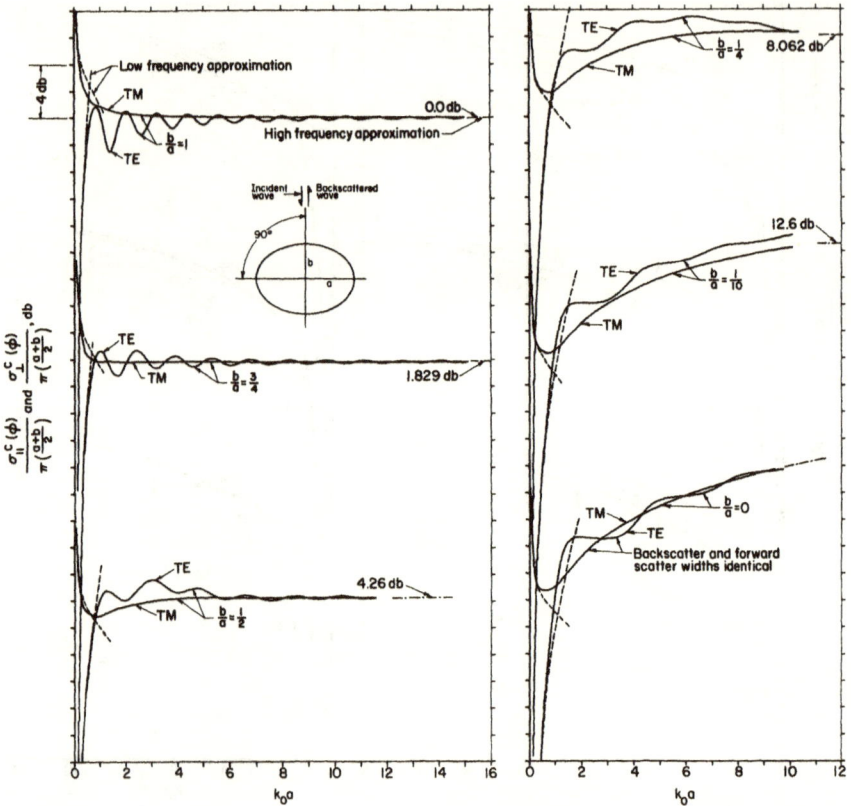

Fig. 4-59a. Exact solutions for backscattering widths of an infinitely long, perfectly conducting elliptic cylinder as a function of size (or frequency) for various axial ratio values.

values of $h_0$ have been computed and are reproduced here (Figures 4-60 to 4-63). All these curves apply to a cylinder with $\epsilon_r = 2$ and $\mu_r = 1$. The backscattering width in the first sets of figures is normalized so that the ordinate represents $k_0 \sigma_\parallel{}^c(\phi, \phi)/4$ and $k_0 \sigma_\perp{}^c(\phi, \phi)/4$. The abscissas represent $k_0 a$, i.e., the major axis of the ellipse relative to wavelength.

It should be noted that these curves, plotted as a function of $k_0 a$, unfortunately do not represent ellipses of constant axial ratio. Each curve of Figures 4-60 to 4-63 represents a constant, $h_0$ (i.e., $h_0 = k_0 a \sqrt{1 - b^2/a^2}$), so that as the abscissa value $k_0 a$ increases, the quantity $\sqrt{1 - b^2/a^2}$ must decrease so that $h_0$ remains constant. Consequently, as the abscissa $k_0 a$ increases, the curve represents a different ellipse whose minor axis, $k_0 b$, is rapidly approaching $k_0 a$; i.e., the ellipse approaches a circle as a function of $k_0 a$. This fact makes it difficult to observe from these figures just how

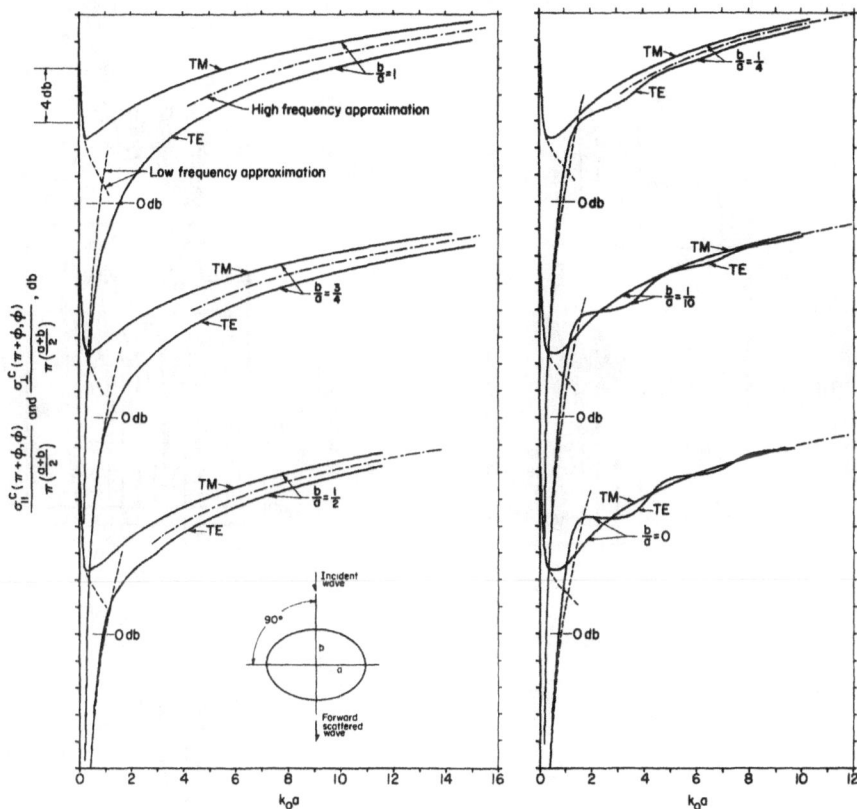

Fig. 4-59b. Exact solutions for forward scattering widths of an infinitely long, perfectly conducting elliptic cylinder as a function of size (or frequency) for various axial ratio values.

the backscattering width of a fixed-shape elliptical cylinder varies as a function of frequency, as was shown in the curves for the perfectly-conducting elliptical cylinder. The reason for this situation was mentioned previously; for the homogeneous cylinder, it is much more involved to compute the series coefficients than for the conducting cylinder. The procedure must be repeated anew for each value of $h_0$. Hence, in general, as few values of $h_0$ as possible are chosen to reduce the computation time.

### 4.4.2.2. Low-Frequency Approximations ($k_0 a < 0.5$, $k_0 b < 0.5$, $k_1 a < 0.5$, $k_1 b < 0.5$)

Fortunately, it is possible to solve for the series solution coefficients in the low-frequency limit. This is done by expanding the Mathieu functions

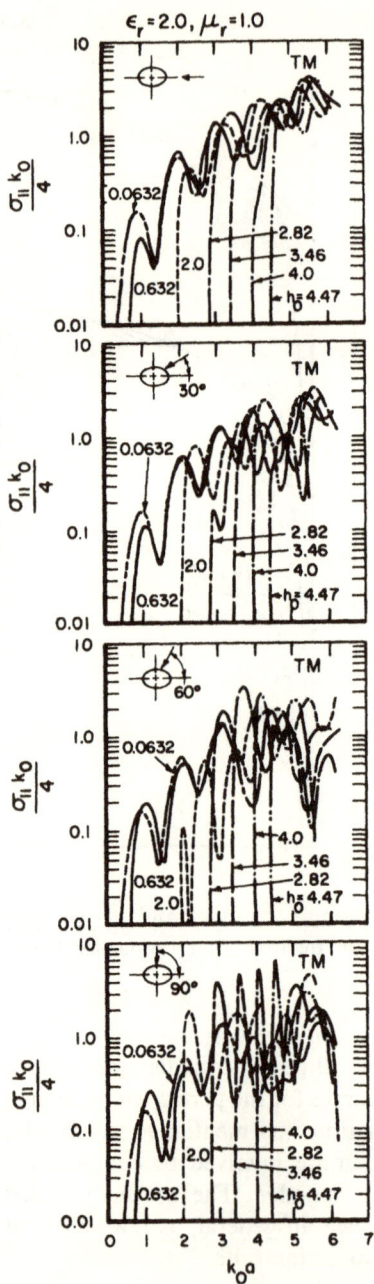

Fig. 4-60. Exact solutions for back-scattering widths of an infinitely long, homogeneous ($\epsilon_r = 2$, $\mu_r = 1$) elliptic cylinder for various constant values of $h_0 = k_0 a(1 - b^2/a^2)^{1/2}$, as a function of $k_0 a$ [after Yeh[37]].

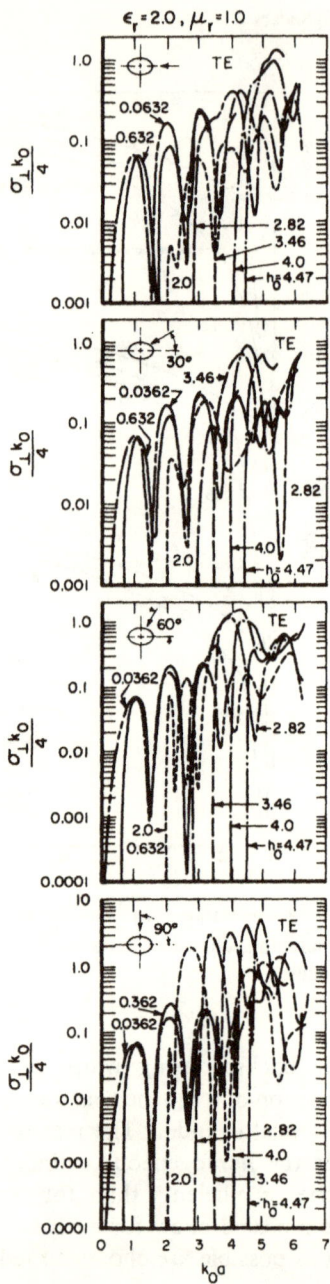

Fig. 4-61. Exact solutions for back-scattering widths of an infinitely long, homogeneous ($\epsilon_r = 2$, $\mu_r = 1$) elliptic cylinder for various constant values of $h_0 = k_0 a(1 - b^2/a^2)^{1/2}$ and as a function of $k_0 a$ [after Yeh[37]].

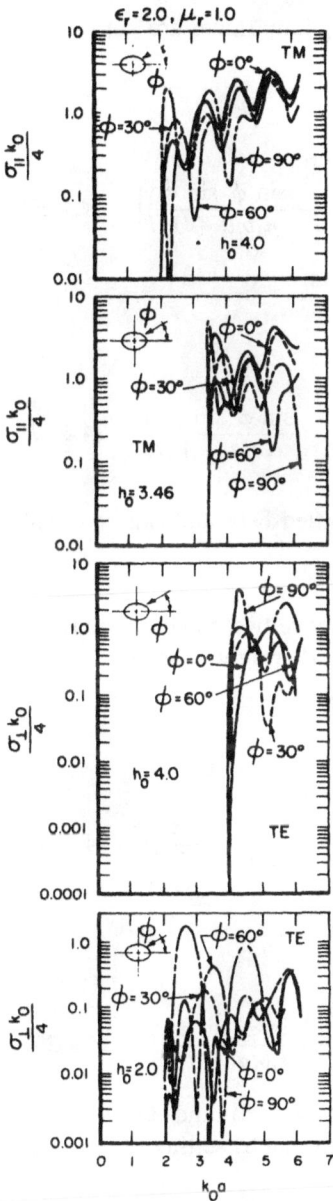

Fig. 4-62. Exact solutions for backscattering widths of an infinitely long, homogeneous ($\epsilon_r = 2$, $\mu_r = 1$) elliptic cylinder for various values of aspect angle, $\phi$, and as a function of $k_0 a$ [after Yeh[37]].

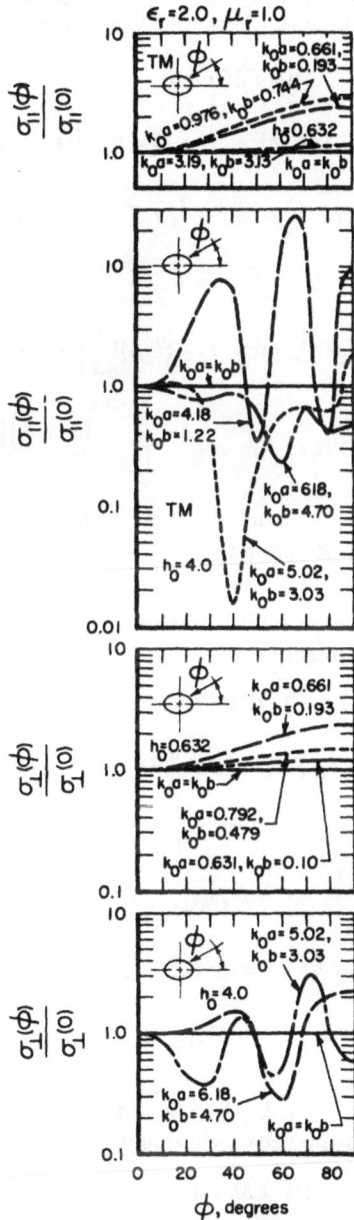

Fig. 4-63. Exact solutions for normalized backscattering widths of an infinitely long, homogeneous ($\epsilon_r = 2$, $\mu_r = 1$) elliptic cylinder for various axial lengths as a function of aspect angle, $\phi$ [after Yeh[37]].

into a power series in $k_0a$, and neglecting higher-order terms. This was done in reference 35, and the resulting scattering amplitudes are

(i) *TM* Case

$$T_\|(\phi', \phi) = i\,\frac{\pi}{4}\,(k_0a)^2\,\frac{b}{a}\,\Big[(\epsilon_r - 1)$$

$$- (\mu_r - 1)\Big(1 + \frac{b}{a}\Big)\Big(\frac{\cos\phi'\cos\phi}{\mu_r + b/a} + \frac{\sin\phi'\sin\phi}{\mu_1 b/a + 1}\Big)\Big] \quad (4.4\text{-}26)$$

(ii) *TE* Case

$$T_\perp(\phi', \phi) = i\,\frac{\pi}{4}\,(k_0a)^2\,\frac{b}{a}\,\Big[(\mu_r - 1)$$

$$- (\epsilon_r - 1)\Big(1 + \frac{b}{a}\Big)\Big(\frac{\cos\phi'\cos\phi}{\epsilon_r + b/a} + \frac{\sin\phi'\sin\phi}{\epsilon_r b/a + 1}\Big)\Big] \quad (4.4\text{-}27)$$

Higher terms of order $(k_0a)^4$ have been omitted here, but can be found in Reference 35.

4.4.2.2.1. *Bistatic Scattering Width.* The above two equations may be substituted into Eq. (4.1-8) to give the bistatic scattering widths, or

(i) *TM* Case

$$\sigma_\|(\phi', \phi) = \pi a\,\frac{\pi}{4}\,(k_0a)^3\,\frac{b^2}{a^2}\,\Big[(\epsilon_r - 1)$$

$$- (\mu_r - 1)\Big(1 + \frac{b}{a}\Big)\Big(\frac{\cos\phi'\cos\phi}{\mu_r + b/a} + \frac{\sin\phi'\sin\phi}{\mu_r b/a + 1}\Big)\Big]^2 \quad (4.4\text{-}28)$$

(ii) *TE* Case

$$\sigma_\perp(\phi', \phi) = \pi a\,\frac{\pi}{4}\,(k_0a)^3\,\frac{b^2}{a^2}\,\Big[(\mu_r - 1)$$

$$- (\epsilon_r - 1)\Big(1 + \frac{b}{a}\Big)\Big(\frac{\cos\phi'\cos\phi}{\epsilon_r + b/a} + \frac{\sin\phi'\sin\phi}{\epsilon_r b/a + 1}\Big)\Big]^2 \quad (4.4\text{-}29)$$

4.4.2.2.2. *Backscattering Width.* Equation (4.4-26) and (4.4-27) for bistatic scattering may be specialized to backscattering by setting $\phi' = \phi$, or

(i) *TM* Case

$$\sigma_\|(\phi, \phi) = \pi a\,\frac{\pi}{4}\,(k_0a)^3\,\frac{b^2}{a^2}\,\Big[(\epsilon_r - 1)$$

$$- (\mu_r - 1)\Big(1 + \frac{b}{a}\Big)\Big(\frac{\cos^2\phi}{\mu_r + b/a} + \frac{\sin^2\phi}{\mu_r b/a + 1}\Big)\Big]^2 \quad (4.4\text{-}30)$$

(ii) *TE* Case

$$\sigma_\perp(\phi, \phi) = \pi a \, \frac{\pi}{4} \, (k_0 a)^3 \, \frac{b^2}{a^2} \left[ (\mu_r - 1) \right.$$

$$\left. - (\epsilon_r - 1) \left( 1 + \frac{b}{a} \right) \left( \frac{\cos^2 \phi}{\epsilon_r + b/a} + \frac{\sin^2 \phi}{\epsilon_r b/a + 1} \right) \right]^2 \quad (4.4\text{-}31)$$

4.4.2.2.3. *Forward Scattering Width.* For forward scattering, setting $\phi' = \pi + \phi$ in Eqs. (4.4-26) and (4.4-27) gives

(i) *TM* Case

$$\sigma_\parallel(\pi + \phi, \phi) = \pi a \, \frac{\pi}{4} \, (k_0 a)^3 \, \frac{b^2}{a^2} \left[ (\epsilon_r - 1) \right.$$

$$\left. + (\mu_r - 1) \left( 1 + \frac{b}{a} \right) \left( \frac{\cos^2 \phi}{\mu_r + b/a} + \frac{\sin^2 \phi}{\mu_r b/a + 1} \right) \right]^2 \quad (4.4\text{-}32)$$

(ii) *TE* Case

$$\sigma_\perp(\pi + \phi, \phi) = \pi a \, \frac{\pi}{4} \, (k_0 a)^3 \, \frac{b^2}{a^2} \left[ (\mu_r - 1) \right.$$

$$\left. + (\epsilon_r - 1) \left( 1 + \frac{b}{a} \right) \left( \frac{\cos^2 \phi}{\epsilon_r + b/a} + \frac{\sin^2 \phi}{\epsilon_r b/a + 1} \right) \right]^2 \quad (4.4\text{-}33)$$

4.4.2.3. *High-Frequency Approximations*

The general behavior of the homogeneous elliptical cylinder at high frequencies is the same as that of the circular cylinder. Rays enter the cylinder, may be reflected from the boundary internally any number of times, and finally emerge in various directions. A modified geometrical-optics technique, as employed with the circular cylinder, has not yet been applied to the homogeneous elliptic cylinder. Without such an approximation, it is not possible to determine a high-frequency limit in general, as was done with the perfectly conducting elliptical cylinder.

An approximation is possible in the high-frequency limit when the cylinder material is sufficiently lossy that rays entering the cylinder are attenuated so that their contribution to the scattered field is negligible. This occurs when $\text{Im}[k_1 a] > 2$ and $\text{Im}[k_1 b] > 2$. When this is true, the only component which can contribute to the scattered field is the ray externally reflected from the cylinder. Then Eqs. (4.4-15) and (4.4-16) may be used for the scattering amplitudes after insertion of the factor $-R_\parallel[(\phi' - \phi)/2]$ into the former, and $+R_\perp[(\phi' - \phi)/2]$ into the latter equation on the right sides; these are the Fresnel reflection coefficients, and are given by Eqs. (7.1-9) and (7.1-10) of Section 7.1. All of the scattering widths given in Section 4.4.1.3 may be used here with the squares of the above reflection coefficients multiplying the results of that section, where

$R_\parallel^2[(\phi' - \phi)/2]$ is used in $\sigma_\parallel^c(\phi', \phi)$, and $R_\perp^2[(\phi' - \phi)/2]$ in $\sigma_\perp^c(\phi', \phi)$. This procedure holds for both the bistatic and the backscattering widths.

The forward scattering widths and patterns in the high-frequency limit are the same as those given in Section 4.4.1.3.3. The Fresnel reflection coefficients do not appear in that case because the forward field depends only upon the width of the geometric shadow and not on the material of the cylinder. Those results for forward scattering are valid in the high-frequency limit regardless of whether or not the cylinder material is lossy, whereas the approximations mentioned in the preceding paragraph are valid only when the cylinder material is sufficiently lossy.

### 4.4.3. Infinitely Long, Perfectly Conducting Elliptical Cylinder at Oblique Incidence

#### 4.4.3.1. Introduction

It was observed for the perfectly conducting circular cylinder that the scattered fields at oblique incidence had the same form as those for normal incidence, except for factors cos $\Psi_i$ appearing at various places. The same holds for perfectly conducting elliptic cylinders, or, for that matter, for infinitely long perfectly conducting cylinders of any shape; the scattered fields and scattering widths for oblique incidence can be written immediately if one knows these quantities at normal incidence.[†] The purpose of this section is to establish the rules for converting the results of Section 4.4.1 for normal incidence to results valid for oblique incidence.

Consider Figure 4-64. A plane wave is incident upon the cylinder at an angle $\pi/2 + \Psi_i$ from the z axis (i.e., the cylinder axis), or $\Psi_i$ from the normal to the cylinder axis. The plane of incidence includes the incident wave propagation direction and the z axis, and this plane lies at an angle, $\phi$, with respect to the $x - z$ plane. The scattering plane is defined by the

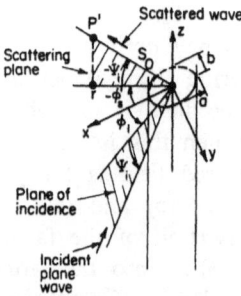

Fig. 4-64. Scattering geometry for an infinitely long elliptic cylinder at oblique incidence.

[†] Unfortunately, the same does not hold true for scattering from cylinders which are not perfectly conducting. Consequently, elliptic cylinders of other material at oblique incidence will not be considered in this chapter.

scattered field propagation direction and the $z$ axis, and this plane lies at an angle, $\phi'$, with respect to the $x - z$ plane.

It must be emphasized that in an arbitrary scattering plane (i.e., any $\phi'$), the scattered field propagates only in the direction $-\Psi_i$ with respect to the normal to the cylinder axis. Any finitely long cylinder always scatters "specularly" in a direction whose angle, $\Psi_s$, with respect to the axis normal, is the negative of the angle of incidence with respect to this normal. Thus, backscattering is possible only at normal incidence ($\Psi_i = \Psi_s = 0$).

Now define the *TM* and *TE* polarization directions as those where the $E$-field vector lies either normal to or within the plane of incidence (or scattering), respectively. Then, for a perfectly conducting, infinitely long cylinder of any cross section, a *TM* incident wave always gives rise only to a *TM* scattered wave, regardless of the positions of the incidence and scattering planes. The same, of course, holds true for *TE* polarization. This result was demonstrated for the circular cylinder in Section 4.2.2. The above statement is true only for perfectly conducting cylinders, however. Hence, it is necessary to resolve the incident field into *TM* and *TE* polarizations, and then find the *TM* and *TE* scattered fields each of these incident components produces, respectively.

Let the scattered field point of interest be $P'$, and let $S_0$ be the slant distance along the scattered field propagation direction from $P'$ to the cylinder axis. Then, the scattering width at oblique incidence is defined in the same manner as it was in Section 4.2 for the circular cylinder, i.e., the ratio of *TM* (or *TE*) scattered power density at $P'$ to the incident *TM* (or *TE*) power density at the cylinder times $2\pi S_0$.

With the preceding paragraphs as background, the following rules are established so that the results of Section 4.4.1 may be used at oblique incidence.

### 4.4.3.2. *Far-Zone Scattered Fields*

Equations (4.2-12), (4.2-15), (4.2-16), and (4.2-19) give the far-zone fields scattered from a perfectly conducting circular cylinder at oblique incidence (angle $\Psi$ there is $\Psi_i$ here.) Those equations may be used here with the following changes: (a) The actual series appearing there is to be replaced by the series defined by $T_\parallel(\phi', \phi)$ of Eq. (4.4-4) for *TM* polarization, and by the series defined by $T_\perp(\phi', \phi)$ of Eq. (4.4-6) for *TE* polarization; and (b) All cylinder dimensions appearing in the Mathieu function series are to be multiplied by $\cos \Psi_i$ (i.e., $h = k_0 a \sqrt{1 - b^2/a^2}$ becomes $k_0 a \cos \Psi_i \sqrt{1 - b^2/a^2}$).

### 4.4.3.3. *Scattering Widths*

All of the closed-form expressions for the scattering widths at normal incidence found in Sections 4.4.1.2 and 4.4.1.3 may be used here to find the scattering widths at oblique incidence, $\sigma_\parallel{}^c(\phi', \phi, \Psi_i)$ and $\sigma_\perp{}^c(\phi', \phi, \Psi_i)$, as follows (these scattering widths at oblique incidence are defined in Section 4.4.3.1.): (a) Anywhere the cylinder dimensions a and b appear, they are to be replaced by $a \cos \Psi_i$ and $b \cos \Psi_i$; and (b) The scattering width for normal incidence result is to be divided by $\cos \Psi_i$ to obtain $\sigma_\parallel{}^c(\phi', \phi, \Psi_i)$, or $\sigma_\perp{}^c(\phi', \phi, \Psi_i)$ for *TM* and *TE* polarizations.

### 4.4.3.4. *Scattering Width Curves*

The scattering width curves at normal incidence (i.e., Figures 4-53 to 4-59) be used as they stand to give normalized scattering widths for oblique incidence as follows: (a) The abscissa of the plots, instead of representing $k_0 a$, should now represent $k_0 a \cos \Psi_i$ for oblique incidence; and (b) The ordinate obtained from the curves now represents $\sigma^c(\phi', \phi, \Psi_i) \cos \Psi_i$.

## 4.5. CYLINDERS OF OTHER CROSS SECTIONS

In this chapter, only circular and elliptic cylinders have been treated explicitly. The reason for this is that for only these two classes of cylinders of finite cross section are scattered field solutions readily obtainable by boundary value techniques. Thus, these two problems have received the "lion's share" of the attention, whereas cylinders of other cross sections have been almost entirely neglected until very recently.

Most other cylinder cross sections are not readily amenable to boundary value solution. However, low-frequency results can be obtained employing scalar potential solutions (see Section 2.2.2.7). The number of two-dimensional cross-section shapes that can be solved by potential techniques is somewhat greater. For instance, Reference 43 treats the potential problem for the rectangular cylinder. However, the results there are in graphical form, and only the total scattering cross section in the low-frequency limit is computed; as such, those results are of no use for the radar cross-section problem treated here. High-frequency scattered fields from rectangular cylinders have been obtained in Reference 44 using the geometrical theory of diffraction. Unfortunately, only the total scattering cross section is presented there.

Reference 45 has considered the potential problem for the more general case of polygonal perfectly conducting cylinders in an attempt to find the low-frequency scattered fields. Although the method presented there appears to be powerful, the results seem to be erroneous. The scattered fields for the *TM* case do not exhibit the expected logarithmic behavior. It may be that those authors considered only the static electric or static magnetic problem, when in reality both must be considered simultaneously for perfectly conducting bodies. Furthermore, the infinite length of the cylinder means that for the *TM* polarization, the correct solution cannot come from a static potential treatment of the problem.

Reference 46 has obtained the total high-frequency cross section for a polygonal cylinder using the geometrical theory of diffraction.

In general, high-frequency scattered fields from a cylinder of arbitrary convex cross section may be obtained using geometrical (or physical) optics and the geometrical theory of diffraction or Fock theory (see Chapter 2 for summaries of these approximations). It is in the low-frequency and resonance regions (i.e., where wavelength is greater than or of the order of the body dimensions) that solution for the scattered field is difficult when the cross-sectional shape is arbitrary. As a consequence, the advent of computers has resulted in a variety of numerical techniques for solving scattering problems in these two frequency regions (see Section 2.2.2.8 for a description of numerical techniques).

For example, Reference 47 computes the scattered field for *TM* polarization at normal incidence from square cylinders, semicircular cylinders, and "*I*" shaped cylinders. This is done for one cylinder size each (at the lower end of the resonance region), and for one incidence aspect. Fortran programs used in generating these programs are also given in that report.

Reference 48 provides a numerical method for computing the *TM* or *TE* scattered fields from an infinitely long, perfectly conducting cylinder of arbitrary cross section. These authors apply the method to the elliptic cylinder to generate scattering widths.[†] The elliptic cylinder results, employed in Section 4.4.1, have been generated from the exact series solution; however, both results show essentially the same behavior.

In general, the few numerically generated curves which are presently available for cylinders of special cross sections, as in Reference 47, are insufficient in detail and quantity to really illustrate the scattering behavior of these cross-section shapes. For that reason, they are not included here.

## 4.6.  REFERENCES

1. Lord Rayleigh, On the Electromagnetic Theory of Light, *Phil. Mag.* **XII**:81 (1881), also appearing in *Scientific Papers* Vol. I, Dover Publications, New York (1964), p. 518.
2. Lord Rayleigh, The Dispersal of Light by a Dielectric Cylinder, *Phil. Mag.* **XXXVI**:365 (1918), also appearing in *Scientific Papers*, Vol. VI, Dover Publications, New York (1964), p. 554.

---

† It may be noted in passing that the numerical method employed in Reference 48 offers faster computer solutions for the elliptic cylinder than does the exact Mathieu function series solution, when the latter is generated and summed by the machine. This is due to the complexity involved in the Mathieu functions and their generation. However, the numerical technique is not exact, and the results begin to deviate significantly from the true solution when the axial ratio is less than $\frac{1}{2}$. This technique is accurate, therefore, only when applied to cylinders of small eccentricity.

3. King, R. W. P., and Wu, T. T., *The Scattering and Diffraction of Waves*, Harvard University Press (1959), p. 22.
4. Harrington, R. F., *Time-Harmonic Electromagnetic Fields*, McGraw-Hill, New York (1961), p. 232.
5. Tang, C. C. H., Backscattering from Dielectric-Coated Infinite Cylindrical Obstacles, *J. Appl. Phys.* **28**:628 (1957).
6. *Handbook of Mathematical Functions with Formulas, Graphs, and Mathematical Tables*, Abramowitz, M., and Stegun, I. A. (Eds.), U. S. Department of Commerce (1965).
7. Jahnke, E., Emde, F., and Lösch, F., *Tables of Higher Functions*, McGraw-Hill, New York (1960).
8. Jones, D. S., *The Theory of Electromagnetism*, MacMillan, New York (1964), p. 465.
9. The Effect of Dielectric Coating on the Radar Cross Section of Typical Re-Entry Vehicles, A. S. Thomas, Inc. (1963), AD 415 801.
10. Beckman, P., and Franz, W., Über die Greensche Function transparenter Zylinder, *Z. Naturforschung* **12a**:257 (1957).
11. Kawano, T., and Peters, L., Jr., An Extension of the Modified Geometrical Optics Methods for Radar Cross Sections of Dielectric Bodies, and An Application of Modified Geometrical Optics Method for Bistatic Radar Cross Sections of Dielectric Bodies, The Antenna Laboratory, The Ohio State University, Reports No. 1116-38,39, November, 1963, January, 1965.
12. Lee, W. C. Y., Peters, L., Jr., and Walter, C. H., Electromagnetic Scattering by Gyrotropic Cylinders with Axial Magnetic Fields, The Antenna Laboratory, The Ohio State University, Report No. 1116-48, May, 1964.
13. Van de Hulst, H. C., *Light Scattering by Small Particles*, John Wiley, New York, (1957).
14. Barrick, D., A Note on Scattering From Dielectric Bodies by the Modified Geometrical Optics Method, *Trans. IEEE* **AP-16** (March 1968).
15. Helstrom, C. W., Scattering from a Cylinder Coated with a Dielectric Material. In Jordan, E. C., *Electromagnetic Theory and Antennas, Part I*, Pergamon Press, New York (1963), p. 133.
16. Kodis, R. D., The Radar Cross Section of a Conducting Cylinder with Dielectric Sleeve at the Optical Limit, In Jordan, E. C., *Electromagnetic Theory and Antennas, Part I*, Pergamon Press, New York (1962), p. 127.
17. Plonus, M. A., Backscattering from a Conducting Cylinder with a Surrounding Shell, *Canad. J. Phys.* **38**:1665 (1960).
18. Wait, J. R., Scattering of a Plane Wave from a Circular Dielectric Cylinder at Oblique Incidence, *Canad. J. Phys.* **33**:189 (1955).
19. Wait, J. R., *Electromagnetic Radiation from Cylindrical Structures*, Pergamon Press, New York (1959), p. 142.
20. Wait, J. R., The Long Wavelength Limit in Scattering from a Dielectric Cylinder at Oblique Incidence, *Canad. J. Phys.* **43**:2212 (1965).
21. Van Vleck, J. H., Bloch, F., and Hamermesh, M., Theory of Radar Reflection from Wires or Thin Metallic Strips, *J. Appl. Phys.* **18**:274 (1947).
22. Siegel, K. M., Far Field Scattering from Bodies of Revolution, *Appl. Sci. Res.* **7**, 293 (1958).
23. Mack, C. L., and Reiffen, B., RF Characteristics of Thin Dipoles, *Proc. IEEE* **52**:533 (1964).
24. Albini, F. A., Radar Cross Section of Wires, Institute for Defense Analyses, *Pen-X Paper* **34** (May, 1965).
25. Tai, C. T., Electromagnetic Back-Scattering from Cylindrical Wires, *J. Appl. Phys.* **23**:909 (1952).
26. Cassedy, E. S., and Fainberg, J., Backscattering Cross Sections of Cylindrical Wires of Finite Conductivity, *IRE Trans. Ant. Prop.* **AP-8**:1 (1960).
27. de Bettencourt, J. T., Bistatic Cross Sections of Cylindrical Wires, Pickard & Burns, Report No. 1, 1961, RADC-TDR-61-285.
28. Weber, J., Scattering of Electromagnetic Waves by Wires and Plates, *Proc. IRE* **43**, 82 (1955).

29. Ufimtsev, P. Ya., Diffraction of Plane Electromagnetic Waves by a Thin Cylindrical Conductor, *Radiotekhnika i Elektronika* **7**:260 (1962).
30. Van de Hulst, H. C., *Light Scattering by Small Particles*, John Wiley, New York (1957), p. 304.
31. Carswell, A. I., Microwave Scattering Measurements in the Rayleigh Region Using a Focused-Beam System, *Canad. J. Phys.* **43**: 1962 (1965).
32. Ufimtsev, P. Ya., Approximate Calculation of the Diffraction of Plane Electromagnetic Waves by Certain Metal Objects. II. The Diffraction by a Disc and by a Finite Cylinder, *J. Tech. Phys.* **28**: 2386 (1958).
33. Ufimtsev, P. Ya., *The Edge Wave Method in Physical Diffraction Theory* "Soviet Radio" Publishing House, Moscow (1962), p. 86.
34. Burke, J. E., and Twersky, V., On Scattering of Waves by an Elliptic Cylinder and by a Semielliptic Protuberance on a Ground Plane, *J. Opt. Soc. Am.* **54**, 732 (1964).
35. Burke, J. E., Low-Frequency Approximations for Scattering by Penetrable Elliptic Cylinders, *J. Acoust. Soc. Am.* **36**:2059 (1964).
36. Yeh, C., The Diffraction of Waves by a Penetrable Ribbon, *J. Math. Phys.* **4**:65 (1963).
37. Yeh, C., Backscattering Cross Section of a Dielectric Elliptical Cylinder, *J. Opt. Soc. Am.* **55**, 309 (1965).
38. Morse, P. M., and Feshbach, H., *Methods of Theoretical Physics*, Vol. II, McGraw-Hill (1953), p. 1407 and 1568.
39. Blanch, G., On the Computation of Mathieu Functions, *J. Math. Phys.* **XXV**:1 (1946).
40. *Tables Relating to Mathieu Functions*, Columbia University Press, New York (1951).
41. Vaynshteyn, L. A., and Fedorov, A. A., Plane and Cylindrical Wave Scattering on an Elliptical Cylinder and the Theory of Diffraction Rays, *Radiotekhnika i Elektronika* **6**: 24 (1961).
42. Kazarinoff, N. D., and Ritt, R. K., Scalar Diffraction by an Elliptic Cylinder, *IRE Trans. Ant. Prop.* **AP-7**: S21 (1959).
43. Mei, K. and Van Bladel, J., Low-Frequency Scattering by Rectangular Cylinders, *IEEE Trans. Ant. Prop.* **AP-11**:52 (1963).
44. Mei, K., Scattering of High-Frequency Waves by Perfectly Conducting Rectangular Cylinders, 1963 IEEE International Convention Record, Part I, 132.
45. Lin Wei-Guan and Pen We-Yen, Low-Frequency Scattering by Polygonal Cylinders, *Acta Physica Sinica* **13**, 1381 (1964).
46. Morse, B. J., Diffraction by Polygonal Cylinders, *J. Math. Phys.* **5**, 199 (1964).
47. Richmond, J. H., Scattering by an Arbitrary Array of Parallel Wires, The Antenna Laboratory, The Ohio State University, Report No. 1522-7 (April 1964), AD 443 833.
48. Mullin, C. R., Sandburg, R., and Velline, C. O., A Numerical Technique for the Determination of Scattering Cross Sections of Infinite Cylinders of Arbitrary Cross Section, *IEEE Trans. Ant. Prop.* **AP-13**:141 (1965).

# Chapter 5
# ELLIPSOIDS AND OGIVES

G. T. Ruck

## 5.1. ELLIPSOIDS[†]

An important shape for which general solutions have not been obtained is the ellipsoid and its special cases, the prolate and oblate spheroid. These shapes are of considerable interest since many more complex geometric shapes can be approximated to first order by an ellipsoid. The available results on the cross sections of ellipsoidal objects are discussed in the following sections.

### 5.1.1. General Ellipsoids

No exact analytical solution to the general problem of the radar cross section of an ellipsoid has as yet been obtained. However, solutions have been obtained in the Rayleigh region, and geometrical optics allows high-frequency approximations to be made.

#### 5.1.1.1. Low-Frequency Solutions

Stevenson[1,2] has obtained the first two terms in a power series expansion of the far zone scattered fields in powers of $k_0$ for an arbitrary ellipsoid, either dielectric or conducting. His results in the general case are quite complex, and his papers should be consulted by anyone interested in applying these.

Consider a perfectly conducting ellipsoid with a plane wave incident, as illustrated in Figure 5-1, with $a$, $b$, $c$, the ellipsoid semi-axes, ordered such that $a \geqslant b \geqslant c$. Stevenson expresses the far-zone scattered field as[‡]

$$E_\theta = H_\phi = \left( \frac{\partial P}{\partial \theta'} + \frac{1}{\sin \theta'} \frac{\partial \bar{P}}{\partial \phi'} \right) \frac{e^{ik_0 r}}{r}, \qquad (5.1\text{-}1)$$

and

$$E_\phi = -H_\theta = \left( \frac{1}{\sin \theta'} \frac{\partial P}{\partial \phi'} - \frac{\partial \bar{P}}{\partial \theta'} \right) \frac{e^{ik_0 r}}{r} \qquad (5.1\text{-}2)$$

---

[†] Throughout Section 5.1, the symbols, $\mu$, $\epsilon$, are to be interpreted as the permittivity and permeability of the body relative to the free space value, $\mu_0$, $\epsilon_0$.

[‡] To simplify the analysis, Stevenson considers a unit incident field, and sets $\epsilon_0 = \mu_0 = 1$.

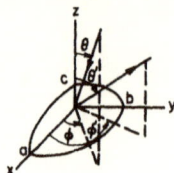

Fig. 5-1. Coordinate system for the general ellipsoid.

with

$$P = k_0^2[K_1\alpha + K_2\beta + K_3\gamma] + k_0^4[L_1\alpha + L_2\beta + L_3\gamma + M_1\alpha^2 + M_2\beta^2$$
$$+ M_3\gamma^2 + N_2\beta\gamma + N_2\alpha\gamma + N_3\alpha\beta - \tfrac{1}{30}(K_1\alpha + K_2\beta + K_3\gamma)$$
$$(a^2\alpha + b^2\beta + c^2\gamma)]$$

(5.1-3)

and analogous to Eq. (5.1-3),

$$\bar{P} = k_0^2[\bar{K}_1\alpha + \cdots + \bar{K}_3\gamma] + k_0^4[\bar{L}_1\alpha + \cdots + \bar{N}_3\alpha\beta$$
$$- \tfrac{1}{30}(\bar{K}_1\alpha + \cdots + \bar{K}_3\gamma)(a^2\alpha^2 + \cdots + c^2\gamma^2)]$$

(5.1-4)

The parameters $\alpha$, $\beta$, $\gamma$ are the direction cosines of the radius vector **r** to the observation point, and in spherical coordinates are

$$(\alpha; \beta; \gamma) = (\sin\theta'\cos\phi'; \sin\theta'\sin\phi'; \cos\theta')$$

(5.1-5)

while the quantities $K_i$, $L_i$, $M_i$, and $N_i$ are defined as follows:[†]

$$K_1 = \tfrac{2}{3}(\epsilon - 1)f_1(\epsilon)\,l_1$$

(5.1-6)

$$L_1 = \frac{f_1(\epsilon)}{15}\left[(\epsilon - 1)\,l_1\,[(\tfrac{1}{3})(6a^2 - b^2 - c^2)\right.$$
$$- (a^2l^2 + b^2m^2 + c^2n^2)] + \epsilon(b^2mn_2 - c^2nm_2)\Big]$$
$$+ \frac{f_1^2(\epsilon)}{15}\,l_1\left[(\epsilon - 1)\left((\epsilon - 2)\,I + \epsilon a^2 I_a - \frac{4a}{bc}\right) + \epsilon^2\mu\frac{(b^2 + c^2)}{abc}\right]$$
$$+ \frac{f_1(\epsilon)\,g_1(\mu)}{15}\left[(I_b + I_c)[(\mu/2)(b^2 + c^2)\right.$$
$$\times (mn_2 - nm_2) - mn_2b^2 - nm_2c^2] - \frac{\epsilon\mu(b^2 - c^2)}{abc}(mn_2 + nm_2)\Big]$$
$$+ \frac{f_1^2(\epsilon)\,g_1(\mu)}{15}\,l_1\left[(I_b - I_c)(\epsilon - 1)\,h_1(\mu) + \epsilon\mu(\epsilon\mu - 2)\frac{(b^2 - c^2)}{abc}\right]$$

(5.1-7)

$$M_1 = \frac{\epsilon - 1}{45Q}\left[(\epsilon - 1)(I_{ab}nn_1 + I_{ac}mm_1 - 2I_{bc}ll_1)\right.$$
$$+ \frac{2\epsilon}{abc(a^2b^2 + b^2c^2 + a^2c^2)}(2a^2ll_1 - b^2mm_1 - c^2nn_1)\Big]$$

(5.1-8)

[†] $N_1$ as given in Example (3), p. 1148 of Stevenson's paper (Reference 2), for a perfectly conducting ellipsoid, should have a factor of 2 in the denominator of the second term.

$$N_1 = -\frac{1}{15}\left[\frac{(\mu-1)f_1(\mu)}{3}(b^2-c^2)\,l_2 - g_1(\epsilon)[(\epsilon/2)(b^2+c^2)(mn_1+nm_1)\right.$$

$$\left. - b^2nm_1 - c^2nm_1] - f_1(\mu)\,g_1(\epsilon)\,l_2\left(\epsilon\mu\frac{(b^2-c^2)}{abc} - h_1(\epsilon)(\mu-1)\right)\right]$$

$$(5.1\text{-}9)$$

where

$$f_1(\epsilon) = \left[(\epsilon-1)\,I_a + \frac{2}{abc}\right]^{-1} \tag{5.1-10}$$

$$g_1(\epsilon) = \left[(\epsilon-1)(b^2+c^2)\,I_{bc} + \frac{2}{abc}\right]^{-1} \tag{5.1-11}$$

$$h_1(\epsilon) = b^2 I_b + c^2 I_c - (\epsilon/2)(b^2+c^2)(I_b-I_c) \tag{5.1-12}$$

$$Q = (\epsilon-1)^2\,(I_{ab}I_{bc} + I_{bc}I_{ca} + I_{ca}I_{ab})$$

$$-\frac{4\epsilon(\epsilon-1)}{abc}\left[\frac{(I_{ab}-I_{ac})}{(c^2-b^2)} + \frac{(a^2+b^2+c^2)}{(a^2b^2+b^2c^2+a^2c^2)}\left(I_{ab}-c^2\frac{(I_{ab}-I_{ac})}{(c^2-b^2)}\right)\right]$$

$$(5.1\text{-}13)$$

and $(l; m; n)$, $(l_1; m_1; n_1)$, $(l_2; m_2; n_2)$ are the direction cosines of the incident-field propagation vector, the incident electric field, and the incident magnetic field, respectively, with respect to the $(x, y, z)$ axis. The quantities $I$, $I_a$, $I_b$ are functions of elliptic integrals, and will be defined and discussed in more detail later. The remaining constants $K_2$, $K_3$, $L_2$, etc., of Eq. (5.1-3) are obtained from $K_1$, $L_1$, $M_1$, and $N_1$ by cyclic permutation of $a$, $b$, $c$; $l$, $m$, $n$; $l_1$, $m_1$, $n_1$; $l_2$, $m_2$, and $n_2$.

The quantities $\bar{K}_i$, $\bar{L}_i$, $\bar{M}_i$, and $\bar{N}_i$ are obtained from the expressions for $K_i$, $L_i$, $M_i$, and $N_i$ by making the substitutions

$$(l_1; m_1; n_1) \rightarrow (l_2; m_2; n_2)$$
$$(l_2; m_2; n_2) \rightarrow -(l_1; m_1; n_1) \tag{5.1-14}$$
$$\epsilon \leftrightarrow \mu$$

For a perfectly conducting ellipsoid, all the above expressions are valid provided $\mu$ is set equal to zero and $\epsilon$ allowed to become infinite.

The elliptic functions $I$, $I_a$, etc., can all be expressed in terms of the three quantities $I_a$, $I_b$, $I_c$ by means of the following relations:

$$I = a^2 I_a + b^2 I_b + c^2 I_c \tag{5.1-15}$$

$$I_{ab} = \frac{(I_a - I_b)}{(b^2-a^2)}, \qquad \text{etc.} \tag{5.1-16}$$

It is apparent that the algebra involved in applying the above results to a general ellipsoid would be somewhat tedious. However, knowing the polarization and direction of the incident field, and the direction of observation, along with $I_a$, $I_b$, $I_c$ and the material properties of the ellipsoid, would allow the coefficients of the $k_0{}^2$ and $k_0{}^4$ terms to be computed in the power series expansions of the scattered fields. To illustrate the approach and to obtain some specific results, the first-order or Rayleigh coefficient will be derived specifically.

From Eqs. (5.1-3) and (5.1-4), the first terms of $P$, $\bar{P}$ are

$$P = k_0{}^2(K_1 \sin \theta' \cos \phi' + K_2 \sin \theta' \sin \phi' + K_3 \cos \theta') \qquad (5.1\text{-}17)$$

and

$$\bar{P} = k_0{}^2(\bar{K}_1 \sin \theta' \cos \phi' + \bar{K}_2 \sin \theta' \sin \phi' + \bar{K}_3 \cos \theta'). \qquad (5.1\text{-}18)$$

Thus, the field expressions (5.1-1) and (5.1-2) become

$$E_\theta = H_\phi = k_0{}^2 \frac{e^{ik_0 r}}{r} (K_1 \cos \theta' \cos \phi' + K_2 \cos \theta' \sin \phi'$$
$$- K_3 \sin \theta' - \bar{K}_1 \sin \phi' + \bar{K}_2 \cos \phi') \qquad (5.1\text{-}19)$$

and

$$E_\phi = -H_\theta = k_0{}^2 \frac{e^{ik_0 r}}{r} (-K_1 \sin \phi' + K_2 \cos \phi' - \bar{K}_1 \cos \theta' \cos \phi'$$
$$- \bar{K}_2 \cos \theta' \sin \phi' + \bar{K}_3 \sin \theta') \qquad (5.1\text{-}20)$$

Now with the direction cosines of the incident electric-field vector being $(l_1 ; m_1 ; n_1)$, and of the incident magnetic-field vector being $(l_2 ; m_2 ; n_2)$, then for an ellipsoid of arbitrary material the $K_i$, $\bar{K}_i$ become, from Eq. (5.1-6),

$$K_1 = \left(\frac{2}{3}\right)(\epsilon - 1)\left[(\epsilon - 1)I_a + \frac{2}{abc}\right]^{-1} l_1$$

$$K_2 = \left(\frac{2}{3}\right)(\epsilon - 1)\left[(\epsilon - 1)I_b + \frac{2}{abc}\right]^{-1} m_1 \qquad (5.1\text{-}21)$$

$$K_3 = \left(\frac{2}{3}\right)(\epsilon - 1)\left[(\epsilon - 1)I_c + \frac{2}{abc}\right]^{-1} n_1$$

and

$$\bar{K}_1 = \left(\frac{2}{3}\right)(\mu - 1)\left[(\mu - 1)I_a + \frac{2}{abc}\right]^{-1} l_2$$

$$\bar{K}_2 = \left(\frac{2}{3}\right)(\mu - 1)\left[(\mu - 1)I_b + \frac{2}{abc}\right]^{-1} m_2 \qquad (5.1\text{-}22)$$

$$\bar{K}_3 = \left(\frac{2}{3}\right)(\mu - 1)\left[(\mu - 1)I_c + \frac{2}{abc}\right]^{-1} n_2$$

The parameters, $I_a$, $I_b$, $I_c$, are definite integrals which can be expressed in terms of elliptic integrals of the first and second kinds.[3,4] The integral expressions for these parameters are

$$I_a = \int_0^\infty \frac{dv}{(a^2 + v)[(a^2 + v)(b^2 + v)(c^2 + v)]^{1/2}}$$

$$I_b = \int_0^\infty \frac{dv}{(b^2 + v)[(a^2 + v)(b^2 + v)(c^2 + v)]^{1/2}} \qquad (5.1\text{-}23)$$

$$I_c = \int_0^\infty \frac{dv}{(c^2 + v)[(a^2 + v)(b^2 + v)(c^2 + v)]^{1/2}}$$

In terms of the elliptic integrals of the first and second kinds, for $a \geqslant b \geqslant c$,

$$I_a = \frac{2}{abc} \frac{\cos \tau \cos \gamma}{\sin^3 \gamma \sin^2 \alpha} [F(\gamma\backslash\alpha) - E(\gamma\backslash\alpha)] \qquad (5.1\text{-}24)$$

$$I_b = \frac{2}{abc} \frac{\cos \tau \cos \gamma}{\sin^3 \gamma \sin^2 \alpha \cos^2 \alpha} \left[ E(\gamma\backslash\alpha) - \cos^2 \alpha F(\gamma\backslash\alpha) - \frac{\sin^2 \alpha \sin \gamma \cos \gamma}{\cos \tau} \right]$$
$$(5.1\text{-}25)$$

$$I_c = \frac{2}{abc} \frac{\cos \tau \cos \gamma}{\sin^3 \gamma \cos^2 \alpha} \left[ \frac{\sin \gamma \cos \tau}{\cos \gamma} - E(\gamma\backslash\alpha) \right] \qquad (5.1\text{-}26)$$

where $F$ and $E$ are incomplete elliptic integrals of the first and second kinds, respectively, and $\gamma$ and $\alpha$ are the amplitude and modular angle, respectively. The functions $F$ and $E$ are tabulated in References 3 and 4. In terms of the semi-axes of the ellipsoid, the angles $\alpha$, $\gamma$, $\tau$ are defined as

$$\cos \gamma = c/a, \qquad \cos \tau = b/a, \qquad \text{and} \quad \sin \alpha = \sqrt{\frac{1 - (b/a)^2}{1 - (c/a)^2}} \qquad (5.1\text{-}27)$$

where $0 \leqslant \gamma$, $\tau$, $\alpha \leqslant \pi/2$. Curves of $abcI_a$, $abcI_b$, $abcI_c$, adopted from Reference 5, are given in Figures 5-2 to 5-4.

For perpendicular polarization, $\mathbf{E}^i = E_\phi{}^i \hat{\boldsymbol{\phi}}$, and $\mathbf{H}^i = H_\theta{}^i \hat{\boldsymbol{\theta}}$, so

$$(l_1 ; m_1 ; n_1)_\perp = (-\sin \phi; \cos \phi; 0) \qquad (5.1\text{-}28)$$

and

$$(l_2 ; m_2 ; n_2)_\perp = (\cos \phi \cos \theta; \sin \phi \cos \theta; - \sin \theta) \qquad (5.1\text{-}29)$$

The bistatic scattered fields are

$$E_{\theta_\perp}^s = H_{\phi_\perp}^s = \left(\frac{2}{3}\right) k_0{}^2 \frac{e^{ik_0 r}}{r}$$

$$\left( (\epsilon - 1) \left[ -\frac{\sin \phi \cos \theta' \cos \phi'}{(\epsilon - 1) I_a + 2/abc} + \frac{\cos \phi \cos \theta' \sin \phi'}{(\epsilon - 1) I_b + 2/abc} \right] \right.$$

$$\left. - (\mu - 1) \left[ \frac{\cos \phi \cos \theta \sin \phi'}{(\mu - 1) I_a + 2/abc} - \frac{\sin \phi \cos \theta \cos \phi'}{(\mu - 1) I_b + 2/abc} \right] \right)$$

$$(5.1\text{-}30)$$

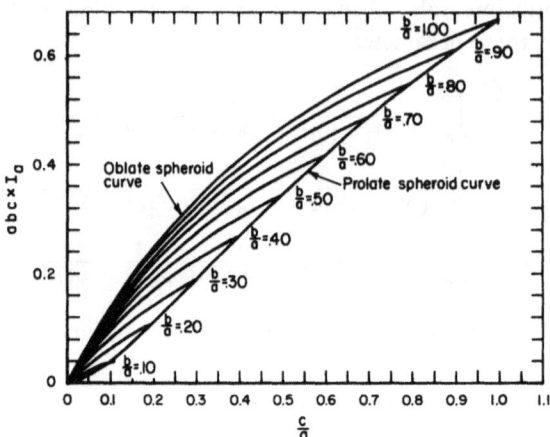

Fig. 5-2.  Ellipsoid coefficient $abcI_a$ versus $b/a$ and $c/a$ [after Osborn[5]].

$$E^s_{\phi\perp} = -H^s_{\theta\perp} = \left(\frac{2}{3}\right) k_0^2 \frac{e^{ik_0r}}{r}$$

$$\times \left( (\epsilon - 1) \left[ \frac{\sin\phi\sin\phi'}{(\epsilon-1)I_a + 2/abc} + \frac{\cos\phi\cos\phi'}{(\epsilon-1)I_b + 2/abc} \right] \right.$$

$$- (\mu - 1) \left[ \frac{\cos\phi\cos\theta\cos\theta'\cos\phi'}{(\mu-1)I_a + 2/abc} \right.$$

$$+ \left. \left. \frac{\sin\phi\cos\theta\cos\theta'\sin\phi'}{(\mu-1)I_b + 2/abc} + \frac{\sin\theta\sin\theta'}{(\mu-1)I_c + 2/abc} \right] \right)$$

$$(5.1\text{-}31)$$

For parallel polarization, $\mathbf{E}^i = E_\theta^i \hat{\theta}$ and $\mathbf{H}^i = -H_\phi^i \hat{\phi}$, so

$$(l_1 \,;\, m_1 \,;\, n_1)_{\parallel} = (\cos\theta\cos\phi \,;\, \cos\theta\sin\phi \,;\, -\sin\theta) \qquad (5.1\text{-}32)$$

and

$$(l_2 \,;\, m_2 \,;\, n_2)_{\parallel} = (\sin\phi \,;\, -\cos\phi \,;\, 0) \qquad (5.1\text{-}33)$$

The bistatic scattered fields are now

$$E^s_{\theta\parallel} = H^s_{\phi\parallel} = \left(\frac{2}{3}\right) k_0^2 \frac{e^{ik_0r}}{r} \left( (\epsilon - 1) \left[ \frac{\cos\theta\cos\phi\cos\theta'\cos\phi'}{(\epsilon-1)I_a + 2/abc} \right. \right.$$

$$+ \left. \frac{\cos\theta\sin\phi\cos\theta'\sin\phi'}{(\epsilon-1)I_b + 2/abc} + \frac{\sin\theta\sin\theta'}{(\epsilon-1)I_c + 2/abc} \right]$$

$$- (\mu - 1) \left[ \frac{\sin\phi\sin\phi'}{(\mu-1)I_a + 2/abc} + \frac{\cos\phi\cos\phi'}{(\mu-1)I_b + 2/abc} \right] \right)$$

$$(5.1\text{-}34)$$

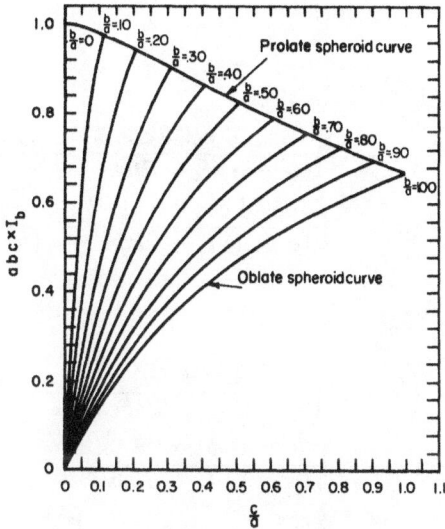

Fig. 5-3.  Ellipsoid coefficient $abcI_b$ versus $b/a$
and $c/a$ [after Osborn[5]].

and

$$E^s_{\phi_\parallel} = -H^s_{\theta_\parallel} = \left(\frac{2}{3}\right) k_0^2 \frac{e^{ik_0 r}}{r}$$

$$\times \left( (\epsilon - 1) \left[ -\frac{\cos\theta \cos\phi \sin\phi'}{(\epsilon - 1) I_a + 2/abc} + \frac{\cos\theta \sin\phi \cos\phi'}{(\epsilon - 1) I_b + 2/abc} \right] \right.$$

$$\left. - (\mu - 1) \left[ \frac{\sin\phi \cos\theta' \cos\phi'}{(\mu - 1) I_a + 2/abc} - \frac{\cos\phi \cos\theta' \sin\phi'}{(\mu - 1) I_b + 2/abc} \right] \right)$$

$$(5.1\text{-}35)$$

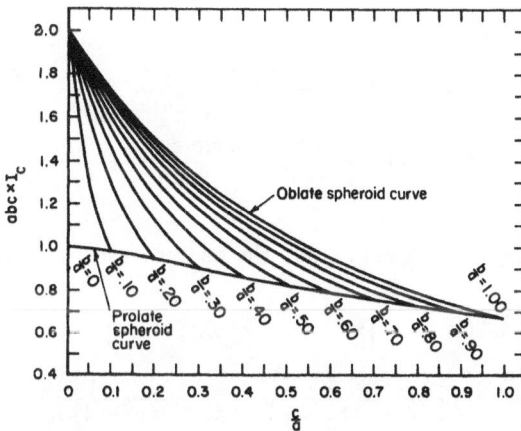

Fig. 5-4.  Ellipsoid coefficient $abcI_c$ versus $b/a$ and $c/a$
[after Osborn[5]].

For backscatter these become

$$E_{\theta_\perp}^s = H_{\phi_\perp}^s = \left(\frac{2}{3}\right) k_0{}^2 \frac{e^{ik_0 r}}{r}$$

$$\times \left( (\epsilon - 1) \left[ \frac{-\sin\phi\cos\theta\cos\phi}{(\epsilon-1)I_a + 2/abc} + \frac{\cos\phi\cos\theta\sin\phi}{(\epsilon-1)I_b + 2/abc} \right] \right.$$

$$\left. - (\mu - 1) \left[ \frac{\cos\phi\cos\theta\sin\phi}{(\mu-1)I_a + 2/abc} - \frac{\sin\phi\cos\theta\cos\phi}{(\mu-1)I_b + 2/abc} \right] \right)$$

$$\tag{5.1-36}$$

$$E_{\phi_\perp}^s = -H_{\theta_\perp}^s = \left(\frac{2}{3}\right) k_0{}^2 \frac{e^{ik_0 r}}{r}$$

$$\times \left( (\epsilon - 1) \left[ \frac{\sin^2\phi}{(\epsilon-1)I_a + 2/abc} + \frac{\cos^2\phi}{(\epsilon-1)I_b + 2/abc} \right] \right.$$

$$- (\mu - 1) \left[ \frac{\cos^2\phi\cos^2\theta}{(\mu-1)I_a + 2/abc} \right.$$

$$\left. \left. + \frac{\sin^2\phi\cos^2\theta}{(\mu-1)I_b + 2/abc} + \frac{\sin^2\theta}{(\mu-1)I_c + 2/abc} \right] \right) \tag{5.1-37}$$

and

$$E_{\theta_\parallel}^s = H_{\phi_\parallel}^s = \left(\frac{2}{3}\right) k_0{}^2 \frac{e^{ik_0 r}}{r} \left( (\epsilon - 1) \left[ \frac{\cos^2\theta\cos^2\phi}{(\epsilon-1)I_a + 2/abc} \right. \right.$$

$$\left. + \frac{\cos^2\theta\sin^2\phi}{(\epsilon-1)I_b + 2/abc} + \frac{\sin^2\theta}{(\epsilon-1)I_c + 2/abc} \right]$$

$$\left. - (\mu - 1) \left[ \frac{\sin^2\phi}{(\mu-1)I_a + 2/abc} + \frac{\cos^2\phi}{(\mu-1)I_b + 2/abc} \right] \right)$$

$$\tag{5.1-38}$$

$$E_{\phi_\parallel}^s = -H_{\theta_\parallel}^s = \left(\frac{2}{3}\right) k_0{}^2 \frac{e^{ik_0 r}}{r}$$

$$\times \left( (\epsilon - 1) \left[ -\frac{\cos\theta\cos\phi\sin\phi}{(\epsilon-1)I_a + 2/abc} + \frac{\cos\theta\cos\phi\sin\phi}{(\epsilon-1)I_b + 2/abc} \right] \right.$$

$$\left. - (\mu - 1) \left[ \frac{\cos\theta\cos\phi\sin\phi}{(\mu-1)I_a + 2/abc} - \frac{\cos\theta\cos\phi\sin\phi}{(\mu-1)I_b + 2/abc} \right] \right)$$

$$\tag{5.1-39}$$

Thus, from Eqs. (5.1-37) and (5.1-38), the backscatter cross sections become

$$\sigma_\perp = \frac{16\pi}{9} k_0{}^4 \left( (\epsilon - 1) \left[ \frac{\sin^2\phi}{(\epsilon-1)I_a + 2/abc} + \frac{\cos^2\phi}{(\epsilon-1)I_b + 2/abc} \right] \right.$$

$$- (\mu - 1) \left[ \frac{\cos^2\phi\cos^2\theta}{(\mu-1)I_a + 2/abc} + \frac{\sin^2\phi\cos^2\theta}{(\mu-1)I_b + 2/abc} \right.$$

$$\left. \left. + \frac{\sin^2\theta}{(\mu-1)I_c + 2/abc} \right] \right)^2 \tag{5.1-40}$$

and

$$\sigma_{\|} = \frac{16\pi}{9} k_0^4 \left( (\epsilon - 1) \left[ \frac{\cos^2 \theta \cos^2 \phi}{(\epsilon - 1) I_a + 2/abc} \right. \right.$$

$$+ \frac{\cos^2 \theta \sin^2 \phi}{(\epsilon - 1) I_b + 2/abc} + \left. \frac{\sin^2 \theta}{(\epsilon - 1) I_c + 2/abc} \right]$$

$$- (\mu - 1) \left[ \frac{\sin^2 \phi}{(\mu - 1) I_a + 2/abc} + \frac{\cos^2 \phi}{(\mu - 1) I_b + 2/abc} \right] \right)^2$$

$$(5.1\text{-}41)$$

For a perfectly conducting ellipsoid $K$, $\bar{K}$ can be obtained from Eqs. (5.1-21) and (5.1-22) by letting $\mu \to 0$ and $\epsilon \to \infty$, giving

$$K_1 = \frac{2}{3I_a} l_1$$

$$K_2 = \frac{2}{3I_b} m_1 \qquad (5.1\text{-}42)$$

$$K_3 = \frac{2}{3I_c} n_1$$

$$\bar{K}_1 = -\frac{2}{3} \frac{l_2}{I_b + I_c}$$

$$\bar{K}_2 = -\frac{2}{3} \frac{m_2}{I_a + I_c} \qquad (5.1\text{-}43)$$

$$\bar{K}_3 = -\frac{2}{3} \frac{n_2}{I_a + I_b}$$

Thus, the bistatic scattered fields from a perfectly conducting ellipsoid are

$$E_{\theta\perp}^s = H_{\phi\perp}^s = \left( \frac{2}{3} \right) k_0^2 \frac{e^{ik_0 r}}{r} \left[ -\frac{\sin \phi \cos \theta' \cos \phi'}{I_a} \right.$$

$$+ \frac{\cos \phi \cos \theta' \sin \phi'}{I_b} + \frac{\cos \phi \cos \theta \sin \phi'}{I_b + I_c}$$

$$\left. - \frac{\sin \phi \cos \theta \cos \phi'}{I_a + I_c} \right] \qquad (5.1\text{-}44)$$

$$E_{\phi\perp}^s = -H_{\theta\perp}^s = \left( \frac{2}{3} \right) k_0^2 \frac{e^{ik_0 r}}{r} \left[ \frac{\sin \phi \sin \phi'}{I_a} + \frac{\cos \phi \cos \phi'}{I_b} \right.$$

$$+ \frac{\cos \phi \cos \theta \cos \theta' \cos \phi'}{I_b + I_c} + \frac{\sin \phi \cos \theta \cos \theta' \sin \theta'}{I_a + I_c}$$

$$\left. + \frac{\sin \theta \sin \theta'}{I_a + I_b} \right] \qquad (5.1\text{-}45)$$

$$E_{\theta_{\|}}^s = H_{\phi_{\|}}^s = \left(\frac{2}{3}\right) k_0{}^2 \frac{e^{ik_0r}}{r} \left[ \frac{\cos\theta\cos\phi\cos\theta'\cos\phi'}{I_a} \right.$$

$$+ \frac{\cos\theta\sin\phi\cos\theta'\sin\phi'}{I_b} + \frac{\sin\theta\sin\theta'}{I_c}$$

$$\left. + \frac{\sin\phi\sin\phi'}{I_b + I_c} + \frac{\cos\phi\cos\phi'}{I_a + I_c} \right] \qquad (5.1\text{-}46)$$

and

$$E_{\phi_{\|}}^s = -H_{\theta_{\|}}^s = \left(\frac{2}{3}\right) k_0{}^2 \frac{e^{ik_0r}}{r} \left[ -\frac{\cos\theta\cos\phi\sin\phi'}{I_a} \right.$$

$$+ \frac{\cos\theta\sin\phi\cos\phi'}{I_b} + \frac{\sin\phi\cos\theta\cos\phi'}{I_b + I_c}$$

$$\left. - \frac{\cos\phi\cos\theta'\sin\phi'}{I_a + I_c} \right] \qquad (5.1\text{-}47)$$

The backscattered fields from a perfectly conducting ellipsoid are

$$E_{\theta_{\perp}}^s = \left(\frac{2}{3}\right) k_0{}^2 \frac{e^{ik_0r}}{r} (\cos\theta\sin\phi\cos\phi)[I_b^{-1} - I_a^{-1} + (I_b + I_c)^{-1} - (I_a + I_c)^{-1}] \qquad (5.1\text{-}48)$$

$$E_{\phi_{\perp}}^s = \left(\frac{2}{3}\right) k_0{}^2 \frac{e^{ik_0r}}{r} \left[ \frac{\sin^2\phi}{I_a} + \frac{\cos^2\phi}{I_b} + \frac{\cos^2\theta\cos^2\phi}{I_b + I_c} \right.$$

$$\left. + \frac{\cos^2\theta\sin^2\phi}{I_a + I_c} + \frac{\sin^2\theta}{I_a + I_b} \right] \qquad (5.1\text{-}49)$$

$$E_{\theta_{\|}}^s = \left(\frac{2}{3}\right) k_0{}^2 \frac{e^{ik_0r}}{r} \left[ \frac{\cos^2\theta\cos^2\phi}{I_a} + \frac{\cos^2\theta\sin^2\phi}{I_b} + \frac{\sin^2\theta}{I_c} \right.$$

$$\left. + \frac{\sin^2\phi}{I_b + I_c} + \frac{\cos^2\phi}{I_a + I_c} \right] \qquad (5.1\text{-}50)$$

and

$$E_{\phi_{\|}}^s = \left(\frac{2}{3}\right) k_0{}^2 \frac{e^{ik_0r}}{r} (\cos\theta\cos\phi\sin\phi)[I_b^{-1} - I_a^{-1} + (I_b + I_c)^{-1} - (I_a + I_c)^{-1}] \qquad (5.1\text{-}51)$$

The backscatter cross sections are

$$\sigma_{\perp} = \frac{16\pi}{9} k_0{}^4 \left[ \frac{\sin^2\phi}{I_a} + \frac{\cos^2\phi}{I_b} + \frac{\cos^2\theta\cos^2\phi}{I_b + I_c} + \frac{\cos^2\theta\sin^2\phi}{I_a + I_c} + \frac{\sin^2\theta}{I_a + I_b} \right]^2 \qquad (5.1\text{-}52)$$

and

$$\sigma_{\prime} = \frac{16\pi}{9} k_0{}^4 \left[ \frac{\cos^2\theta\cos^2\phi}{I_a} + \frac{\cos^2\theta\sin^2\phi}{I_b} + \frac{\sin^2\theta}{I_c} + \frac{\sin^2\phi}{I_b + I_c} + \frac{\cos^2\phi}{I_a + I_c} \right]^2 \qquad (5.1\text{-}53)$$

## 5.1.1.2. *High-Frequency Solutions*

The high-frequency cross section of a perfectly conducting ellipsoid can be approximated by geometrical optics, which for the bistatic case gives

$$\sigma = \frac{4\pi a^2 b^2 c^2}{\left[a^2(\sin\theta\cos\phi + \sin\theta'\cos\phi')^2 + b^2(\sin\theta\sin\phi + \sin\theta'\sin\phi')^2 + c^2(\cos\theta + \cos\theta')^2\right]^2}$$

(5.1-54)

and for backscatter gives

$$\sigma = \frac{\pi a^2 b^2 c^2}{[a^2\sin^2\theta\cos^2\phi + b^2\sin^2\theta\sin^2\phi + c^2\cos^2\theta]^2}$$ (5.1-55)

For the general ellipsoid, the physical optics integrals cannot be evaluated exactly analytically, and a stationary phase evaluation yields the geometrical optics results given above.

For a dielectric ellipsoid no high-frequency results are known to exist, except for the Rayleigh–Gans weak scatterer case.[6]

## 5.1.1.3. *Resonance-Region Solutions*

At the present time there is no known analytical technique for determining the radar cross section of a general, perfectly conducting ellipsoid in the resonance region.

For dielectric ellipsoids, no resonance region results other than for the Rayleigh–Gans weak scatterer case[6] are known to have been obtained.

### 5.1.2. Prolate Spheroids

A considerable amount of research has been undertaken to investigate the scattered fields from prolate spheroidal objects, both in the vector and scalar cases. An excellent and comprehensive summary of the general status of research on scattering and diffraction by prolate spheroids has been given by Sleator.[7] It should be consulted by anyone seriously contemplating work in this area.

Formally exact solutions to the problem of the scattering of a plane electromagnetic wave by a prolate spheroid can be obtained in terms of a pair of Debye potentials or their associated Hertz vectors,[8] or in terms of vector wave functions analogous to Hanson's vector wave functions[9] developed for the sphere problem. Unlike the sphere, however, imposition of the electromagnetic boundary conditions at the surface of a spheroid does not yield specific expressions for the expansion coefficients in terms of spheroidal functions. Instead, the best that can be done is to obtain an infinite number of equations for the infinite set of coefficients. Numerical results can be obtained by truncation, i.e., taking the first $n$ equations of each set and solving for the first $n$ unknown coefficients. In general, a large-scale digital computer is required to accomplish this.

Such a procedure was first applied by Schultz[10] to a plane wave incident along the major axis of a perfectly conducting spheroid. Numerical results based on Schultz's work were obtained by Siegel, et al.,[11] for a 10 : 1 axial ratio prolate spheroid over a $k_0a$ range from 0.1 to approximately 6, where $a$, was the semimajor axis. The results of this numerical computation are shown in Figure 5-5. Generalization of Schultz's solution to the case of a plane wave incident from an arbitrary direction has been carried out by Reitlinger;[12] however, no numerical results are available due to the greatly increased complexity over the axially incident case.

### 5.1.2.1. Low-Frequency Solutions

Stevenson's results for the general ellipsoid can be applied to the prolate spheroid by setting $b = c$. In this event, the coefficients, $I_a$, $I_b$, $I_c$, introduced in Section 5.1.1, simplify considerably, with the elliptic functions degenerating to logarithmic functions.

In the general bistatic case, the first-order, or Rayleigh far-zone scattered fields, are given by Eqs. (5.1-30), (5.1-31), and (5.1-34), (5.1-35), with $b = c$.

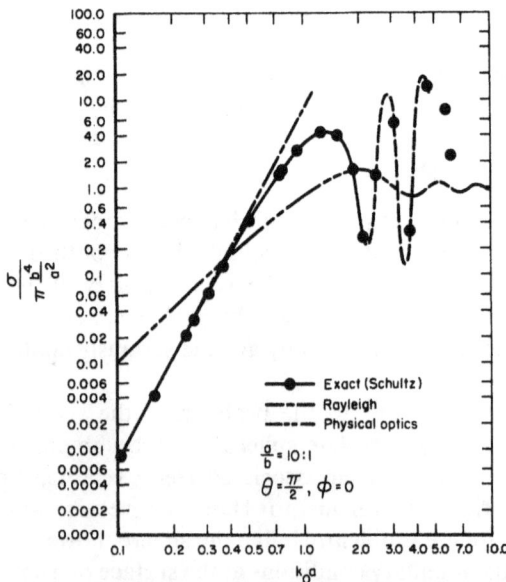

Fig. 5-5. Backscatter cross section of a 10:1 perfectly conducting prolate spheroid for a plane wave incident along the major axis [from Siegel et al.[11]].

Thus,

$$E_{\theta_\perp}^s = H_{\phi_\perp}^s = \left(\frac{2}{3}\right) k_0^2 \frac{e^{ik_0 r}}{r}$$

$$\times \left( (\epsilon - 1) \left[ \frac{\cos\phi \cos\theta' \sin\phi'}{(\epsilon - 1) I_b + 2/ab^2} - \frac{\sin\phi \cos\theta' \cos\phi'}{(\epsilon - 1) I_a + 2/ab^2} \right] \right.$$

$$\left. - (\mu - 1) \left[ \frac{\cos\phi \cos\theta \sin\phi'}{(\mu - 1) I_a + 2/ab^2} - \frac{\sin\phi \cos\theta \cos\phi'}{(\mu - 1) I_b + 2/ab^2} \right] \right)$$

$$(5.1\text{-}56)$$

$$E_{\phi_\perp}^s = -H_{\theta_\perp}^s = \left(\frac{2}{3}\right) k_0^2 \frac{e^{ik_0 r}}{r}$$

$$\times \left( (\epsilon - 1) \left[ \frac{\sin\phi \sin\phi'}{(\epsilon - 1) I_a + 2/ab^2} + \frac{\cos\phi \cos\phi'}{(\epsilon - 1) I_b + 2/ab^2} \right] \right.$$

$$- (\mu - 1) \left[ \frac{\cos\phi \cos\theta \cos\theta' \cos\phi'}{(\mu - 1) I_a + 2/ab^2} \right.$$

$$\left. \left. + \frac{\sin\phi \cos\theta \cos\theta' \sin\phi'}{(\mu - 1) I_b + 2/ab^2} + \frac{\sin\theta \sin\theta'}{(\mu - 1) I_b + 2/ab^2} \right] \right)$$

$$(5.1\text{-}57)$$

and

$$E_{\theta_\parallel}^s = \left(\frac{2}{3}\right) k_0^2 \frac{e^{ik_0 r}}{r} \left( (\epsilon - 1) \left[ \frac{\cos\theta \cos\phi \cos\theta' \cos\phi'}{(\epsilon - 1) I_a + 2/ab^2} \right. \right.$$

$$\left. + \frac{\cos\theta \sin\phi \cos\theta' \sin\phi'}{(\epsilon - 1) I_b + 2/ab^2} + \frac{\sin\theta \sin\theta'}{(\epsilon - 1) I_b + 2/ab^2} \right]$$

$$\left. - (\mu - 1) \left[ \frac{\sin\phi \sin\phi'}{(\mu - 1) I_a + 2/ab^2} + \frac{\cos\phi \cos\phi'}{(\mu - 1) I_b + 2/ab^2} \right] \right)$$

$$(5.1\text{-}58)$$

$$E_{\phi_\parallel}^s = \left(\frac{2}{3}\right) k_0^2 \frac{e^{ik_0 r}}{r}$$

$$\times \left( (\epsilon - 1) \left[ \frac{\cos\theta \sin\phi \cos\phi'}{(\epsilon - 1) I_b + 2/ab^2} - \frac{\cos\theta \cos\phi \sin\phi'}{(\epsilon - 1) I_a + 2/ab^2} \right] \right.$$

$$\left. - (\mu - 1) \left[ \frac{\sin\phi \cos\theta' \cos\phi'}{(\mu - 1) I_a + 2/ab^2} - \frac{\cos\phi \cos\theta' \sin\phi'}{(\mu - 1) I_b + 2/ab^2} \right] \right)$$

$$(5.1\text{-}59)$$

For backscatter, setting $\theta = (\pi/2)$ for simplicity, these become

$$E_{\theta_\perp}^s = 0, \qquad E_{\phi_\parallel}^\theta = 0$$

$$E_{\phi_\perp}^s = \left(\frac{2}{3}\right) k_0^2 \frac{e^{ik_0 r}}{r} \left( (\epsilon - 1) \left[ \frac{\sin^2\phi}{(\epsilon - 1) I_a + 2/ab^2} + \frac{\cos^2\phi}{(\epsilon - 1) I_b + 2/ab^2} \right] \right.$$

$$\left. - (\mu - 1) \left[ \frac{1}{(\mu - 1) I_b + 2/ab^2} \right] \right)$$

$$(5.1\text{-}60)$$

and

$$E_{\theta_{\parallel}}^{s} = \left(\frac{2}{3}\right) k_0^2 \frac{e^{ik_0 r}}{r} \left(\frac{(\epsilon - 1)}{(\epsilon - 1) I_b + 2/ab^2}\right.$$
$$\left. - (\mu - 1) \left[\frac{\sin^2 \phi}{(\mu - 1) I_a + 2/ab^2} + \frac{\cos^2 \phi}{(\mu - 1) I_b + 2/ab^2}\right]\right) \quad (5.1\text{-}61)$$

In the case of a perfectly conducting spheroid, setting $\mu = 0$ and $\epsilon \to \infty$ in the above expressions gives the scattered fields. Then the low-frequency radar cross sections for a perfectly conducting prolate spheroid are

$$\sigma_{\perp} = \frac{16\pi}{9} k_0^4 \left[\frac{1}{I_a + I_b} + \frac{\cos^2 \phi}{I_b} + \frac{\sin^2 \phi}{I_a}\right]^2 \quad (5.1\text{-}62)$$

and

$$\sigma_{\parallel} = \frac{16\pi}{9} k_0^4 \left[\frac{1}{I_b} + \frac{\cos^2 \phi}{I_a + I_b} + \frac{\sin^2 \phi}{2 I_b}\right]^2 \quad (5.1\text{-}63)$$

In arriving at the above equations, the following identity was used

$$I_a + I_b + I_c = 2/abc \quad (5.1\text{-}64)$$

For the prolate spheroid, the parameters $I_a$, $I_b$ become

$$I_a = \left(\frac{2}{ab^2}\right)\left(\frac{1}{e^2 - 1}\right)\left[\frac{e}{2(e^2 - 1)^{1/2}} \ln\left(\frac{e + (e^2 - 1)^{1/2}}{e - (e^2 - 1)^{1/2}}\right) - 1\right] \quad (5.1\text{-}65)$$

and

$$I_b = \left(\frac{2}{ab^2}\right)\left(\frac{e}{2(e^2 - 1)}\right)\left[e - \frac{1}{2(e^2 - 1)^{1/2}} \ln\left(\frac{e + (e^2 - 1)^{1/2}}{e - (e^2 - 1)^{1/2}}\right)\right], \quad (5.1\text{-}66)$$

where $e = (a/b)$. Curves of $I_a$ and $I_b$ are given in Figures 5-2 and 5-3, labeled "prolate spheroid."

In Figures 5-6 to 5-9, curves of the axial radar cross section versus $k_0 a$ are given for several $(a/b)$ ratios. These curves, taken from Reference 13, compare the exact solution computed using Schultz's technique with the Rayleigh results [Eqs. (5.1-62) and (5.1-63)], and with Stevenson's third-order solutions.

### 5.1.2.2. High-Frequency Solutions

5.1.2.2.1. *Perfectly Conducting Spheroids.* The geometrical-optics result for the cross section of a prolate spheroid can be obtained from Eq. (5.1-54) for the ellipsoid by setting $b = c$. Thus the bistatic optical cross section is

$$\sigma = \frac{4\pi a^2 b^4}{\left[[a^2(\sin\theta\cos\phi + \sin\theta'\cos\phi')^2 + b^2(\sin\theta\sin\phi + \sin\theta'\sin\phi')^2 + b^2(\cos\theta + \cos\theta')^2]^2\right]} \quad (5.1\text{-}67)$$

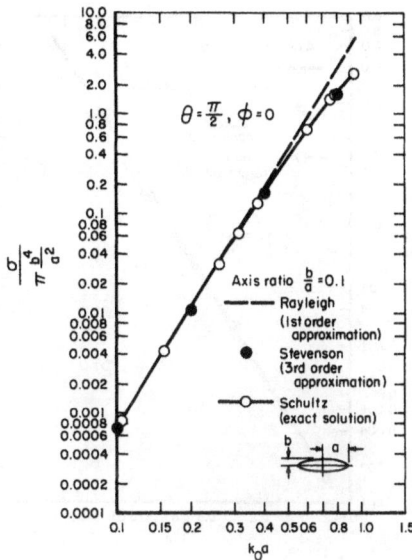

Fig. 5-6. Low-frequency backscatter cross section of a perfectly conducting prolate spheroid for a plane wave incident along the major axis and $b/a = 0.1$ [from Mathur and Mueller[13]].

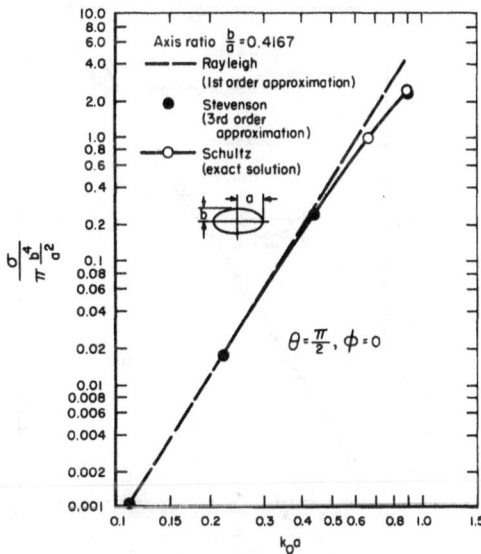

Fig. 5-7. Low-frequency backscatter cross section of a perfectly conducting prolate spheroid for a plane wave incident along the major axis and $b/a = 0.4167$ [from Mathur and Mueller[13]].

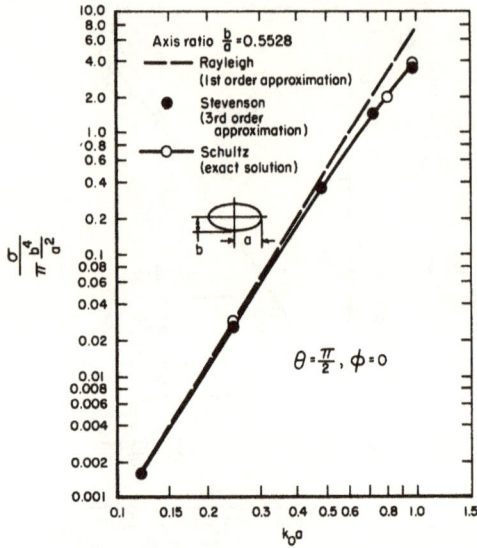

Fig. 5-8. Low-frequency backscatter cross section of a perfectly conducting prolate spheroid for a plane wave incident along the major axis and $b/a = 0.5528$ [from Mathur and Mueller[13]].

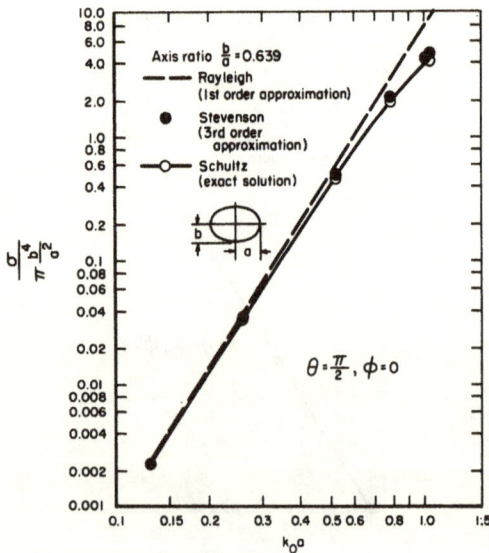

Fig. 5-9. Low-frequency backscatter cross section of a perfectly conducting prolate spheroid for a plane wave incident along the major axis and $b/a = 0.639$ [from Mathur and Mueller[13]].

and the backscatter cross section is, setting $\theta = (\pi/2)$,

$$\sigma = \frac{\pi a^2 b^4}{[a^2 \cos^2 \phi + b^2 \sin^2 \phi]^2} \qquad (5.1\text{-}68)$$

This is plotted versus $b/a$ in Figure 5-10 for several angles of incidence.
For incidence along the major axis, the physical-optics backscatter cross section can be evaluated analytically giving

$$\sigma_{\text{p.o.}}(0, 0) = \frac{\pi b^4}{a^2} \left[ 1 - \frac{\sin(2k_0 a)}{k_0 a} + \frac{\sin^2(2k_0 a)}{k_0^2 a^2} \right] \qquad (5.1\text{-}69)$$

while for incidence along a minor axis, the physical-optics integral can also be evaluated analytically giving

$$\sigma_{\text{p.o.}}(0, \pi/2) = \pi a^2 \left[ 1 - \frac{\sin(2k_0 b)}{k_0 b} + \frac{\sin^2(2k_0 b)}{k_0^2 b^2} \right] \qquad (5.1\text{-}70)$$

Levy and Keller[14] have obtained the backscattered field for a perfectly conducting prolate spheroid with a plane wave incident along the major axis,

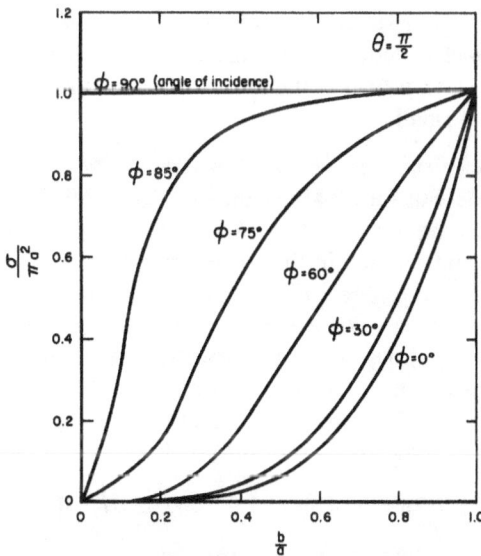

Fig. 5-10. Geometrical-optics cross section of a perfectly conducting prolate spheroid versus $b/a$ for several angles of incidence.

using the geometrical theory of diffraction. The backscatter cross section predicted by Levy and Keller is

$$\sigma = \frac{\pi b^4}{a^2} \left| \left[ 1 - \frac{\pi^2 k_0^{1/3} a^{5/3}}{6^{1/3} b^{4/3}} \exp(i\pi/3 + i2k_0 a) \right. \right.$$

$$\left. \left. \left( \frac{\exp(ik_0^{1/3}\tau_1 Q)}{3A'^2} - \frac{\exp(ik_0^{1/3}\tau_2 Q)}{q_2 A^2} \right) \right] \right|^2 \qquad (5.1\text{-}71)$$

where

$$Q = \frac{(ab)^{2/3}}{(a^2 - b^2)^{1/2}} \int_{\pi/2}^{3\pi/2} \left( \frac{a^2}{a^2 - b^2} - \cos^2 n \right)^{-1/2} dn \qquad (5.1\text{-}72)$$

and

$$\tau_i = 6^{-1/3} e^{i\pi/3} q_i \qquad (5.1\text{-}73)$$

Levy and Keller give the numerical values of the constants, $q_i$, $A$, and $A'$, as

$$q_1 = 3.3721$$

$$q_2 = 1.4693$$

$$A = 1.1668$$

$$A' = -1.0590$$

Apparently, no numerical results have as yet been obtained from the above expression, and its validity is not known. From the general assumptions made in the geometrical theory of diffraction, however, it is felt that the results should be valid provided the radius of curvature at any point on the spheroid is not too small.

5.1.2.2.2. *Dielectric Spheroids.* Low-frequency cross sections for dielectric spheroids can be obtained directly from Stevenson's results, the first terms of which are given by Eqs. (5.1-56) to (5.1-59) in the bistatic case, and Eqs. (5.1-60) and (5.1-61) in the backscatter case.

The only known, general, high-frequency or resonance-region result available for a dielectric prolate spheroid is that obtained by Thomas,[15] using a geometrical optics technique appropriate to homogeneous, non-absorptive dielectric scatterers (see Section 3.3). Thomas computed the axially incident backscatter cross section for a dielectric prolate spheroid with $\mu_r = 1$, $\epsilon_r = 1.8$, and an axial ratio $(a/b)$ of 1.35, and obtained

$$\sigma = \pi b^2 \left| R(0) \right|^2 \left| 0.74 + [5.77 - 5.73(b/\lambda_0)^{1/2}] e^{i7.24(k_0 b)} \right|^2 \qquad (5.1\text{-}74)$$

where $R(0)$ is the normal incidence, planar interface, reflection coefficient given by Eqs. (7.1-9) and (7.1-10) of Section 7.1. A comparison of some measured cross sections with the above expression is given in Figure 5-11.[15]

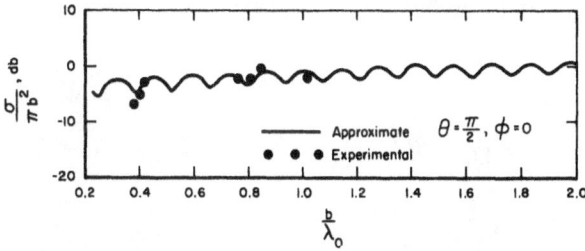

Fig. 5-11. Resonance-region backscatter cross section of a dielectric prolate spheroid with $\mu_r = 1$, $\epsilon_r = 1.8$, and $a/b = 1.35$ for incidence along the major axis [from Thomas[15]].

### 5.1.2.3. Resonance-Region Solutions

Excellent approximations to the backscatter cross section of a perfectly conducting prolate spheroid, for a plane wave incident from an arbitrary direction in the $\theta = \pi/2$ plane, have been obtained by Moffatt[16] using a modification of Kennaugh's impulse-response technique (which is discussed briefly in Section 2.2), designated by Moffatt as the "ramp-response approach."

This ramp-response technique gives, for the backscatter cross section,

$$\sigma = \pi a^2 [E_R{}^2 + E_I{}^2], \tag{5.1-75}$$

where

$$
\begin{aligned}
E_R = \Big( & A[\cos(\omega\tau_0 T_1) + \omega\tau_0 T_1 \sin(\omega\tau_0 T_1) - 1] \\
& - \frac{B}{2}\Big[2T_1 \cos(\omega\tau_0 T_1) + \Big(\omega\tau_0 T_1{}^2 - \frac{2}{\omega\tau_0}\Big)\sin(\omega\tau_0 T_1)\Big] \\
& - \frac{C(\omega\tau_0)^2}{\beta^2 - (\omega\tau_0)^2}[\beta\cos(\omega\tau_0 T_2) + \omega\tau_0 \sin(\omega\tau_0 T_1)] \\
& + \frac{\beta C(\omega\tau_0)^2}{[\alpha^2 + \beta^2 - (\omega\tau_0)^2]^2 + [2\alpha\omega\tau_0]^2}[(\alpha^2 + \beta^2 - \omega^2\tau_0{}^2)\cos(\omega\tau_0 T_2) \\
& - 2\alpha\omega\tau_0 \sin(\omega\tau_0 T_2)]\Big)
\end{aligned}
\tag{5.1-76}
$$

and

$$
\begin{aligned}
E_I = \Big( & A[2k_0 b T_1 \cos(2k_0 b T_1) - \sin(2k_0 b T_1)] \\
& + \frac{B}{2}\Big[2T_1 \sin(2k_0 b T_1) - \frac{1}{k_0 b} - \cos(2k_0 b T_1) \\
& \Big(2k_0 b T_1{}^2 - \frac{1}{k_0 b}\Big)\Big] - \frac{(2k_0 b)^2}{(\beta^2 - 4k_0{}^2 b^2)}[2k_0 b \cos(2k_0 b T_1) \\
& - \beta \sin(2k_0 b T_2)] - \frac{\beta C(2k_0 b)^2}{(\alpha^2 + \beta^2 - 4k_0{}^2 b^2)^2 + (4\alpha k_0 b)^2} \\
& \times [4\alpha k_0 b \cos(2k_0 b T_2) + (\alpha^2 + \beta^2 - 4k_0{}^2 b^2)\sin(2k_0 b T_2)]\Big)
\end{aligned}
\tag{5.1-77}
$$

The parameters are defined as follows:

$$A = \frac{ab}{b^2 \cos^2 \phi + a^2 \sin^2 \phi}, \qquad B = \frac{ab}{2[b^2 \cos^2 \phi + a^2 \sin^2 \phi]^{3/2}} \frac{\omega}{k_0}$$

$$C = -AT_1 + B(T_1^2/2), \qquad T_1 = A/B, \qquad \tau_0 = 2k_0 b/\omega$$

$$\alpha = \left[ \frac{\beta C}{T_1 C/2 - BT_1^3/12 - K/2 + C/\beta} - \beta^2 \right]^{1/2}$$

$$\beta = \frac{\pi/2}{T_2 - A/B}, \qquad K_{\perp, \parallel} = \left[ \frac{\sigma_{\perp, \parallel}}{\pi k_0^2 b^4} \right]^{1/2} \qquad (5.1\text{-}78)$$

where $\sigma_{\perp, \parallel}$ is given by Eqs. (5.1-62) and (5.1-63). For perpendicular polarization,

$$T_2 = \frac{-\sin^2 \phi \cos^2 \phi (b^2 - a^2)^2}{(b^2 \cos^2 \phi + a^2 \sin^2 \phi)^{1/2} (b^2 \sin^2 \phi + a^2 \cos^2 \phi)}$$

$$+ (b^2 \cos^2 \phi + a^2 \sin^2 \phi)^{1/2} + \frac{\pi}{2} \left( \frac{C_1^2 + C_2^2}{2} \right)^{1/2} \qquad (5.1\text{-}79)$$

while for parallel polarization,

$$T_2 = \frac{\sin \phi \cos \phi (b^2 - a^2)}{2(a^2 \cos^2 \phi + b^2 \sin^2 \phi)^{1/2}}$$

$$+ \frac{1}{2} (b^2 \cos^2 \phi + a^2 \sin^2 \phi)^{1/2} + \frac{\pi}{2} \left( \frac{a^2 + b^2}{2} \right)^{1/2}$$

$$+ \frac{1}{2} \left| \frac{-\sin \phi \cos \phi (b^2 - a^2)}{(a^2 \cos^2 \phi + b^2 \sin^2 \phi)^{1/2}} - (b^2 \cos^2 \phi + a^2 \sin^2 \phi)^{1/2} \right|$$

$$\qquad (5.1\text{-}80)$$

where

$$C_1^2 = a^2 \left( 1 - \frac{\sin^2 \phi \cos^2 \phi (b^2 - a^2)^2}{a^2 b^2} + \left[ \frac{\sin^2 \phi \cos^2 \phi (b^2 - a^2)(a^2 - b^2)}{a^2 b^2 (b^2 \cos^2 \phi + a^2 \sin^2 \phi)^{1/2}} \right]^2 \right.$$

$$\left. \times \left[ \frac{b^2 \sin^2 \phi + a^2 \cos^2 \phi}{a^2 b^2} \right]^{-1} \right) \qquad (5.1\text{-}81)$$

and

$$C_2^2 = \frac{a^2 b^2 - \sin^2 \phi \cos^2 \phi (b^2 - a^2)^2}{b^2 \sin^2 \phi + a^2 \cos^2 \phi}$$

$$\times \left[ \frac{\sin^2 \phi \cos^2 \phi (b^2 - a^2)^2}{(b^2 \cos^2 \phi + a^2 \sin^2 \phi)^{1/2} (b^2 \sin^2 \phi + a^2 \cos^2 \phi)} \right]^2 \qquad (5.1\text{-}82)$$

In Figures 5-12 to 5-15 are given curves of $\sigma$ versus $k_0 b$, as computed by Eq. (5.1-75), as measured for both polarizations, and at angles of incidence of 0, 30, 60, and 90°.[16] The agreement between computed and measured points

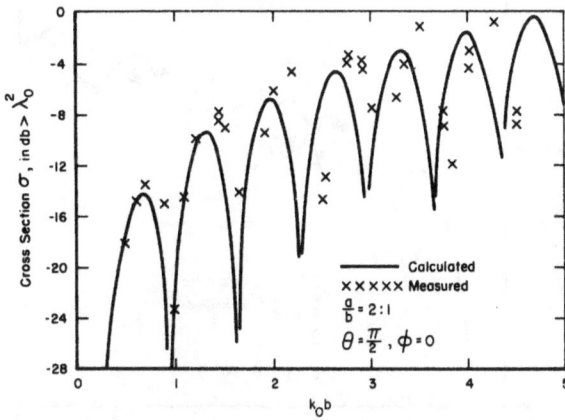

Fig. 5-12. Comparison of the experimental backscatter cross section for a perfectly conducting 2:1 prolate spheroid with theoretical calculations using the ramp-response technique for $\varphi = 0$ [from Moffatt[16]].

is quite good, in general, although the positions of the maxima and minima are shifted somewhat for the computed curves, and the agreement becomes poorer at higher frequencies. In obtaining an approximation to the ramp response for the spheroid, Moffatt made some rather rough approximations. It is expected that a more sophisticated analysis would yield still better

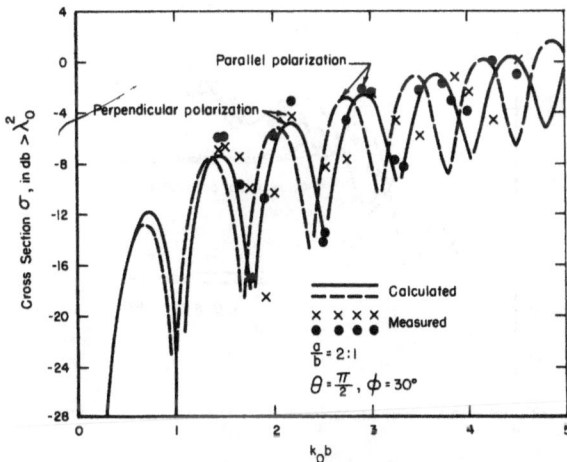

Fig. 5-13. Comparison of the experimental backscatter cross section for a perfectly conducting 2:1 prolate spheroid with theoretical calculations using the ramp-response technique for $\varphi = 30°$ [from Moffatt[16]].

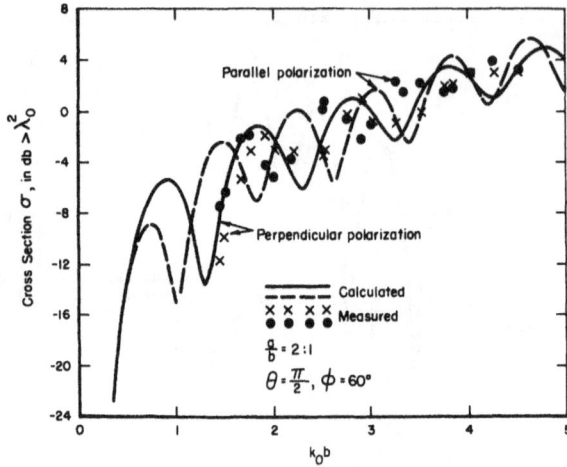

Fig. 5-14. Comparison of the experimental backscatter
cross section for a perfectly conducting 2:1 prolate spheroid
with theoretical calculations using the ramp-response technique for $\varphi = 60°$ [from Moffatt[16]].

agreement with experiment. Nevertheless, in view of the fact that no other
analytical technique has, as yet, yielded results of any kind for the oblique-
incidence backscatter from a perfectly conducting spheroid in the resonance
region, Moffatt's work is outstanding.

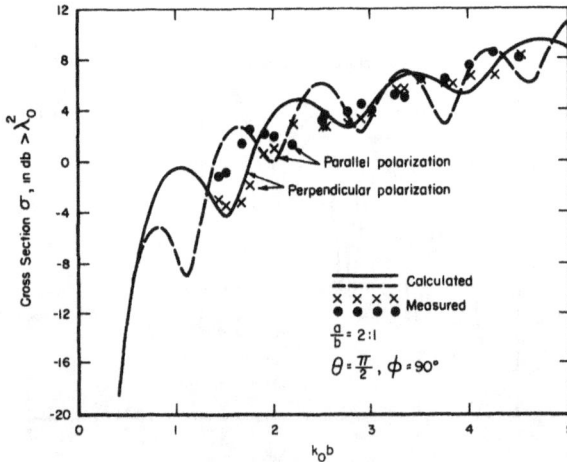

Fig. 5-15. Comparison of the experimental backscatter
cross section for a perfectly conducting 2:1 prolate spheroid
with theoretical calculations using the ramp-response technique for $\varphi = 90°$ [from Moffatt[16]].

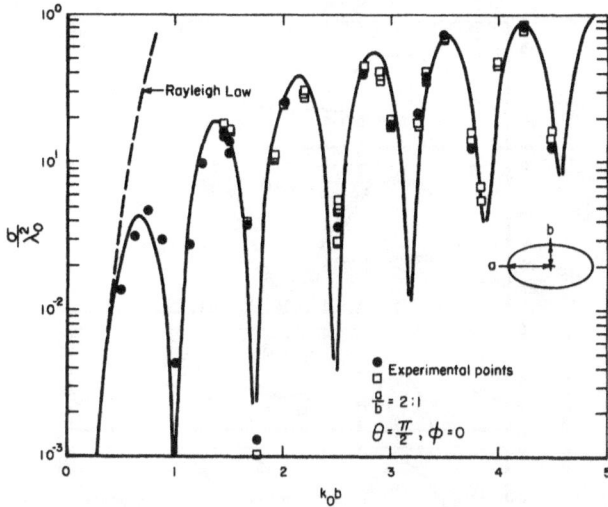

Fig. 5-16. Comparison of the experimental backscatter cross section for a perfectly conducting 2:1 prolate spheroid with theoretical calculations using the impulse-response technique for $\varphi = 0$ [from Moffatt and Kennaugh[17]].

An example of the results obtainable with a more sophisticated application of the impulse response technique is given in Figure 5-16. Here the backscatter cross section for a plane wave incident along the major axis of a conducting prolate spheroid, as computed by Kennaugh and Moffatt,[17] using the impulse response technique, is compared with experiment.

In addition, in Figures 5-17 and 5-18, measured cross sections[16] for 2 : 1 axial ratio prolate spheroids, as a function of the angle of incidence, are shown for a number of spheroids of different sizes.

### 5.1.2.4. *Other Approximations*

A number of other approximate solutions have been obtained for special conditions on the spheroid shape or material. For example, for spheroids with small eccentricity, an approximate solution can be obtained by treating the spheroid as a shape perturbation of a sphere. This process was carried out by Mushiake[18] for an arbitrarily polarized plane wave incident on a conducting spheroid of small eccentricity in an arbitrary direction.

For the opposite case of large eccentricity, traveling wave theory can be used for a particular polarization and range of aspect angles to predict the backscatter cross section of conducting spheroids.

In the case of weak scatters, i.e., dielectric spheroids with $\epsilon_r$ or $\mu_r \approx 1$, various perturbation techniques have been employed to obtain approximate results.[19,20]

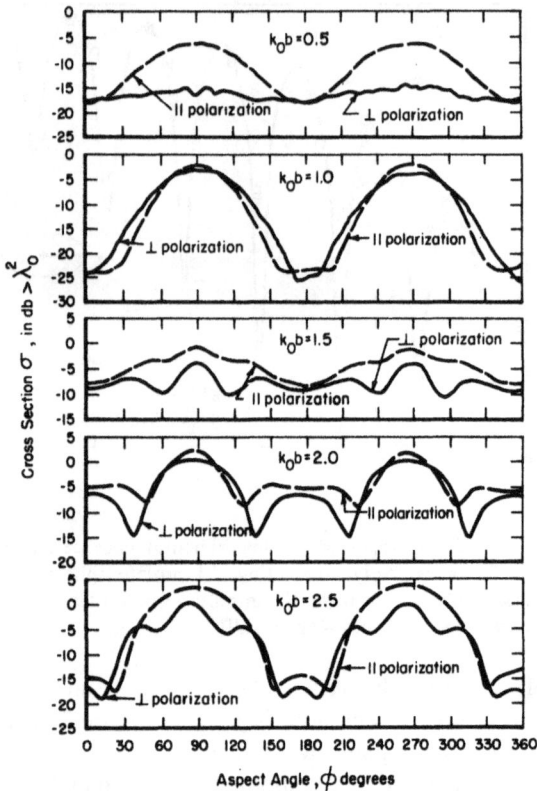

Fig. 5-17. Experimental angular backscatter cross-section patterns for a 2:1 perfectly conducting prolate spheroid and $k_0b = 0.5$, 1.0, 1.5, 2.0, and 2.5 [after Moffatt[16]].

### 5.1.3. Oblate Spheroids

Formally exact solutions for the scattered electromagnetic fields from a perfectly conducting oblate spheroid can be obtained in terms of a pair of Debye potentials, or a set of vector wave functions in a fashion analogous to the prolate spheroid problem. Again, as in the prolate spheroid case, the imposition of the electromagnetic boundary conditions at the spheroid surface results in a doubly infinite set of equations for the unknown expansion coefficients, which can be truncated and presumably solved by numerical methods using a large digital computer. Rauch[21] has formulated a solution for the electromagnetic fields, scattered from a perfectly conducting oblate spheroid illuminated by an axially incident plane wave, in terms of vector wave functions. No numerical results are known to have been computed using Rauch's solution.

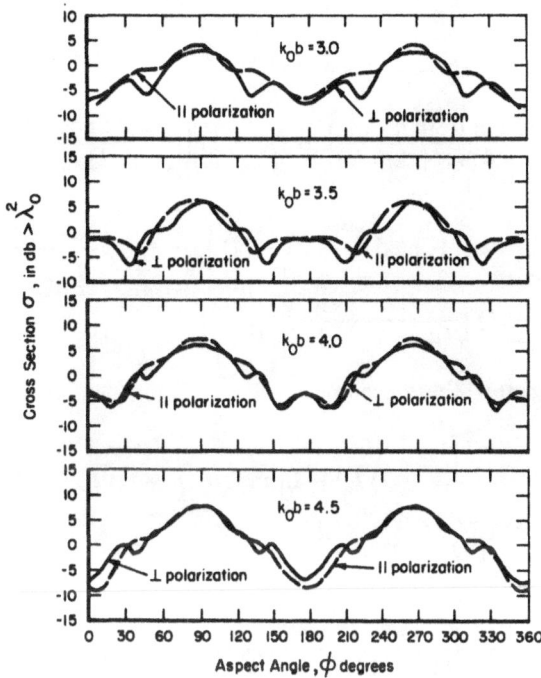

Fig. 5-18. Experimental angular backscatter cross-section patterns for a 2:1 perfectly conducting prolate spheroid and $k_0 b = 3.0$, 3.5, 4.0, and 4.5 [after Moffatt[16]].

### 5.1.3.1. *Low-Frequency Solutions*

Stevenson's general ellipsoid results can be applied to the oblate spheroid by setting $a = b > c$. In the general bistatic case, setting $a = b$ and $\phi = 0$ for simplicity gives, from Eqs. (5.1-30), (5.1-31) and (5.1-34), (5.1-35), the Rayleigh scattered fields:

$$E_{\theta_\perp}^s = H_{\phi_\perp}^s = \left(\frac{2}{3}\right) k_0^2 \frac{e^{ik_0 r}}{r} \left( (\epsilon - 1) \left[ \frac{\cos \theta' \sin \phi'}{(\epsilon - 1) I_a + 2/(a^2 c)} \right] \right.$$

$$\left. - (\mu - 1) \left[ \frac{\cos \theta \sin \phi'}{(\mu - 1) I_a + 2/(a^2 c)} \right] \right) \qquad (5.1\text{-}83)$$

$$E_{\phi_\perp}^s = - H_{\theta_\perp}^s = \left(\frac{2}{3}\right) k_0^2 \frac{e^{ik_0 r}}{r} \left( (\epsilon - 1) \left[ \frac{\cos \phi'}{(\epsilon - 1) I_a + 2/(a^2 c)} \right] - (\mu - 1) \right.$$

$$\times \left[ \frac{\cos \theta \cos \theta' \cos \phi'}{(\mu - 1) I_a + 2/(a^2 c)} \right] - (\mu - 1) \left[ \frac{\sin \theta \sin \theta'}{(\mu - 1) I_c + 2/(a^2 c)} \right] \right)$$

$$(5.1\text{-}84)$$

and

$$E_{\theta_\parallel}^s = H_{\phi_\parallel}^s = \left(\frac{2}{3}\right) k_0^2 \frac{e^{ik_0 r}}{r} \left( (\epsilon - 1) \left[ \frac{\cos\theta \cos\theta' \cos\phi'}{(\epsilon - 1) I_a + 2/(a^2 c)} \right] + (\epsilon - 1) \right.$$
$$\left. \left[ \frac{\sin\theta \sin\theta'}{(\epsilon - 1) I_c + 2/(a^2 c)} \right] - (\mu - 1) \left[ \frac{\cos\phi'}{(\mu - 1) I_a + 2/(a^2 c)} \right] \right)$$

(5.1-85)

$$E_{\phi_\parallel}^s = - H_{\theta_\parallel}^s = - \left(\frac{2}{3}\right) k_0^2 \frac{e^{ik_0 r}}{r} \left( (\epsilon - 1) \left[ \frac{\cos\theta \sin\phi'}{(\epsilon - 1) I_a + 2/(a^2 c)} \right] \right.$$
$$\left. - (\mu - 1) \left[ \frac{\cos\theta' \sin\phi'}{(\mu - 1) I_a + 2/(a^2 c)} \right] \right)$$

(5.1-86)

For backscatter, these become

$$E_{\phi_\perp}^s = \left(\frac{2}{3}\right) k_0^2 \frac{e^{ik_0 r}}{r} \left[ \frac{(\epsilon - 1)}{(\epsilon - 1) I_a + 2/(a^2 c)} - \frac{(\mu - 1) \cos^2\theta}{(\mu - 1) I_a + 2/(a^2 c)} \right.$$
$$\left. - \frac{(\mu - 1) \sin^2\theta}{(\mu - 1) I_c + 2/(a^2 c)} \right]$$

(5.1-87)

and

$$E_{\theta_\parallel}^s = \left(\frac{2}{3}\right) k_0^2 \frac{e^{ik_0 r}}{r} \left[ \frac{(\epsilon - 1) \cos^2\theta}{(\epsilon - 1) I_a + 2/(a^2 c)} + \frac{(\epsilon - 1) \sin^2\theta}{(\epsilon - 1) I_c + 2/(a^2 c)} \right.$$
$$\left. - \frac{(\mu - 1)}{(\mu - 1) I_a + 2/(a^2 c)} \right]$$

(5.1-88)

For a perfectly conducting spheroid, setting $\mu = 0$ and $\epsilon \to \infty$, in the above expressions, gives the scattered fields. Thus, the radar cross sections of a perfectly conducting oblate spheroid at low frequencies are

$$\sigma_\perp = \frac{16\pi}{9} k_0^4 \left[ \frac{1 + (1/2)\sin^2\theta}{I_a} + \frac{\cos^2\theta}{I_a + I_c} \right]^2$$

(5.1-89)

$$\sigma_\parallel = \frac{16\pi}{9} k_0^4 \left[ \frac{\cos^2\theta}{I_a} + \frac{\sin^2\theta}{I_c} + \frac{1}{I_a + I_c} \right]^2$$

(5.1-90)

In Figure 5-19 (from Reference 13), curves of the backscatter cross section for a plane wave, axially incident on a perfectly conducting oblate spheroid, are given for several axial ratios. Both the Rayleigh results from Eqs. (5.1-89) and (5.1-90) and Stevenson's full third-order results are shown for comparison.

For an oblate spheroid, $I_a$ and $I_c$ are given by

$$I_a = I_b = \left(\frac{2}{a^2 c}\right) \frac{1}{2(e^2 - 1)} \left( e^2 (e^2 - 1)^{-1/2} \sin^{-1} \left[ \frac{(e^2 - 1)^{1/2}}{e} \right] - 1 \right)$$

(5.1-91)

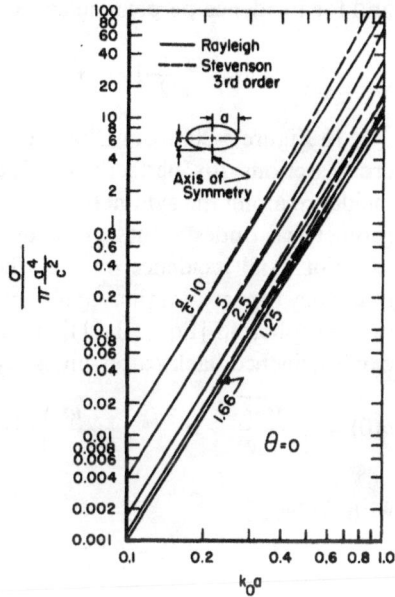

Fig. 5-19. Low-frequency backscatter cross section of a perfectly conducting oblate spheroid for a plane wave incident along the axis of symmetry and $a/c = 10, 5, 2.5, 1.66,$ and 1.25 [from Mathur and Mueller[13]].

and

$$I_c = \left(\frac{2}{a^2 c}\right) \frac{e^2}{(e^2 - 1)} \left(1 - (e^2 - 1)^{-1/2} \sin^{-1}\left[\frac{(e^2 - 1)^{1/2}}{e}\right]\right) \qquad (5.1\text{-}92)$$

with $e = a/c$. Curves of $I_a$, $I_c$ are given in Figures 5-2, and 5-4, labeled "oblate spheroid."

### 5.1.3.2. High-Frequency Solutions

The geometrical-optics cross section for a perfectly conducting oblate spheroid is given by Eq. (5.1-54), with $a = b$ and $\phi = 0$; thus for the bistatic case,

$$\sigma = \frac{4\pi a^4 c^2}{[a^2(\sin\theta + \sin\theta'\cos\phi')^2 + a^2(\sin\theta'\sin\phi')^2 + c^2(\cos\theta + \cos\theta')^2]^2} \qquad (5.1\text{-}93)$$

and for backscatter,

$$\sigma = \frac{\pi a^4 c^2}{[a^2\sin^2\theta + c^2\cos^2\theta]^2} \qquad (5.1\text{-}94)$$

As in the case of the prolate spheroid, the physical-optics integral can be evaluated analytically for two directions of incidence, along the symmetry axis and perpendicular to it. Thus, the axial-incidence physical-optics backscatter cross section is

$$\sigma_{\text{p.o.}}(0) = \frac{\pi a^4}{c^2}\left[1 - \frac{\sin 2k_0 c}{k_0 c} + \frac{\sin 2k_0 c}{(k_0 c)^2}\right] \qquad (5.1\text{-}95)$$

and for incidence perpendicular to the axis,

$$\sigma_{p.o.}\left(\frac{\pi}{2}\right) = \pi c^2 \left[1 - \frac{\sin\left(2k_0 a\right)}{k_0 a} + \frac{\sin^2\left(2k_0 a\right)}{(k_0 a)^2}\right] \tag{5.1-96}$$

In Figure 5-20, from Olte and Silver,[22] are shown measured backscatter cross sections for perfectly conducting oblate spheroids for the cases of incidence along the symmetry axis and along the major axis. In addition, the geometrical optics values given by Eq. (5.1-94) are shown for comparison.

For axial incidence on a perfectly conducting oblate spheroid, a high-frequency approximation analogous to Levy and Keller's result for the prolate spheroid [Eq. (5.1-71)] can be obtained using Fock theory. Thus the axial-incidence backscatter cross section is

$$\sigma(0) = \left| \frac{\sqrt{\pi}\, a^2}{c}\, e^{-2ik_0 c} + \frac{ia\sqrt{\pi}}{2}\left(\frac{k_0 c^2}{2a}\right)^{1/3} [\hat{q}(\xi) - \hat{p}(\xi)]\, e^{i2k_0 aE(\epsilon^2)} \right|^2 \tag{5.1-97}$$

with

$$\xi = 2\left(\frac{k_0 c^2}{2a}\right)^{1/3} K(\epsilon^2), \qquad \epsilon^2 = \frac{a^2 - c^2}{a^2}$$

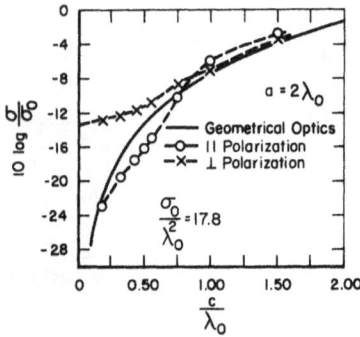

(a) Incidence Along Major Axis

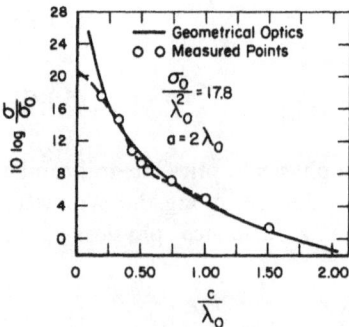

(b) Incidence Along Minor Axis

Fig. 5-20. Experimental backscatter cross sections for perfectly conducting oblate spheroids and incidence along the major and minor axes [from Olte and Silver[22]].

In these equations, $E$ and $K$ are complete elliptic integrals of the first and second kinds, respectively.[4]

No results are available for dielectric oblate spheroids.

### 5.1.3.3. Resonance-Region Solutions

No results are known of in the resonance region for oblate spheroids, although the impulse approximation technique could be applied to conducting oblate, as well as prolate spheroids. In addition, the numerical techniques developed for general perfectly conducting bodies of revolution can, of course, be applied to oblate spheroids.

## 5.2. OGIVES

Considering an ogive to be a body of revolution formed by rotating a convex arc around a chord or symmetry axis, there are two types of ogives commonly encountered in the literature. One is formed by rotating a circular arc around its chord, and the other by rotating a parabolic arc around its chord. There are no exact analytical solutions available for either of these bodies and, in fact, only a few approximate results are available.

In the low-frequency region, Siegel[23] has obtained an approximate expression for the backscatter cross section, at axial incidence, of a perfectly conducting ogive as[†]

$$\sigma = (4/\pi) k_0^4 V^2 [1 + e^{-\tau}/\pi\tau]^2 \tag{5.2-1}$$

For the parabolic ogive, $\tau$ and $V$ are given by

$$\tau = \tfrac{4}{5}(a/b) \tag{5.2-2}$$

and

$$V = \tfrac{16}{15}\pi a b^2 \tag{5.2-3}$$

while for the circular ogive,

$$\tau = \left(\tfrac{3}{2}\right) \frac{\sin\alpha - \alpha\cos\alpha - (1/3)\sin^3\alpha}{(1 - \cos\alpha)^3} \tag{5.2-4}$$

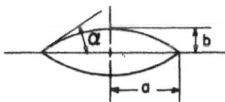

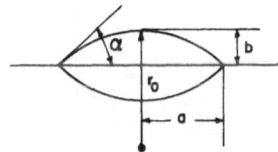

Fig. 5-21. Parabolic ogive dimensions.         Fig. 5-22. Circular ogive dimensions.

[†] See Eqs. (2.2-292) to (2.2-294) and the associated discussion in Section 2.2.2.7.

and

$$V = 2\pi r_0^2 [\sin \alpha - \alpha \cos \alpha - \tfrac{1}{3}\sin^3 \alpha]. \qquad (5.2\text{-}5)$$

The ogive dimensions, $a$, $b$, $\alpha$, and $r_0$, are illustrated in Figures 5.21 and 5-22.

At high frequencies, stationary-phase evaluation of the physical-optics integral gives, for the circular ogive,

$$\sigma(\theta) = \frac{\lambda_0^2 \tan^4 \alpha}{16\pi \cos^6 \theta \, (1 - \tan^2 \alpha \, \tan^2 \theta)^3}, \qquad 0 \leqslant \theta < (90° - \alpha) \qquad (5.2\text{-}6)$$

$$\sigma(90 - \alpha) = \frac{b^2}{4\pi \tan^2 (\alpha/2)}, \qquad \theta = 90° - \alpha \qquad (5.2\text{-}7)$$

$$\sigma(\theta) = \pi r_0^2 \left[ 1 - \frac{r_0 - b}{r_0 \sin \theta} \right], \qquad (90° - \alpha) < \theta \leqslant 90° \qquad (5.2\text{-}8)$$

For axial incidence, the physical-optics integral for the circular ogive can be integrated exactly giving

$$\sigma(0) = \frac{\lambda_0^2 \tan^4 \alpha}{16\pi} \left[ 1 + \frac{2 \cos^2 \alpha \cos 2k_0 a}{1 + \cos \alpha} + \frac{\cos^4 \alpha}{(1 + \cos \alpha)^2} \right] \qquad (5.2\text{-}9)$$

For long, thin bodies of revolution, Adachi[24] has derived a modified physical-optics formulation which predicts the axial-incidence backscatter cross section quite well (Adachi's technique is discussed in more detail in

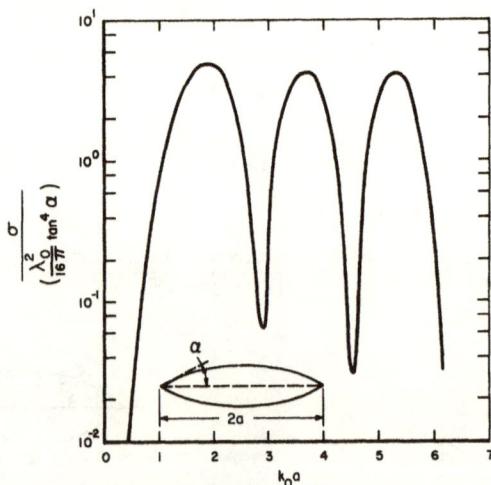

Fig. 5-23. Theoretical axial-incidence backscatter cross section for a perfectly conducting parabolic ogive, Adachi's method.

Chapter 8). For a long, thin parabolic ogive, Adachi gives, as the axial cross section,

$$\sigma(0) = \frac{\lambda_0^2 \tan^4 \alpha}{16\pi} \left| (1 - e^{i4k_0a}) - \frac{3}{2k_0a}(1 + e^{i4k_0a}) - \frac{3}{4(k_0a)^2}(1 - e^{i4k_0a}) \right|^2$$

$$(5.2\text{-}10)$$

In Figure 5-23, this is plotted versus $k_0a$, with $\sigma(0)$ normalized with respect to the geometrical-optics value:

$$\frac{\lambda_0^2 \tan^4 \alpha}{16\pi} \qquad (5.2\text{-}11)$$

Adachi's expression for the parabolic ogive, Eq. (5.2-10), reduces in the high-frequency limit to

$$\sigma(0) = \frac{\lambda_0^2 \tan^4 \alpha}{8\pi}(1 + \cos 4k_0a) \qquad (5.2\text{-}12)$$

and in the low-frequency limit to

$$\sigma(0) = \frac{4}{\pi}k_0^4 V^2 \qquad (5.2\text{-}13)$$

the Rayleigh result for long, thin bodies. Thus, one might expect it to be valid in the resonance region, as well as at high and low frequencies. To the extent that the body can be considered thin, this is apparently true, as illustrated by Figure 5-24. In this figure, the experimental nose-on radar cross section of a 20° circular ogive is compared with the theoretical cross section obtained by numerically integrating Adachi's expression. It is apparent that good agreement is obtained, with the largest deviation from the experimental results occurring at the highest frequencies.

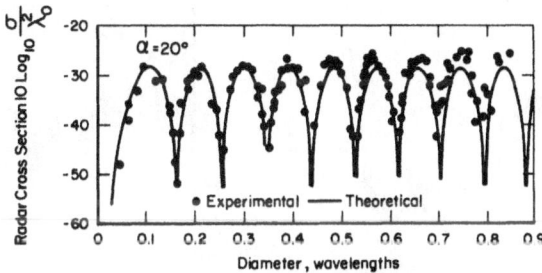

Fig. 5-24. Comparison of theoretical results using Adachi's method with experimental measurements for the axial-incidence backscatter cross section of a perfectly conducting 20° circular ogive.

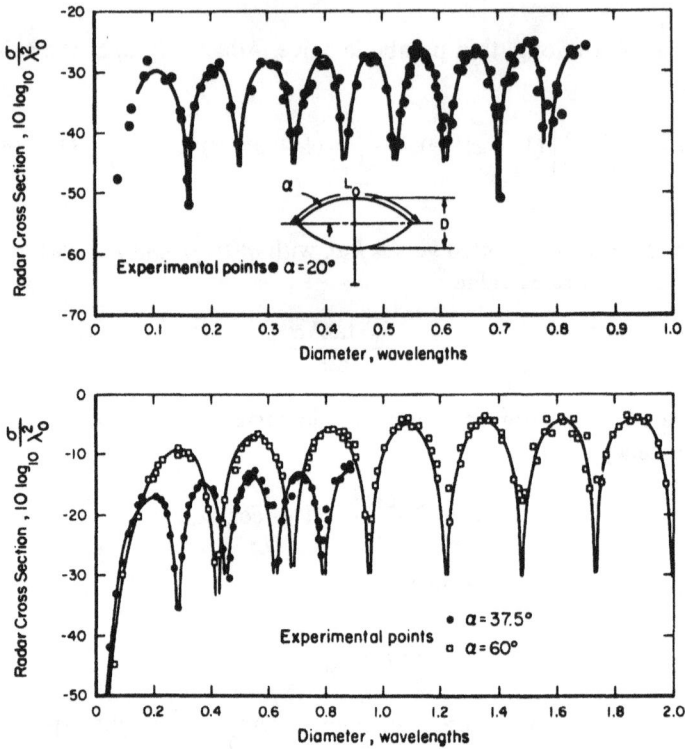

Fig. 5-25. Measured axial-incidence backscatter cross section for perfectly conducting circular ogives with nose angles of 20, 37.5, and 60° [from Blore and Royer[26]].

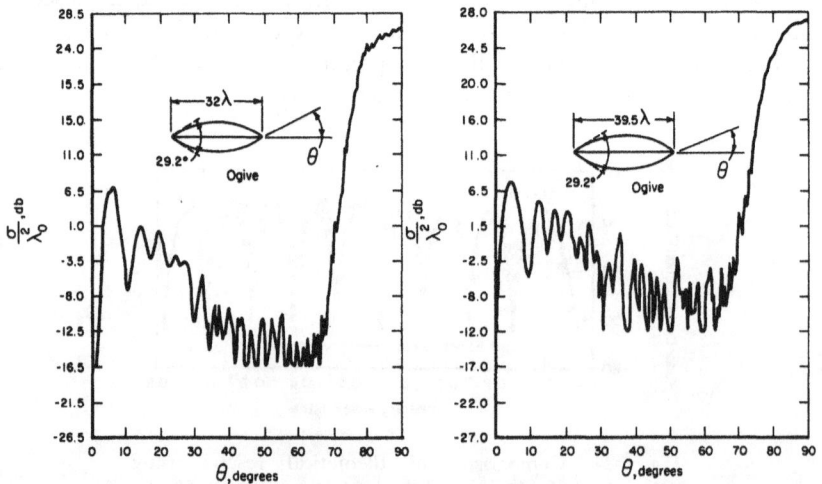

Fig. 5-26. Backscatter cross section for parallel polarization of perfectly conducting circular ogives versus the angle of incidence, with $\alpha = 14.6°$ and $2a = 32\lambda_0$, and $39.5\lambda_0$.

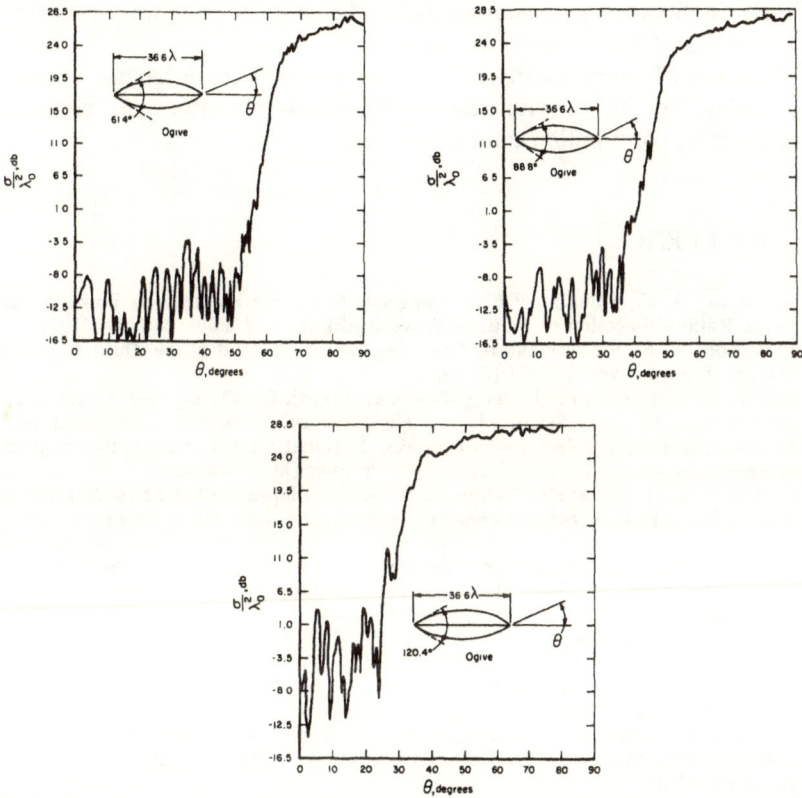

Fig. 5-27. Backscatter cross section for parallel polarization of perfectly conducting circular ogives versus the angle of incidence, with $2a = 36.6\lambda_0$ and $\alpha = 30.7, 44.4,$ and $60.2°$.

For long thin ogives and parallel polarization, Peter's[25] traveling wave theory, discussed in Section 2.3.3, can be used to predict the cross section for incidence other than nose-on.

In Figure 5-25a, b are presented the results of an extensive set of measurements by Blore and Royer[26] of the nose-on cross section of 20, 37.5, and 60° circular ogives. From the results of these measurements, Blore[27] has obtained an empirical expression for the nose-on radar cross section of a circular ogive as follows:

$$\sigma = \frac{\pi \lambda_0^2 (2\alpha)^{4.3}}{2 \cdot 10^{10}} \left\{ 1 + \sin \left[ 2\pi \left( \frac{4k_0 a \alpha}{\pi \sin 2\alpha} - 1.25 \right) \right] \right\} \qquad (5.2-14)$$

In Figures 5-26, and 5-27 are given measured curves of the backscatter cross section for parallel polarization versus the angle of incidence for several perfectly conducting circular ogives. Figure 5-26 compares two ogives of different lengths and the same nose angle. The large traveling wave contribution at angles of 5 to 8° is clearly apparent.

Figure 5-27 compares three ogives of the same length having different nose angles. For ogives with nose angles of this magnitude, no clear-cut traveling wave contribution can be observed.

## 5.3. REFERENCES

1. Stevenson, A. F., Solution of Electromagnetic Scattering Problems as Power Series in the Ratio (Dimension of Scatterer/Wavelength), *J. Appl. Phys.* **24**:1134 (1953).
2. Stevenson, A. F., Electromagnetic Scattering by an Ellipsoid in the Third Approximation, *J. Appl. Phys.* **24**:1143 (1953).
3. Jahnke, E., and Emde, F., *Tables of Functions*, Fourth Ed. Dover, New York (1945).
4. Abramowitz, M., and Stegun, I. A., *Handbook of Mathematical Functions with Formulas, Graphs, and Mathematical Tables*, National Bureau of Standards Applied Mathematics Series—55, U.S. Government Printing Office (1965).
5. Osborn, J. A., Demagnetizing Factors of the General Ellipsoid, *Phys. Rev.* **67**:351 (1945).
6. Van De Hulst, H. C., *Light Scattering by Small Particles*, John Wiley, New York (1957), p. 93.
7. Sleator, F. B., Studies in Radar Cross Sections XLIX—Diffraction and Scattering by Regular Bodies III: The Prolate Spheroid, The University of Michigan–Radiation Laboratory, Report No. 3648-6-T, February, 1964.
8. Van Bladel, J., *Electromagnetic Fields*, McGraw-Hill, New York (1964), p. 210.
9. Stratton, J. A., *Electromagnetic Theory*, McGraw-Hill, New York (1941), p. 393.
10. Schultz, F. V., Scattering by a Prolate Spheroid, The University of Michigan, Report No. UMM-42 (March, 1950).
11. Siegel, K. M., Schultz, F. V., Gere, B. H., and Sleator, F. B., The Theoretical and Numerical Determination of the Radar Cross Section of a Prolate Spheroid, *Trans. IRE* **AP-4**:266 (1956).
12. Reitlinger, N., Scattering of a Plane Wave Incident on a Prolate Spheroid at an Arbitrary Angle, The University of Michigan–The Radiation Laboratory, Unpublished Memo 2686-506-M, 1957.
13. Mathur, P. N., and Mueller, E. A., Radar Back-Scattering from Non-Spherical Scatterers, Part I, Cross-Sections of Conducting Prolates and Spheroidal Functions, Part II, Cross-Sections from Non-Sphrerical Raindrops, Report of Investigation No. 28, State Water Survey Division, State of Illinois, AD 76451, 1955.
14. Levy, B. R., and Keller, J. B., Diffraction by a Spheroid, *Canad. J. Phys.* **38**:128 (1960).
15. Thomas, D. T., Scattering by Plasma and Dielectric Bodies, The Antenna Laboratory –The Ohio State University, Report No. 1116-20, AD 286854, (August 1962).
16. Moffatt, D. L., Electromagnetic Scattering by a Perfectly Conducting Prolate Spheroid, The Antenna Laboratory–The Ohio State University, Report No. 1774-11 (September, 1965).
17. Moffatt, D. L., and Kennaugh, E. M., The Axial Echo Area of a Perfectly Conducting Prolate Spheroid, *IEEE Trans. Antennas and Propagation* **AP-13**:401 (1965).
18. Mushiake, Y., Backscattering for Arbitrary Angles of Incidence of a Plane Electromagnetic Wave on a Perfectly Conducting Spheroid with Small Eccentricity, *J. Appl. Phys.* **27**:1459 (1956).
19. Montroll, W., and Hart, R. W., Scattering of Plane Waves by Soft Obstacles II Scattering by Cylinders, Spheroids, and Disks, *J. Appl. Phys.* **22**:1278 (1951).
20. Ikeda, Y., Extension of the Rayleigh–Gans Theory, IN Kerker, M. (Ed.), *Electromagnetic Scattering*, Pergamon Press, New York (1963), 47.
21. Rauch, L. M., Studies in Radar Cross-Sections—IX, Willow Run Research Center–The University of Michigan, UMM-116 (October 1953).

22. Olte, A., and Silver S., New Results in Backscattering from Cones and Spheroids, *Trans. IRE* **AP-7** (Special Supplement), S61 (1959).
23. Siegel, K. M., Far Field Scattering from Bodies of Revolution, *Appl. Sci. Res.* **7**:293 (1959).
24. Adachi, S., The Nose-on Echo Area of Axially Symmetric Thin Bodies Having Sharp Apices, The Antenna Laboratory–The Ohio State University, Report No. 925-1 (March 1960), AD 240 651.
25. Peters, L., Memorandum on the Echo Area of Ogives, The Antenna Laboratory–The Ohio State University, Report No. 601-7 (January 1956).
26. Blore, W. E., and Royer, G. M., The Radar Cross-Section of Bodies of Revolution, Defense Research Telecommunications Establishment (Canada), Report No. 1105 (March 1963), AD 406 112.
27. Blore, W. E., The Radar Cross Section of Ogives, Double-Backed Cones, Double-Rounded Cones, and Cone Spheres, *IEEE Trans. Antennas and Propagation* **AP-12**:582 (1964).

*Chapter 6*

# CONES, RINGS, AND WEDGES

W. D. Stuart


## 6.1. INTRODUCTION

This chapter is concerned with radar scattering, primarily monostatic, from some simple, singular shapes. A singular shape is defined as one which has at least one point where the radius of curvature of the body approaches zero with respect to a wavelength. Some of the theoretical aspects of singular points and edges are considered in Chapter 2. In his excellent book, Jones[1] devotes an entire chapter to the theoretical aspects of this problem for perfectly conducting objects.

The first topic to be considered is scattering from a semi-infinite cone. It is the simplest such shape, being singular at only one point. Within certain limitations, solutions can be obtained for the semi-infinite cone and related to other more practical shapes such as the finite cone and the ogive. The contribution to the backscattered field due to the tip is approximately the same for all three of the shapes for the same angle at the tip. The ogive is discussed in Chapter 5.

The second shape considered is the finite cone, a shape of considerable practical importance and theoretical difficulty. Its significance is such that much of the chapter is devoted to this shape, and the discussion on the finite cone terminated with a portion of a sphere, is continued in Chapter 8. The finite cone may have singularities both at the tip and at the termination at the base of the cone. (Blunting the cone tip into a frustum is also briefly considered.) The effects of the base termination may sometimes be considered in terms of scattering from an equivalent loop, which is the next topic discussed.

An infinite line singularity such as the tip of a wedge is also discussed in this chapter. Some theoreticians have linked their approaches to the problem of scattering from a finite flat-based cone with the problem of scattering from a wedge. The wedge is also discussed in Chapter 2, as is the half-plane, a special case of the wedge.

## 6.2.  SEMI-INFINITE CONES

### 6.2.1.  Perfectly Conducting Cones

#### 6.2.1.1. *Exact Solution for Axial Incidence*

The problem of scattering from an elliptical semi-infinite cone with a plane wave incident axially has been solved by Kraus and Levine.[2,3] The circular cone is a special case of the elliptical cone; however, its solution is not readily obtainable from the elliptic cone solution. The solution of Kraus and Levine is in terms of series of products of Lamé functions. The Lamé differential equation has periodic and nonperiodic solutions. The doubly periodic Lamé functions have been the subject of numerous investigations. However, the simply periodic and nonperiodic Lamé functions have not been studied extensively. Because of the complexity of the Kraus and Levine solution, and because it is in terms of functions for which information is not readily available, it is not included here; in all the discussions which follow, the cones are considered to have circular cross sections.

No solution exists for the general case of a plane wave arbitrarily incident on a perfectly conducting semi-infinite cone, although solutions to the scalar problems are available.[4] The radar cross section of a semi-infinite perfectly conducting circular cone for plane-wave incidence along the axis of symmetry, that is, $\mathbf{E}^i = \hat{\mathbf{y}} \exp(-ik_0 z)$, has been found by Hansen and Schiff.[5] This configuration is illustrated in Figure 6-1. Because of the special symmetry of the cone for this type of illumination, there is no polarization dependence. A detailed discussion of the derivation of this solution is available in Mentzer's treatise.[6]

The bistatic scattered field is expressed in the form of an infinite series of Legendre functions of noninteger order as follows:

$$E_\theta{}^s = - \frac{\cos \phi' \exp(+ik_0 r)}{ik_0 r} \sum_{n=0}^{\infty} \left[ \frac{\nu_n(\nu_n + 1) \exp(-i\nu_n \pi)}{2 \int_{\cos\theta'}^{1} [P_{\nu_n}^1(\cos \theta')]^2 \, d(\cos \theta')} \right.$$

$$\left. - \frac{\mu_n(\mu_n + 1) \exp(-i\mu_n \pi)}{2 \int_{\cos\theta'}^{1} [P_{\mu_n}^1(\cos \theta')]^2 \, d(\cos \theta')} \right] + O\left(\frac{1}{r^2}\right)$$

$$(6.2\text{-}1)$$

$$E_\phi{}^s = + \frac{\sin \phi' \exp(+ik_0 r)}{ik_0 r} \sum_{n=0}^{\infty} \left[ \frac{\mu_n(\mu_n + 1) \exp(-i\mu_n \pi)}{2 \int_{\cos\theta'}^{1} [P_{\mu_n}^1(\cos \theta')]^2 \, d(\cos \theta')} \right.$$

$$\left. - \frac{\nu_n(\nu_n + 1) \exp(-i\nu_n \pi)}{2 \int_{\cos\theta'}^{1} [P_{\nu_n}^1(\cos \theta')]^2 \, d(\cos \theta')} \right] + O\left(\frac{1}{r^2}\right) \qquad (6.2\text{-}2)$$

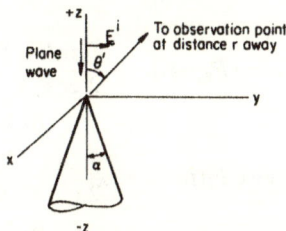

Fig. 6-1.   Cone coordinate system.

The backscattered radar cross section is found by substituting the scattered field from Eq. (6.2-1) with $\theta' = 0$, $\phi' = \pi/2$ into the expression

$$\sigma(0) = 4\pi \lim_{r \to \infty} r^2 \mid E_\theta{}^s / E_\theta{}^i \mid^2$$

thus giving

$$\sigma(0) = \frac{4\pi}{k_0{}^2} \left| \sum_{n=0}^{\infty} \left[ + \frac{\nu_n(\nu_n + 1) \exp(-i\nu_n\pi)}{2 \int_{\cos\theta'}^1 [P_{\nu_n}^1(\cos\theta')]^2 \, d(\cos\theta')} \right. \right.$$
$$\left. \left. - \frac{\mu_n(\mu_n + 1) \exp(-i\mu_n\pi)}{2 \int_{\cos\theta'}^1 [P_{\mu_n}^1(\cos\theta')]^2 \, d(\cos\theta')} \right] \right|^2 \qquad (6.2\text{-}3)$$

The integrals in the denominator of Eqs. (6.2-1), (6.2-2), and (6.2-3) can be evaluated as

$$\int_1^{\cos\theta'} [P_{\nu_n}^1(\cos\theta')]^2 \, d(\cos\theta') = \frac{1 - \cos^2\theta'}{2\nu_n + 1} \left[ \frac{dP_{\nu_n}^1(\cos\theta')}{d(\cos\theta')} \right] \left[ \frac{dP_{\nu_n}^1(\cos\theta')}{d\nu_n} \right]$$
$$(6.2\text{-}4)$$

$$\int_1^{\cos\theta'} [P_{\mu_n}^1(\cos\theta')]^2 \, d(\cos\theta') = - \frac{1 - \cos^2\theta'}{2\mu_n + 1} P_{\mu_n}^1(\cos\theta') \frac{\partial^2 P_{\mu_n}^1(\cos\theta')}{\partial(\cos\theta') \, \partial\mu_n}$$
$$(6.2\text{-}5)$$

where $\nu_n$ and $\mu_n$ are the $n$th zeroes of $dP_\mu{}^1/d\theta' \mid_{\theta' = \pi - \alpha}$ and $P_\nu{}^1(\cos\theta') \mid_{\theta' = \pi - \alpha}$. Approximate values for $\nu$ and $\mu$ have been reported by Siegel et al.[7] and recently recalculated to higher accuracy by Schultz et al.[8] The evaluation of the series in Eq. (6.2-1) is difficult because difficulties arise in the convergence of the series for certain regions of cone angle $2\alpha$ and observation angle $\theta'$.

The difficulties with the convergence of the series can be avoided with an integral representation of the field. Goryainov[9] has obtained bistatic far-field results for $0 \leqslant \theta' < \pi - 2\alpha$ of the form

$$E_\theta{}^s = (e^{ik_0 r}/ik_0 r) \cos \phi' L_1(\theta', \alpha) \qquad (6.2\text{-}6)$$

and

$$E_\phi{}^s = (-e^{ik_0 r}/ik_0 r) \sin \phi' L_2(\theta', \alpha) \qquad (6.2\text{-}7)$$

where

$$L_1 = \frac{d}{d\theta'} \int_0^\infty \nu \sinh \pi\nu \, P_1(i\nu) \, d\nu - \frac{1}{\sin \theta'} \int_0^\infty \nu \sinh \pi\nu \, P_2(i\nu) \, d\nu + \frac{\tan^2(\alpha/2)}{2 \cos^2(\theta'/2)}$$

$$(6.2\text{-}8)$$

$$L_2 = \frac{1}{\sin \theta'} \int_0^\infty \nu \sinh \pi\nu \, P_1(i\nu) \, d\nu - \frac{d}{d\theta'} \int_0^\infty \nu \sinh \pi\nu \, P_2(i\nu) \, d\nu + \frac{\tan^2(\alpha/2)}{2 \cos^2(\theta'/2)}$$

$$(6.2\text{-}9)$$

and

$$P_1(i\nu) = \frac{[P_{i\nu-1/2}^1(\cos \alpha)][P_{i\nu-1/2}^1(\cos \theta')]}{[P_{i\nu-1/2}^1(\cos \alpha)](\nu^2 + 1/4) \cosh \nu\pi} \qquad (6.2\text{-}10)$$

$$P_2(i\nu) = \frac{[(\partial/\partial\alpha) \, P_{i\nu-1/2}^1(\cos \alpha)][P_{i\nu-1/2}^1(\cos \theta')]}{[(\partial/\partial\alpha) \, P_{i\nu-1/2}^1(-\cos \alpha)](\nu^2 + 1/4) \cosh \nu\pi} \qquad (6.2\text{-}11)$$

The integration path in the integral for $L_1$, $L_2$ is from zero to infinity along the positive imaginary axis in the $\nu$-plane. In Figures 6-2, and 6-3, curves of $L_1$ and $L_2$, as computed by Goryainov, are given.

Schensted,[10] Siegel et al.,[11] and Felsen[12] have considered Eqs. (6.2-1) and (6.2-3) using special summation techniques. By means of their techniques, approximate expressions for the backscattered field for small-angle and

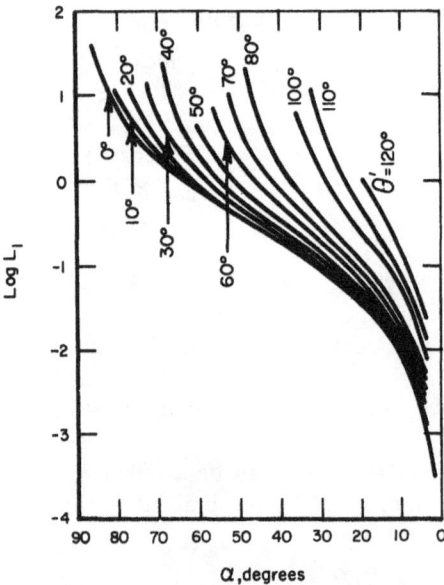

Fig. 6-2. Evaluation of Goryainov's $L_1(\theta', \alpha)$ function [after Goryainov[9]].

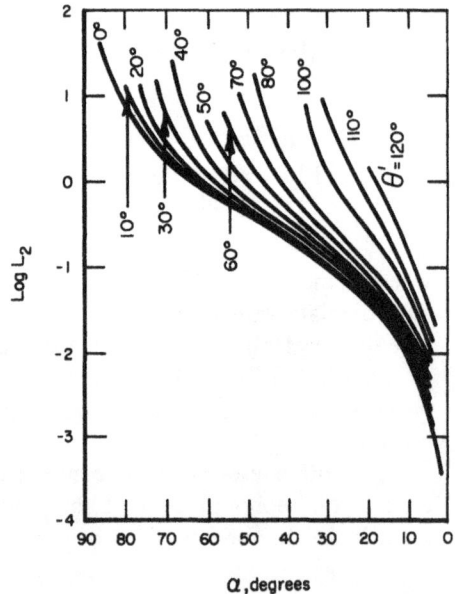

Fig. 6-3. Evaluation of Goryainov's $L_2(\theta', \alpha)$ function [after Goryainov[9]].

large-angle cones can be obtained. The results for the backscattered radar cross sections are

$$\sigma(0) = (\lambda_0^2/4\pi)(1 - \cos\alpha)^2\,[1 + 3(1 - \cos\alpha)] \qquad (6.2\text{-}12)$$

for $\alpha \approx 0$, and

$$\sigma(0) = (\lambda_0^2/16\pi\,\cos^4\alpha)(1 - 2\cos^2\alpha) \qquad (6.2\text{-}13)$$

for $\alpha \approx \pi/2$.

### 6.2.1.2. Approximate Solutions

Since the exact solutions are difficult to evaluate, requiring approximations for numerical results, it is appropriate to consider approximate solutions to the basic scattering problem. Felsen[12] has examined the bistatic scattered far field for a source of the form

$$\mathbf{E}^i = \hat{\phi}' \exp[-ik_0 r(\sin\theta\,\sin\theta'\,\cos(\phi - \phi') + \cos\theta\,\cos\theta']$$

With the approximations $k_0 r \gg 1$, $\theta + \theta' < \pi - 2\alpha$, $\alpha$ small, he obtains

$$E_\theta{}^s = \frac{\exp(ik_0 r)}{ik_0 r} \sin(\phi - \phi')\left(\frac{\alpha}{2}\right)^2 \left[\frac{\sec^2(\theta/2) + \sec^2(\theta'/2)}{\cos\theta + \cos\theta'}\right.$$
$$\left. + \frac{\sin^2\theta + \sin^2\theta'}{2\cos^2(\theta/2)\cos^2(\theta'/2)(\cos\theta + \cos\theta')^2}\right] \qquad (6.2\text{-}14)$$

$$E_\phi{}^s = \frac{\exp(ik_0 r)}{ik_0 r}\left(\frac{\alpha}{2}\right)\left\{\frac{2\sin\theta\,\sin\theta'}{(\cos\theta + \cos\theta')^3} + \cos(\phi - \phi')\right.$$
$$\times\left[\frac{\sec^2(\theta/2) + \sec^2(\theta'/2)}{\cos\theta + \cos\theta'} + \frac{\sin^2\theta + \sin^2\theta'}{2\cos^2(\theta/2)\cos^2(\theta'/2)(\cos\theta + \cos\theta')^2}\right.$$
$$\left.\left. + \frac{16\sin^2(\theta/2)\,\sin^2(\theta'/2)}{(\cos\theta + \cos\theta')^3}\right]\right\} \qquad (6.2\text{-}15)$$

For the case of backscattering, $\mathbf{E}^i = \hat{\phi}'\exp(-ik_0 r)$, and the $\theta$ component of the scattered field goes to zero. The $\phi$ component reduces to the form

$$E_\phi{}^s = \frac{\exp(ik_0 r)}{ik_0 r}\left(\frac{\alpha}{2}\right)^2\left[\frac{3 + \cos^2\theta}{4\cos^3\theta}\right] \qquad (6.2\text{-}16)$$

where $\theta$ is the angle between the axis and the incident-field direction.

Another approximation which is commonly used is the physical-optics approximation; it is not obvious that a physical-optics approximation is justifiable here because of the singularity at the tip. Goryainov[9] reports that he has considered the difference between the true field as obtained from Eqs. (6.2-6) and (6.2-7) and the physical-optics field and found that the difference is small when $\theta'$ is near 0, and that the error increased uniformly away from the backscatter direction. The error never exceeded 10% for

$\theta' < (\pi - 2\alpha)$. The physical-optics bistatic scattered fields for axial incidence, as a function of $\theta'$, given by Goryainov are

$$E_\theta{}^s(\theta') = \frac{\exp(ik_0 r)}{ik_0 r} L_p(\theta') \cos \phi' \qquad (6.2\text{-}17)$$

and

$$E_\phi{}^s(\theta') = - \frac{\exp(ik_0 r)}{ik_0 r} L_p(\theta') \sin \cdot \phi' \qquad (6.2\text{-}18)$$

where

$$L_p(\theta') = \left| \frac{-\sin^2 \alpha \cos \alpha}{4 \cos(\theta'/2)[\cos(\alpha + \theta'/2) \cos(\alpha - \theta'/2)]^{3/2}} \right| \qquad (6.2\text{-}19)$$

The physical-optics, axial-incidence bistatic cross section as a function of $\theta'$ is also given by Siegel et al.[13] as

$$\sigma(\theta) = \frac{\lambda_0{}^2 \tan^4 \alpha}{16\pi} \frac{2[1 + \cos 2\alpha]^3}{[1 + \cos \theta'][\cos \theta' + \cos 2\alpha]^3} \qquad (6.2\text{-}20)$$

where $0 \leqslant \theta' < \pi - 2\alpha$. This is identical to the bistatic cross section obtainable from Goryainov's equations.

Similarly, the Luneburg–Kline expansion cannot in principle be legitimately applied for scattering from a cone since the radius of curvature at the nose of the cone is zero. However, if one carries out the Luneburg–Kline expansion for axial incidence, then Eq. (6.2-20) is the first term of the expression obtained,[13] or

$$\sigma(\theta') = \frac{\lambda_0{}^2 \tan^4 \alpha}{16\pi} \frac{2(1 + \cos 2\alpha)}{(1 + \cos \theta')(\cos 2\alpha + \cos \theta')} \left[ \left( \frac{1 + \cos 2\alpha}{\cos 2\alpha + \cos \theta'} \right)^2 \right.$$
$$\left. + 2 \left( \frac{1 + \cos 2\alpha}{\cos 2\alpha + \cos \theta'} \right) \left( \frac{1 - \cos \theta'}{1 + \cos \theta'} \right) \cos 2\psi + \left( \frac{1 - \cos \theta'}{1 + \cos \theta'} \right)^2 \right] \qquad (6.2\text{-}21)$$

where $\psi$ is the angle between the plane of polarization of the incident field and the plane containing the axis of the cone and the receiver. Moreover, Eq. (6.2-21) predicts that the cross section is a maximum in the plane of polarization as is expected. Also, the cross section increases as a function of the separation between transmitter and receiver in the expected manner.

In Figure 6-4, the physical-optics approximation is compared to the maximum cross section ($\psi = 0$) and to the minimum cross section ($\psi = \pi/2$) predicted by Eq. (6.2-21). For convenience, the normalized cross section, $16\pi\sigma/\lambda_0{}^2 \tan^4 \alpha$, is plotted against $\theta'$. It is seen that the physical-optics prediction of Eq. (6.2-20) lies approximately halfway between the maximum and minimum curves from the Luneburg–Kline method.

For axial-incidence backscatter ($\theta' = 0$), Eqs. (6.2-17) and (6.2-18) from

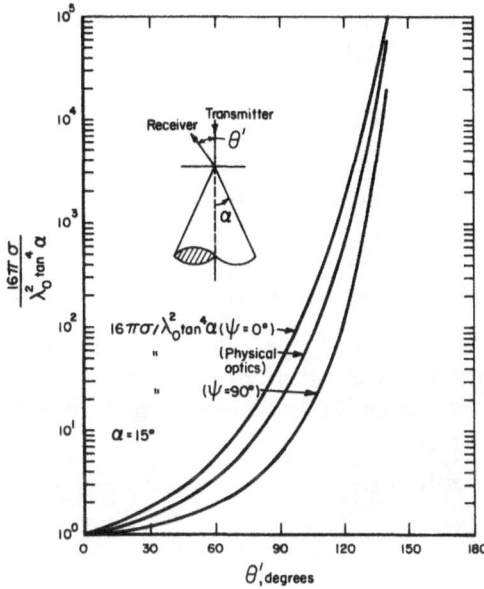

Fig. 6-4. Comparison of the radar cross section of a semi-infinite cone as predicted by the physical-optics approximation with the maximum and minimum values predicted by the Luneburg–Kline expansion [after Siegel et al.[13]].

Goryainov, Eq. (6.2-20) from Siegel, and the Luneburg–Kline series result, Eq. (6.2-21), all reduce to the familiar equation of Spencer[14]

$$\sigma(0) = (\lambda_0^2/16\pi) \tan^4 \alpha \qquad (6.2\text{-}22)$$

If the approximation is made in Eq. (6.2-22) that $\tan \alpha$ is equal to $\alpha$, then

$$\sigma(0) = (\lambda_0^2/16\pi) \alpha^4 \qquad (6.2\text{-}23)$$

with the limitation that $\alpha \ll 1$, $k_0 r \gg 1$.

Equation (6.2-23) was also obtained by Felsen.[12,15–17] He used a transmission line formulation to derive Green's function for wide- and narrow-angle cones. The cone surface is considered to constitute the wall of a conical waveguide in which transmission is in the radial direction. For a narrow-angle cone, he obtained a scattered field of

$$\mathbf{E}^s = -\hat{\mathbf{x}} \frac{i e^{i k_0 z}}{k_0 z} \left(\frac{\alpha}{2}\right)^2 \qquad (6.2\text{-}24)$$

for an incident field polarized in the $x$ direction and incident as in Figure 6-1.

The resulting backscatter cross section is given by Eq. (6.2-23). For a wide-angle cone for the same incident field, the result is

$$\mathbf{E}^s = -\hat{\mathbf{x}} \frac{i e^{i k_0 z}}{4 k_0 z [(\pi/2) - \alpha]^2} \qquad (6.2\text{-}25)$$

The radar cross section obtained from this equation is

$$\sigma(0) = \frac{\lambda_0{}^2}{16\pi} \frac{1}{[(\pi/2) - \alpha]^4} \qquad (6.2\text{-}26)$$

where $\alpha \approx \pi/2$, $k_0 r \gg 1$, and $\sqrt{k_0 r} \cos \alpha \gg 1$. Equations (6.2-23) and (6.2-26) are the first terms of equations reported by Felsen[12,15-17] and by Siegel et al.[11] Felsen, from the transmission line formulation, obtains

$$\sigma(0) = \frac{\lambda_0{}^2}{16\pi \cos^4 \alpha} (1 - 4\cos^2 \alpha + \cdots) \qquad (6.2\text{-}27)$$

for $\alpha \approx \pi/2$; and

$$\sigma(0) = \frac{\lambda_0{}^2 \sin^4(\alpha/2)}{4\pi} \left[ 1 + \left( 4 \ln \frac{1}{\sin^2(\alpha/2)} - 2 \right) \sin^2(\alpha/2) + \cdots \right] \qquad (6.2\text{-}28)$$

for $\alpha \approx 0$. Siegel et al., by means of manipulation of divergent series, obtain a factor of two instead of the four which multiplies the $\cos^2 \alpha$ term in Eq. (6.2-27), they also obtain a factor of six which multiplies the $\sin^2(\alpha/2)$ term in Eq. (6.2-28), instead of the $4 \ln[1/\sin^2(\alpha/2)]$.

The various approximations are compared in Figure 6-5 for cone half-angles of 20° or greater. Curve $E$ is the physical-optics approximation of

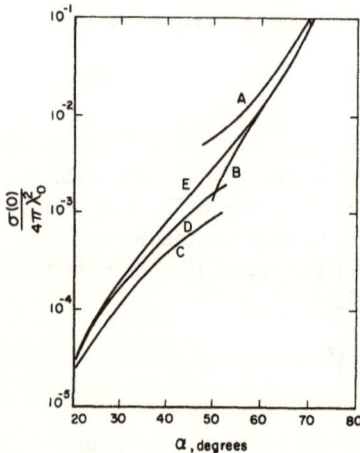

Fig. 6-5. Comparison of axial backscatter from semi-infinite cones of various sized half angles as computed by five approximate methods [after Kleinman and Senior[4]].

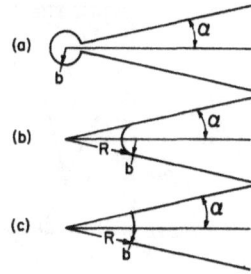

Fig. 6-6. Various blunted tip configurations: (a) ball-point tip; (b) rounded tip; and (c) concave tip.

Eq. (6.2-22). It is bounded below at large cone angles by curve $B$, computed from Eq. (6.2-13), an approximate solution to the exact scattering equation. Curve $E$ lies below the Felsen approximation of curve $A$, computed from Eq. (6.2-26). The small-cone approximate calculation of the exact scattering equation is given in Curve $D$, computed from Eq. (6.2-12). This curve is seen to lie between the physical-optics approximation of Curve $E$ and the Felsen small-angle approximation, Curve $C$, computed from Eq. (6.2-23). Thus, the physical-optics approximation appears to give reasonable results over the entire range of cone angles.

### 6.2.2. Blunted-Tip Effects

Felsen[15,16] and Weiner and Borison[18] have considered the effects of rounding the tips of semi-infinite perfectly conducting cones. Felsen, by a transmission-line analogy, found the Green's function for a sphere on the end of a semi-infinite cone, as the Green's function for a sphere minus a correction term for the cone. Weiner and Borison considered the three cases illustrated in Figure 6-6. The upper cone is the same as that considered by Felsen. In the second case, the cone is rounded off with a portion of a sphere serving as the termination. The third cone has had its tip removed to form a concave termination. Calculations by eigenfunction expansions were performed for a plane wave incident along the symmetry axis (nose-on incidence). For the first case, the cone with a ball on the tip, the Hansen–Schiff summation is included in the resulting series. The approximation was made that $k_0 b \alpha \ll 1$, and the series was summed on a digital computer. Figure 6-7 shows $\sigma/\pi b^2$ as a function of $k_0 b$ for a cone with a 10° half-angle. The curve appears to be a combination of two curves, one for backscattering from a sphere and the other for tip scattering from a semi-infinite cone. Figure 6-8 illustrates the region of merger of the two curves for $\alpha = 5, 8, 10,$ and 12°. The rounded- and concave-tip cone solutions involved expanding the fields in two separate regions and matching coefficients. The limit of validity of approximations, such as truncating the resulting infinite matrix and terminating the series, can only be determined by actual numerical

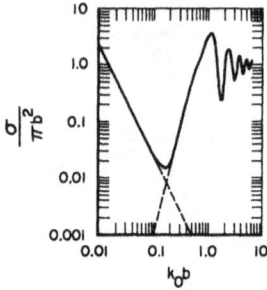

Fig. 6-7. Axial backscatter cross section for ball-point tip as a function of $k_0 b$ for $\alpha = 10°$; dashed curves are for a pointed cone and for an isolated sphere [after Weiner and Borison[18]].

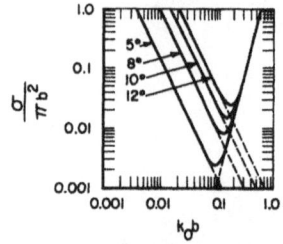

Fig. 6-8. Axial backscatter cross section for ball-point tip as a function of $k_0 b$ for $\alpha = 5, 8, 10$, and $12°$; dashed curves are for pointed cones and for an isolated sphere [after Weiner and Borison[18]].

computations. Figure 6-9 presents the results of the computation for $\alpha = 10°$, and Figure 6-10 for $\alpha = 5, 8, 10$, and $12°$. Also shown in Figure 6-9 are the physical-optics approximations for the rounded-tip and concave-tip cones. The concave-tip, axial-incidence, physical-optics backscatter cross section is given by

$$\sigma = 4\pi k_0^2 R^4 \left| \frac{(2ik_0 R - 1) \exp(2ik_0 R)}{(2k_0 R)^2} - \frac{(2ik_0 R \cos \alpha - 1) \exp(2ik_0 R \cos \alpha)}{(2k_0 R \cos \alpha)^2} \right|^2$$

(6.2-29)

where $R$, as defined in Figure 6-6, is the maximum distance from the projected tip of the cone to the blunt tip.

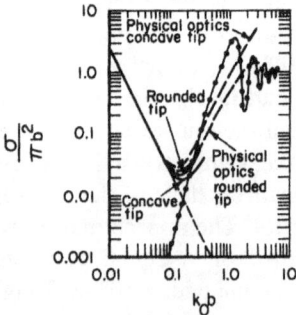

Fig. 6-9. Axial backscatter cross section for concave and rounded tips for $\alpha = 10°$; also shown are curves predicted by the physical-optics approximation and the curves for a pointed cone and for an isolated sphere [after Weiner and Borison[18]].

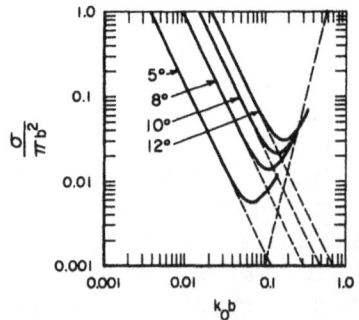

Fig. 6-10. Axial backscatter cross section for a rounded tip as a function of $k_0 b$ for $\alpha = 5, 8, 10$, and $12°$; also shown are the curves for pointed cones and for an isolated sphere [after Weiner and Borison[18]].

Expanding Eq. (6.2-29) about $\alpha = 0$ for small $k_0 b \alpha$, Weiner and Borison obtain

$$\sigma \approx (\pi b^4/4k_0^2) + (k_0 b)^2 \, (\pi b^2) \qquad (6.2\text{-}30)$$

which is the sum of the cross section of a pointed cone and a flat plate of radius $b$.

The physical-optics approximation for the rounded-tip cone is, for axial incidence,

$$\sigma = \pi b^2 \left[ 1 - \frac{\sin[2k_0 b(1 - \sin \alpha)]}{k_0 b \cos^2 \alpha} + \frac{1 + \cos^4 \alpha}{4(k_0 b)^2 \cos^4 \alpha} - \frac{\cos[2k_0 b(1 - \sin \alpha)]}{2(k_0 b)^2 \cos^2 \alpha} \right]$$

$$(6.2\text{-}31)$$

The agreement with the exact theory is quite good. For incidence other than nose-on, but for $\theta \leqslant \alpha$ so that the entire surface is illuminated, the physical-optics approximation for the rounded-tip cone gives

$$\frac{\sigma}{\pi b^2} = \frac{1 + \theta^2}{4(k_0 b)^2} \, [A_1 + A_2 \cos[2k_0 \cos \theta (1 - \sin \alpha)]$$

$$+ A_3 \sin[2k_0 \cos \theta (1 - \sin \alpha)]], \qquad k_0 b \theta \ll 1 \qquad (6.2\text{-}32)$$

where

$$A_1 = 2 + 2\alpha^2 - 2\theta^2 + \alpha^4 - \alpha^2\theta^2 + \tfrac{1}{2}\theta^4 + 2\alpha^4\theta^2 + 4(k_0 b)^2 - 2(k_0 b)^2\,\theta^2$$
$$- 8(k_0 b)^3\,\theta^2 + (k_0 b)^2\,\theta^4 + 6(k_0 b)^2\,\alpha^2\theta^2 + 8(k_0 b)^3\,\theta^4 + 13(k_0 b)^4\,\theta^4$$

$$A_2 = -2 - 2\alpha^2 + 2\theta^2 + \alpha^2\theta^2 - \tfrac{1}{2}\theta^4 - 6(k_0 b)^2\,\theta^2 + 8(k_0 b)^4\,\theta^3 + 3(k_0 b)^2\,\theta^4$$

$$A_3 = -4[1 + \alpha^2 - \tfrac{1}{2}\theta^2 + 3(k_0 a \theta)^2][k_0 b - k_0 b \theta^2 - (k_0 b \theta)^2] - 4(k_0 b \theta)^3$$

The results of a numerical evaluation of Eq. (6.2-32) for $\alpha = 15°$, for several values of $k_0 b$, and for $\theta$ from $0°$ to $14°$, are given in Figure 6-11.

In summary, the physical-optics approximation appears to be a good approximation for calculating the radar cross section of small-angle, blunted-tip cones, with the approximation apparently better for smooth-tip cone joints.

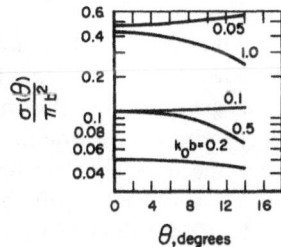

Fig. 6-11. Backscatter cross section with nonaxial incidence on rounded-tip cone for $\alpha = 15°$ and several values of $k_0 b$ [after Weiner and Borison[18]].

### 6.2.3.  Coated and Dielectric Semi-Infinite Cones

Überall[19] has presented four approximate methods for analyzing the axial-incidence backscatter for a perfectly conducting, semi-infinite cone coated with a weak scattering layer (complex relative dielectric constant and permeability with magnitudes near unity) of thickness $\delta$. The backscattered field and the cross section are expressed as

$$\mathbf{H}^s(0) = i(4k_0 r)^{-1} e^{ik_0 r} \mathbf{H}^i \tan^2 \alpha \, | \, S \, | \qquad (6.2\text{-}33)$$

and

$$\sigma(0) = \frac{\lambda_0^2}{16\pi} \tan^4 \alpha \, | \, S \, |^2 \qquad (6.2\text{-}34)$$

where $S$ is a parameter to be evaluated.

A straightforward physical-optics approximation results in the following expression for $S$:

$$S_{\text{p.o.}} = [\tfrac{1}{2}][e^{-2ik_0\delta\csc\alpha}][[m^2 \sin\alpha \cos(k_0 q\delta) + i\mu_r q \sin(k_0 q\delta)]$$
$$\times \ m^2 \sin\alpha \cos(k_0 q\delta) - i\mu_r q \sin(k_0 q\delta)]^{-1} + [q \cos(k_0 q\delta)$$
$$+ \ i\mu_r \sin\alpha \sin(k_0 q\delta)][q \cos(k_0 q\delta) - i\mu_r \sin\alpha \sin(k_0 q\delta)]^{-1}]$$
$$(6.2\text{-}35)$$

where

$$m = \sqrt{\mu_r \epsilon_r},$$
$$q = (m^2 - \cos^2 \alpha)^{1/2}$$

A Born-type approximation (see Section 2.2.1.2) for $S$ gives

$$S_B = e^{ik_0\delta[(m-1)\csc\alpha - \sin\alpha]}[\cos(mk_0\delta \sin\alpha) - (i\mu_r/m)\sin(mk_0\delta \sin\alpha)]^{-1}$$
$$\times \ [1 + (1 - \epsilon_r)(i\mu_r/m)\sin(mk_0\delta \sin\alpha) \, e^{-ik_0\delta(m\csc\alpha + \cos\alpha\cot\alpha)}]$$
$$(6.2\text{-}36)$$

A more precise method of determining $S$ combines the physical-optics approximation for the perfectly conducting cone with the eikonal method of Saxon and Schiff[20] for the dielectric coating. The results of this method are

$$S_S = S_B + \{[(m - 1)/2] \csc\alpha(\sin\alpha + \csc\alpha)[1 - \cos(mk_0\delta \sin\alpha)]$$
$$\times \ \exp[-ik_0\delta(m\csc\alpha + \cos\alpha\cot\alpha)] - [(m - 1)/(m + 1)^2]$$
$$\times \ [1 - \exp(-2imk_0\delta \cos\alpha\cot\alpha)]\}. \qquad (6.2\text{-}37)$$

It is seen that Eq. (6.2-37) includes the Born results of Eq. (6.2-36) as the first few terms. The results of the determination of cross section with these two methods are shown in Figure 6-12. The Born-type approximation exaggerates the diffraction minima. Both equations become less reliable as the cone angle becomes smaller. Variations of normalized cross section versus thickness of coating for cone half-angles of $\alpha = 30, 50,$ and $70°$ and $m = 1.33$ and 1.67 are shown in Figure 6-13, as computed from Eq. (6.2-37).

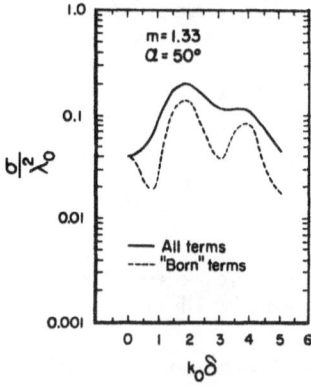

Fig. 6-12. Comparison of the axial backscatter radar cross section of a coated cone as a function of coating thickness as computed by Eq. (6.2-37) and by the "Born" equation (6.2-36) [after Überall[19]].

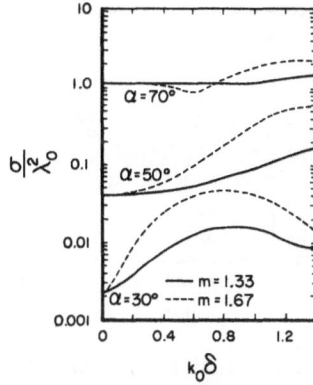

Fig. 6-13. Axial backscatter radar cross section of a coated cone as a function of coating thickness for three sizes of cone half-angle and two values of the refractive index of the coating [after Überall[19]].

The previous approximations neglected refraction in the dielectric coating. Überall derives an exact eikonal Green's function which allows the refraction effects to be included. The result of combining this technique with the physical-optics approximation for the metallic cone is

$$S_R = a_1/b_1 + a_2/b_2 \qquad (6.2\text{-}38)$$

where

$$a_1 = \tfrac{1}{2}m\{[\exp(-2ik_0\delta \csc \alpha)][(q - \sin \alpha) \cos(k_0\delta q) \\ + i(1 - 1/\epsilon_r) q \sin(k_0q\delta)] - q \exp[ik_0\delta(q - 2 \sin \alpha)]\}$$

$$a_2 = (1/2) \sin \alpha\{\exp(-2ik_0\delta \csc \alpha)[q(q - \sin \alpha) \cos(k_0q\delta) \\ + im^2(1 - 1/\epsilon_r) \sin(k_0q\delta)] \\ + \exp[ik_0\delta(q - 2 \sin \alpha)] q[\tfrac{1}{2}(q - \sin \alpha)[1 - \exp(2ik_0\delta \cos \alpha \cot \alpha)] \\ - (q + \cos \alpha \cot \alpha)]\}$$

$$b_1 = -m \sin \alpha \cos(k_0q\delta) + i\frac{\mu_r q \sin(k_0q\delta)}{m}$$

$$b_2 = -q \cos(k_0q\delta) + i\mu_r \sin \alpha \sin(k_0q\delta)$$

$$q = (m^2 - \cos^2 \alpha)^{1/2}$$

$$m = (\mu_r\epsilon_r)^{1/2}$$

At large cone angles for the weak-scattering case, Eqs. (6.2-37) and (6.2-38)

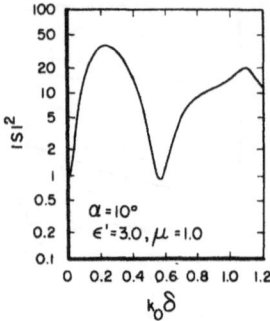

Fig. 6-14. Axial backscatter from a coated cone as a function of coating thickness [after Überall[19]].

give essentially the same results. Equation (6.2-38), however, gives better results for small angles and for large dielectric constants.

All the expressions for $S$ [Eqs. (6.2-35) to (6.2-38)] reduce to unity if $\delta = 0$, or if $\mu_r$ and $\epsilon_r = 1$. Thus, as the dielectric coating approaches free space, Eq. (6.2-34) approaches Eq. (6.2-22). For a cone of half-angle $\alpha = 10°$ and a coating with a purely real dielectric constant ($\epsilon_r = 3$, $\mu_r = 1$), the values of $|S|^2$, as determined by Eq. (6.2-38), are shown in Figure 6-14. The result is a rapidly varying function of $k_0\delta$ which shows pronounced diffraction maxima and minima.

In general, it is seen that even a weakly scattering coating can have considerable effect upon the cross section of a semi-infinite cone. The effect of layers whose relative dielectric parameters are not near unity awaits solution.

Little theoretical work has been done on the scattering from a semi-infinite dielectric cone. Felsen[21,22] has considered the case of a semi-infinite cone of material having a surface impedance which varies with distance from the cone apex. He considers four types of ring sources, electric or magnetic currents flowing in radial or azimuthal directions. No numerical results are known to have been obtained for this problem, and due to the special nature of the sources, the scattered field is not easily related to the plane-wave radar cross section.

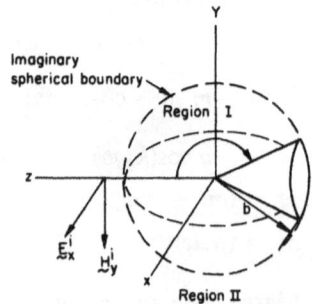

Fig. 6-15. Spherically-capped right-circular cone configuration.

## 6.3. FINITE CONES

### 6.3.1. Metallic Cones

#### 6.3.1.1. *Theoretical Solutions*

The only finite cone for which an "exact" solution has been obtained is the spherical sector, i.e., the spherically-capped, right circular cone, as illustrated in Figure 6-15. A spherical coordinate system is used, centered at the apex of the cone, and having its polar axis coinciding with the axis of the cone. It is seen that the spherical sector is the intersection of the sphere $r = b$ and the cone $\theta = \theta_0$. The wave equations are separable, and solutions exist for each of the surfaces. Thus, by forcing the solutions to agree on the portion of the sphere that is free space, by applying the radiation condition at infinity, and by applying the proper boundary conditions on the metal cone, one obtains a system of linear equations with as many unknowns as there are equations. Unfortunately, the system is infinite and the answer can only be approximated by a finite system. Two cases of this nature have been examined. Northover[23] considered scattering from a perfectly conducting cone when the transmitting antenna is an electric dipole oriented along the axis of symmetry of the cone. Rogers, Schultz, *et al.*[8,24-27] of Purdue University have examined the problem of a plane wave incident along the axis of symmetry, which is the limiting case of an infinitely distant transverse dipole.

Northover's solution is not known to have been reduced to numerical results. However, the group at Purdue University has obtained approximate answer by truncating the infinite set of equations. In doing this, it was necessary to consider up to one hundred equations. They express the back-scattering cross section as a function of two parameters, $c_n$ and $d_n$, which must be determined, or

$$\sigma(0) = \frac{\lambda_0^2}{4\pi} \left| \sum_n i^n n(n + 1)(c_n - id_n) \right|^2 \qquad (6.3\text{-}1)$$

The $c_n$'s are all interdependent, as are the $d_n$'s. This interdependence takes the form of an infinite set of complex equations involving the roots of the Legendre function with the number of equations always equal to the number of $c$'s (or $d$'s). The validity of arbitrarily terminating this set with a finite number of equations has received some consideration by Rogers and Schultz.[24] They found that the values obtained for $c_1$, $c_2$, and $c_3$ by using three equations were within 2% of the values obtained when seven equations were used. An examination of the ratios $|c_1/c_2|$ and $|d_1/d_2|$ for $k_0 a = 0.1$ yielded values of 200 and 140, respectively, indicating rapid convergence of the series for small cones. The calculated value of the radar cross section is extremely sensitive to small errors in the roots of the Legendre function

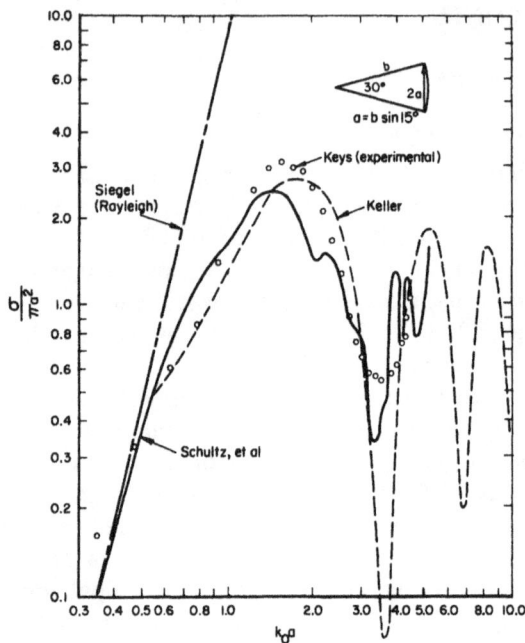

Fig. 6-16. Axial backscatter radar cross section for cone-spherical segment as obtained by several methods [after Schultz et al.[26]].

and in order to obtain accuracy, it was necessary for Schultz et al.[8] to reevaluate the roots.

In Figure 6-16, the results of the Purdue calculations are compared to the Rayleigh approximation of Siegel discussed in the following section, to the geometric theory of diffraction results of Keller, discussed in Section 6.3.1.2.2, and to some experimental results of Keys.[†] It is seen that for $1 \lesssim k_0 a \lesssim 3$, the "exact" solution (with the approximations required to compute numerical answers) does not appear to agree as well with the experimental results as does Keller's theory. On the other hand, the experimental results cannot be said to be exact. With the continual improvement of computers perhaps a more precise evaluation of this method can be accomplished soon.

### 6.3.1.2. Approximate Solutions

From the previous discussion it is seen that exact theoretical values for the radar cross section of finite cones are not available in general. Since the cone is an important shape, it is essential to have analytical estimates of its

---

† These measurements were taken from unpublished data on flat-backed cones furnished to Purdue University by Keys. It is said [26] that these measurements were indistinguishable from measurements of a spherically-capped cone of the same dimensions.

radar scattering characteristics. There are many ways of estimating cone cross sections, each with varying accuracy for different situations. These approximations have been divided into four classes for this discussion. First, when either the cone is small or the frequency is sufficiently low so that the cone is small compared to the radar wavelength. Then the opposite extreme, the high-frequency approximation, is considered; i.e., the cone is large compared to the wavelength. The high-frequency approximation is applied to the bistatic as well as the monostatic case. The region between these approximations is called the resonance region, where the cone and the radar wavelength are comparable in size. The cross section in this region is more difficult to approximate. In the first two sections only cw or monochromatic scattering is discussed. In the final section the time-domain response and the pulse response of a cone are examined.

6.3.1.2.1. *Rayleigh Region.* If the exact solution is known for a body such as a sphere, a series expansion in powers of $k_0$ can be made, the first term of which gives the scattering from a body very small in terms of the wavelength. The resulting equation for the cross section (the well-known Rayleigh law) is expressed as

$$\sigma = 4\pi(k_0)^4 |f_0|^2 \qquad (6.3\text{-}2)$$

where $f_0$ is the coefficient of the leading term of the series expansion. This coefficient is given for the sphere in Chapter 3 and for the ellipsoid in Chapter 5. Since the exact results for a terminated cone cannot be readily expressed in an appropriate series form, the question arises as to how to evaluate $f_0$. One technique,[28] valid for any body which can be represented as the intersection of a finite number of regions within each of which the electrostatic Green's function can be determined, was used by Darling[29] for a spherically-capped cone, and $f_0$ was expressed in terms of infinite matrices. In order to use this method, it will be necessary to investigate on a large computer the convergence of the answer obtained by truncating the matrices.

As discussed in Section 2.2.2.7, Siegel[30-34] has proposed an approximate method of estimating $f_0$ for bodies of revolution. He expresses $f_0$ as the product of the volume and a dimensionless correction factor $F$ which accounts for the approximate dependence of the cross section on the shape. The expression for $F$ is

$$F = 1 + \frac{\exp(-\tau)}{\pi\tau} \qquad (6.3\text{-}3)$$

where $\tau$ is taken as a measure of the body elongation. In order that the cross section approach that of a disc as $h$ approaches zero, $\tau$ is set equal to $h/4a$ for a right circular cone of altitude $h$ and radius $a$. The equation for the cross section then reduces to

$$\sigma = \frac{4\pi}{9} k_0^4 a^4 h^2 \left[1 + \frac{4a \exp(-h/4a)}{\pi h}\right]^2 \qquad (6.3\text{-}4)$$

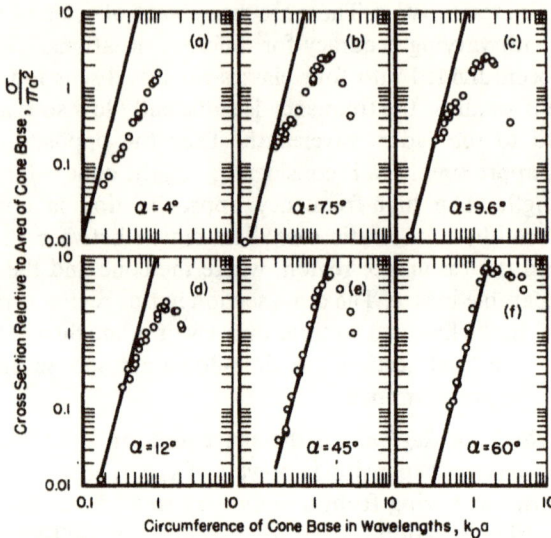

Fig. 6-17.  Theoretical and experimental axial backscatter radar cross section as a function of size of cone base for several cone half-angles [after Brysk et al.[32]].

The agreement between this equation and experimental results is illustrated in Figure 6-17, adapted from Reference 32. It is seen that the agreement is best for the moderate- and larger-angle cones. For cones of very large angle, one might expect the Rayleigh curve for a disc, discussed in Chapter 7, to fit the results. The agreement between Eq. (6.3-4) and experimental data is not as good for cones of smaller half-angle, since they are not really in the Rayleigh region. As the half-angle decreases, the boundary between the Rayleigh region and the resonance region moves to smaller bodies. Equation (6.3-4) is a valid approximation only if the product of the slant height of the cone and the wave number is very much less than unity, or $\sqrt{h^2 + a^2} \ll \lambda_0/2\pi$.

It should also be noted that, to the order of this approximation, Eq. (6.3-4) predicts that the nose-on and base-on cross section of the cone should be the same.[31] The experimental data plotted in Figures 6-18 and 6-19 show that the nose-on and base-on cross sections converge as the Rayleigh region is approached. Figure 6-18[33] gives the ratio in decibels of the cross section of the base to the cross section of the nose, as determined from experimental measurements of Keys and Primich,[35] Honda, Silver, and Clapp,[36] and August and Angelakos.[37] It is seen that at small $k_0a$, the ratio is indeed approaching 0 db. Figure 6-19 gives the results of experiments at the University of Michigan showing that the separate curves of nose-on and base-on cross section do converge for small $k_0a$.[33]

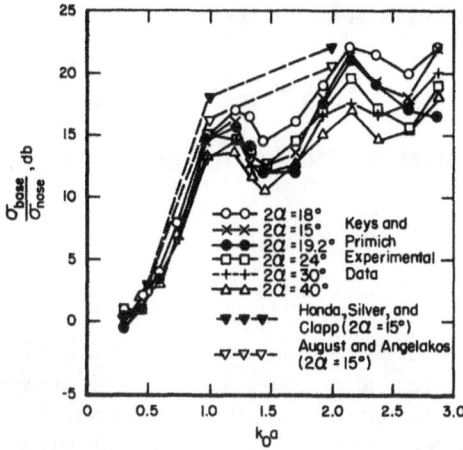

Fig. 6-18. Ratio of base backscatter cross section to axial backscatter cross section as a function of base size for cones with various half-angles [after Siegel[33]].

Using the same techniques as were used to generate Eq. (6.3-4), Siegel[34] has arrived at the Rayleigh cross section of a cone-hemispheroid as

$$\sigma = \frac{4}{\pi} k_0{}^4 \left( \frac{1}{3} \pi a^2 h + \frac{2}{3} \pi a^2 b \right)^2 \left[ 1 + \frac{\exp[-(h + 2b)/4a]}{\pi(h + 2b)/4a} \right]^2 \quad (6.3\text{-}5)$$

where $b$ is the semimajor axis of the spheroid, and $a$ is the radius of the base of the cone and the semiminor axis of the spheroid. If $b$ is set equal to zero, Eq. (6.3-5) reduces to (6.3-4).

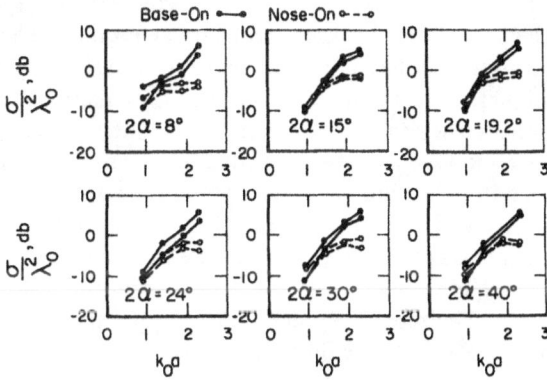

Fig. 6-19. Comparison of experimental base and axial backscatter cross sections for various cones [after Siegel[33]]

For a cone-hemisphere, $b = a$, thus

$$\sigma = \frac{4}{\pi} k_0^4 \left(\frac{1}{3} \pi a^2 h + \frac{2}{3} \pi \dot{a}^3\right)^2 \left[1 + \frac{\exp[-(h + 2a)/4a]}{\pi(h + 2a)/4a}\right]^2 \quad (6.3\text{-}6)$$

Equation (6.3-6) can be rewritten in terms of the cone half-angle, $\alpha$, as

$$\sigma = \frac{4\pi}{9} k_0^4 a^6 C^2 \left[1 + \frac{\exp[-C/4]}{\pi C/4}\right]^2 \quad (6.3\text{-}7)$$

where

$$C = [1/\tan(\alpha/2)] + 2 = (h + 2a)/a$$

A generalization of Siegel's method is discussed in Chapter 8. The exact Rayleigh results for general spheroids and ellipsoids are given in Chapter 5. From these, approximations for cones of various sizes and shaped can be made.

The Rayleigh cross section can be used not only to predict the cross section in the Rayleigh region, but also to aid in predicting the resonance-region cross section. This is done by using the Rayleigh coefficient together with high-frequency approximations to predict the time response of a radar target to an impulsive plane wave. This technique appears to have some potential, and is discussed in more detail in Section 6.3.1.2.2.2.

Excellent reviews of Rayleigh scattering have been written by Kleinman[38] and Kleinman and Senior.[4]

6.3.1.2.2. *The High-Frequency and Resonance Regions.* At the other extreme from the Rayleigh region is the region where the observing wavelength is very short with respect to the dimensions of the scatterer. In this region, the shape of the scatterer is of vital importance, as contrasted to the Rayleigh region where variations in shape that keep the volume constant are scarcely noticeable. The frequency range between the Rayleigh region and the high-frequency region is usually characterized by large oscillations appearing in curves of the backscattering cross section versus frequency. Predictions of scattering behavior in this frequency range have been attempted by many techniques, most involving extension of predictions from the high-frequency region or the Rayleigh region. One of the best approximations in the resonance region seems to be generated by impulse-response techniques (see Section 2.2.2.4), which are discussed in Section 6.3.1.2.2.2. The Rayleigh region extensions largely involve finding the second or third terms in the expansion of the cross section in powers of the wave number $k_0$. Depending upon the radius of convergence of the series expansion, these extensions may be of little use. The remaining techniques require the extension of high-frequency approximations to the limit of their accuracy.

6.3.1.2.2.1. *Continuous Wave Cross Sections*

6.3.1.2.2.1.1. *Physical Optics.* Perhaps the most commonly used approximation is that of physical optics, discussed in Section 2.2.2.3. Strictly speaking, physical optics is not valid near a discontinuity such as the tip of a cone; however, as indicated in Section 6.2.1.2, the physical-optics approximation, when applied to a semi-infinite cone, gives results that agree reasonably well with the exact solution. Crispin et al.,[39] using physical optics, have obtained the general expression for the bistatic cross section from a flat-base cone when the transmitter is located on the axis of symmetry. The resulting cross section is, for $0 \leqslant \theta' < \pi - 2\alpha$,

$$\sigma(\theta') = \left[ \frac{4\pi h^2 \sin^2 \alpha \tan^2 \alpha}{[\sin^2 \theta' \sin^2 \alpha - \cos^2 \alpha(1 + \cos^2 \theta')^2]^2} \right]$$
$$\times [\cos^2 \alpha(1 + \cos \theta')^2 J_0^2(kh \sin \theta' \tan \alpha)$$
$$+ \sin^2 \theta' \sin^2 \alpha J_1^2(kh \sin \theta' \tan \alpha)] \qquad (6.3\text{-}8)$$

Although this equation was derived under the assumption that $\theta' \neq 0$, it reduces to the physical-optics equation for backscattering when $\theta' = 0$. For a flat-back cone of height $h$ and half-angle $\alpha$, the backscatter cross section from Eq. (6.3-8) is given by

$$\sigma(0) = \frac{\lambda_0^2}{\pi} \left| -\frac{i}{4} \tan^2 \alpha[1 + (2ik_0h - 1) \exp(2ik_0h)] \right|^2 \qquad (6.3\text{-}9)$$

The first term is the tip scattering

$$\sigma = (\lambda_0^2/16\pi) \tan^4 \alpha \qquad (6.3\text{-}10)$$

which is the same as obtained from Eq. (6.2-22) for scattering from a semi-infinite cone. The second term is the physical-optics expression for an edge singularity. It is not very accurate. A more accurate estimate, obtained by other techniques, will be given later.

Kleinman and Senior[4] point out that if $\tan \alpha$ is replaced by $a/h$ in Eq. (6.3-9), the exponential function expanded for small argument, and then $h$ is allowed to go to zero, Eq. (6.3-9) goes into the physical-optics expression for rear-aspect incidence on a flat-back cone.

Figure 6-20 illustrates the three types of terminations for cones commonly considered other than flat back. The lower cone is commonly designated the cone-sphere although all these cones are in reality cone-spheres. When the term cone-sphere is used in this book without further qualifications, the lower cone is assumed. In the lower cone, the first derivative of the body contour is matched at the cone-sphere join. The second cone is representative of a discontinuity in which less of the sphere is present than in the cone-sphere case. The case in which the center of the sphere is at the apex or tip of the cone is the case considered by Schultz et al. discussed in Section 6.3.1.1. The upper cone in Figure 6-20 is representative of the case where a larger sphere is attached to the cone than for the cone-sphere. With the physical-optics

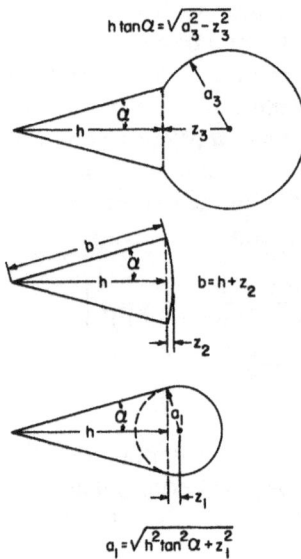

Fig. 6-20.  Various cone-sphere combinations.

approximation, no differences between flat-back cones and other terminations are predicted, unless a portion of the termination or cap is in the illuminated region. If the sphere has its radius on the axis of symmetry at a distance $z$ from the origin, it must have a radius equal to $[(h \tan \alpha)^2 + z_1]^{1/2}$ to achieve a join with the cone such that, at most, the first derivative is discontinuous. In this case, Kleinman and Senior[4] give the axial incidence backscatter cross section as

$$\sigma(0) = \frac{\lambda_0^2}{\pi} \left| -\frac{i}{4} \tan^2 \alpha - \frac{i}{4} [2ik_0(h \tan^2 \alpha - z_1) - \sec^2 \alpha] \exp(2ik_0 h) \right.$$
$$\left. - \frac{i}{4} \exp[2ik_0(z_1 + h)] \right|^2 \qquad (6.3\text{-}11)$$

The $\exp[2ik_0(z_1 + h)]$ term is a spurious term, generated because the physical-optics approximation predicts a current discontinuity at the shadow boundary. This term should, in general, be disregarded.

Neglecting the tip scattering and rewriting the second term in terms of $\delta$, the angle between the tangents to the cone and the sphere at their junction, one obtains

$$\sigma(0) = \frac{\lambda_0^2}{\pi} \left| \left( \frac{k_0 h}{2} \tan \alpha \tan \delta + \frac{i}{4} \right) \sec^2 \alpha \exp(2ik_0 h) \right|^2 \qquad (6.3\text{-}12)$$

which is the cross section attributable to the join. For the cone-sphere, $\delta = 0$; thus the cross section attributable to the join is

$$\sigma(0) = \frac{\lambda_0^2}{\pi} \left| \frac{i}{4} \sec^2 \alpha \exp(2ik_0 h) \right|^2 \qquad (6.3\text{-}13)$$

This has the same wavelength dependence as tip scattering; however, except for cones with $\alpha$ very large, it is much larger than tip scattering.

Dawson et al.[40-42] arrived at similar results for the effects of terminations, assuming the surface of the cone consisted of a series of small wedges. For a flat-back cone they give

$$\sigma(0) = \frac{\pi^3 a^2}{(\alpha + 3\pi/2)^2} \csc^2 \left( \frac{4\pi^2}{3\pi + 2\alpha} \right) \tag{6.3-14}$$

For a cone-sphere, the results are the same as with physical optics. They also applied wedge theory to backscattering from a cone-cylinder with $\alpha = 30°$, $a = 0.905\lambda$, and an overall length of cone plus cylinder varying between 5.2 and 5.5. The scattering portions of the body were assumed to consist of two edges; one at the junction of the cone and cylinder and one at the rear of the cylinder. The scattered fields from these edges were assumed to be independent and related in phase by the free space propagation time. The comparison with theory is shown in Figure 6-21. It appears that an additional phase correction is needed to produce good agreement.

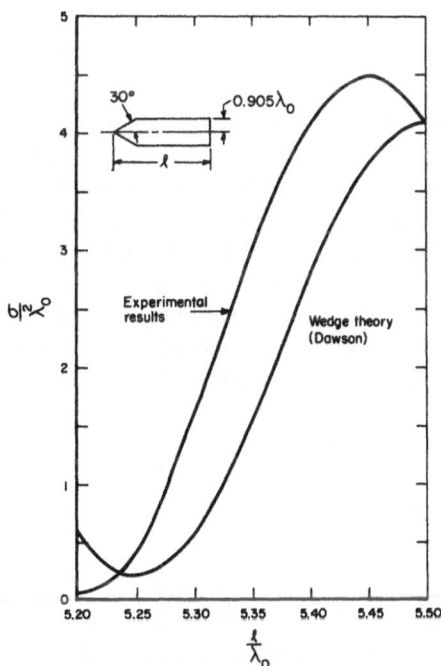

Fig. 6-21. Comparison of theoretical and experimental axial backscatter cross section of a cone cylinder [after Dawson et al.[40]].

For angles of incidence away from axial, the physical-optics approximation should be applied only to the determination of the specular return. At other angles of incidence, the spurious return from the shadow boundary makes calculations tedious and inaccurate. For incidence near broadside to the cone, Kleinman and Senior[4] give for backscatter,

$$\sigma(\theta) = \frac{\lambda_0^2}{\pi} \left| \frac{1}{4} \exp\left[i\frac{\pi}{4} + 2ik_0h\cos(\theta+\alpha)\sec\alpha\right]\left(\frac{k_0h\tan\alpha}{\pi\sin\theta}\right)^{1/2} \tan(\theta+\alpha) \right.$$

$$\left. \times \left[1 - F(\sqrt{2k_0h}\cos(\theta+\alpha)\sec\alpha)\right] \right|^2 \tag{6.3-15}$$

where

$$F(\tau) = \frac{\exp(-i\tau^2)}{\tau}\int_0^\tau e^{ix^2}\,dx \tag{6.3-16}$$

and $F(\tau)$ is related to the Fresnel integral.[43] For small $\tau$,

$$F(\tau) = 1 - \tfrac{2}{3}i\tau^2 + O(\tau^4) \tag{6.3-17}$$

Substituting Eq. (6.3-17) into (6.3-15), as $\theta$ approaches $(\pi/2) - \alpha$, Eq. (6.3-15) reduces to

$$\sigma\left(\frac{\pi}{2} - \alpha\right) = \frac{\lambda_0^2}{\pi}\left| \frac{k_0h}{3}\sqrt{\frac{k_0h\sin\alpha}{\pi}}\sec^2\alpha \right|^2 \tag{6.3-18}$$

This is equivalent to scattering from a thick cylinder whose length is the slant height of the cone, and whose radius is $\tfrac{4}{5}a\sec\alpha$. Other choices of cylinder dimensions can give the same cross section, and no choice seems to give the side lobes accurately. The asymmetrical behavior of the side lobes observed experimentally can be seen from a detailed examination of Eq. (6.3-15).[†]

Pannell, Rheinstein, and Smith[44] define the dimensions of equivalent cylinders of a cone sphere as having a length equal to the slant height ($h\sec\alpha$) of the cone, and a radius such that the volume of the cylinder is equal to that of the cone. The specular return from such a cylinder under the physical-optics approximation is given by

$$\sigma = k_0h^2\sec^2\alpha\{(a^2/3)\cos\alpha\}^{1/2} \tag{6.3-19}$$

Such a cylinder would have a specular lobe of width $\Delta\psi$ given by

$$\Delta\psi = \frac{\lambda_0}{h\sec\alpha} \tag{6.3-20}$$

Figures 6-22 and 6-23 show that there is good agreement between the measured data and Eqs. (6.3-19) and (6.3-20), respectively. Pannell *et al.* also

[†] See Figures 6-47 and 6-48 for experimental examples of this asymmetrical behavior.

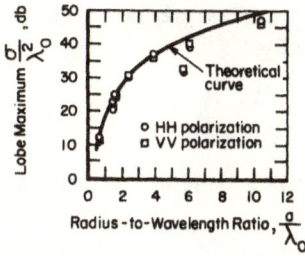

Fig. 6-22. Specular-lobe maximum
cross section area for cone-spheres
[after Pannell *et al.*[44]].

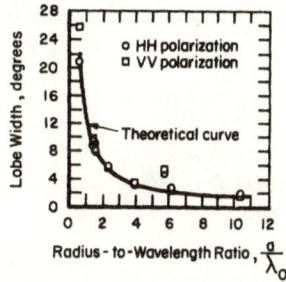

Fig. 6-23. Specular-lobe width
for return from cone-spheres
[after Pannell *et al.*[44]].

compared the measured scattering from the spherical cap with the geometrical-optics scattering from a sphere given by

$$\frac{\sigma}{\lambda_0^2} = \frac{(k_0 a)^2}{4\pi} \qquad (6.3-21)$$

It is seen from Figure 6-24 that the measured values bracket the values
predicted by Eq. (6.3-21), with the greatest deviation being approximately 2 db.

Burke, Mower, and Twersky[45] have derived results for scattering by
finite, conducting cones whose slant height $h$ and base radius $a$ are large
compared to the observing wavelength. The surface fields are approximated
in the integral representation by geometrical-optics values. Thus the approximation is a form of physical optics. The bistatic cross section is given by

$$\sigma(\hat{\mathbf{k}}_0^{\,i}, \hat{\mathbf{r}}') = \frac{4\pi}{k_0^2} \mid \mathbf{F}(\hat{\mathbf{k}}_0^{\,i}, \hat{\mathbf{r}}') \mid^2 \qquad (6.3-22)$$

where $\hat{\mathbf{r}}'$ is a unit vector pointing from the tip of the cone in the direction of
the receiving radar.

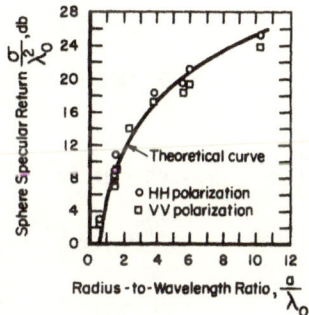

Fig. 6-24. Comparison of measured backscatter cross
section of the spherical cap of a cone-sphere with return
predicted by geometrical optics [after Pannell *et al.*[44]].

For an incident electric field polarized in an arbitrary $\hat{e}_1$ direction, the scattering amplitude function, $F_e(\hat{k}_0{}^i, \hat{r}')$, is given by

$$F_e(\hat{k}_0{}^i, \hat{r}') = k_0\{[(\hat{k}_0{}^i \times \hat{e}_1) \times \hat{n}] - \hat{r}'[\hat{r}' \cdot (\hat{k}_0{}^i \times \hat{e}_1) \times \hat{n}]\}$$
$$\{\tfrac{2}{3}(k_0 a/2\pi D)^{1/2} \, l \, \exp(-i\pi/4) \, I[k_0 a(P - D)]\} \quad (6.3\text{-}23)$$

where $\hat{n}$ is the unit vector normal to the cone surface lying in the scattering plane,

$$D = [(\sin\theta + \cos\phi \sin\theta')^2 + \sin\phi \sin\theta]^{1/2} \quad (6.3\text{-}24)$$

$$P = \cot\alpha(\cos\theta - \cos\theta') \quad (6.3\text{-}25)$$

$$I(x) = \frac{3}{2}\int_0^1 t^2 e^{ixt}\, dt = \frac{3}{2ix}\left[\exp(ix) - \frac{1}{2}\int_0^1 \frac{\exp(ixt)}{t^{1/2}}\, dt\right] \quad (6.3\text{-}26)$$

The function $I(x)$ above is similar to the complex Fresnel integral,[43] and as $x \to 0$,

$$I(x) \approx 1 + i(\tfrac{3}{5})\, x - \tfrac{3}{14}x^2 + \cdots \quad (6.3\text{-}27)$$

As $x \to \infty$,

$$I(x) \approx \frac{3}{2}\frac{\exp ix}{ix} - \frac{3}{4x}\left(\frac{\pi}{ix}\right)^{1/2} + \frac{3}{4x^2}\exp ix + \cdots \quad (6.3\text{-}28)$$

For an incident magnetic field polarized in an arbitrary direction $\hat{e}_2$,

$$F_m(\hat{k}_0{}^i, \hat{r}') = k_0[\hat{r}' \times (\hat{e}_2 \times \hat{n})]\{\tfrac{2}{3}(k_0 a/2\pi D)^{1/2} \, lI[k_0 a(P - D)]\} \quad (6.3\text{-}29)$$

If $\hat{r}'$ is in the plane defined by the direction of incidence and the cone axis, and if $\hat{e}_2 = \hat{y}$, Eqs. (6.3-23) and (6.3-29) reduce to

$$F_e = -\frac{2lk_0}{3}\exp\left(-i\frac{\pi}{4}\right)\left(\frac{a}{\lambda_0(\sin\theta + \sin\theta')}\right)^{1/2}\sin(\theta' - \alpha)$$
$$\times I[k_0 l \cos(\theta + \alpha) - k_0 l \cos(\theta' - \alpha)](\hat{e}_3 \sin\tau - \hat{n}\cos\tau) \quad (6.3\text{-}30)$$

and

$$F_m = \frac{2lk_0}{2}\exp\left(-i\frac{\pi}{4}\right)\left(\frac{a}{\lambda_0(\sin\theta + \sin\theta')}\right)^{1/2}\sin(\theta' - \alpha)$$
$$\times I[k_0 l \cos(\theta + \alpha) - k_0 l \cos(\theta' - \alpha)]\,\hat{y} \quad (6.3\text{-}31)$$

where $\hat{e}_3$ is the direction of the reflecting generator and

$$\hat{k}_0{}^i \cdot \hat{e}_3 = \cos\tau$$

If $\hat{e}_1 = \hat{y}$, then Eqs. (6.3-23) and (6.3-29) reduce to

$$F_e = -\frac{2lk_0}{3}\exp\left(-i\frac{\pi}{4}\right)\left(\frac{a}{\lambda_0(\sin\theta + \sin\theta')}\right)^{1/2}\sin(\theta + \alpha)$$
$$\times I[k_0 l \cos(\theta + \alpha) - k_0 l \cos(\theta - \alpha)]\,\hat{y} \quad (6.3\text{-}32)$$

and

$$\mathbf{F}_m = -\frac{2lk_0}{3} \exp\left(-i\,\frac{\pi}{4}\right)\left(\frac{a}{\lambda_0(\sin\theta + \sin\theta')}\right)^{1/2} \sin(\theta + \alpha)$$

$$\times\, I[k_0l\cos(\theta + \alpha) - k_0l\cos(\theta - \alpha)](\hat{\mathbf{e}}_3\sin\tau - \hat{\mathbf{n}}\cos\tau) \quad (6.3\text{-}33)$$

6.3.1.2.2.1.2. *Born Approximation.* Hodge and Schultz[46] attempt to improve the physical-optics approximation by use of the Born approximation.[47] They consider the case of scattering from a cone whose slant height $l$ is much larger than the observing wavelength. The type of termination does not effect the cross section to the order of this approximation. The first Born iteration assumes the scattered field is related to the incident field as

$$\mathbf{H}_1^s(\mathbf{r}) = \frac{1}{4\pi} \iint_s [\hat{\mathbf{n}}(\mathbf{r}') \times \mathbf{H}^i(\mathbf{r}')] \times \nabla'\psi_0(\mathbf{r}, \mathbf{r}')\, ds' \quad (6.3\text{-}34)$$

where $\psi_0(\mathbf{r}, \mathbf{r}')$ is the free-space Green's function defined by Eq. (2.2-21), and $\hat{\mathbf{n}}$ is the outward unit vector normal to the surface at the point $\mathbf{r}'$. It is seen that $\mathbf{H}_1^s(\mathbf{r})$ is one half the usual physical-optics scattered field. The second Born iteration replaces the incident field with the field determined by Eq. (6.3-34), giving

$$\mathbf{H}_2^s(\mathbf{r}) = \frac{1}{4\pi} \iint_s [\hat{\mathbf{n}}(\mathbf{r}') \times \mathbf{H}_1^s(\mathbf{r}')] \times \nabla'\psi_0(\mathbf{r}, \mathbf{r}')\, ds' \quad (6.3\text{-}35)$$

Hodge and Schultz express $\mathbf{H}_2^s(\mathbf{r})$ as the sum of two magnetic fields, $\mathbf{H}_1^s(\mathbf{r})$ and a correction term. The surface integral for $\mathbf{H}_1^s(\mathbf{r})$, for the case of the incident field polarized perpendicular to the plane of incidence, can be partially evaluated analytically, giving

$$\mathbf{H}_1^s(\mathbf{r}) = -\frac{iH^i e^{ik_0 r}}{2r}\,\hat{\mathbf{y}} \int_0^l k_0[\cos u \sin \alpha J_0(2k_0r'\sin u\sin\alpha)$$

$$-\, i\sin u\cos\alpha J_1(2k_0r'\sin u\cos\alpha)][\exp(i2k_0r'\cos u\cos\alpha)]\sin\alpha\,r'\,dr'$$

$$(6.3\text{-}36)$$

where

$$\cos u = \hat{\mathbf{k}}_0^i \cdot \hat{\mathbf{z}}$$

This remaining integral must be evaluated numerically. The correction term is expanded in terms of spherical eigenfunctions. Again one integral could be evaluated analytically. However, the resulting expression is too complex for inclusion here. This expression was evaluated at Purdue University for axial incidence, a simplification which reduces the number of infinite summations from four to two. The double infinite series was evaluated numerically by considering successively larger square arrays up to one hundred elements. The radar cross section had not yet converged to a steady value at this time, thus limiting the usefulness of this technique. The results of this calculation are

shown in Figure 6-25, along with the physical-optics approximation of
Eq. (6.3-34) and the numerical approximation to the exact solution of a cone
spherical segment as generated by Schultz et al.,[26] and discussed in
Section 6.3.1.1. It is seen that the second Born approximation gives values
an order of magnitude larger than the physical-optics approximation,
providing better agreement with the previous curve of Schultz; however,
there is essentially no agreement in the fine structure. Whether use of still
more terms would give better agreement is unknown.

   6.3.1.2.2.1.3. *Cylindrical Current Approximation.* Another approximation for determining the backscattering radar cross section of bodies of
circular geometrical cross section, but for which the radius of the circle
varies along the axis, is the cylindrical current method of Dawson and
Turner.[40-42] The assumption is made that the surface current density at
any point on the body is the same as that which would exist on an infinitely
long, perfectly conducting circular cylinder with the same incident field. The
radius of the cylinder is the radius of the cone at the particular point along
a perpendicular to the axis of the cone at which measurement of the current

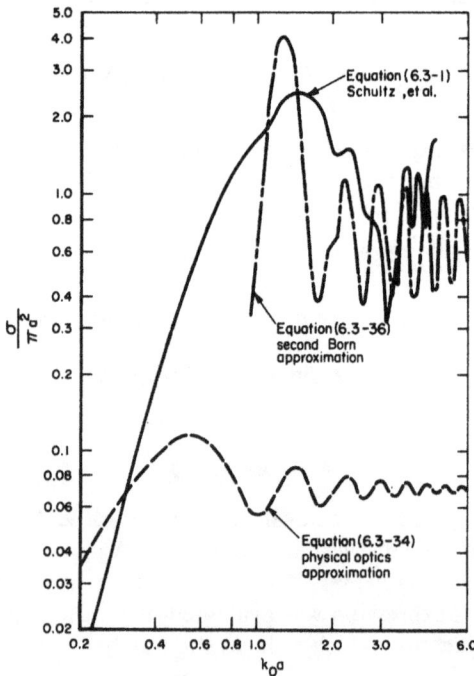

Fig. 6-25.  Comparison of axial backscatter cross
section of a cone of 15° half-angle as predicted by
three techniques [after Hodge and Schultz[46]].

is desired. The cross section is determined by summing the contributions from the individual portions of the infinite cylinder.

For the case where the electric vector is perpendicular to the body axis (perpendicular polarization), the scattered field is, for $\theta > \alpha$,

$$E^s = \frac{iE^i}{\pi r'} \exp(ik_0 r') \int_0^{L_s} S_1 \exp(-2ik_0 z \cos \theta) \, dz \qquad (6.3\text{-}37)$$

where $z$ is the distance along the cone axis measured from the tip of the cone, $L_s$ is the distance to the shadow boundary, and

$$S_1 = \sum_{-\infty}^{\infty} \frac{(-1)^n J_n'(k_0 r_c \sin \theta)}{H_n^{(1)'}(k_0 r_c \sin \theta)} \qquad (6.3\text{-}38)$$

where $r_c$ is the radius of the equivalent cylinder. The radar cross section is

$$\sigma(\theta) = \frac{4}{\pi} \left| \int_0^{L_s} S_1 \exp(-2ik_0 z \cos \theta) \, dz \right|^2 \qquad (6.3\text{-}39)$$

For the case where the magetic vector of the incident field is perpendicular to the body axis (parallel polarization), Eqs. (6.3-37) and (6.3-39) apply, except $S_1$ is replaced by $S_2$ where

$$S_2 = \sum_{-\infty}^{\infty} (-1)^n \frac{J_n(k_0 r_c \sin \theta)}{H_n^{(1)}(k_0 r_c \sin \theta)} \qquad (6.3\text{-}40)$$

In Figures 6-26 and 6-27 the theoretical results from Eqs. (6.3-37) to (6.3-40) are compared with experimental results for two polarizations.[41] In both figures it is seen that the results agree well for $\theta = 55°$ to $\theta = 110°$. Since this formulation neglects traveling waves and creeping waves, and edge effects are not taken into account properly, it would be expected to be valid only for the specular echo. A heuristic combination of cylindrical current theory with surface wave and edge effect theories should produce better agreement.

Using the cylindrical current approximation, Dawson and Turner have approximated the radar cross section of a truncated cone with a flat base for perpendicular polarization as

$$\sigma = \frac{\pi \sin \theta \cos^2 \alpha \sin \alpha}{16 k_0^2 \cos^3(\theta + \alpha)} \left| K_1 \exp\left(-\frac{\pi i}{2} K_1^2\right) \right.$$
$$\left. - K_2 \exp\left(-\frac{\pi i}{2} K_2^2\right) - C(K_1) + iS(K_1) + C(K_2) - iS(K_2) \right|^2 \qquad (6.3\text{-}41)$$

where

$$K_1^2 = (4/\pi) k_0 a (\sin \theta - \cot \alpha \cos \theta) \qquad (6.3\text{-}42)$$

$$K_2^2 = (4/\pi) k_0 (a - L_s \tan \alpha)(\sin \theta - \cot \alpha \cos \theta) \qquad (6.3\text{-}43)$$

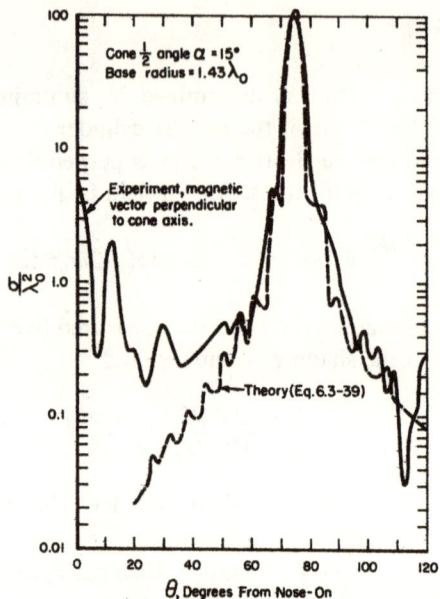

Fig. 6-26. Comparison of backscatter from a flat-backed cone as a function of aspect angle as predicted by cylindrical current theory with measured results [after Dawson et al.[41]].

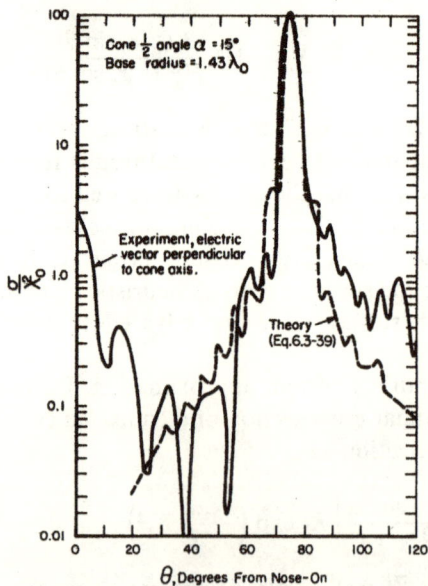

Fig. 6-27. Comparison of backscatter from a flat-backed cone as a function of aspect angle as predicted by cylindrical current theory with measured results [after Dawson et al.[41]].

and $C(x)$, $S(x)$ are the Fresnel integrals.[43] For the case of a pointed cone, Eq. (6.3-41) reduces to

$$\sigma = \lambda_0 a \left[ \frac{1 + \dfrac{C^2(\xi) + S^2(\xi)}{\xi^2} - \dfrac{2}{\xi}\left[ C(\xi) \cos\left(\dfrac{\pi}{2}\,\xi^2\right) + S(\xi) \sin\left(\dfrac{\pi}{2}\,\xi^2\right)\right]}{8\pi \sin\theta \tan^2\alpha(\cot\theta \cot\alpha - 1)} \right]$$

(6.3-44)

where

$$\xi^2 = (4/\pi)\, k_0 a \sin\theta(\cot\theta \cot\alpha - 1)$$

At an angle $\theta = \pi/2 - \alpha$, the radar looks normal to the side of the cone. For this case, Eq. (6.3-41), for a truncated cone, simplifies to

$$\sigma = \tfrac{4}{9} k_0 \cos\alpha \cot^2\alpha[a^{3/2} - a_1^{3/2}]^2$$

(6.3-45)

where $a_1$ = radius of small end of truncated cone. For a pointed cone, $a_1 = 0$, and Eq. (6.3-45) reduces to

$$\sigma = \frac{8\pi a^3}{9\lambda_0} \cos\alpha \cot^2\alpha$$

(6.3-46)

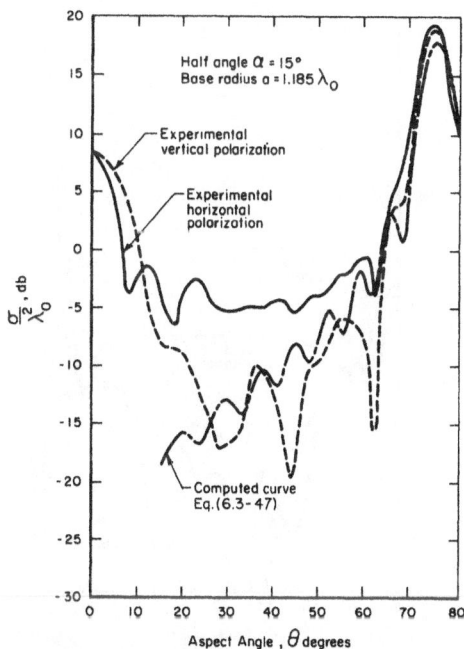

Fig. 6-28. Comparison of backscatter from a flat-backed cone as a function of aspect angle as predicted by cylindrical current theory with measured results [after Melling and Clark[48]].

Equations (6.3-45) and (6.3-46) can only be expected to give good results for specular scattering when the cones are several wavelengths in extent.

Melling and Clark[48] evaluated the integrand obtained in the cylindrical current method for a flat-base cone by the stationary phase method and obtained the first two terms. The resulting expression is independent of polarization, or

$$\frac{\sigma}{\lambda_0^2} = \frac{k_0 a \sin^2 \theta}{16\pi^2(\cos \theta - \sin \theta \tan \alpha)^2} \left| 1 + \frac{\sqrt{\pi}}{2} \frac{\exp[i(A + 3\pi/4)]}{A^{1/2}} \right|^2 \quad (6.3\text{-}47)$$

where

$$A = 2k_0 a \left[ \frac{\cos \theta - \sin \theta \tan \alpha}{\tan \alpha} \right] \quad (6.3\text{-}48)$$

and $\theta > \alpha, \neq \pi/2 - \alpha$.

Figures 6-28 and 6-29 compare the results of computing the cross section of 9.6° and 15° half-angle cones by Eq. (6.3-47) with measurements taken by Keys and Primich.[35] Considering the polarization independence of the

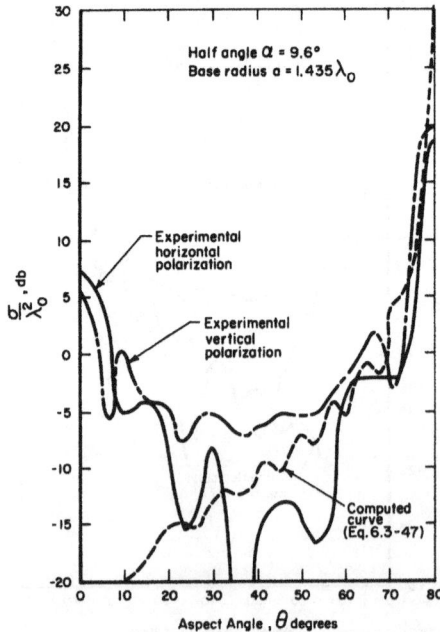

Fig. 6-29. Comparison of backscatter from a flat-backed cone as a function of aspect angle as predicted by cylindrical current theory with measured results [after Melling and Clark[48]].

theory, there is good agreement from 25 to 78° in aspect angle. No comparisons were made beyond 80°.

The expression for the cross section of a cone-sphere obtained by the cylindrical current technique is very complex, and is available in the literature.[41,48] Figure 6-30 compares the results with experiment for a cone-sphere with $\alpha = 15°$ and sphere radius equal to $7.15\lambda_0$. The experimental results[41] are said to be accurate to $\pm 3°$ in angle $\theta$, and $\pm 10\%$ in magnitude. Also shown on the figure is a point predicted by the wedge approximation and by the physical approximation. The experimental results can be approximated grossly up to $\theta = 55°$ by $\sigma = .01\lambda_0 a$, which can be obtained by crudely approximating the terms in Dawson's formulation.[41]

6.3.1.2.2.1.4. *Geometrical Diffraction.* One of the better high-frequency approximations is the geometrical theory of diffraction developed by J. B. Keller. The basic theory is summarized in Section 2.2.2.2. The optical concept of a ray is extended to include reflected, refracted, and diffracted rays. Tips, edges, and other singularities in body shape give rise to diffracted rays.

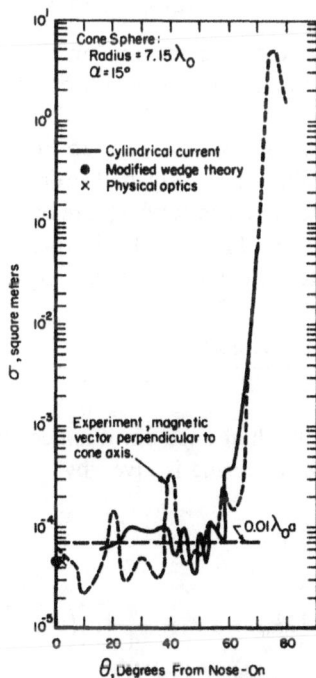

Fig. 6-30. Comparison of backscatter from a cone-sphere as a function of aspect angle as predicted by cylindrical current theory with results obtained by several other techniques [after Dawson et al.[41]].

The value of the diffracted field created by these singularities is determined by evaluation of diffraction coefficients which are found from asymptotic expansions of the exact solutions for canonical bodies, such as infinite wedges, spheres, etc. The geometrical theory of diffraction, which has been applied to flat-back cones[49-52] and cone-spheres,[49,50] provides good agreement with experimental results. Although the technique is a high-frequency approximation, the theoretical results agree reasonably well with experimental results for bodies of a few wavelengths in dimension.

For the case of a flat-backed cone, Keller's result as corrected by Bechtel[52] is, for axial incidence,

$$
\frac{\sigma}{\lambda_0^2} = \frac{1}{\pi} \left[ \frac{k_0 a \sin(\pi/n)}{n} \right]^2 \left| \left[ \cos\frac{\pi}{n} - \cos\frac{3\pi}{n} \right]^{-1} \right.
$$
$$
\left. + \left( \sin\frac{\pi}{n} \right) [n\sqrt{\pi k_0 a}]^{-1} \left[ \cos\frac{\pi}{n} - \cos\frac{3\pi}{2n} \right]^{-2} \exp i \left[ 2k_0 a - \frac{\pi}{4} \right] \right|^2
$$

$$(6.3\text{-}49)$$

where
$$
n = (3/2) + (\alpha/\pi)
$$

Equation (6.3-49) predicts that the maxima in the oscillation of the back-scattered cross section for axial incidence will occur at $k_0 a = (m + \frac{5}{8})$, and that the minima will occur at $k_0 a = (m + \frac{9}{8})$ where $m = 0, 1, 2,...$

For large values of $k_0 a$, the second term in Eq. (6.3-49) is much smaller than the first, which implies that only first-order diffraction is important since the second term arises from rays which are diffracted at one edge, cross the base of the cone, and are diffracted back toward the radar at the other edge. If the second-order terms can be neglected, Eq. (6.3-49) can be simplified to

$$
\frac{\sigma}{\lambda_0^2} = \frac{1}{\pi} \left[ \frac{k_0 a \sin(\pi/n)}{n} \right]^2 \left[ \cos\frac{\pi}{n} - \cos\frac{3\pi}{n} \right]^{-2} \qquad (6.3\text{-}50)
$$

For aspect angles other than nose-on, it is necessary to consider several cases. For $0 < \theta < \alpha$, the scattering is given by

$$
\frac{\sigma_\parallel^\perp(\theta)}{\lambda_0^2} = \frac{k_0 a}{4\pi^2} \left( \frac{\sin(\pi/n)}{n} \right)^2 \frac{1}{\sin\theta}
$$
$$
\times \left| \exp\left[ -i\left( 2k_0 a \sin\theta - \frac{\pi}{4} \right) \right] \left[ \left( \cos\frac{\pi}{n} - 1 \right)^{-1} \right.\right.
$$
$$
\mp \left( \cos\frac{\pi}{n} - \cos\frac{3\pi - 2\theta}{n} \right)^{-1} \right] + \exp\left[ i\left( 2k_0 a \sin\theta - \frac{\pi}{4} \right) \right]
$$
$$
\left. \times \left[ \left( \cos\frac{\pi}{n} - 1 \right)^{-1} \mp \left( \cos\frac{\pi}{n} - \cos\frac{3\pi + 2\theta}{n} \right)^{-1} \right] \right|^2 \qquad (6.3\text{-}51)
$$

where the superscript $\perp$ indicates polarization perpendicular to the plane determined by the vector, $\hat{\mathbf{k}}_0{}^i$, and the axis of the cone ($z$ axis), and requires the upper signs in Eq. (6.3-51) to be used. The subscript 11 goes with the lower signs and indicates polarization parallel to the plane of $\hat{\mathbf{k}}_0{}^i$ and $z$.

When the cone is viewed along a direction normal to the generator of the cone, that is, $\theta = (\pi/2) - \alpha$, the geometrical theory of diffraction fails. Bechtel makes an asymptotic expansion of the physical-optics expression and obtains an expression equivalent to Eq. (6.3-18), or

$$\frac{\sigma}{\lambda_0{}^2} = \frac{8}{9}\pi\left(\frac{a}{\lambda_0}\right)^3 [\sin^2\alpha\cos\alpha]^{-1} \tag{6.3-52}$$

This expression is valid only for $\theta = (\pi/2) - \alpha$. For other angles of $\theta$ between $\alpha$ and $\pi/2$, geometrical theory of diffraction predicts

$$\frac{\sigma_{\parallel}{}^{\perp}(\theta)}{\lambda_0{}^2} = \frac{k_0 a}{4\pi^2}\left(\frac{\sin(\pi/n)}{n}\right)^2 \frac{1}{\sin\theta}\left[\left(\cos\frac{\pi}{n} - 1\right)^{-1}\right.$$
$$\left.\mp\left(\cos\frac{\pi}{n} - \cos\frac{3\pi - 2\theta}{n}\right)^{-1}\right]^2 \tag{6.3-53}$$

For $\pi/2 < \theta < \pi$, the corresponding equation is

$$\frac{\sigma_{\parallel}{}^{\perp}(\theta)}{\lambda_0{}^2} = \frac{k_0 a}{4\pi^2}\left(\frac{\sin(\pi/n)}{n}\right)^2 \frac{1}{\sin\theta}\left| \exp\left[-i\left(2k_0 a\sin\theta - \frac{\pi}{4}\right)\right]\right.$$
$$\times\left[\left(\cos\frac{\pi}{n} - 1\right)^{-1} \mp \left(\cos\frac{\pi}{n} - \cos\frac{3\pi - 2\theta}{n}\right)^{-1}\right]$$
$$+ \exp\left[i\left(2k_0 a\sin\theta - \frac{\pi}{4}\right)\right]\left[\left(\cos\frac{\pi}{n} - 1\right)^{-1}\right.$$
$$\left.\mp\left(\cos\frac{\pi}{n} - \cos\frac{2\theta - \pi}{n}\right)^{-1}\right]\bigg|^2 \tag{6.3-54}$$

When the base of the cone is viewed specularly, i.e., $\theta = \pi$, Bechtel states that the radar cross section is given very well by the physical-optics expression for specular scattering from a disc, or

$$\frac{\sigma(\pi)}{\lambda_0{}^2} = \frac{(k_0 a)^4}{4\pi} \tag{6.3-55}$$

Keller[51] has compared the results of his theory with the measured data of Keys and Primich[35] with good agreement for a number of cases. Bechtel[52,53] points out that the agreement is even better if the experimental accuracy is taken into account. The measured data are claimed to be valid within $\pm 2$ db. Bechtel replots several of the measured curves of Keys and Primich showing the 4 db spread of the data. Curves for parallel and perpendicular polarization

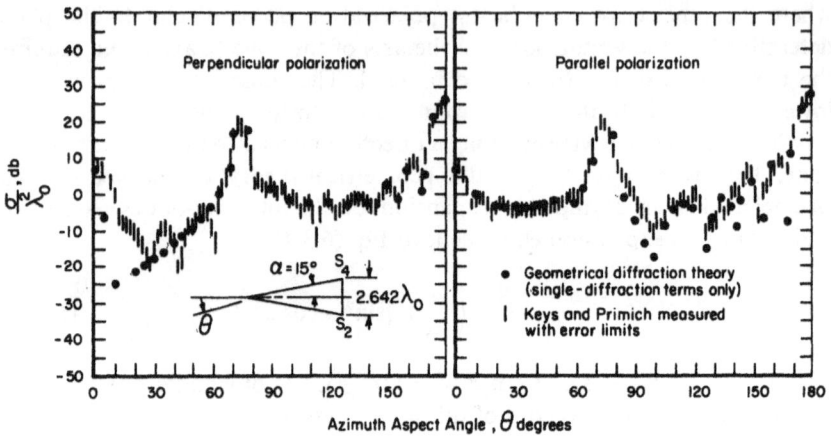

Fig. 6-31. Comparison of backscatter cross section as a function of aspect angle as predicted by geometrical theory of diffraction with measured results [after Bechtel[52]].

for $\alpha = 15°$ and $\alpha = 4°$ cones are shown in Figures 6-31 and 6-32, respectively. The base diameter of the cones is $2.64\lambda_0$ ; thus $k_0 a = 8.3$.

The agreement for parallel polarization and for perpendicular polarization at aspect angles greater than about 40° is seen to be quite good. Some of the measured nulls are not as deep as the nulls predicted by theory, but this may be due to experimental error because of scattering from the target-support structures and other background clutter. The poor agreement for perpendicular polarization at aspect angles less than 40° has not yet been explained.

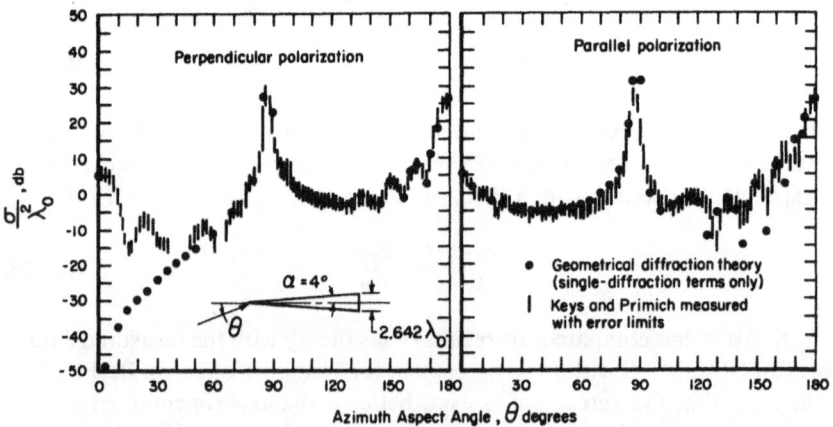

Fig. 6-32. Comparison of backscatter cross section as a function of aspect angle as predicted by geometrical theory of diffraction with measured results [after Bechtel[52]].

Since the geometrical theory of diffraction states that the radar cross section is proportional to the sum of all the rays diffracted in the direction of the receiving antenna, and that each ray originates on some discontinuity of the target, it is possible to think of the cross section as an infinite series. Each term in the series is the amplitude, $\sqrt{\sigma_j}$, of the field radiated from the $j$th scattering center associated with an appropriate phase angle $\rho_j$. The cross section is then written as

$$\sigma = \left| \sum_j \sqrt{\sigma_j}\, e^{i\rho_j} \right|^2 \tag{6.3-56}$$

Let the coordinate system be placed so that $\hat{k}_0{}^i$ is in the $\phi = 0$ plane at an angle $\theta$ from the body axis. Bechtel and Ross[53] show that for axially asymmetric edge diffraction the amplitude may be written as

$$\sqrt{\sigma_j}\Big|_\parallel^\perp = \frac{\sin \pi/n}{n} \sqrt{\frac{a \csc \psi_1}{k_0 \cos(\beta/2)}} \left[ \left( \cos \frac{\pi}{n} - \cos \frac{\pi + 2\beta}{n} \right)^{-1} \right.$$
$$\left. \mp \left( \cos \frac{\pi}{n} - \cos \frac{\beta_1}{n} \right)^{-1} \right], \tag{6.3-57}$$

where $\beta$ is the angle between $\hat{k}_0{}^i$ and the desired $\hat{k}_0{}^s$, $\beta_1$ is the projection of $\beta$ on the $\phi = 0$ plane, $\psi_1 = \theta + \beta_1/2$, $n = (3/2) + (\alpha/\pi)$, and $a$ is the distance to the axis of symmetry of the target. In the case of the scattering from the base of a cone, $a$ is equal to the radius of the base as before. The phase associated with each amplitude is given by

$$\rho_j = [-(2k_0 \cos(\beta/2)(a \sin \theta + h \cos \theta) + \pi/4] \tag{6.3-58}$$

where $h$ is the distance along the axis of symmetry from the base to the phase reference. Choosing the tip of the cone as the phase reference makes $h$ equal to the height of the cone as before. Bechtel and Ross[53] show the application of the concept in Eq. (6.3-49), where only two terms are used in the series. Figure 6-33 illustrates the cross section contributions from the two scattering centers at the base of the cone. Special considerations must be taken at angles of 0, $\pi/2$, and $\pi$. For example, the curves may be smoothly joined with the value predicted by physical optics at the specular flare, and with the value predicted by scattering from a disc for the base.

They also applied the scattering center concept to a doubly truncated cone.(frustum). The angle of the frustum is the same as the half-angle of the cone that would be formed if the sides of the frustum were extended until they met. This angle $\alpha$ is related to height, $h_f$, of the frustum and the radii of its two truncated ends, by

$$\alpha = \tan^{-1}\left( \frac{a_2 - a_1}{h_f} \right) \tag{6.3-59}$$

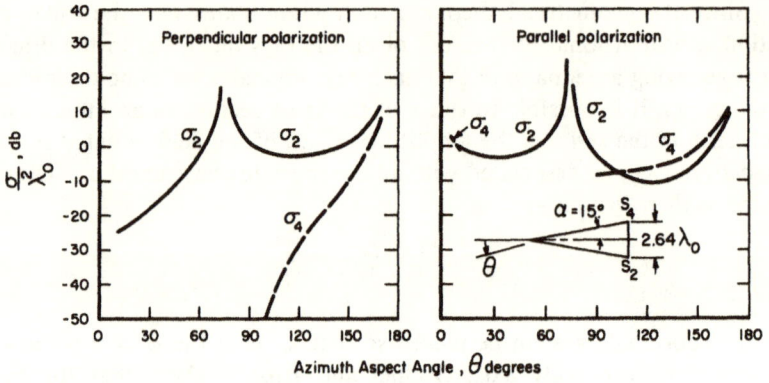

Fig. 6-33. Backscatter radar cross sections of scattering centers on cone [after Bechtel and Ross[53]].

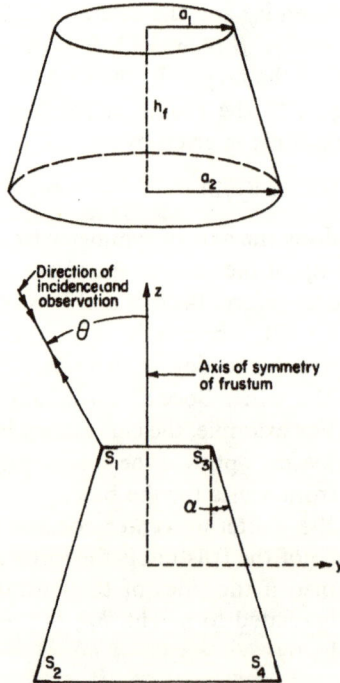

Fig. 6-34. Scattering centers of frustum.

The frustum has wedge-like scattering centers, $S_1$, $S_2$, $S_3$, and $S_0$, as illustrated in Figure 6-34. The center, $S_1$, is assumed to contribute for $\theta < \pi - \alpha$, $S_2$ for $\theta > \alpha$, $S_3$ for $\theta < \pi/2$, and $S_4$ for $\pi < \theta < \alpha$. The backscattered cross section of the frustum is

$$\sigma(\theta) = |\sqrt{\sigma_1}\, e^{i\rho_1} + \sqrt{\sigma_2}\, e^{i\rho_2} + \sqrt{\sigma_3}\, e^{i\rho_3} + \sqrt{\sigma_4}\, e^{i\rho_4}|^2 \qquad (6.3\text{-}60)$$

The values of $\sigma_j$ and $\rho_j$ are determined from Eqs. (6.3-57) and (6.3-58), respectively. For example, the interior wedge angle at $S_1$ is $\pi/2 + \alpha$; therefore, the backscattering from $S_1$ is given by

$$\sqrt{\sigma_{1_1}^\perp} = \frac{\sin\{\pi/[(3/2) - (\alpha/\pi)]\}}{(3/2) - (\alpha/\pi)} \sqrt{\frac{a_1 \csc \theta}{k_0}} \left\{ \left[ \cos\left( \frac{\pi}{(3/2) - (\alpha/\pi)} \right) \right.\right.$$
$$\left.\left. - \cos\left( \frac{\pi + 2\theta}{(3/2) - (\alpha/\pi)} \right) \right]^{-1} \mp \left[ \cos\left( \frac{\pi}{(3/2) - (\alpha/\pi)} \right) - 1 \right]^{-1} \right\} \quad (6.3\text{-}61)$$

Figure 6-35 shows the results of evaluating Eq. (6.3-60) for parallel polarization for a frustum of dimensions $a_1 = 2.057$ cm, $a_2 = 5.753$ cm, $h_f = 20.945$ cm, and $\lambda_0 = 0.861$ cm. Thus, $\alpha = 10.03°$. Also shown in the

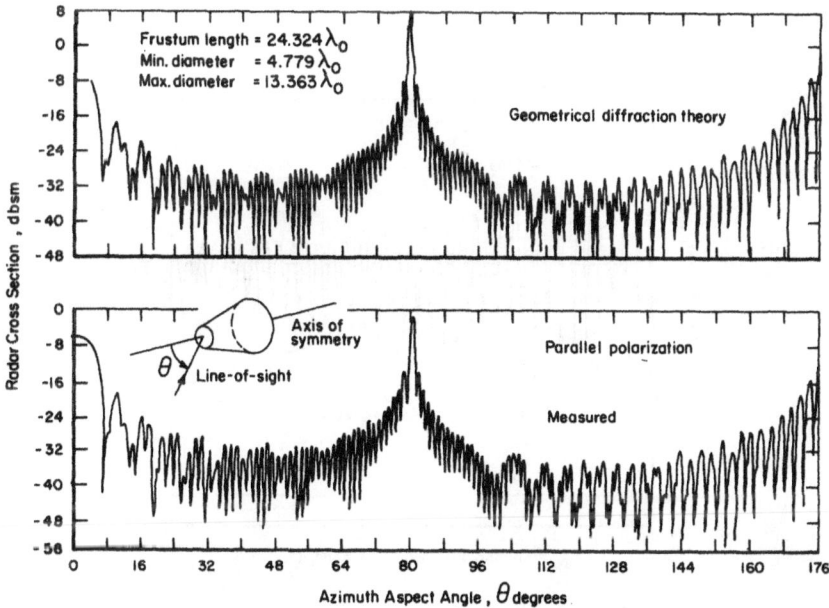

Fig. 6-35. Comparison of computed and measured radar cross section of conical frustum [after Bechtel and Ross[53]].

figure is the measured cross section for parallel polarization. The agreement is very good. Similar agreement was reached for perpendicular polarization, as shown in Reference 53.

*6.3.1.2.2.1.5. Superposition of Scattering Components (Cone-Sphere).* Returning to the cone-sphere, the nose-on backscatter may be attributed to four scattering components. These contributors include tip scattering; scattering from the join between the cone and the sphere; creeping waves; and for long thin cones, traveling waves. The relative magnitude of the contributions for nose-on backscatter from a 15° half-angle cone-sphere with a sphere radius varying between 0 and 10 wavelengths is illustrated in Figure 6-36 from Reference 44. Since a 15° half-angle cone is not a long thin body, the traveling wave component is not shown. Neglecting the traveling wave, Pannell, Rheinstein, and Smith[44] use the physical-optics expression for the tip and join return, and determine the creeping wave contribution from the exact solution for a sphere (see Chapter 3). For axial incidence, the cross section is

$$\sigma(0) = \frac{\lambda_0^2}{\pi} \left| -\frac{i}{4} \left[ \tan^2 \alpha \exp(-2ik_0 a \csc \alpha) \right] \right.$$

$$\left. - \sec^2 \alpha \exp(-2ik_0 a \sin \alpha) + F^c(0) \right|^2 \qquad (6.3\text{-}62)$$

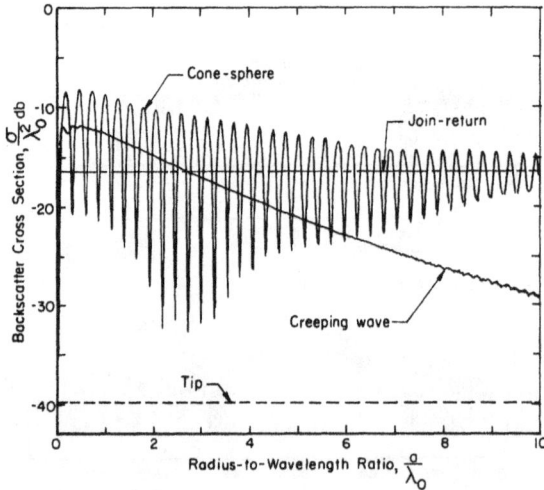

Fig. 6-36. Predicted axial backscatter cross section of the scattering components for a cone-sphere of 15° half-angle and their sum [after Pannell *et al.*[44]].

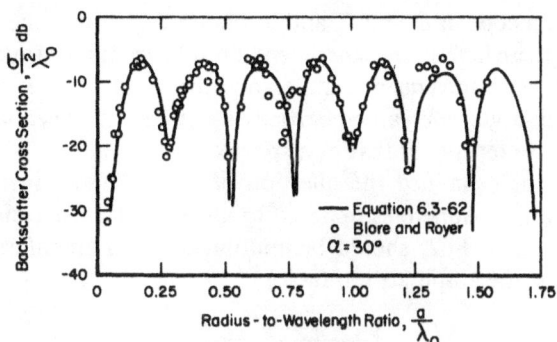

Fig. 6-37. Comparison of predicted and measured axial backscatter cross section of a cone-sphere of 30° half-angle [after Pannell *et al.*[44]].

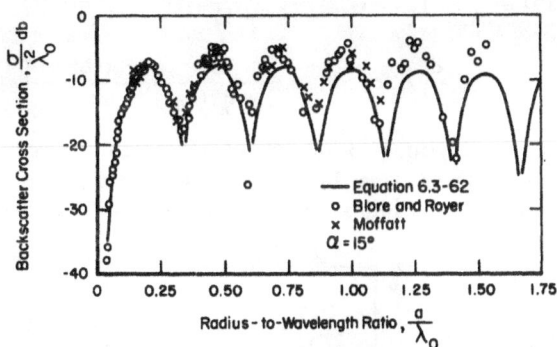

Fig. 6-38. Comparison of predicted and measured axial backscatter cross section of a cone-sphere of 15° half-angle [after Pannell *et al.*[44]].

where $F^c(0)$ is the creeping wave contribution and is[†]

$$
\begin{aligned}
F^c(0) = & \left(\frac{k_0 a}{2}\right)^{4/3} \exp\left[i\pi\left(k_0 a - \frac{2}{3}\right)\right] \frac{1}{\beta_c[\text{Ai}(-\beta_c)]^2} \\
& \times \left[1 + \frac{8\beta_c}{15}\left(1 + \frac{9}{32\beta_c^3}\right)\left(\frac{2}{k_0 a}\right)^{2/3} \exp(i\pi/3)\right] \\
& \times \exp\left[-\beta_c\pi\left(\frac{k_0 a}{2}\right)^{1/3} \exp(-i\pi/6) - \frac{\beta_c^2 \pi}{60}\left(1 - \frac{9}{\beta_c^3}\right)\right. \\
& \times \left.\left(\frac{2}{k_0 a}\right)^{1/3} \exp(i\pi/6)\right]
\end{aligned}
$$

(6.3-63)

where $\beta_c = 1.01879...$, and $\text{Ai}(-\beta_c) = 0.53565...$ .

Equation (6.3-62) has been plotted by Pannell *et al.*, and is shown in

[†] See Eq. (3.2-8) of Chapter 3, where the coefficients have been evaluated numerically.

Figure 6-37 for a cone of $\alpha = 30°$, and in Figure 6-38 for a cone of $\alpha = 15°$. Also shown in these figures are experimental results from Blore and Royer[4,54] and Moffatt.[55] The agreement is quite good for smaller values of $a/\lambda_0$. The disagreement at larger $a/\lambda_0$ can be partially attributed to the enhancement of the creeping wave term over that of a sphere.

Senior[56] has examined the question of the difference in the creeping wave field for a cone-sphere as opposed to a sphere. He determined that the $F^c(0)$ term in Eq. (6.3-62) should be multiplied by an amplification factor. Senior arrives at the amplification factor

$$\left| \frac{2}{T_2[(\pi/2) - \alpha]} \right| \tag{6.3-64}$$

from considerations of the currents on the cone. The value of $T_2$ is given by

$$T_2(\theta) = \frac{1}{k_0 a} \sum_{n=1}^{\infty} (-i)^{n+1} \frac{2n+1}{n(n+1)} \left[ \left( \frac{\partial [k_0 a h_n^{(1)}(k_0 a)]}{\partial k_0 a} \right)^{-1} \right.$$
$$\left. \times \frac{\partial}{\partial \theta} P_n'(\cos \theta) - \frac{i}{k_0 a h_n^{(1)}(k_0 a)} \frac{P_n'(\cos \theta)}{\sin \theta} \right] \tag{6.3-65}$$

With this correction to Eq. (6.3-62), better agreement between theory and measurements was obtained as illustrated in Figure 6-39. Many measurements were taken for each point. The points indicate the mean of the mea-

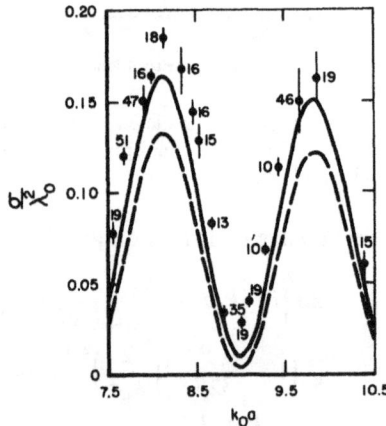

Fig. 6-39. Axial backscatter cross section of a cone-sphere of $12\frac{1}{2}°$ half-angle with base radius 4.519 cm; dashed curve computed from Eq. (6.3-62); solid curve contains correction term of Eq. (6.3-64); points indicate mean of measurements with numerals indicating number of measured values and lines indicating the standard deviations [after Senior[56]].

surements; the length of the vertical lines indicate the standard deviations; and the numerals indicate the number of measurements taken for this value of $k_0 a$. The dashed curve is computed from Eq. (6.3-62), and the solid curve is computed from this equation with $S$ multiplied by the correction factor, Eq. (6.3-64).

In general, reasonably good nose-on cross section estimates can be made for a cone-sphere for all frequencies. The Rayleigh curve agrees fairly well at lower frequencies, and intersects the high-frequency approximation obtained by summing the various scattering components in the neighborhood of $k_0 a = 1$. The impulse-response technique, discussed in Section 6.3.1.2.2.2, predicts behavior similar to that predicted by this section. At very high frequencies there are a variety of approximations that give good results.

6.3.1.2.2.2. *Impulse Approximation.* Previous sections of this chapter have considered the radar cross section of various types of cones at single frequencies. Radar signals are generally composed of many frequencies. In order to find the actual response one could form a weighted sum or integral over the responses at the frequencies comprising the waveforms. On the other hand, it might be advantageous to find the response to various waveforms. The impulse response of a target, $F_I(t)$, defines its scattering characteristics for a radar signal which is impulsive in time. The theoretical details of finding the impulse response are discussed in Section 2.2.2.4. The impulse response of a cone sphere has been discussed by Kennaugh[58-60] and Moffatt[55] for backscattering with axial incidence. As a first approximation they used the physical-optics result for an infinite cone, combined with the exact short-pulse response[62] of a sphere, minus the physical-optics scattering from the portion of the sphere blocked by the cone. Zero interaction was assumed at the cone-sphere junction. However, this formulation does predict a contribution from the cone-sphere junction, since the first return from the sphere is added at the transit time corresponding to the time of scattering from the junction. The first two moment conditions (repeated from Section 2.2.2.4 for convenience)

$$\int_{-\infty}^{\infty} F_I(t)\, dt = 0 \qquad (6.3\text{-}66)$$

and

$$\int_{-\infty}^{\infty} t F_I(t)\, dt = 0 \qquad (6.3\text{-}67)$$

were satisfied. No attempt was made to estimate Rayleigh-region backscatter. Thus, the third moment condition

$$\int_{-\infty}^{\infty} t^2 F_I(t)\, dt = -2K_R \qquad (6.3\text{-}68)$$

could not be imposed.

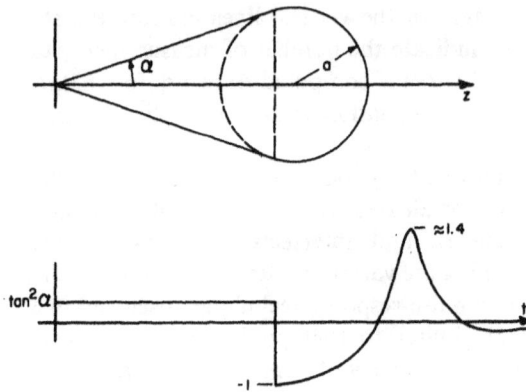

Fig. 6-40.  Predicted cone-sphere impulse response [after Moffatt[55]].

Figure 6-40 illustrates the predicted time-domain response. Figure 6-41 compares the radar cross section predicted by the impulse response with measured data taken at The Ohio State University.[55] The experimental data have an accuracy limit of $\pm 0.5$ db. The agreement is quite good, suggesting that interpolation is, perhaps, easier in the time domain than in the frequency domain.

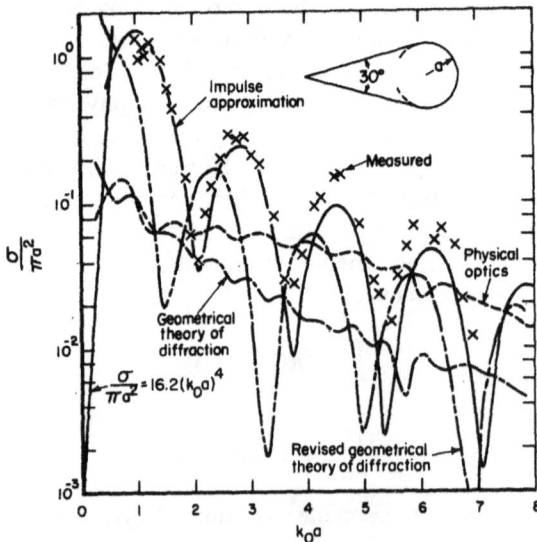

Fig. 6-41.  Comparison of axial backscatter radar cross section of a cone-sphere between measured values and values predicted by impulse approximation and by other techniques [after Moffatt[55]].

Siegel[59] has calculated the Rayleigh region cross section for this shape as $16.2(k_0 a)^4$. Siegel's Rayleigh curve is also included in Figure 6-41. Presumably, the Rayleigh curve and the impulse response would join smoothly if the third moment condition were satisfied. Also included in Figure 6-41 are the curves predicted by physical optics and by the geometrical theory of diffraction. The physical-optics curve neglects the creeping waves thus, the curve fails to predict the cancellation and enhancement effect caused by the interference between the creeping waves and the other contributors. The geometrical theory of diffraction curve neglects the contribution from the discontinuity in the second derivative of the curve of body shape at the cone-sphere junction. The curve in Figure 6-41, labeled revised geometrical theory of diffraction, represents Moffatt's attempt[55] to include a contribution from the junction discontinuity in the geometrical theory of diffraction. From the impulse analysis, the contribution from the cone-sphere junction referenced to the tip is, approximately,

$$\mathbf{E}^s = -\frac{ia\sec^2\alpha}{4k_0 a}\frac{\exp(ik_0 r)}{r}\exp(i2k_0 a\cot\alpha\cos\alpha)\,\mathbf{E}^i \qquad (6.3\text{-}69)$$

Moffatt adds this term to Keller's expression,[50] giving

$$\mathbf{E}^s = \frac{a\exp(ik_0 r)}{2r}\left[-\frac{2i}{k_0 a}\left(\frac{\alpha}{2}\right)^2 + \frac{(k_0 a)^{1/3}}{2}\exp[ik_0 a(\pi + 2\csc\alpha)]\right]$$
$$\times\left[\frac{2\pi^2}{6^{1/3}q_m}\frac{\exp(i\pi/3)}{A(q_m)^2}\exp\left(-\frac{\exp(-i\pi/6)}{6^{1/3}}q_m\pi(k_0 a)^{1/3}\right)\right.$$
$$\left. + \frac{\sec^2\alpha}{i2k_0 a}\exp[i2k_0 a\cot\alpha\cos\alpha]\right]\mathbf{E}^i \qquad (6.3\text{-}70)$$

where

$$q_m = 1.469354$$
$$A(q_m) = 1.16680$$

The revised geometrical theory of diffraction still does not appear to agree with the experimental data as well as the impulse approximation. It appears that the creeping wave contributions are not predicted correctly.

The impulse response clearly illustrates the contributions from the different scattering mechanisms.[61] In Figure 6-42a and b are shown the theoretical axial return in the time domain as computed by the impulse approximation for a cone-sphere of half-angle $\alpha = 15°$, and with a sphere of $k_0 a = 21.8$ for sine-squared envelope input pulse durations of 0.577 and 1.0 sphere transit times, respectively. In Figure 6-43a and b, the sphere has a value of $k_0 a = 4.5$, and the pulse durations are 1.396 and 2.793 sphere transit times. The contributions of the tip, the junction discontinuity, and the creeping waves are clearly evident. These predicted results compare very well with the short pulse measurements of Alongi, Kell, and Newton.[63]

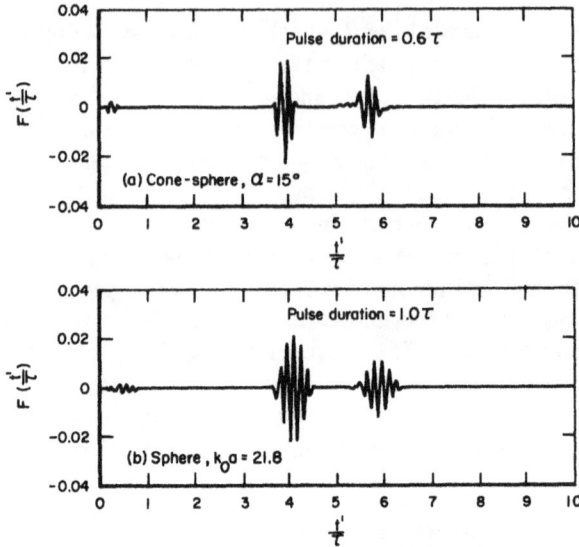

Fig. 6-42. Scattered-field waveform for a cone-sphere of 15° half-angle with sphere circumference $21.8\lambda_0$, sine-squared pulse, and sphere-diameter transit time equal to $\tau$ [after Kennaugh and Moffatt[61]].

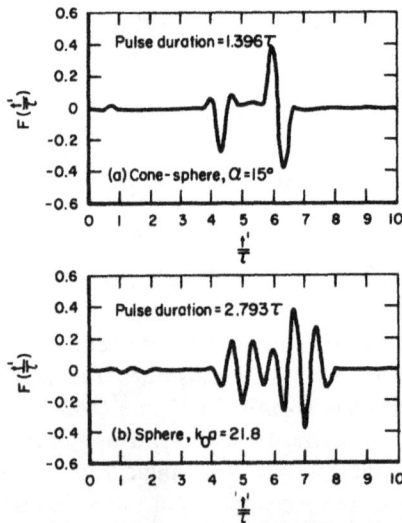

Fig. 6-43. Scattered field waveform for a cone-sphere of 15° half-angle with sphere circumference $4.5\lambda_0$, sine-squared pulse, and sphere-diameter transit time equal to $\tau$ [after Kennaugh and Moffatt[61]].

The impulse approximation predicts that changing the cone angles will have little effect on the magnitude of the nose-on return. It is stated[55] that the variations of axial echo area of an $\alpha = 7.5°$ cone-sphere would differ little from that of the $\alpha = 15°$ cone-sphere, illustrated in Figure 6-41. The maxima and minima would shift slightly in position because of the shift in the location of the cone-sphere junction.

Moffatt[55] also considered the effects of modifying the shape of the cone-sphere. The shapes considered and the predicted time-response junctions are illustrated in Figure 6-44. A cone-transition-sphere was considered where the transition section matched the geometrical cross-sectional area function and its first two derivatives at both the cone and the sphere. The functional form of the geometrical area of the transition section was assumed to be a fifth-order polynominal with unknown coefficients. The requirement that the area be monotonically increasing at the two boundaries was used to identify the six coefficients. The decrease in radar cross section caused by eliminating the junction contribution is more than compensated by the increase in cross section caused by the transition element. It is suggested that as the transition

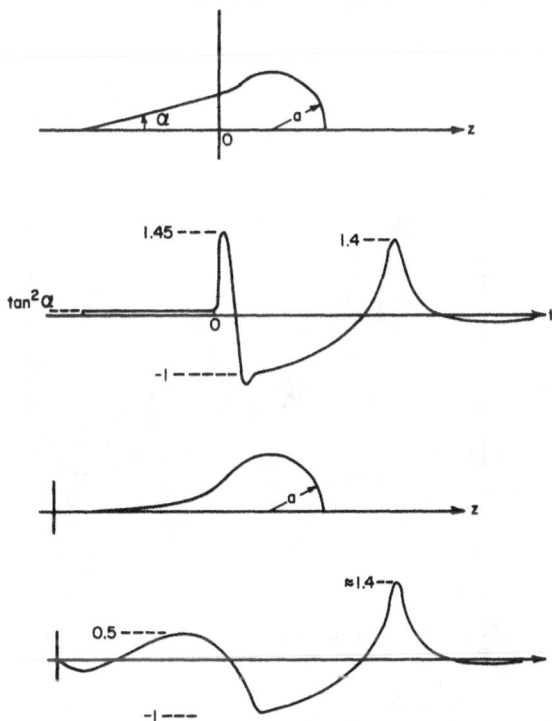

Fig. 6-44. Impulse time response of a cone-transition-sphere and of a polynomial sphere [after Moffatt[55]].

element increases in length, the time response would become smooth so that for very long transition sections the cross section would decrease. This agrees with the results obtained for a polynomial sphere. The fifth-order polynomial is assumed to have zero for the second-, first-, and zeroth-order coefficients. The resulting shape is basically the same as when the transition section in the cone-transition-sphere is lengthened. The resultant radar cross section decreases slightly. The predicted radar cross sections as a function of $k_0a$ for these shapes are given in Figure 6-45. In calculating these curves it was assumed that the change in the forward portion of the body did not significantly alter the current distribution on the spherical base. Recent work by Senior[56] on the enhancement of creeping waves casts some doubt on the validity of this assumption.

The impulse approximation technique is not limited to finding the response for impulsive excitation. Response waveforms for step and ramp excitations can also be defined. The response to a step-function excitation is denoted $F_u(t)$, and is given by the integral of $F_I(t)$ as shown:

$$F_u(t) = \int_0^b F_I(t') \, dt' \qquad (6.3\text{-}71)$$

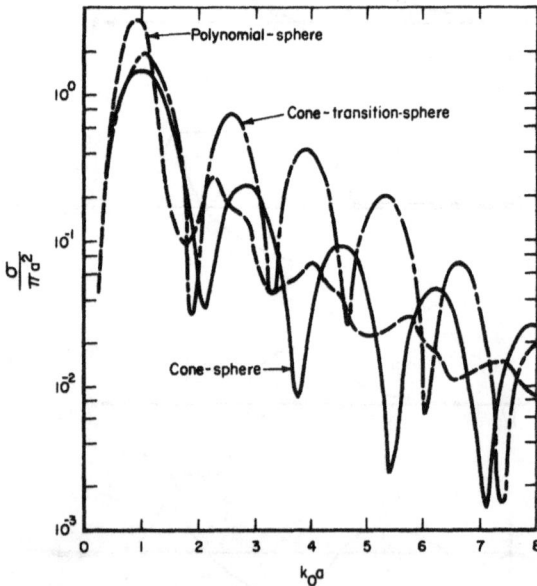

Fig. 6-45. Axial backscatter cross section of a cone-sphere, cone-transition-sphere and polynomial sphere as predicted by the impulse approximation [after Moffatt[55]].

The ramp response $F_R(t)$ is then the integral of the step response as shown:

$$F_R(t) = \int_0^t F_u(t')\, dt' \tag{6.3-72}$$

The ramp response automatically satisfies the first two moment conditions, and if set proportional to the Rayleigh coefficient, it satisfies the third moment condition. The ramp response is considerably smoother in the time domain than is the impulse response, and thus might be easier to estimate.

The theoretical backscatter ramp response of a cone sphere to a synthesized sawtooth waveform is illustrated in Figure 6-46. The sawtooth waveform is a periodic succession of ramp functions of alternating polarity. The waveform was synthesized by combining five frequencies in the harmonic ratios of $1 : 3 : 5 : 7 : 9$ with their appropriate magnitudes. That these five might be sufficient was first demonstrated by comparing the response of a perfectly conducting sphere to a waveform with sixty-four terms to the response to the five-term waveform. Only minor differences were noted.

Kennaugh and Moffatt[61] suggest that the ramp response is related to the physical shape and composition of the radar target. The area under the ramp response waveform is proportional to the Rayleigh scattering coefficient which is proportional to the volume of the scatterer. With the physical-optics approximation, the ramp response is a maximum at the shadow boundary where it is proportional to the maximum projected cross-sectional area. The impulse approximation is certain to be the subject of much future research.

### 6.3.1.3. Measured Cross Sections

6.3.1.3.1. *Introduction.* This section summarizes measurements on right circular metallic cones with various terminations. Section 6.3.1.3.2 discusses flat-backed cones; Section 6.3.1.3.3 discusses cones terminated with

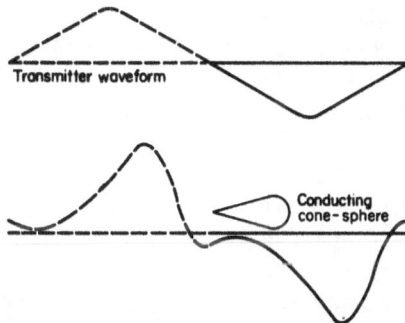

Fig. 6-46. Input waveform and approximate ramp response.

portions of a sphere; and Section 6.3.1.3.4 discusses other terminations. Other measurements of metallic cones have been presented in the previous sections of this chapter for comparison with theoretical computations.

6.3.1.3.2. *Flat-Backed Cones.* One of the most extensive set of measurements of right circular metal cones has been published in two reports by Keys and Primich.[35,64] These measurements are often the "standard" against which theoretical calculations are compared. The measurements were made at a wavelength of 8.6 mm, using cw and short-pulse radars. Measurements with the cw system were made in the near field. At a distance of 35 wavelengths, from a 10-wavelength aperture, the power and phase variation across a model with a maximum dimension of approximately four wavelengths was found to be 2 db and 5°, respectively. Larger models were measured with the pulse radar at a distance of 300 wavelengths. It is said that a comparison of model measurements showed no significant difference. The overall accuracy is said to be $\pm 2$ db for points above the background signal level, which was approximately 30 db below a square wavelength.

The models measured were made of aluminum machined to a tolerance of $\pm 0.001$ inches with a surface finish of 12 microinches r.m.s. The first report[35] contains 168 curves. Curves are given for horizontal and vertical polarizations for cones of six different half angles and fourteen different base diameters. The second report[64] contains 140 curves obtained by measuring, for both polarizations, the same fourteen base diameters and five different half-angles. The base diameters measured (in wave-lengths) were 0.297, 0.445, 0.592, 0.742, 0.980, 1.21, 1.33, 1.45, 1.69, 1.93, 2.16, 2.37, 2.64, and 2.87. The half-angles $\alpha$ were 4, 7.5, 9.6, 12, 15, and 20° in the first report; and 30, 37.5, 45, 52.5, and 60° in the second report. It is stated that the total cone angle $(2\alpha)$ on each of the cones was accurate to within one minute.

Since this data is so extensive and is widely available, it is not reproduced here. Portions of it, or data extracted from these reports, have been used extensively in the previous sections.

A less readily available set of measurements has been made by Radiation Incorporated for the Lincoln Laboratory of the Massachusetts Institute of Technology.[65,66] A cone of 20° half-angle, 2.374 meters in height, was measured at six frequencies and four polarizations. The $k_0 a$ of the cones lie between the approximate values of 2.64 and 21.8. The measured data are reproduced in Figures 6-47 through 6-50. Based on data from References 65, 66 Figure 6-47 illustrates the return at the six frequencies, 200, 400, 600, 800, 1000, and 1200 Mcps, for parallel polarization both transmitted and received. Parallel polarization is defined in this case as the electric field vector being parallel to the plane containing the symmetry axis of the cone and the radar antenna. Figure 6-48 illustrates the behavior for perpendicular polarization, which is defined as the electric field vector perpendicular to the plane of the symmetry axis and the radar. At nose-on incidence ($\theta = 0°$), or base-on ($\theta = 180°$), there should be no difference between the cross sections at the two polarizations, which is, in fact, observed upon comparing the six cases of Figure 6-47 with the respective cases in Figure 6-48.

Another comparison of interest is that of the cross section at normal incidence to the sides of the cone, with the values predicted by Eqs. (6.3-18) or (6.3-46). The measured cross section at 70° varies between 10 db at 200 Mcps for perpendicular polarization to 20 db at 1000 Mcps for parallel

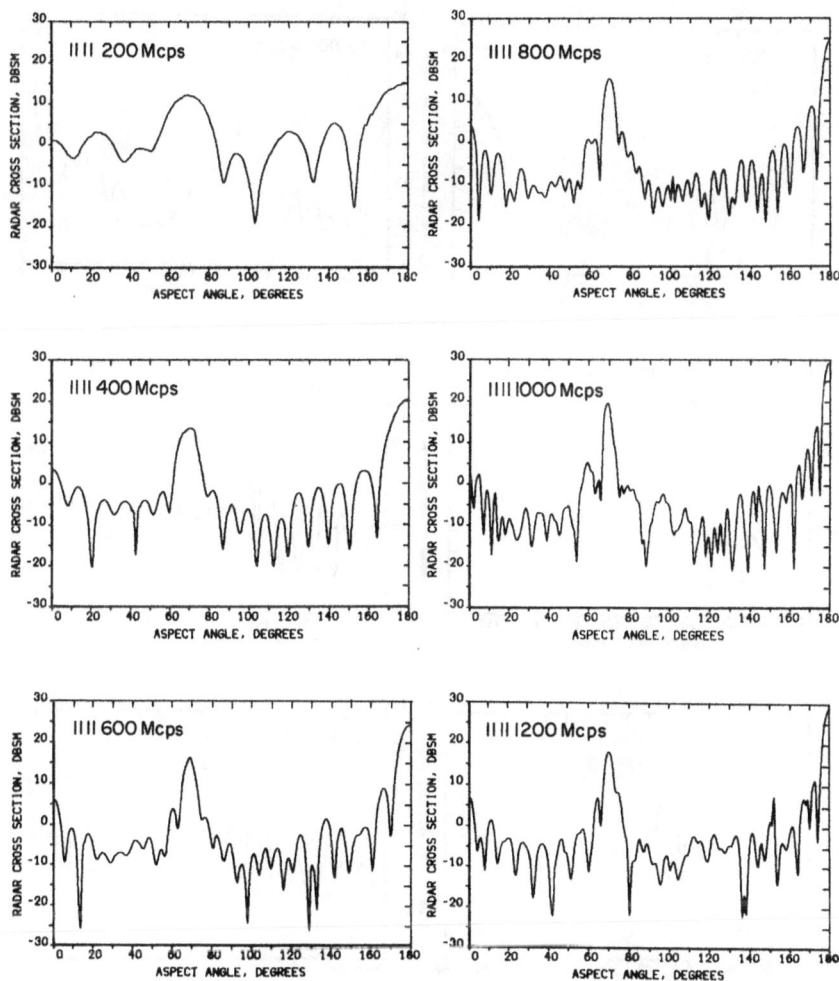

Fig. 6-47. Backscatter cross section of flat-backed cone of 20° half-angle, 2.374 m in length, for parallel polarization.

polarization. Equation (6.3-18) predicts 9.9 db at 200 Mcps, 16.9 db at 1000 Mcps, and 17.6 db at 1200 Mcps.

Figure 6-49 illustrates the behavior of the cone for right circular polarization transmitted and left circular received at all frequencies. Figure 6-50 illustrates the behavior for left circular transmitted and left circular received. It is seen that the signal level is considerably less in the left–left case, which is as it should be. The cone does depolarize to some extent, but it should be a poor reflector for cross-polarization. Obviously, it does not have zero reflection as is predicted by the unmodified physical-optics approximation.

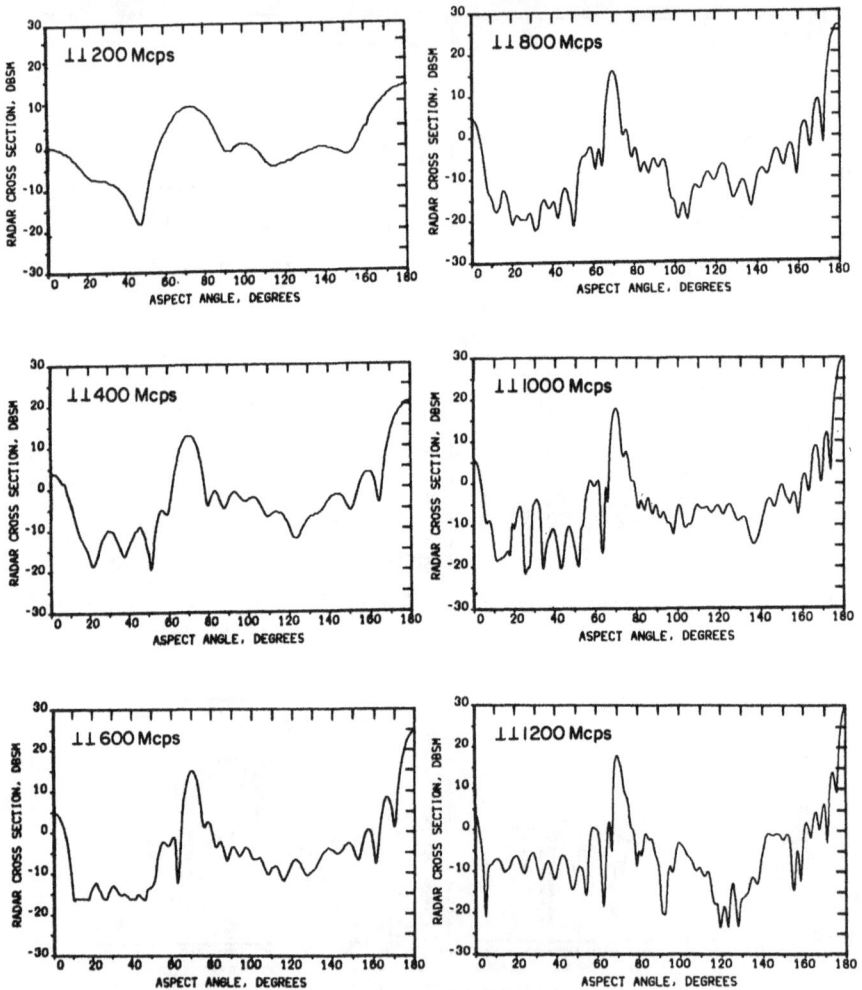

Fig. 6-48.   Backscatter cross section of flat-backed cone of 20° half-angle, 2.374 m in length, for perpendicular polarization.

The range used for the circular polarization measurements was equal to or greater than $2D^2/\lambda_0$, where $D$ is the maximum dimension of the cone. For the linear polarization measurements the range was equal to or greater than $4D^2/\lambda_0$, except for the measurement at 1200 Mcps. At this frequency, the range for linear polarization was $2.8D^2/\lambda_0$. It is said that these ranges proved to be satisfactory experimentally. A general discussion of range criteria in radar cross section measurements is given in Chapter 11.

6.3.1.3.3. *Cone-Sphere.* Pannell, Rheinstein, and Smith[44] have compiled a set of measurements on cone-spheres. They report backscatter

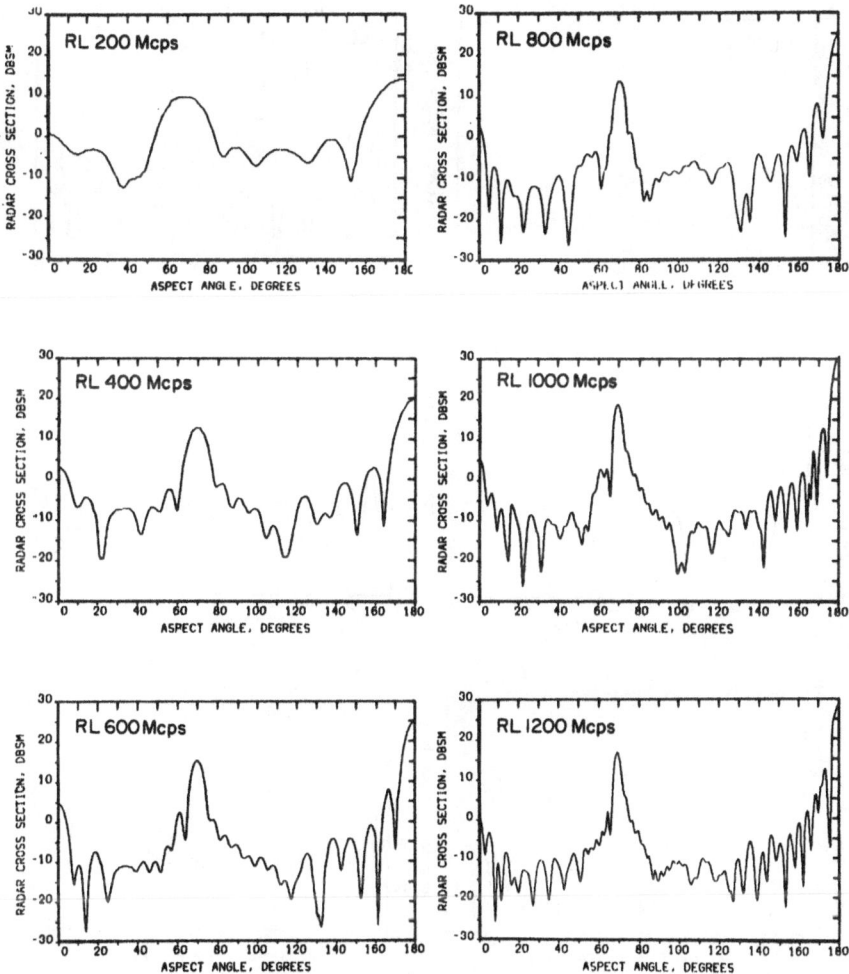

Fig. 6-49. Backscatter cross section of flat-backed cone of 20° half-angle, 2.374 m in length, for circular polarization.

measurements for both parallel and perpendicular polarization were made for $\alpha = 12.5°$ cones, whose base radius $a$, measured in wavelengths, varied from 0.60 to 10.43. Radiation, Inc., Melbourne, Florida, measured an aluminum cone with $a/\lambda_0 = 1.57$; a silver-coated epoxy cone with $a/\lambda_0 = 3.91$; and a silver-coated balsa cone with $a/\lambda_0 = 0.60$, 1.45, 6.09, and 10.43. The Defense Research Telecommunications Establishment, Ottawa, Canada, measured magnesium cones with $a/\lambda_0 = 1.47$ and 5.70. The USAF facility "RAT SCAT," Holloman AFB, New Mexico, measured a silver-coated balsa cone with $a/\lambda_0 = 2.4$. Figure 6-51 compares, for perpendicular polarization,

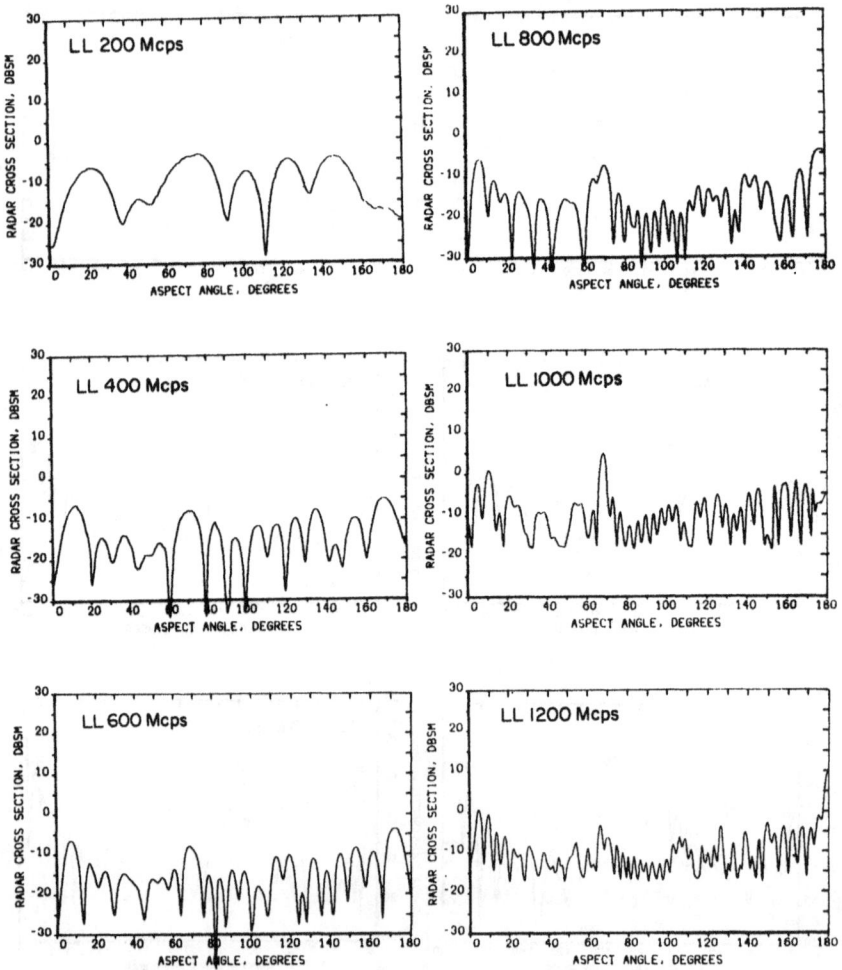

Fig. 6-50. Backscatter cross section of flat-backed cone of 20° half-angle, 2.374 m in length, for cross-polarized circular polarization.

the silver-coated balsa cone with an $a/\lambda_0 = 1.45$, measured at 1335 Mcps, with the magnesium cone with an $a/\lambda_0 = 1.47$, measured at 9000 Mcps. It is seen that there is essentially no difference for aspect-angles near base-on.

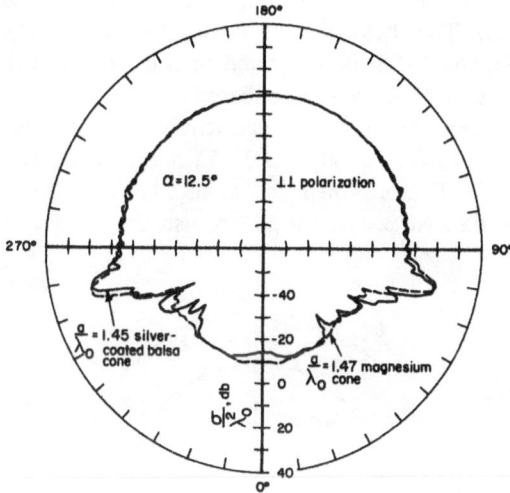

Fig. 6-51. Backscatter cross section of two similar cone-spheres for perpendicular polarization [after Pannell *et al.*[(44)]].

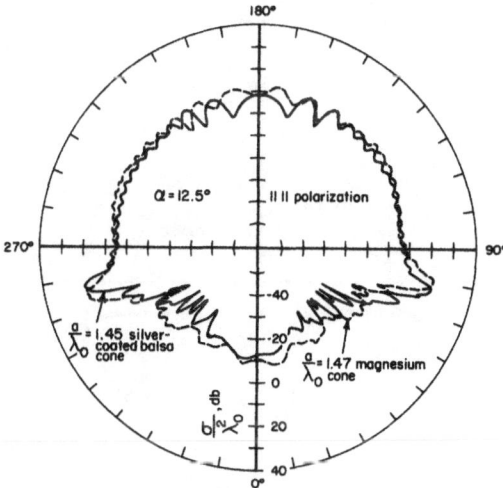

Fig. 6-52. Backscatter cross section of two similar cone-spheres for parallel polarization [after Pannell *et al.*[(44)]].

However, there is approximately 5 db difference for axial incidence. For parallel polarization, from Figure 6-52, it is seen that the differences are greater than for perpendicular polarization.

For large cones, differences between the two polarizations are small. For example, see Figure 6-53 which presents both polarizations for the $a/\lambda_0 = 10.4$ cone. The behavior of cones for a progression of sizes, $a/\lambda_0 = 0.60$, 3.91, and 6.09, is illustrated in Section 8.4.1; a section which continues the discussion of the cone-sphere.

Blore[54] has reported the measurement of backscattering from cone spheres with $\alpha = 7.5$, 15, 20, 30, and 37.5°, made at the Defense Research Telecommunications Establishment in Canada. His curves are reproduced in Section 8.3.1. The data for $\alpha = 15$ and 30° are also given in Section 6.3.1.2.2.1.5,

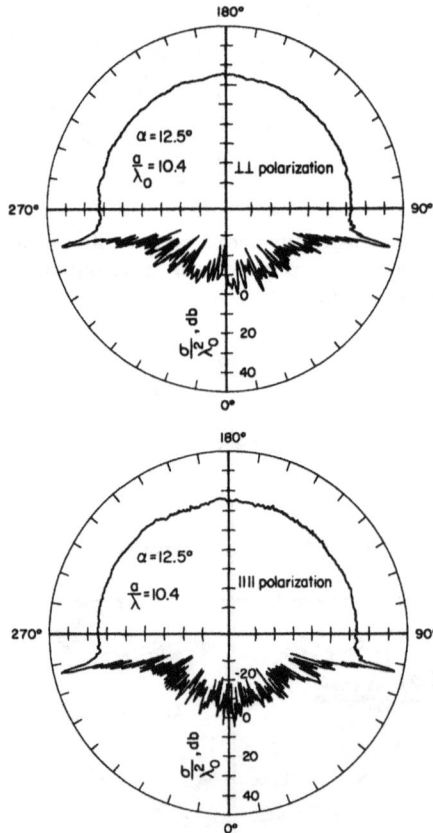

Fig. 6-53. Backscatter cross sections of a large cone-sphere at both perpendicular and parallel polarizations [after Pannell *et al.*[44]].

along with the measured data of Moffatt[55] for $\alpha = 15°$. The data of Moffatt and Blore agree quite well and can be predicted reasonably well theoretically, as discussed in Section 6.3.1.2.2.1.5.

Figure 6-54 (from Reference 54) compares the backscattered radar cross section of a set of $\alpha = 20°$ cones. Flat-backed cones, double-backed cones, cone-spheres, and double-rounded cones are compared, along with an ogive of $\alpha = 20°$. The bodies have identical projected areas at all diameters, have identical length-to-diameter ratios, and identical nose angles. The volumes differ by less than a factor of two between the largest and the smallest. However, as seen in the figure, the radar cross sections are different by several orders of magnitude.

6.3.1.3.4. *Other Terminations.* An experimental investigation was made at the University of Michigan[67] to determine the effect of discontinuities in the shadow region on the near-nose cross section of a cone sphere. A cone of half angle $\alpha = 12.5°$ was made out of rolled aluminum stock, which was machined for a good surface finish. The radius of the sphere (1.779 inches) and the frequency (9300 Mcps) were such that $k_0 a = 8.808$. For a fin entirely in the shadow region, it was found that the fin shape was not critical as long as its outline was smooth and its curvatures gentle. The fins were cut from 0.016-inch aluminum sheet with a circular bottom edge to attach to the sphere and with a parabolic upper edge. The maximum width was 2.75 inches on all fins. Figure 6-55 shows the effects of varying the height of the fins. The curves are smoothed fits of the experimental data. The number-two fin had a height at the center of $\frac{1}{4}$ inch. The height increased in $\frac{1}{8}$-inch steps for each larger fin number. Larger fins than those illustrated are said to have repeated essentially the same curves, with fin numbers 2 and 9 being quite similar.

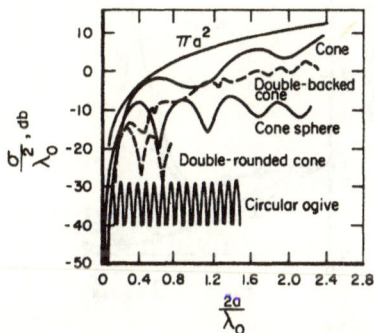

Fig. 6-54. Comparison of backscatter cross section of various objects with a half-angle of 20° [after Blore[54]].

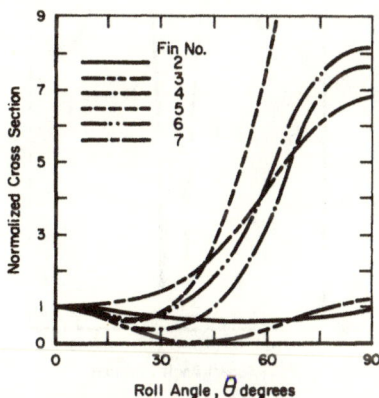

Fig. 6-55. Normalized backscatter cross section as a function of roll angle for several fin sizes [after Knott and Senior[67]].

From Figure 6-55, it is seen that none of the fins have any effects when they are oriented perpendicular to the incident electric vector. The fins reduce the cross section for some roll angles, and increase them for others. The increase is far greater than what would occur if only the phase of the creeping wave is being changed. It was suggested that enhancement of the creeping wave took place.

The radar cross section within $\pm 40°$ of nose-on was measured for fin number 3 for a roll angle of 40°. The data are presented in Figure 6-56. It is seen that the fin reduced the cross section considerably over a limited range of aspect angle.

If it were desired to reduce the cross section for a range of roll angles, several fins would be required. However, together they would look like a corner reflector and perhaps give considerable reflection at some roll angles. In order to prevent the corner reflector effect, a portion of the area of a number 4 fin was removed and tests were made.

A centrally-located slot was cut in several number 4 fins. The sides were normal to the spherical surface, and the size of the slot was designated by the angle subtended by the cuts at the center of the sphere. The results are illustrated in Figure 6-57. It is seen that the smallest return was observed with a slot whose width was 52°. This is equivalent to the removal of over 80 % of the fin area.

However, when two such 52° slots were installed at right angles, only a

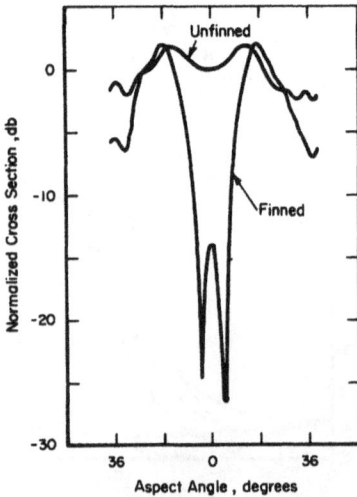

Fig. 6-56. Normlaized backscatter cross section as a function of aspect angle for a roll angle of 40° [after Knott and Senior[67]].

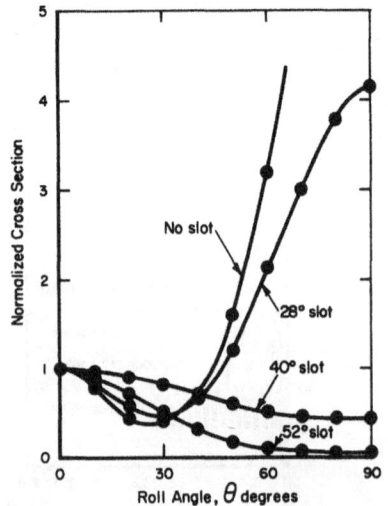

Fig. 6-57. Normalized backscatter cross section as a function of roll angle for different slot widths in a number 4 fin [after Knott and Senior[67]].

small reduction in nose-on cross section was obtained at all roll angles. There is not yet a satisfactory explanation for this.

Eberle and St. Clair[57] have reported an extensive set of model measurements made at The Ohio State University. Bistatic aspect angle patterns at bistatic angles of $\beta = 0$, 30, 60, 90, 120, and 140° were made on all double-body combinations of cones of $\alpha = 10$, 20, and 30°, rounded 20° cones, prolate spheroids, oblate spheroids, and spherical caps (except the combination of a spherical cap terminated with a spherical cap). The diameter of the target at the join was 4.27 cm. The rounded-tip cone has a nose radius of 6.4 mm. The prolate spheroid had a major radius of 4.65 cm, and the oblate spheroid had a minor radius of 1.475 cm. All measurements were made at a wavelength of 3.2 cm. Several of their figures are reproduced in Section 8.1.1.3, the section discussing experimental data from measurements of complex bodies.

Backscatter measurements on cones with their base edges rounded off and terminated with other similar cones have been reported by Blore.[54,68] Selected curves from his report are also reproduced in Section 8.1.1.3.

Extensive measurements have been reported by Lincoln Laboratory[69] on a variety of other terminations. These data are discussed in Section 8.4.1, which deals with the control of radar cross section by shaping.

### 6.3.2. Finite Dielectric and Dielectric-Coated Cones

#### 6.3.2.1. *Approximate Solutions*

The complexity of the theoretical problem of a dielectric coating on a cone is such that very little in the way of analytical results are available. Überall[19] has considered the effect of a dielectric coating on a semi-infinite cone. His work is discussed in Section 6.2.3.

The same problem is encountered with a dielectric cone. When the boundary conditions appropriate to a dielectric boundary are used, the complexity is such that very few solutions have been obtained.

For bodies large with respect to a wavelength, application of a modified form of physical optics, as discussed in Section 2.2.2.3, gives

$$\sigma_{\text{Dielectric}} = |R|^2 \sigma_{\text{p.c.}} \qquad (6.3\text{-}73)$$

where $R$ is the reflection coefficient of a semi-infinite planar dielectric surface, provided the Leontovich boundary condition[70] is valid at the cone surface.

In the resonance region, one must depend upon the experimental results for dielectric cone cross sections, and these are also rather limited.

#### 6.3.2.2. *Experimental Results*

At the University of California[71] monostatic measurements have been made on cones of 7.5° half-angle with relative dielectric constants of 5, 10, 15, and 20. Three different base diameters were used, i.e., 1.267, 1, and 0.619 inches. All the measurements were made at 9333 Mcps, for which the base diameters are $\lambda_0$, $0.79\lambda_0$, and $\lambda_0/2$, respectively. The backscattering from metallic cones of these sizes was also measured. These results are shown in Figures 6-58 through 6-63. These figures show that the metallic cones have, in general, three primary lobes, two specular reflections from the sides and one from the base. The dielectric cones show more lobes. There is also a large reflection at nose-on aspect for the dielectric cones, whereas it is very small for the metallic cones. In some cases, the reflection from the nose-on aspect for the dielectric cones exceeded the reflection for the base-on aspect. The results indicate that considerable difference exists between metallic cones and lossless dielectric cones for dielectric constants as high as 20.

In Table 6-1, the ratio of the cross sections of dielectric and metallic cones at incidence normal to the cone side and normal to the base are compared with the square of the semi-infinite, flat-plate reflection coefficient.

**Table 6-1**

**Comparison of Measured and Estimated Radar Cross-Section Ratios for Dielectric and Metallic Cones of $7\frac{1}{2}°$ Half-Angle**

| Cone base diameter | $\epsilon_r$ | $\lvert R \rvert^2$ | Measured ratios of dielectric to metallic cross section | |
|---|---|---|---|---|
| | | | Specular flare | Base |
| $\lambda_0$ | 20 | 0.402 | 0.32 | 0.25 |
| | 15 | 0.348 | 0.03–0.16 | 0.18 |
| | 5 | 0.146 | 0.32–0.5 | 0.03 |
| $0.79\lambda_0$ | 20 | 0.402 | 0.2 | 0.25 |
| | 15 | 0.348 | 0.78–1.0 | 0.71 |
| | 10 | 0.270 | 0.2–0.25 | 0.49 |
| | 5 | 0.146 | 0.1–0.32 | 0.11 |
| $0.5\lambda_0$ | 20 | 0.402 | 0.32 | 1.41 |
| | 15 | 0.348 | 3.2 | 1.58 |
| | 10 | 0.270 | 0.32 | 0.06 |
| | 5 | 0.146 | 0.32 | 0.16 |

It is seen that there is little agreement. The reflections for the back surfaces of the cone are apparently not taken into account by the use of the flat-plate reflection coefficient. This poor agreement should, in fact, be expected, since for lossless dielectric cones, the Leontovitch boundary condition [and, hence, Eq. (6.3-73)] is not valid. For larger cones with loss, there should be better agreement.

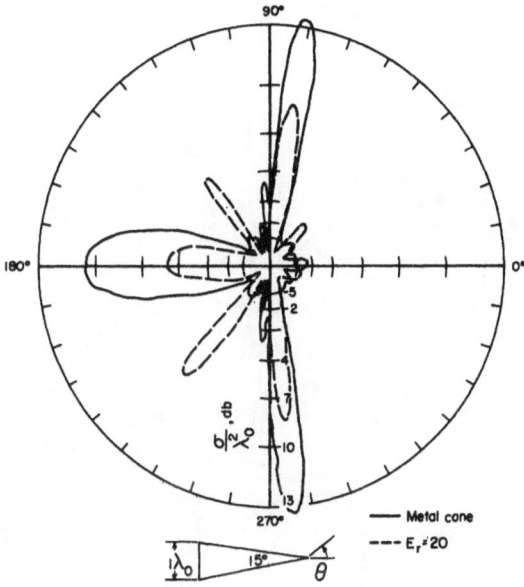

Fig. 6-58. Comparison of backscatter cross section of metallic and dielectric flat-back cones [after Pon and Angelakos[71]].

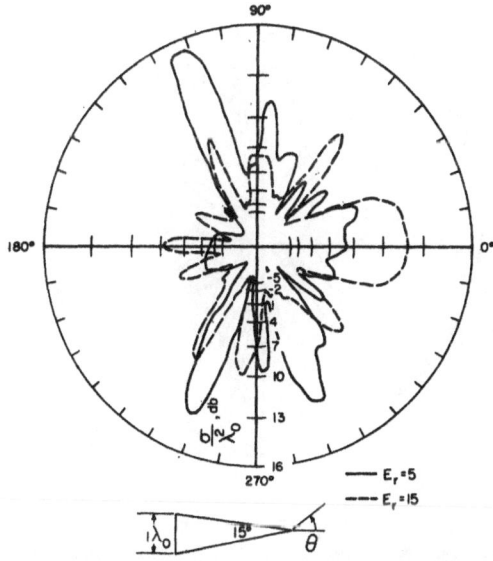

Fig. 6-59. Comparison of backscatter cross section of flat-back cones of different dielectric constants [after Pon and Angelakos[71]].

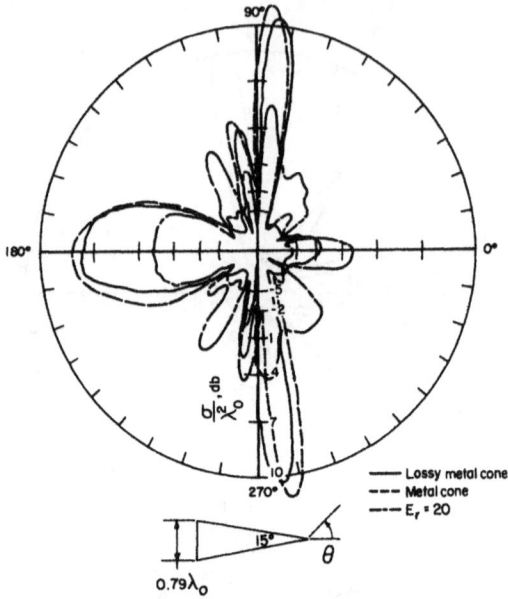

Fig. 6-60. Comparison of backscatter cross section of metallic and dielectric flat-back cones [after Pon and Angelakos[71]].

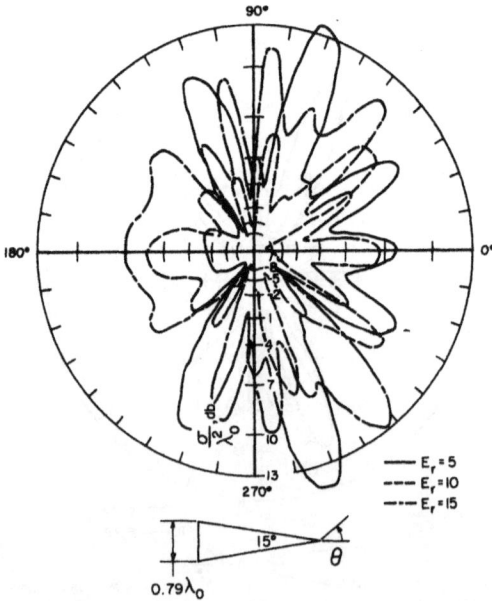

Fig. 6-61. Comparison of backscatter cross section of flat-back cones of different dielectric constants [after Pon and Angelakos[71]].

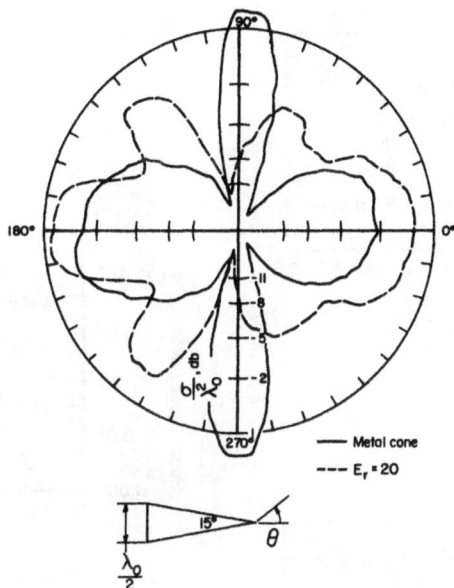

Fig. 6-62. Comparison of backscatter cross section of metallic and dielectric flat-back cones [after Pon and Angelakos[71]].

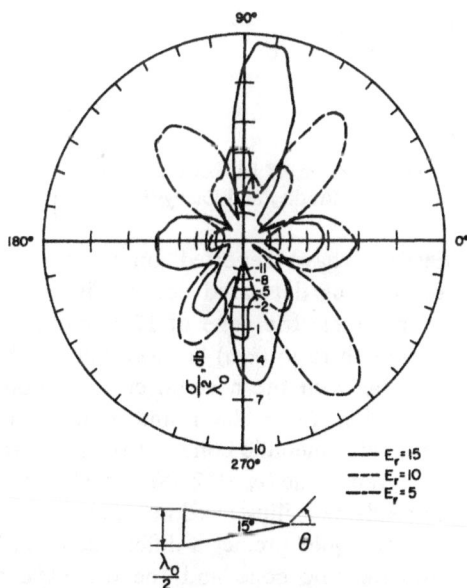

Fig. 6-63. Comparison of backscatter cross section of flat-back cones of different dielectric constants [after Pon and Angelakos[71]].

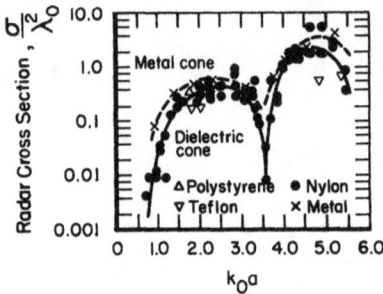

(a) Nose-on

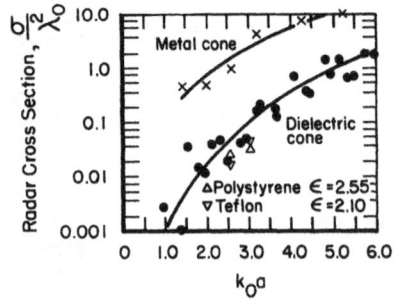

(b) Broadside

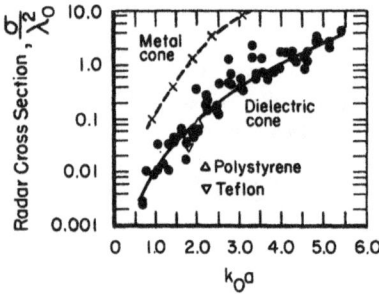

(c) Base-on

Fig. 6-64.  Backscatter cross section of flat-back dielectric cone of $12\frac{1}{2}°$ half-angle for perpendicular polarization [after Bradley and Eastly[72]].

Bradley and Eastly[72] have reported on measurements made at the Boeing Company on various flat-based cones. Figure 6-64 illustrates the radar cross section for a dielectric cone of 12.5° half-angle with the base-circumference-to-wavelength ratio $(k_0 a)$ between 0.8 and 5.3. Figure 6-64(a) shows the comparison between the nose-on cross section of the dielectric cone constructed of nylon $(\epsilon_r = 2.85)$ and data taken by Keys and Primich[35] for the same size metallic cone. Also shown are a few measurements taken of a polystyrene cone $(\epsilon_r = 2.55)$ and a Teflon cone $(\epsilon_r = 2.10)$. The cross section at broadside is illustrated in Figure 6-64(b). It is interesting to note that Eq. (6.3-73) would predict a difference of 11.8 db between the cross sections for the metallic cone and the dielectric cone. The actual difference varies between 11.6 db for $k_0 a = 2$ and 10.5 db for $k_0 a = 5$. The base aspect cross section is illustrated in Figure 6-64(c).

Figure 6-65 shows the variation of nose-on backscatter cross section

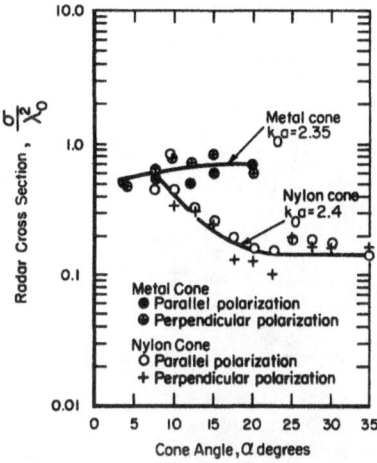

Fig. 6-65. Axial backscatter cross section of flat-back metallic and dielectric cone as a function of cone angle [after Bradley and Eastly[72]].

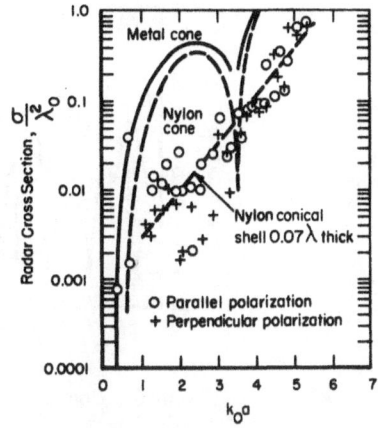

Fig. 6-66. Axial backscatter cross section of a thin-cone shell of $12\frac{1}{2}°$ half-angle [after Bradley and Eastly[72]].

for various cone angles. The nylon cone is seen to have a nose-on radar cross section which is less than that of a similar metallic cone for all half-angles greater than 7.5°.

Figure 6-66 compares a nylon conical shell 0.07 dielectric wavelengths thick with comparable solid dielectric and metallic cones of 12.5° half-angle.

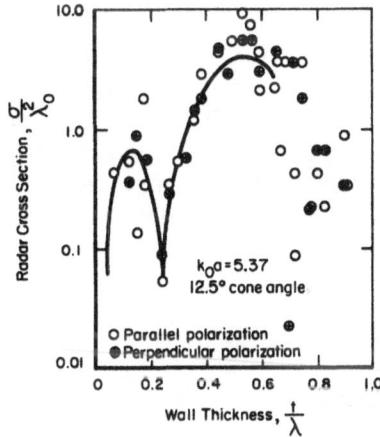

Fig. 6-67. Axial backscatter cross section of a cone shell versus shell electrical thickness in wavelengths [after Bradley and Eastly[72]].

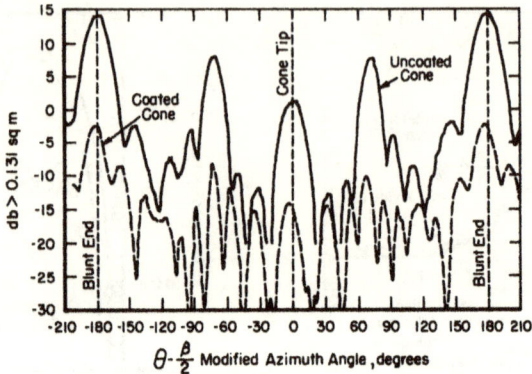

Fig. 6-68. Comparison of cross sections of coated and uncoated metallic flat-back cones of 19.25° half-angle and $k_0a$ of 14.34 for a fixed bistatic angle of 40° [after Garbacz and Moffatt[73]].

The shell gives a lower cross section for most values of $k_0a$. In Figure 6-67, the behavior of the nose-aspect cross section of a nylon shell for various wall thicknesses is shown for a cone of 12.5° half-angle with $k_0a = 5.37$.

In summary, the scattering behavior of dielectric cones has not been well established, either theoretically or experimentally. The radar cross section at nose-on aspect may be either larger or smaller than the cross section of a similar metallic cone.

A similar status exists for the scattering behavior of dielectric coated metal cones, in that very few theoretical or experimental results exist. Garbacz and Moffatt[73] have compared the bistatic scattering from coated and uncoated right circular cones with $k_0a$ of 14.34 and half-angle $\alpha = 19.25°$ at 9375 Mcps. This comparison for a bistatic angle of 40° is illustrated in Figure 6-68. The approximate material parameters of the coating were $\mu_r = 1.5 + i1.5$ and $\epsilon_r = 8 + i0.5$. The thickness of the coating is 0.077 inch. It is seen that the coating reduced the bistatic tip scattering, specular flare, and base scattering by 14 to 18 db. For large bistatic angles (in the neighborhood of 140° or larger), Garbacz and Moffatt reported enhancement of the uncoated cone cross section.

## 6.4. WIRE RINGS OR LOOPS

### 6.4.1. Theoretical Solutions

A theoretically exact solution for the scattered field from a perfectly-conducting circular loop or torus has been obtained by Weston[74] by using solutions[75–79] of the wave equation in toroidal coordinates. In order to

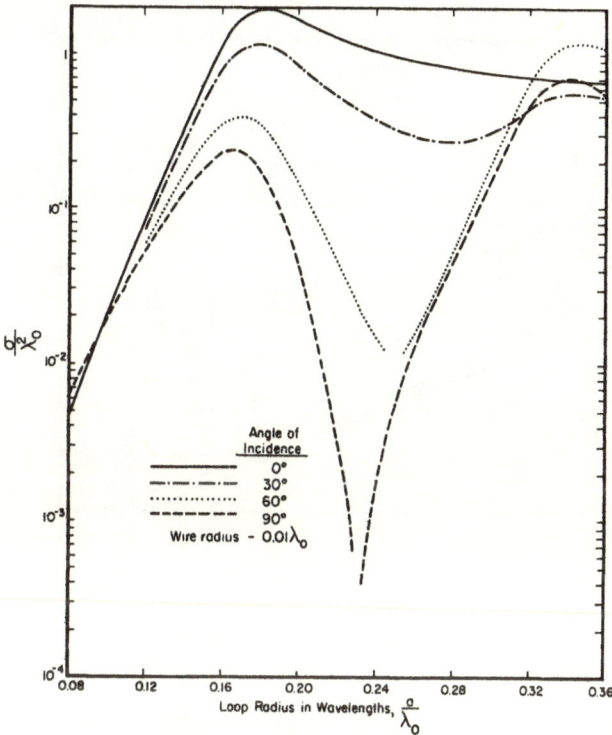

Fig. 6-69. Backscatter cross section of thin circular loops at various angles of incidence as a function of loop radius, perpendicular polarization [after Weston[74]].

compute the echo area numerically, he neglected all terms of order $(b/a)^2$ and $[k_0 d/(a/b)]^2$ and higher; where $b$ is the radius of the wire, $a$ is the radius of the loop, and $d$ is the distance to the observation point from the center of the loop.[†] Higher-order terms can be obtained by a proper combination of the vector wave functions described in his dissertation.[75]

Weston calculated the echo area for four angles of incidence, 0, 30, 60, and 90°; for three cases of wire radius, $b = 0.01\lambda_0$, $0.005\lambda_0$, and $0.0008\lambda_0$; for a range of loop radius, $a$, roughly between $0.06\lambda_0$ and $0.38\lambda_0$; and for two types of polarization, perpendicular to the plane of incidence (in the plane of the ring), and parallel to the plane of incidence. His results are shown in Figures 6-69 through 6-73. It is apparent from the figures that there is a

---

† The approximations used by Weston to obtain numerical results are valid only for thin-wire loops, and yield expressions equivalent to those obtained by Kouyoumjian using a variational approach. Kouyoumjian's work is discussed in Section 6.4.2.

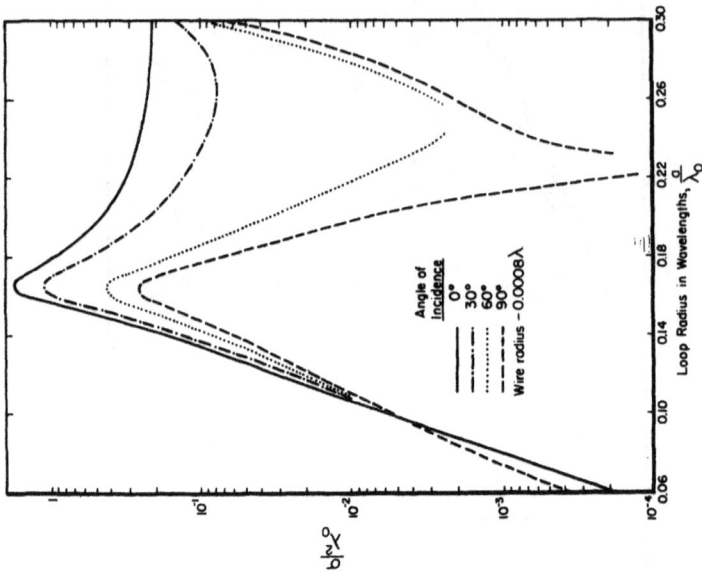

Fig. 6-71. Backscatter cross section of thin circular loops at various angles of incidence as a function of loop radius, perpendicular polarization [after Weston[74]].

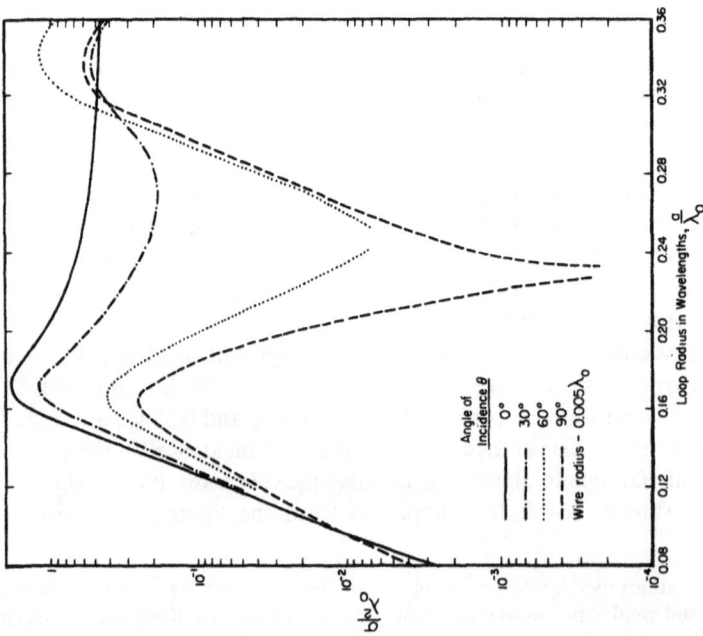

Fig. 6-70. Backscatter cross section of thin circular loops at various angles of incidence as a function of loop radius, perpendicular polarization [after Weston[74]].

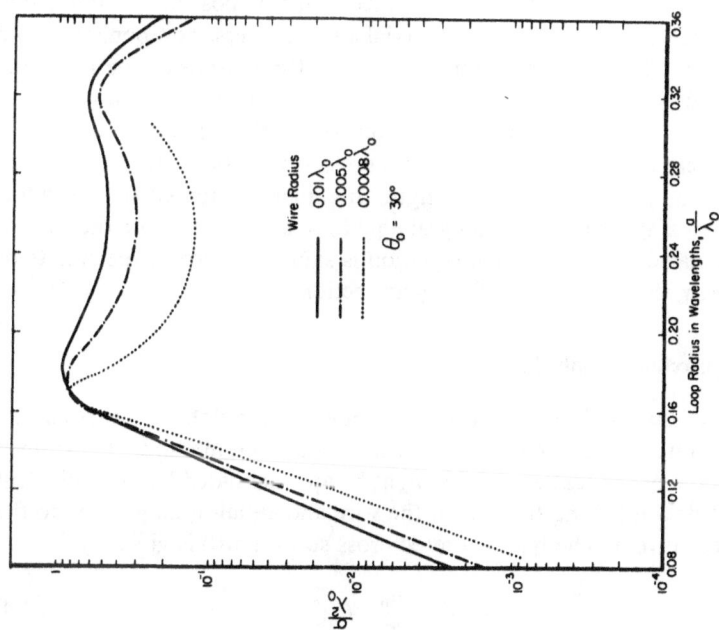

Fig. 6-73.  Backscatter cross section of thin circular loops of various wire sizes at 60° angle of incidence as a function of loop radius, parallel polarization [after Weston[74]].

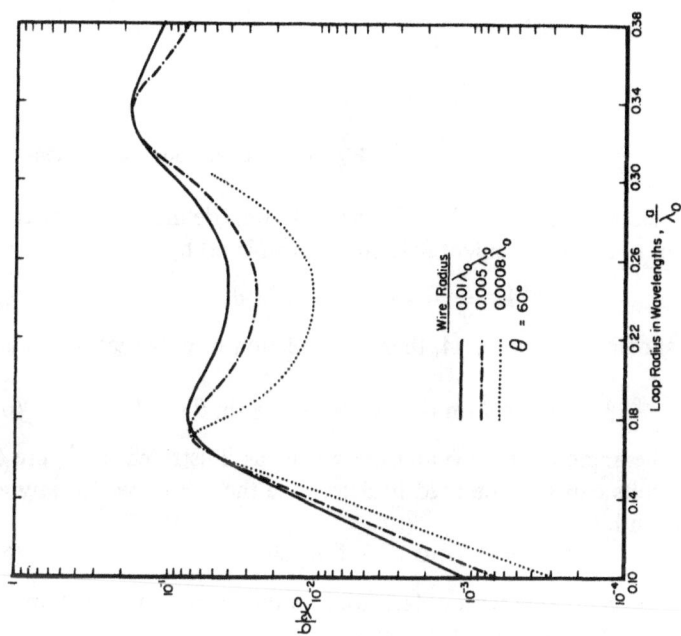

Fig. 6-72.  Backscatter cross section of thin circular loops of various wire sizes at 30° angle of incidence as a function of loop radius, parallel polarization [after Weston[74]].

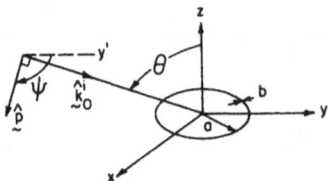

Fig. 6-74.  Loop coordinate system.

resonant peak near $k_0 a = 1$ for all cases. If one defines the direction of propagation, $\hat{\mathbf{k}}_0{}^i$, of the incident field (in the $xz$-plane) by

$$\hat{\mathbf{k}}_0{}^i = -\hat{\mathbf{x}} \sin \theta - \hat{\mathbf{z}} \cos \theta \qquad (6.4\text{-}1)$$

where $\theta$ is shown in Figure 6-74, then the direction of polarization $\hat{\mathbf{p}}$ may be written as

$$\hat{\mathbf{p}} = \hat{\mathbf{x}} \cos \theta \sin \psi + \hat{\mathbf{y}} \cos \psi - \hat{\mathbf{z}} \sin \theta \sin \psi \qquad (6.4\text{-}2)$$

where $\psi$ is the angle between $\hat{\mathbf{p}}$ and the $y$ axis, as illustrated in Figure 6-75. Weston's results can then be used to determine the echo area for any angle of polarization by

$$\sigma = \sigma_\perp \cos^2 \psi + \sigma_\parallel \sin^2 \psi \qquad (6.4\text{-}3)$$

where $\sigma_\perp$ is the echo area with polarization in the plane of the ring and $\sigma_\parallel$ is the echo area with parallel polarization.

King and Wu[80] suggest the possibility of separating the effect of the shape of the cross section of the write from the effect of the shape of the loop for the case of normal incidence. It is stated that it is possible to derive two simultaneous integral equations in the axial and transverse components of the surface current. In the limit of infinite radius of curvature of the loop, the equations become uncoupled and the equations are identical to those for an infinite cylinder, with **E** and **H** parallel to the axis of the cylinder, respectively. When the linear dimension of the physical cross section of the wire is required to be small compared to the wavelength and small compared to the radius of the loop, a single simplified integral equation is obtained. For the case of wire of circular cross section, the equation is said to reduce to the results of Kouyoumjian, discussed in the following section.

### 6.4.2.  Approximate Solutions and Experimental Results

Prior to the analysis of Weston, Kouyoumjian[81–83] considered the scattering from a small ring by means of the variational method. He assumed that the source of the scattered field might be approximated by an equivalent line current flowing along the axis of the wire and equal in magnitude to the total surface current. The backscattered cross section $\sigma(0)$ is given by

$$\sigma(0) = \frac{\pi \mu_0}{\lambda_0{}^2 \epsilon_0} \, |\, U \,|^2 \qquad (6.4\text{-}4)$$

where $U$ is the far-zone amplitude of the backscattered field for either polarization. The backscattered cross section received by an antenna with polarization $p$ is

$$\sigma(0) = \frac{\pi\mu_0}{\lambda_0^2\epsilon_0} \mid \hat{\mathbf{p}} \cdot \tilde{U} \cdot \hat{\mathbf{p}} \mid^2 \tag{6.4-5}$$

where

$$\tilde{U} = \begin{bmatrix} U_y & 0 \\ 0 & U_\theta \end{bmatrix}$$

At the surface of the loop ($xy$-plane), where the incident field is ($TE$ or $\perp$ case),

$$\mathbf{E}^i = \hat{\mathbf{y}}\exp(ik_0 a \sin\theta \cos\phi)$$

Kouyoumjian obtains a cross section in square wavelengths of

$$\frac{\sigma(\theta)}{\lambda_0^2} = 2\pi(k_0 a)^6 \left| \frac{J_1^2(\beta)}{k_0^2 a^2 K_1} + \sum_{n=1}^\infty \frac{(-1)^n [J_{n-1}(\beta) - J_{n+1}(\beta)]^2}{k_0^2 a^2 [K_{n-1} + K_{n+1} - 2n^2 K_n]} \right|^2 \tag{6.4-6}$$

where

$$K_n = \int_0^\pi \frac{\cos nx \exp\{-ika[2(1-\cos x) + (b/a)^2]^{1/2}\}}{[b^2/2a^2 + (1-\cos x)]^{1/2}}\, dx$$

$$\approx \sqrt{2}\left[\ln(8a/b) - 2\left(1 + \frac{1}{2} + \frac{1}{2^2} + \cdots + \frac{1}{2^{n-1}}\right)\right]$$

$$- \frac{\pi}{2}\int_0^{2k_0 a} \Omega_{2n}(x) + iJ_{2n}(x)\, dx + O(b/a) \tag{6.4-7}$$

with $\beta = k_0 a \sin\theta$, $\Omega_{2n}(x)$ the Lommel–Weber[43] functions, and the series $\sum_{l=0}^{n-1} 2^{-l}$ is taken to be zero for $n = 0$.[†] For an incident field,

$$\mathbf{E}^i = \hat{\mathbf{\theta}}\exp(ik_0 a \sin\theta \cos\phi),$$

($TM$ or $\parallel$ case), Kouyoumjian obtains

$$\frac{\sigma(\theta)}{\lambda_0^2} = 2n(k_0 a)^6 \cos^4\theta \left| \sum_{n=1}^\infty \frac{(-1)^n [J_{n-1}(\beta) + J_{n+1}(\beta)]^2}{(k_0 a)^2 [K_{n-1} + K_{n+1} - 2n^2 K_n]} \right|^2 \tag{6.4-8}$$

For the broadside case with $\theta = 0$, Eqs. (6.4-6) and (6.4-8) reduce to

$$\frac{\sigma(0)}{\lambda_0^2} = \frac{2\pi(k_0 a)^6}{\mid k_0^2 a^2 [K_0 + K_2 - 2K_1]\mid^2} \tag{6.4-9}$$

For a large loop, $k_0 a \gg 1$, and for $\theta = 0$,

$$\frac{\sigma(0)}{\lambda_0^2} \approx \frac{\pi(k_0 a)^2}{\pi^2 + [2\ln(2/k_0 b)]^2} \tag{6.4-10}$$

[†] The function $K_n$ of Eq. (6.4-7) is tabulated in Reference 82.

For a small loop, $k_0 a \ll 1$, the backscattering may be represented by the fields of an electric and magnetic dipole. The far-zone amplitude for the $TE$ case is given by

$$U_y = \frac{i\pi^2 a^2 (k_0 a)(2 + \sin^2 \theta)}{\sqrt{\mu_0/\epsilon_0}[\ln(8a/b) - 2]} \qquad (6.4\text{-}11)$$

The radar cross section seen by a radar polarized in the $y$ direction is, therefore,

$$\sigma_y = (\pi^3/4)\, a^2 (k_0 a)^4 \, (2 + \sin \theta)^2 [\ln(8a/b) - 2]^{-2} \qquad (6.4\text{-}12)$$

For the $TM$ case, there is only an equivalent electric dipole which lies in the plane of the loop and is parallel to the plane of incidence. Thus, for $k_0 a \ll 1$,

$$U_\theta = i \frac{2\pi^2 a^2 (k_0 a) \cos^2 \theta}{\sqrt{\mu_0/\epsilon_0}[\ln(8a/b) - 2]} \qquad (6.4\text{-}13)$$

The resulting cross section is

$$\sigma_\theta = \pi^3 a^2 (k_0 a)^4 \cos^4 \theta / [\ln(8a/b) - 2]^2 \qquad (6.4\text{-}14)$$

Experimental values are compared with Kouyoumjian's computations in Figures 6-75 and 6-76 for $b = 0.005\lambda_0$ and $b = 0.008\lambda_0$, and for a range of loop radius from $0.06\lambda_0$ to $0.40\lambda_0$. These measurements are for the broadside aspect, and, in addition, for the thicker wire for edge-on aspect. The agreement between the theoretical and experimental results is good, and they also agree well with Weston's results discussed previously. In Figure 6-76, the increasing discrepancy for larger loops may be attributed to the need for more than three terms in the calculation of $\sigma(0)$.

If wire thicknesses of less than $0.0008\lambda_0$ were to be investigated, perhaps the surface impedance should be taken into account. Kouyoumjian reported[82] that experimental results indicated that the finite conductivity caused a 2 % reduction in cross section over the calculated values at the point of largest effect (resonance) for the thinner loop. For the thicker loop, varying the conductivity is said to have created negligible effects.

By using the formula of Chu[84] for the cross section of a thin wire, Crispin, et al.[39] have obtained the following expression for the backscatter cross section of circular wire loops:

$$\frac{\sigma(0)}{\pi a^2} = \frac{(\pi/2)^2 + [\ln(85/\gamma\pi)]^2}{(\pi/2)^2 + [\ln(\lambda_0/\gamma\pi b)]^2} \qquad (6.4\text{-}15)$$

where $\gamma = 1.78$. Equation (6.4-15) can be used to extend Kouyoumjian's results to large loops as shown in Figure 6-77. It can also be used to illustrate the variation of cross section with wire radius, as shown in Figure 6-78.

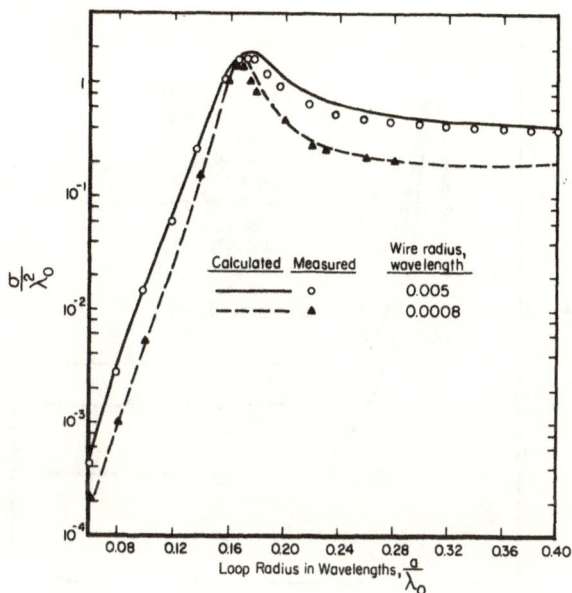

Fig. 6-75. Comparison of calculated and measured backscatter cross section of thin circular loops at the broadside aspect [after Kouyoumjian[82]].

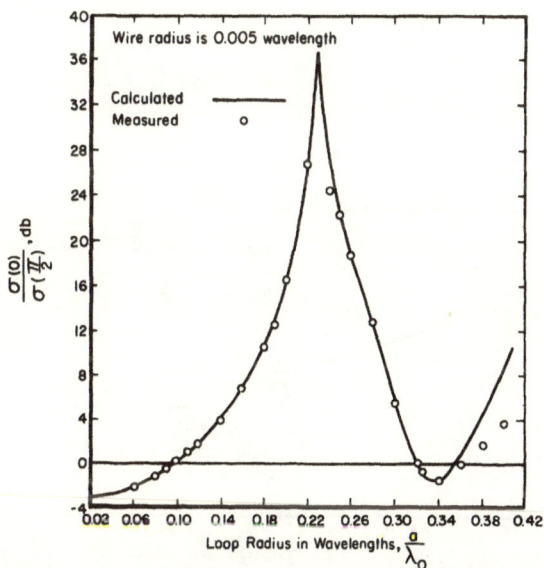

Fig. 6-76. The ratio of the backscatter cross section of a thin circular loop at the broadside aspect to the backscatter cross section at the edge aspect [after Kouyoumjian[82]].

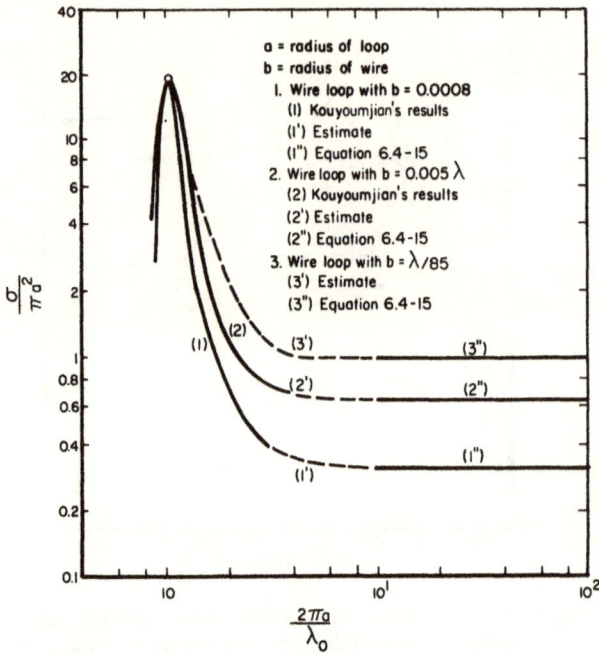

Fig. 6-77. Backscatter cross section of wire loops at broadside aspect [after Crispin et al.[39]].

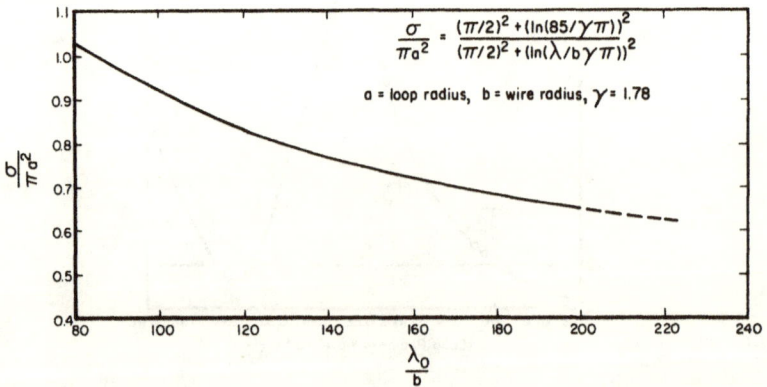

Fig. 6-78. The backscatter cross section of a wire loop at broadside aspect as a function of the wire radius [after Crispin et al.[39]].

Crispin, *et al.*[39] by use of the physical-optics approximation, have also determined the cross section for different polarizations. In the notation used, $\sigma(RT)$ is the cross section at the polarization of the receiver due to the field transmitted by the given polarization of the transmitter. Let the incident field direction, $\hat{k}_0{}^i$, lie in the $xz$-plane. Let $\psi$ be the angle between the polarization vector and the $y$-axis. The two orthogonal polarization directions are given by[†]

$$\hat{p}(A) = \hat{x} \cos \theta \sin \psi + \hat{y} \cos \psi + \hat{z} \sin \theta \sin \psi \qquad (6.4\text{-}16)$$

and

$$\hat{p}(B) = \hat{x} \cos \theta \cos \psi - \hat{y} \sin \psi + \hat{z} \sin \theta \cos \psi \qquad (6.4\text{-}17)$$

The right and left directions of circular polarization are given by

$$\hat{p}(R) = \frac{\exp(-i\psi)}{\sqrt{2}} [\hat{x}(i \cos \theta) + \hat{y} + \hat{z}(i \sin \theta)] \qquad (6.4\text{-}18)$$

and

$$\hat{p}(L) = \frac{\exp(+i\psi)}{\sqrt{2}} [-\hat{x}(i \cos \theta) + \hat{y} - \hat{z}(i \sin \theta)] \qquad (6.4\text{-}19)$$

For these polarizations, the following relations have been obtained for the radar cross sections:

$$\sigma(AA) = \pi a^2 \,|[(\sin^2 \psi \cos^2 \theta + \cos^2 \psi) J_0(2k_0 a \sin \theta)$$
$$+ (\sin^2 \psi \cos^2 \theta - \cos^2 \psi) J_2(2k_0 a \sin \theta)]|^2 \qquad (6.4\text{-}20)$$

$$\sigma(BB) = \pi a^2 \,|[(\cos^2 \theta \cos^2 \psi - \sin^2 \psi) J_2(2k_0 a \sin \theta)$$
$$+ (\cos^2 \theta \cos^2 \psi + \sin^2 \psi) J_0(2k_0 a \sin \theta)]|^2 \qquad (6.4\text{-}21)$$

$$\sigma(AB) = \pi a^2 \sin^2 \psi \cos^2 \psi \,|[(1 + \cos^2 \theta) J_2(2k_0 a \sin \theta)$$
$$- \sin^2 \theta \, J_0(2k_0 a \sin \theta)]|^2 \qquad (6.4\text{-}22)$$

$$\sigma(AR) = \frac{\pi a^2}{2} \,|(i \cos^2 \theta \sin \psi - \cos \psi) J_2(2k_0 a \sin \theta)$$
$$+ (i \cos^2 \theta \sin \psi + \cos \psi) J_0(2k_0 a \sin \theta)|^2 \qquad (6.4\text{-}23)$$

$$\sigma(BR) = \frac{\pi a^2}{2} \,|(i \cos^2 \theta \cos \psi + \sin \psi) J_2(2k_0 a \sin \theta)$$
$$+ (i \cos^2 \theta \cos \psi - \sin \psi) J_0(2k_0 a \sin \theta)|^2 \qquad (6.4\text{-}24)$$

$$\sigma(RR) = \frac{\pi a^2}{4} \,|(-\cos^2 \theta - 1) J_2(2k_0 a \sin \theta) + (1 - \cos^2 \theta) J_0(2k_0 a \sin \theta)|^2$$
$$(6.4\text{-}25)$$

where $(R)$ indicates right circular polarization.

---

[†] Note that $\psi = 0$ corresponds to $\perp$ or *TE* polarization, and $\psi = \pi/2$ to *TM* or $\|$ polarization in terms of the notation previously used.

One sees that Eq. (6.4-25) is independent of $\psi$, as it should be. Further, we see from Eq. (6.4-22) that for $\psi = 0$, $\sigma(AB) = 0$, and from Eqs. (6.4-20) through (6.4-24) that $\sigma(AR) = \sigma(AA)$ and $\sigma(BB) = \sigma(BR)$.

The problem of the radar cross section of a dielectric ring does not appear to have been treated to any extent in the literature. Philipson[85] who appears to be interested primarily in radome effects, has considered the scattering from a thin dielectric ring under two assumptions: (1) the field incident on the ring is given by the free-space field of course, and (2) the total field, incident plus scattered, tends asymptotically to the incident field as the radial thickness of the ring approaches zero. His resulting formulas were applied to the case in which the incident field is generated by a dipole antenna uniaxial with the ring. Very good agreement between his theory and experimental results were obtained.

### 6.4.3. Discussion

The problem of scattering from a torus, as given by Weston, is one of the few problems that has an exact theoretical solution. However, the computation is difficult to program and expensive to compute so that only limited data exist, and these are for loops whose radius is large with respect to the size of the wire. No numerical data appear to exist for loops made of wire of other than circular cross section or of dielectric material. The approximations for the cross section of a loop are also difficult to utilize, in general, if one is interested in the resonance region. The effect of surface impedance appears to be minor for the size of wire considered. Rectangular loops appear to be the only shape other than circular to have been considered.[86] The approximate solutions for this case agree fairly well with measurements, but again are very complicated.

Figure 6-79 compares the scattering from a wire ring with that from a dipole of identical wire radius, and length equal to the diameter of the loop.[82] From this figure the loop appears to be a more efficient scatterer than the dipole. However, considering the volume of material required to produce a given cross section, rings are seen to be much less efficient than are dipoles. In the event that the dipole were to be used at its first resonant peak, one would need a wire about $0.47\lambda_0$ long. Looking at the same cross section on the loop curve, one would require a loop of approximately $0.31\lambda_0$ in diameter. The same loop, however, has better behavior for radar signals of a slightly higher frequency, since the loop cross section is still increasing while the dipole is at its peak. Also shown in Figure 6-79 are curves for the radar cross section of an infinitesimally thin plate and a sphere, both having the same diameter as the loop.

The primary interest in scattering from a loop, other than as an object of theoretical study, is the modeling of more complex shapes by breaking them

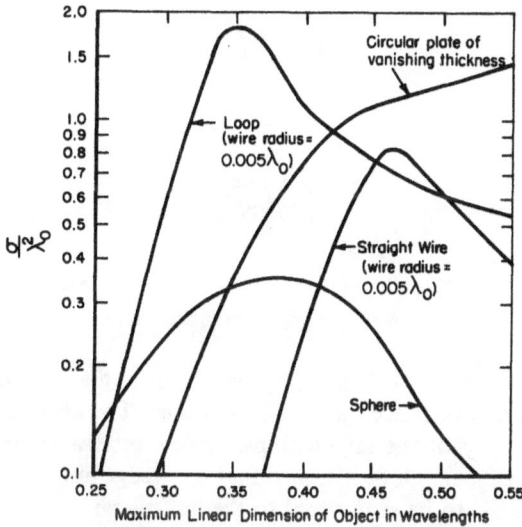

Fig. 6-79. Comparison of the broadside backscatter cross section of several objects in the region of first resonance [after Kouyoumjian[82]].

into simple scatterers. For example, the scattering from the join between a finite cone and its base may in some cases approximate scattering from a ring. Applications to more complex shapes are discussed in Chapter 8.

## 6.5. WEDGES AND HALF-PLANES

The problem of diffraction from a perfectly conducting infinite wedge and its zero-angle limit, the half-plane, is a classic problem. It was examined in the late nineteenth century and early twentieth century by Sommerfeld, Poincaré, MacDonald, Carslaw, Bromwich, and others. Discussions of the history of the development of wedge solutions are found in References 1 and 87 to 89. The fields diffracted from an infinite perfectly conducting wedge are discussed in Section 2.3.1. In this section, infinite wedges are considered in more detail and finite wedges are briefly examined.

Throughout this section only the exterior wedge problem is considered. That is, the interior angle of the wedge ($2\alpha$) is always less than or equal to $\pi$ radians. The edge of the wedge is assumed to lie along the $z$-axis and to extend to infinity in both the plus and minus $z$ directions. The wedge is, unless otherwise specified, located symmetrically with respect to the $x$-axis, as illustrated in Figure 6-80. The faces of the wedge are located at $\phi = \alpha$ and $\phi = -\alpha$. The total interior angle of the wedge is thus $2\alpha$ with $0 \leqslant \alpha \leqslant \pi/2$.

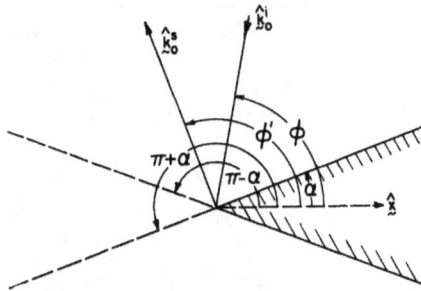

Fig. 6-80.  Wedge coordinate system.

The source is located at a distance $r_0$ and at an angle, $\phi$. The observation point is located at a distance $r$ and at an angle $\phi'$. The distance $r_0$ and $r$ tend to infinity for many of the applications. Unless otherwise specified, in any reference to a wedge it will be assumed that the wedge angle may be zero thus allowing the results to be specialized to the half-plane.

### 6.5.1.  Perfectly Conducting Wedges and Half-Planes

#### 6.5.1.1.  *Infinite Wedges*

Two cases, with the incident electric field vector parallel to the edge and perpendicular to the edge of an infinitely thin, perfectly conducting half-plane, or a nonzero-angle perfectly conducting wedge, were considered by Sommerfeld, as discussed in detail by Baker and Copson.[87] Sommerfeld's method was extended, and the solution simplified by MacDonald[92] and Carslaw.[93] Pauli[94] has obtained results in a form more suitable for numerical evaluation.

For a plane wave incident perpendicular to the edge of the wedge, with

$$E^i_{z\perp} = \exp(-ik_0 x \cos \phi) \tag{6.5-1}$$

or

$$H^i_{z\parallel} = \exp(-ik_0 x \cos \phi) \tag{6.5-2}$$

Pauli writes the total fields as [Eqs. (2.3-33) and (2.3-34)]

$$E_z^T = U(r, \psi_1) - U(r, \psi_2) \tag{6.5-3}$$

$$H_z^T = U(r, \psi_1) + U(r, \psi_2) \tag{6.5-4}$$

where $\psi_1 = \phi - \phi'$ and $\psi_2 = \phi + \phi' - 2\alpha$. The functions $U(r, \psi)$ depend upon whether or not one face of the wedge is shadowed with respect to an optical ray path from the transmitter, and whether the receiver is in the illuminated or shadow region. This function is divided by Pauli into two parts, or

$$U(r, \psi) = v_A + v_B \tag{6.5-5}$$

The quantity $v_A$ represents either the incident or the geometrically-reflected field, and is given by

$$v_A = \begin{cases} \exp[-ik_0 r \cos(\psi + 2\pi n N)], & -\pi < \psi + 2\pi n N < \pi \\ 0, & \text{otherwise,} \end{cases} \quad N = 0, \pm 1, \pm 2, \cdots \quad (6.5\text{-}6)$$

where

$$n = 2(1 - \alpha/\pi) \quad (6.5\text{-}7)$$

The quantities $v_B(r, \psi_1)$, $v_B(r, \psi_2)$, when properly combined in either Eq. (6.5-3) or (6.5-4), give the diffracted field. A general asymptotic form for $v_B$ is

$$v_B \approx [2\pi + 2\pi \cos \psi]^{-1/2} \left[ \frac{1}{n} \sin \frac{\pi}{n} \right] \exp i(k_0 r + \pi/4)$$

$$\times \left( \sum_{m=0}^{\infty} \frac{(-i)^m \, \Gamma(m + \frac{1}{2})}{\pi k_0 r} A_{2m}(\psi) \, S_m[k_0 r(1 + \cos \psi)] \right) \quad (6.5\text{-}8)$$

where

$$S_m(w) = \exp(\psi i w) \, w^m \int_{\sqrt{w}}^{\infty} \exp(i\tau^2) \, \tau^{-2m} \, d\tau \quad (6.5\text{-}9)$$

$$A_0(\psi) = \frac{1 + \cos \psi}{\cos(\pi/n) - \cos(\psi/n)}, \qquad A_1(\psi) \approx 0 \quad (6.5\text{-}10)$$

and

$$A_2(\psi) = \frac{1}{4} A_0(\psi) - \frac{1}{n^2} \cos\left(\frac{\psi}{n}\right) \left[ \frac{1 + \cos \psi}{[\cos(\psi/n) - \cos(\psi/n)]^2} \right]$$

$$+ \frac{(2/n^2)(\sin^2 \psi)(1 + \cos \psi) - [\cos(\pi/n) - \cos(\psi/n)]^2}{[\cos(\pi/n) - \cos(\psi/n)]^3}$$

$$(6.5\text{-}11)$$

At the boundary of the shadow or of the reflected wave

$$\frac{A_0(\psi)}{[2 + 2 \cos \psi]^{1/2}} \bigg|_{\psi = \pm(\pi \mp \Delta)} = \mp \frac{1}{2} \frac{n}{\sin(\pi/n)} \quad (6.5\text{-}12)$$

and

$$A_2(\psi = \pm \pi) = \frac{\cos(\pi/n)}{2 \sin^2(\pi/n)} \quad (6.5\text{-}13)$$

where the $\Delta$ signifies an arbitrarily small number. The asymptotic series of Eq. (6.5-8) is generally approximated by its first term. For large values of $k_0 r$, Eq. (6.5-8) becomes

$$v_B \approx \frac{\exp[ik_0 r + i(\pi/4)]}{(2\pi k_0 r)^{1/2}} \left[ \frac{1/n \sin(\pi/n)}{\cos(\pi/n) - \cos(\psi/n)} \right] \left[ 1 + O\left(\frac{1}{k_0 r}\right) \right] \quad (6.5\text{-}14)$$

It is seen that the combinations of $v_0(\psi_1)$ and $v_0(\psi_2)$, from Eq. (6.5-14), gives the edge diffraction coefficients used in Sections 2.3.1.1, 6.3.1.2.2.1.4, and in the next two chapters in the discussion of scattering from ribbons and from complex bodies.

At the shadow boundary, Eq. (6.5-8) reduces to

$$v_B = \mp \frac{1}{2} \exp(ik_0 r) - \frac{i}{2n} \frac{\cot(\pi/n) \exp(ik_0 r)}{(2k_0 r)^{1/2}} + \cdots \qquad (6.5\text{-}15)$$

The upper (lower) sign in Eq. (6.5-15) is used when approaching the shadow boundary from the illuminated (shadowy) side. The discontinuity of $v_B$ on the shadow boundary is compensated by the discontinuity of $v_A$, so that the sum is continuous and is given by

$$U_{\text{shadow}} = \tfrac{1}{2} \exp(ik_0 r) + O(1/\sqrt{r}) \qquad (6.5\text{-}16)$$

which, for large $r$, is one-half of the incident field on the illuminated side of the shadow boundary.

For the half-plane, $\alpha = 0$; thus, from Eq. (6.5-7), $n = 2$. In this case, all the $A_{2m}$ vanish identically for $m > 0$, and Eq. (6.5-8) reduces to

$$v_B = \frac{1}{k_0 r \sqrt{\pi}} \exp\left[i \frac{3\pi}{4} - ik_0 r \cos(\phi + \phi')\right] \int_{\sqrt{k_0 r[1+\cos(\phi+\phi')]}}^{\infty} \exp(i\tau^2) d\tau$$

$$(6.5\text{-}17)$$

Oberhettinger[88,95] has shown that the integral obtained in formulating the scattering problem can be rearranged in such a manner that, after applying the Watson transform, a different series is obtained. It consists of a Fresnel integral as the leading term, and an asymptotic series in inverse powers of $k_0 r$. The Fresnel integral form is identical to the Pauli results for half-plane diffraction. Thus Pauli's form has a leading term dependent upon the wedge angle, whereas Oberhettinger's results do not.[†] Values of $v_B$ are given by Oberhettinger for plane, cylindrical, and spherical incident waves. The plane wave value is

$$v_B = -\frac{1}{4\alpha} \sin\left[\frac{\pi}{2\alpha}(\pi - \phi' - \phi)\right] \int_0^{\infty} \frac{\exp(k_0 r \cosh x)\, dx}{\cosh(\pi x/2\alpha) - \cos[(\pi/2\alpha)(\pi - \phi' - \phi)]}$$

$$-\frac{1}{4\alpha} \sin\left[\frac{\pi}{2\alpha}(\pi + \phi' - \phi)\right] \int_0^{\infty} \frac{\exp(k_0 r \cosh x)\, dx}{\cosh(\pi x/2a) - \cos[(\pi/2a)(\pi + \phi' - \phi)]}$$

$$(6.5\text{-}18)$$

The integrals in Eq. (6.5-18) are seen to be of the form

$$I(\delta, \alpha) = -\frac{1}{4\alpha} \sin\left[\frac{\pi\delta}{2\alpha}\right] \int_0^{\infty} \frac{\exp(k_0 r \cosh x)\, dx}{\cosh(\pi x/2\alpha) - \cos(\pi\delta/2\alpha)} \qquad (6.5\text{-}19)$$

---

[†] Note added in proof: Hutchins[128] has shown that Pauli's and Oberhettinger's results are both special cases of a more general solution.

with $\delta = (\pi \mp | \phi' - \phi|)$. An asymptotic expansion of $I(\delta, \alpha)$ is

$$I(\delta, \alpha) \approx - \frac{1}{\sqrt{\pi}} \exp\left(ik_0r \cos\delta - i\frac{\pi}{4}\right) (\text{sgn } \delta)\, S\left[(2k_0r)^{1/2}\left| \sin\frac{\delta}{2}\right|\right]$$

$$- (k_0r)^{-1/2} \exp\left(ik_0r + i\frac{\pi}{4}\right) \sum_{m=0}^{\infty} i^{-m}\Gamma\left(m + \frac{1}{2}\right)$$

$$\times [A_m(\delta, \alpha) - A_m(\delta, 2\pi) - A_m(2\pi - \delta, 2\pi)](k_0r)^{-m} \quad (6.5\text{-}20)$$

where $S(z)$ is the Fresnel integral, or

$$S(z) = \int_z^\infty \exp(-it^2)\, dt \quad (6.5\text{-}21)$$

The $A_m(\delta, \alpha)$ are found from a Taylor series expansion, about $t = 0$, of a portion of the integral $I(\delta, \alpha)$; that is,

$$\sum_{m=0}^{\infty} A_m(\delta, \alpha)\, t^m = \frac{\sin(\pi\delta/\alpha)}{4\alpha(t+2)^{1/2}} \left\{\cosh\left[\frac{\pi}{2\alpha} \ln[1 + t + (t^2 + 2t)^{1/2}]\right]\right.$$

$$\left. - \cos\left(\frac{\pi\delta}{2\alpha}\right)\right\}^{-1} \quad (6.5\text{-}22)$$

The first three values of $A_m(\delta, \alpha)$ are

$$A_0(\delta, \alpha) = \frac{\cot(\pi\delta/4\alpha)}{4\alpha\sqrt{2}}, \quad (6.5\text{-}23)$$

$$A_1(\delta, \alpha) = \frac{\cot(\pi\delta/4\alpha)}{-16\alpha\sqrt{2}}\left[1 + \frac{\pi^2}{2\alpha^2}\left(\sin\frac{\pi\delta}{4\alpha}\right)^{-2}\right] \quad (6.5\text{-}24)$$

$$A_2(\delta, \alpha) = \frac{\cot(\pi\delta/4\alpha)}{8\alpha\sqrt{2}}\left[\frac{3}{8} + \left(\frac{5\pi^2}{24\alpha^2} - \frac{\pi^4}{48\alpha^4}\right)\left(\sin\frac{\pi\delta}{4\alpha}\right)^{-2}\right] \quad (6.5\text{-}25)$$

The sum of $A_m(2\pi - \delta, 2\pi)$ and $A_m(\delta, 2\pi)$, needed in Eq. (6.5-20), is given by

$$A_m(2\pi - \delta, 2\pi) + A_m(\delta, 2\pi) = \frac{(-1)^m}{2\pi} 2^{-m-1/2}\left(\sin\frac{\delta}{2}\right)^{-2m-1} \quad (6.5\text{-}26)$$

For $(kr)^{1/2} \sin(\delta/2) \gg 0$, Eq. (6.5-20) reduces to

$$I(\delta, \alpha) \approx - \frac{\exp(ik_0r + i\pi/4)}{(k_0r)^{1/2}} \sum_{m=0}^{\infty} i^m \frac{A_m(\delta, \alpha)\, \Gamma(m + \frac{1}{2})}{(k_0r)^m} \quad (6.5\text{-}27)$$

Hansen and Schiff[96] derived an eigenfunction series for wedge scattering similar in form to the Mie series for a sphere. A more general form of their result is available in several texts; for example, Harrington[97] considers the fields generated by electric- and magnetic-current filaments located at an arbitrary distance and at an arbitrary angle from the wedge. The edge of the wedge is located along the z-axis. The total field for plane-wave incidence is obtained by letting the distance between the source and the wedge go to infinity. The total field for a perpendicular polarized incident field of amplitude $E^i$ is given by

$$E^T_{z\perp} = \frac{\pi E^i}{\pi - \alpha} \sum_{\nu} \epsilon_{\nu} \exp(-i\nu\pi/2) \, J_{\nu}(k_0 r) \sin[\nu(\phi' - \alpha)] \sin[\nu(\phi - \alpha)]$$

(6.5-28)

where

$$\nu = \frac{n\pi}{(2\pi - \alpha)}, \qquad n = 0, 1, 2,...$$

and $\epsilon_{\nu}$ is the Neumann number. It is equal to 1 for $\nu = 0$, and equal to 2 for $\nu > 0$. The Neumann number is included in this equation to make the electric field solution symmetrical to the magnetic field solution although the $n = 0$ term is zero. For a magnetic field with amplitude $H^i$, polarized parallel to the z-axis, the total field is given by

$$H^T_{z\parallel} = \frac{\pi H^i}{\pi - \alpha} \sum_{\nu} \epsilon_{\nu} \exp(-i\nu\pi/2) \, J_{\nu}(k_0 r) \cos[\nu(\phi' - \alpha)] \cos[\nu(\phi - \alpha)]$$

(6.5-29)

The series of Eqs. (6.5-28) and (6.5-29) are of use in determining the diffraction coefficients as discussed in Chapter 2. With some manipulation, Eqs. (6.5-27) and (6.5-28) can be shown to be equivalent to Eqs. (2.3-38) and (2.3-39).

For $\alpha = 0$, Eqs. (6.5-28) and (6.5-29) (Harrington's equations for a wedge) reduce to the equations for the total field for a plane wave incident on a half-plane. These are as follows:

$$E^T_{z\perp} = E^i \sum_{m=0}^{\infty} \epsilon_m (-i)^{m/2} J_{m/2}(k_0 r) \sin\frac{m\phi'}{2} \sin\frac{m\phi}{2} \qquad (6.5-30)$$

and

$$H^T_{z\parallel} = H^i \sum_{m=0}^{\infty} \epsilon_m (-i)^{m/2} J_{m/2}(k_0 r) \cos\frac{m\phi'}{2} \cos\frac{m\phi}{2} \qquad (6.5-31)$$

Numerous other investigators have obtained special forms for the scattered fields from particular wedges, such as half-planes or right-angle wedges. However, many of these results will reduce to those given here which appear to be the most useful for numerical evaluations (see also reference 128).

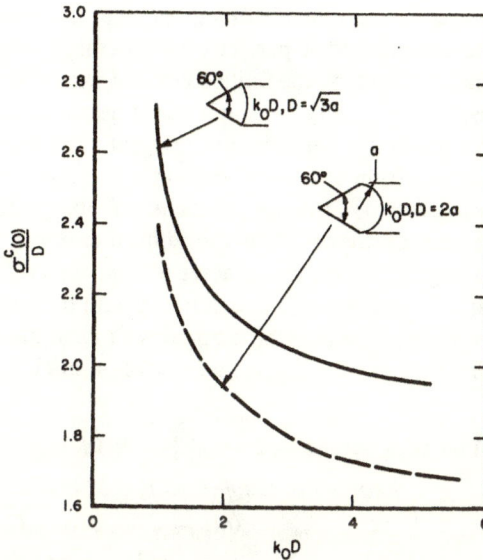

Fig. 6-81. Normalized axial backscatter radar cross section of infinite cylinders whose geometrical cross sections are finite wedges [after Schultz et al.[8]].

### 6.5.1.2. Finite Wedges

Schultz et al.[8] have considered the finite two-dimensional wedge, that is, the case of scattering from an infinite cylinder whose cross section is a finite wedge. The incident electric field is assumed parallel to the cylinder axis. The boundary conditions for a perfect conductor are matched at a finite number of points to evaluate the coefficients of a series expansion of the field. The number of points is chosen so that adjacent points are separated by a distance less than $0.1\lambda_0$. The series were evaluated with a digital computer for two cases. The half-angle of the wedge was chosen as $30°$. In one case, the wedge was terminated by hemicylinder of radius $a$. The length of the side of the wedge was thus $2a$. In the second case, the wedge was terminated by less than a hemicylinder. The length of the side of the wedge was $a\sqrt{3}$. The radar cross section of the second case is larger than for the first case, as illustrated in Figure 6-81. This agrees with what would be predicted by an analysis of the join return, i.e., the smoother the join, the less the return.

### 6.5.2. Imperfectly Conducting and Dielectric Wedges

The primary papers on the diffraction of a plane electromagnetic wave by a wedge with a large but finite conductivity are those of Jones and Pidduck,[98] Felson,[99,100] Senior,[90,101,102] Williams,[91,103] and Malyuzhinets.[104–110]

Jones and Pidduck[98] adopted approximate boundary conditions at the faces of the wedge and, using Green's functions, expressed the solution as a perturbation of the solution to a perfectly conducting wedge. An integral form for the correction caused by the finite conductivity was found. However, the integral obtained is not readily amenable to numerical evaluation.

Felsen[99] also obtained a perturbation integral. For an electric current source, a series representation of the correction term is given which is rapidly convergent when either the source or the point of observation is near the wedge apex. The case of plane wave diffraction requires an asymptotic evaluation of the integral, which yields a set of plane waves reflected from the wedge faces with reflection coefficients appropriate to the composition of the wedge, plus a cylindrical wave which appears to emanate from the apex.

The total far field due to an electric line source parallel to the edge of the wedge is

$$
\begin{aligned}
E_z(r, \phi, \phi') = {} & \exp[-ik_0 r \cos(\phi - \phi')] + R_\perp(\phi') \\
& \times \exp[-ik_0 r \cos(\phi + \phi')]\, \epsilon(\phi + \phi') \\
& + \exp[-ik_0 r \cos(4\alpha - \phi - \phi')]\, R_\perp(2\alpha - \phi') \\
& \times \epsilon(4\alpha - \phi - \phi') + \exp[-ik_0 r \cos(4\alpha - \phi - \phi') \\
& \times \{R_\perp(2\alpha + \phi')\, R_\perp(\phi')\, \epsilon(\phi - \phi' + \pi - 4\alpha) \\
& + \epsilon(\phi' - \phi + \pi - 2\alpha)\, R_\perp(2\alpha - \phi')/R_\perp(4\alpha - \phi')\} \\
& + \frac{\sin(\pi^2/2\alpha)}{2\alpha} \left[ \frac{\pi}{2k_0 r} \right]^{1/2} \\
& \times \left[ \left( 1 - \frac{1 + R_\perp(\phi)}{2 \sin \phi}\, a_1 \right) a_2 \right] \exp\left( ik_0 r + i\,\frac{\pi}{4} \right)
\end{aligned}
$$

$$(6.5\text{-}32)$$

where

$$
a_1 = 2 \csc \alpha \cos\left( \frac{\phi - \phi'}{2} \right) \cos\left[ \frac{2\alpha - (\phi - \phi')}{2} \right] \tag{6.5-33}
$$

$$
a_2 = \frac{1}{\cos(\pi^2/2\alpha) - \cos[(\pi/2\alpha)(\phi - \phi')]} - \frac{1}{\cos(\pi^2/2\alpha) - \cos[(\pi/2\alpha)(\phi + \phi')]}
$$

$$(6.5\text{-}34)$$

for $\pi/2 > \alpha > \pi/4$, and

$$
\epsilon(\tau) = \begin{array}{ll} 1, & 0 \leqslant \tau \leqslant \pi \\ 0, & \text{otherwise} \end{array}
$$

The parameter $R_\perp(\beta)$ equals the reflection coefficient of a plane wave incident at an angle $\beta$ on a planar semi-infinite medium of the wedge material. The faces of the wedge are located at $\phi = 0$ and $\phi = 2\alpha$. The plane-wave terms in Eq. (6.5-32) represent, respectively, the incident wave and the waves caused by simple or multiple reflections from the faces of the wedge at the angle of incidence. For the case where $\alpha < \pi/4$, the solution would contain more reflected plane waves but the same diffraction term.

Felsen, Jones, and Pidduck assume that the solution can be expanded in a series of the inverse powers of the complex refractive index of the material comprising the wedge. Senior[101] states that in the case of a metallic half-plane this expansion is valid only for the field with the magnetic vector parallel to the edge. Senior uses the physical approximation of impedance-type boundary conditions on the faces of the wedge, and finds a difference equation for a function related to the Laplace transform of the field with respect to the radial distance from the edge. This is solved exactly to give an expression for the total field valid for any angle of wedge. The resulting expression is quite complex and therefore is not included here. The expression, when evaluated for $\alpha = 0$, gives the same result as Senior obtained earlier by a different technique for diffraction by a half plane. In a later work, Senior[102] analyzed diffraction at an oblique angle on a highly conducting half-plane. The analysis requires the solution of complex Wiener–Hopf integral equations for the Fourier transforms of four current distributions.

Williams[91] obtained an exact solution for the diffraction of the field of a line current by an imperfectly conducting right-angled wedge. In a later paper,[103] he states that the method used is not capable of generalization to a wedge of arbitrary angle, and he uses a different technique to obtain the diffraction of an $E_\perp$ plane wave by an imperfectly conducting wedge of external angle equal to $p\pi/2q$, where $p$ is an odd integer and $p$ and $q$ are integers which have no common divisior. The original boundary-value problem is reduced to the solution of an ordinary difference equation. A radiation condition is imposed, and a complete solution of the difference equation is found as a contour integral of a sum of products of double gamma functions. The solution is considerably simplified if only wedges of external angle equal to $p\pi/2q$ are considered. This is not a serious limitation since any desired wedge angle may be approximated to any degree of accuracy by suitable choice of $p$ and $q$. Even after the angular simplification the resulting solution is rather complex, requiring evaluation of several Fresnel integrals; therefore, the complete asymptotic solution is not included here. If the additional assumption that the observation is far from the shadow lines is made, the results are further simplified, with the diffracted field $E_{z\perp}^d$ given by

$$E_{z\perp}^d = \sqrt{\frac{2\pi}{r}} \Bigg\{ w_1(\beta)\, w_3(\phi + \pi)\, w_2(\beta)$$

$$- \frac{i\sin(2q\pi/p) - w_2(\beta)\exp[-i(2q/p)(\phi' + \phi)]\sin[(q/p)(\phi' - \phi)]}{\cos[(2q/p)(\phi' - \phi)] - \cos[(2q/p)\,\pi]}$$

$$+ \frac{i\sin(2q\pi/p) - w_2(\beta)\exp[-i(2q/p)(\phi' - \phi)]\sin[(q/p)(\phi' - \phi)]}{\cos[(2q/p)(\phi' + \phi)] - \cos[(2q/p)\,\pi]} \Bigg\}$$

$$\times \left[ \frac{w_3(\phi')}{w_3(\phi + \pi)} \exp i\left(k_0 r - \frac{\pi}{4}\right)\right] \qquad (6.5\text{-}35)$$

where

$$\beta = -i \ln \left[ \frac{1 + (1 - S^2)^{1/2}}{S} \right] \tag{6.5-36}$$

$$S = \left[ \mu_r \epsilon_r \left( 1 + i \frac{\sigma}{\omega \epsilon_0} \right) \right]^{-1/2} \tag{6.5-37}$$

$$w_1(\beta) = \frac{1}{4\pi - 4\alpha} \left\{ \frac{\exp[-i(2q/p)(\phi + \pi/2)]}{\exp[-i(2q/p)\beta] - \exp[-i(2q/p)(\phi + \pi/2)]} \right.$$
$$\left. - \frac{\exp[-(2q/p)(\pi/2 - \phi)]}{\exp[-i(2q/p)\beta] - \exp[-i(2q/p)(\pi/2 - \phi)]} \right\} \tag{6.5-38}$$

$$w_2(\beta) = - \frac{\left[ \begin{array}{c} i \sin[(2q/p)\,\pi]\{\exp[-i(2q/p)(\phi - 1)]\cos[(2q/p)(\beta + \pi/2)] \\ - \cos[(2q/p)(\beta - \pi/2)]\exp(4\pi i q/p)\} \end{array} \right]}{\left[ \begin{array}{c} \exp[-i(2q/p)(2\phi' - 3\pi/2)]\sin[(2q/p)(\phi' + \beta - \pi/2)] \\ \times \sin[(q/p)(\phi' - \beta - \pi/2)]\cos[(q/p)(\phi' - \beta + \pi/2)] \end{array} \right]} \tag{6.5-39}$$

$$w_3(\tau_1) = \frac{w_4(\tau_1 - \beta)\, w_4(\tau_1 - 2\pi - \beta)}{w_4(\tau_1 - \pi + p\pi/2q - \beta)\, w_4(\tau_1 - \pi + p\pi/2q + \beta)} \tag{6.5-40}$$

$$w_4(\tau_2) = \frac{|\cos(\tau_2/2)|\exp[-ik_0 r\cos(\tau_2 + i\pi/4)]}{\cos[(2q/p)\,\pi] - \cos[(2q/p)\,\tau_2]} \int_{\sqrt{k_0 r(1+\cos\tau_2)}}^{\infty} e^{-it^2}\, dt \tag{6.5-41}$$

Williams compared his theoretical results with experimental results for a wedge of $2\alpha = 16°$ and $S = 0.205 - i0.146$, and found good agreement in the shadow region and poor agreement in the illuminated region. He attributed the poor agreement to the experimental difficulty of distinguishing between incident and diffracted fields. He also stated that for perfect conductivity his results agree with those of Pauli discussed in the previous section.

Malyuzhinets[104-110] had previously given a general solution to the problem of diffraction of a plane wave at an impedance wedge. His results are similar to those obtained later by Williams and by Senior, but are expressed in terms of functions whose tabulation is not readily available outside the U.S.S.R. In addition to the various plane waves and cylindrical waves discussed in the previous solutions, he also evaluates surface waves traveling along the wedge surfaces. These surface waves are further discussed by Bobrovnikov and Kislitsyna.[111]

Lebedev and Skal'skaya[112] use an integral transformation of a special form which leads to a functional equation. Leontovich impedance boundary conditions are assumed. The conductivity is high but finite. Their solution is for wedges with angles equal to $(p/2q)\,\pi$. They apply their general form to a rectangular wedge $(2\alpha = \pi/2)$, and indicate a form of the solution applicable for numerical evaluation.

The problem of a wedge with little or no conductivity is also exceedingly complex. Radlow[113] has considered diffraction by a right-angled dielectric wedge by a generalization of the function-theoretical method of Wiener and Hopf (see Section 2.2.1.3). His results, however, are in the form of an integral solution which must be numerically integrated and, thus, are very difficult to utilize.

The Fresnel diffraction by a transparent half-plane of finite thickness has been discussed by Khomazyuk.[114] His results contain the solutions for diffraction from an opaque half-plane, plus additional terms due to waves passing through the half-plane.

The three-dimensional problem of plane wave scattering by an absorbing screen, located on one side of a perfectly conducting, right-angled wedge, has been discussed by Karal and Karp.[115] Felsen[21,22,100] has discussed diffraction by a wedge with a linearly varying surface impedance. The classic Sommerfeld black screen[87] is a half-plane with a surface impedance being a function of position in such a manner that its physical realization is questionable. Felsen and Marcuvitz[116,117] have formulated the diffraction pattern of such a black screen, and Marcinkowski[118,119] has published theoretical data comparing the diffraction patterns of the Sommerfeld black screen and the black screen with a constant surface impedance.

## 6.5.3.  Experimental Results

Measured results simulating the scattering behavior of an infinite wedge are difficult to obtain because of the doubly infinite nature of the wedge. Using a large finite wedge creates a problem in that if one wants to avoid the end effects of the finite termination, one must usually make the measurements in the near field. Some of the resulting problems with near-field measurements are discussed in Chapter 11. Row[120,121] managed to avoid these difficulties by making measurements in a parallel-plate region. A scattering obstacle is placed between two very large parallel metallic plates, and is illuminated by plane or cylindrical waves. The scattering obstacle is imaged in the parallel plate, thus simulating an infinite body. Row measured the total field scattered by infinite metallic wedges of interior angle $2\alpha = \pi/2$, $\pi/4$, and 0 radians, and compared the results with the theoretical expression for perpendicular incidence. The zero radian case corresponds to the half-plane. The agreement was quite good in all cases.

For the half-plane, Row considered the effects of the finite thickness of the metallic sheets used in the experiments. He measured various thicknesses and found that an increase in the thickness does affect the scattered wave, however, the effect is insignificant for thicknesses less than $\lambda_0/10$.

Sletten[122] also made measurements between parallel plates for wedges of $2\alpha = 20, 30, 45, 60, 70$, and $80°$. His measurements agreed quite well with theoretical predictions.

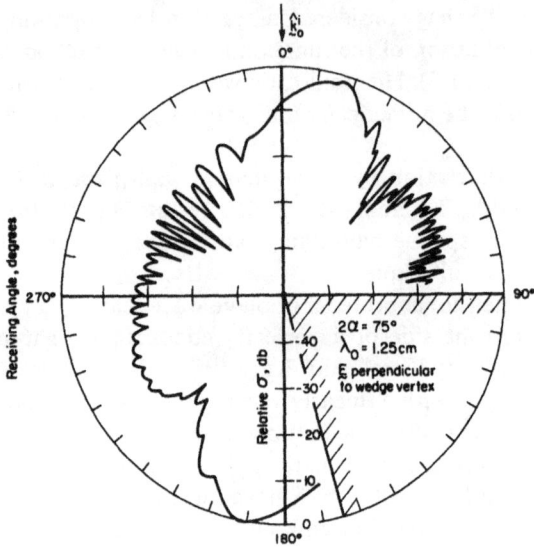

Fig. 6-82.   Measured bistatic diffraction pattern of 75°
wedge, perpendicular polarization [after Hirt[123]].

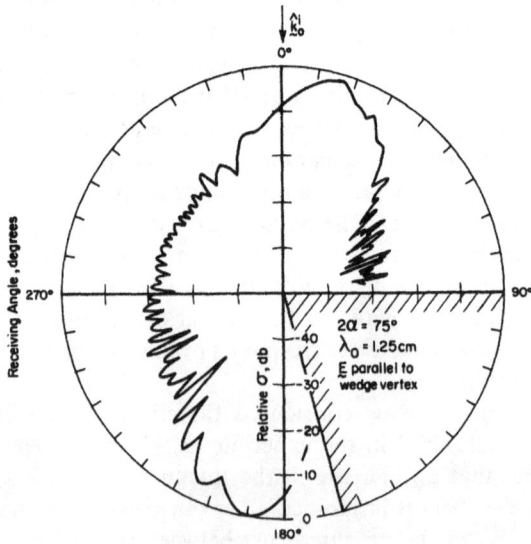

Fig. 6-83.   Measured bistatic diffraction pattern of 75°
wedge, parallel polarization [after Hirt[123]].

Near-field measurements of large finite wedges simulating infinite wedges have been made at the University of Texas.[123-126] Hirt[123] measured the diffraction pattern of a $2\alpha = 75°$ wedge with the direction of propagation of the transmitted energy perpendicular to one face. Measurements were made at a distance of 5 m at a wavelength of 1.25 cm. Figure 6-82 gives the results for parallel polarization, and Figure 6.83 gives the results for perpendicular polarization. The measurements were made on a rooftop range with some experimental error being introduced because of wind effects on the receiving system.

Watson and Horton[124] have measured the diffraction pattern of a hollow wedge with interior angle $2\alpha = 22\frac{1}{2}°$ made of two sheets of copper $30 \times 60 \times \frac{1}{16}$ inches thick. The measurements were made at a wavelength of 3.2 cm. Thus, the dimensions of the sheet were approximately 23.8, 47.6, and $0.05\lambda_0$. The angle of incidence was $67\frac{1}{2}°$ from the wedge face, as indicated in Figure 6-84. The bistatic scattering patterns were measured with a bisectorial horn at a distance of 38.1 cm from the tip of the wedge. Figure 6-84a is for the incident $E$-vector parallel to the edge of the wedge ($\perp$ polarization). Figure 6-84b shows the results for the incident $E$-vector perpendicular to diffracting edge (parallel polarization). The solid lines in Figure-6-84a and b are the measured results. The dotted lines and the points are results predicted by Eq. (6.5-14), except for the regions near the shadow line (180° in the figure) and near the line of geometric reflection (45° in the figure). For these regions,

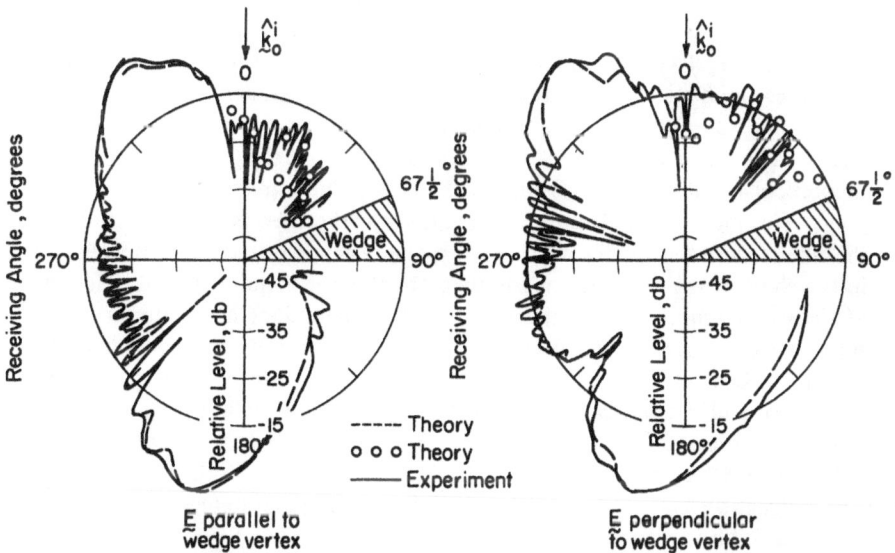

Fig. 6-84. Comparison of experimental and theoretical bistatic diffraction patterns for $22\frac{1}{2}°$ wedge [after Watson and Horton[124]].

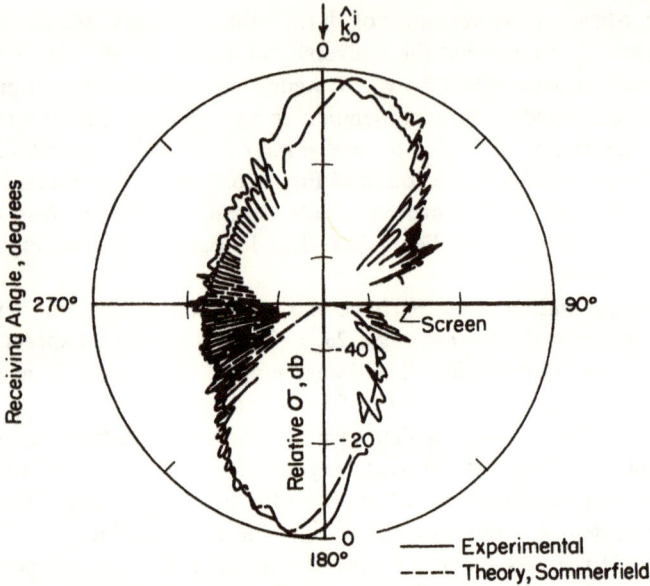

Fig. 6-85.  Comparison of experimental and theoretical bistatic diffraction patterns for semi-infinite screen for perpendicular polarization [after Horton and Watson[125]].

the more accurate asymptotic series of Eq. (6.5-8) was used. The agreement between theory and experiment is seen to be good.

Horton and Watson[125] also have measured the diffraction pattern of a single sheet of copper, thus simulating a semi-infinite plane. Figure 6.85 illustrates the bistatic results for the electric vector parallel to the diffracting edge for a wave incident at an angle of $\pi/2$ radians from the plane of the screen. The solid line is the measured result, and the dotted line is the theoretical result predicted from Eq. (6.5-16). The agreement is seen to be good. The measurements are not as regular as the theoretical predictions in the fine structure near tip-on from the plane. The measured curve is irregular in the shadow region, whereas the theoretical curve is quite smooth. Neither of these differences are unexpected since the plane does have supports and other background to give clutter, and the shadow region behavior is perhaps the most difficult to predict theoretically. Reference 125 also contains measured and theoretical curves for perpendicular polarization and for both polarizations for an angle of incidence of $67\frac{1}{2}°$ from the screen, along with measured curves for the quarter-infinite plane for both polarizations. The diffraction pattern of the quarter-infinite plane is quite similar to that of a semi-infinite plane.

Fainberg[127] has reported bistatic measurements for parallel polarizations of scattering from a 30° metallic wedge and from a metal plate which was

$\frac{3}{4}$ wavelength thick. The measurements were made indoors at 70 Gcps. The measurements agreed quite well with theoretical predictions made by use of the geometrical theory of diffraction discussed in Section 6.3.1.2.2.1.4. This agreement is a verification of the theoretical work of Pauli discussed in Section 6.5.1.1.

## 6.6. REFERENCES

1. Jones, D. S., *The Theory of Electromagnetism*, MacMillan, New York (1964).
2. Kraus, L., and Levine, L. M., Diffraction by an Elliptic Cone, New York University, Research Report No. EM-156 (March 1960), AF 19(604)-5238, AFCRL-TN-60-195.
3. Kraus, L., and Levine, L. M., Diffraction by an Elliptic Cone, *Comm. Pure Applied Math.* **XIV**:49 (1961).
4. Kleinman, R. E., and Senior, T. B. A., Studies in Radar Cross Sections, XLVIII—Diffraction and Scattering by Regular Bodies—II: the Cone, University of Michigan, Scientific Report No. 5 (January 1963), AF 19(604)-6655.
5. Hansen, W. W., and Schiff, L. I., Theoretical Study of Electromagnetic Waves Scattered from Shaped Metal Surfaces, Stanford University, Quarterly Report No. 4 (September 1948), W28-099-333, AD 46568.
6. Mentzer, J. R., *Scattering and Diffraction of Radio Waves*, Pergamon Press, London (1955).
7. Siegel, K. M., Crispin, J. W., Kleinman, R. E., and Hunter, H. E., The Zeros of $P'_{ni}(X_i)$ of Non-integral Degree, *J. Math. Phys.* **XXXI**:170 (1952).
8. Schultz, F. V., Bolle, D. M., and Schindler, J. K., The Scattering of Electromagnetic Waves by Perfectly Reflecting Objects of Complex Shapes, Final Report from Purdue University to the Air Force Cambridge Research Center (January 1963), AF 19(604)-4051, AFCRL-63-319.
9. Goryainov, A. S., Diffraction of a Plane Electromagnetic Wave Propagating Along the Axis of a Cone, *Radio Eng. Electronic Phys.* **6**:39 (1961).
10. Schensted, C. E., Application of Summation Techniques to the Radar Cross Section of a Cone, presented at the McGill Symposium on Microwave Optics, Air Force Cambridge Research Center, Report No. TR-59-118(II), AD 211 500, 1959.
11. Siegel, K. M., *et al.*, Studies in Radar Cross-Sections, IV—Comparison Between Theory and Experiment of the Cross-Section of a Cone, University of Michigan, Report No. UMM-92 (February 1953), AF 30(602)-9.
12. Felsen, L. B., Backscattering from Wide-Angle and Narrow-Angle Cones *J. Appl. Phys.* **26**:138 (1955). See also Backscattering from a Semi-infinite Cone, Microwave Research Institute, Polytechnic Institute of Brooklyn, Brooklyn, New York, Electrophysics Group Memo, No. 43 (July 1958).
13. Siegel, K. M., *et al.*, Bistatic Radar Cross Sections of Surfaces of Revolution, *J. Appl. Phys.* **26**:297 (1955).
14. Spencer, R. C., Back Scattering from Conducting Surfaces, Air Force Cambridge Research Labs., E5070 (April 1951).
15. Felsen, L. B., Plane Wave Scattering by Small-Angle Cones, Microwave Research Institute, Research Report No. R-362-54, PIB-296 (February 1954), AF 19(604)-890.
16. Felsen, L. B., Plane Wave Scattering by Small-Angle Cones, *IRE Trans. on Antennas and Propagation* **AP-5**:121 (1957).
17. Felsen, L. B., Asymptotic Expansion of the Diffracted Wave for a Semi-infinite Cone, *IRE Trans. on Antennas and Propagation* **AP-5**:402 (1957).
18. Weiner, S. D., and Borison, S. L., Radar Scattering from Blunted Cone Tips, *IEEE Trans. on Antennas and Propagation* **AP-14**:774 (1966).
19. Überall, H., Radar Scattering from Coated Perfect Conductors: Application to the Semi-infinite Cone and Use of Exact Eikonal, *Radio Science Journal of Research NBS/USNC-URSI* **68D**:749 (1964).

20. Saxon, D. S., and Schiff, L. I., Theory of High-Energy Potential Scattering, *Il Nuovo Cimento* **VI**:614 (1957).
21. Felsen, L. B., Electromagnetic Properties of Wedges and Cone Surfaces with a Linearly Varying Surface Impedance, *IRE Trans. on Antennas and Propagation* **AP-7**:S231 (1959).
22. Felsen, L. B., On the Electromagnetic Properties of Wedges and Cones with a Linearly Varying Surface Impedance, Polytechnic Institute of Brooklyn, Research Report No. R-736-59, PIB-664 (April 1960), AF 19(604)-4143, AFCRL-TN-60-978.
23. Northover, F. H., The Diffraction of Electromagnetic Waves Around a Finite, Perfectly Conducting Cone, *Quart. J. Mech. Appl. Math.* **XV**:1 (1962).
24. Rogers, C. C., and Schultz, F. V., The Scattering of a Plane Electromagnetic Wave by Finite Cone, Purdue University, ERD-TN-60-765, (August 1960), AF 19(604)-4051.
25. Rogers, C. C., Schindler, J. K., and Schultz, F. V., The Scattering of a Plane Electromagnetic Wave by a Finite Cone. In Jordan, E. C. (Ed.), *Electromagnetic Theory and Antennas*, Pergamon Press, New York (1963), p. 67.
26. Schultz, F. V., *et al.*, The Theoretical and Numerical Determination of the Radar Cross Section of a Finite Cone, Purdue University, Scientific Report No. 1 (August 1964), AF 19(628)-1691.
27. Schultz, F. V., Ruckgaber, G. M., Schindler, J. K., and Rogers, C. C., The Theoretical and Numerical Determination of the Radar Cross Section of a Finite Cone, *Proc. IEEE* **53**:1065 (1965).
28. Darling, D. A., and Senior, T. B. A., Low-Frequency Expansions for Scattering by Separable and Nonseparable Bodies, *J. Acoust. Soc. Am.* **37**:228 (1965).
29. Darling, D. A., Some Relations Between Potential Theory and the Wave Equation, University of Michigan, Report No. 2871-5T (December 1960), AF 19(604)-4993, AD 258 307.
30. Siegel, K. M., Far-Field Scattering from Bodies of Revolution, *Appl. Sci. Res.* **7B**:293 (1959).
31. Siegel, K. M., Goodrich, R. F., and Weston, V. H., Comments on Far-Field Scattering from Bodies of Revolution, *Appl. Sci. Res.* **8B**:8 (1959).
32. Brysk, H., Hiatt, R. E., Weston, V. H., and Siegel, K. M., The Nose-on Radar Cross Sections of Finite Cones, *Canad. J. Phys.* **37**:675 (1959).
33. Siegel, K. M., The Quasi-static Radar Cross Sections of Complex Bodies of Radiation. *In* Langer, R. E. (Ed.), *Electromagnetic Waves*, The University of Wisconsin Press, Wisconsin (1962), p. 181.
34. Siegel, K. M., Low-Frequency Radar Cross-Section Computations, *Proc. IEEE* **51**:232 (1963).
35. Keys, J. E., and Primich, R. I., The Radar Cross Section of Right Circular Metal Cones. 1, Defense Research Board, DRTE Report No. 1010 (May 1959), AD 217 921.
36. Honda, J. S., Silver, S., and Clapp, F. D., Scattering of Microwaves by Figures of Revolution, University of California Electronics Research Laboratory, Issue No. 282, Series No. 60 (March, 1959).
37. August, G., and Angelakos, D. J., Back Scattering from Cones, University of California, Report No. 95 (September 1959), N70nr-29529, AD 230 508.
38. Kleinman, R. E., The Rayleigh Region, *Proc. IEEE* **53**:848 (1965).
39. Crispin, J. W., Jr., Goodrich, R. F., and Siegel, K. M., A Theoretical Method for the Calculation of the Radar Cross Sections of Aircraft and Missiles, The University of Michigan, Report No. 2591-1-H (July, 1959), AF 19(604)-1949. AFCRC-TN-59-774, AD 227695.
40. Dawson, T. W. G., Miller, J. G. W., and Turner, W. R., Calculation of the Head-on Radar Echoes from Bodies of Circular Cross Section Using Wedge Theory, Royal Aircraft Establishment, Technical Note No. RAD.787 (September 1960), AD 250 351.
41. Dawson, T. W. G., and Turner, W. R., Calculation of Radar Echoing Areas by the Cylindrical Current Method, Royal Aircraft Establishment, Technical Note No. RAD.788 (September 1960), AD 250 352.

42. Turner, W. R., Further Calculations of Radar Echoing Areas by the Cylindrical Current Method, Royal Aircraft Establishment, Technical Note No. RAD. 792 (December 1960), AD 251 697.
43. Abramowitz, M. A., and Stegun, I. A., *Handbook of Mathematical Functions with Formulas, Graphs, and Mathematical Tables*, U. S. Government Printing Office (1964).
44. Pannell, J. H., Rheinstein, J., and Smith, A. F., Radar Scattering from a Conducting Cone-Sphere, Lincoln Laboratory–Massachusetts Institute of Technology, Technical Report No. 349 (March 1964), AF 19(628)-500.
45. Burke, J. E., Mower, L., and Twersky, V., Elementary Results for Scattering by Large Cones, *J. Opt. Soc. Am.* **52**:1093 (1962).
46. Hodge, D. B., and Schultz, F. V., The Born Approximation Applied to Electromagnetic Scattering from a Finite Cone, Purdue University, Scientific Report No. 2 (September 1964), AF 19(628)-1691.
47. Schiff, L. I., *Quantum Mechanics*, McGraw-Hill, New York (1949), p. 159.
48. Melling, W. P., and Clark, J. J., Radar Cross Section Minimization Study (Project ZEPHYR–Final Report), Part II, Cornell Aeronautical Laboratory, Inc., Report No. UB-1473-P-2 (July 1961), DA-30-069-ORD-3113.
49. Keller, J. B., Backscattering from a Finite Cone, New York University Research Report No. EM-127 (February 1959), AF 19(604)-1717, AD 211 808.
50. Keller, J. B., Backscattering from a Finite Cone, *IRE Trans. Antennas and Propagation* **AP-8**:175 (1960).
51. Keller, J. B., Backscattering from a Finite Cone—Comparison of Theory and Experiment, *IRE Trans. Antennas and Propagation* **AP-9**:411 (1961).
52. Bechtel, M. E., Application of Geometric Diffraction Theory to Scattering from Cones and Disks, *Proc. IEEE* **53**:877 (1965).
53. Bechtel, M. E., and Ross, R. A., Radar Scattering Analysis, Cornell Aeronautical Laboratory, Inc., Report No. ER/RIS-10 (August 1966), AD 488 833.
54. Blore, W. E., The Radar Cross Section of Ogives, Double-Backed Cones, Double-Rounded Cones, and Cone-Spheres, *IEEE Trans. Antennas and Propagation* **AP-12**: 582 (1964).
55. Moffatt, D. L., Low Radar Cross Sections—The Cone-Sphere, The Ohio State University, Report No. 1223-5 (May 1962), AF 33(616)-8039.
56. Senior, T. B. A., The Backscattering Cross Section of a Cone-Sphere, *IEEE Trans. Antennas and Propagation* **AP-13**:271 (1965).
57. Eberle, J. W., and St. Clair, R. W., Echo Area of Combinations of Cones, Spheroids, and Hemispheres as a Function of Bistatic Angle and Target Aspect, The Antenna Laboratory—The Ohio State University, Report No. 1073-1 (June 1960), AF 19(604)-6157.
58. Kennaugh, E. M., and Moffatt, D. L., On the Axial Echo Area of the Cone-Sphere Shape, *Proc. IRE* **50**:199 (1962).
59. Siegel, K. M., Kennaugh, E. M., and Moffatt, D. L., Radar Cross Section of a Cone-Sphere, *Proc. IEEE* **51**:231 (1963).
60. Kennaugh, E. M., and Moffatt, D. L., Transient and Impulse Response Approximations, *Proc. IEEE* **53**, 893 (1965).
61. Kennaugh, E. M., and Moffatt D. L., The Use of Transient and Impulse Response Approximations in Electromagnetic Scattering Problems, The Ohio State University, Scientific Report No. 3 (February 1966), AF 19 (628)-4002.
62. Kennaugh, E. M., The Scattering of Short Electromagnetic Pulses by a Conducting Sphere, *Proc. IRE* **49**:380 (1961).
63. Alongi, A. V., Kell, R. E., and Newton, D. J., A High-Resolution X-Band FM/CW Radar for RCS Measurements, *Proc. IEEE* **53**:1072 (1965).
64. Keys, J. E., and Primich, R. I., The Radar Cross Section of Right Circular Metal Cones. II, Defense Research Board, DRTE Report No. 1023 (August 1959), AD 227 166.
65. Radar Reflectivity Measurements of Geometric Shapes, Radiation Inc., Report No. 1125-2 to MIT-LL (December 1957), AF 19(122)-958, Subcontract No. 80.

66. Radar Reflectivity Measurements of Geometric Shapes with Circular Polarization, Radiation Inc., Report No. 11 to MIT-LL (August 1958), AF 19(122)-458, Subcontract No. 80.
67. Knott, E. F., and Senior, T. B. A., The Effect of Fins on the Nose-on Cross Section of Cone-Sphere, *IEEE Trans. Antennas and Propagation* **AP-11**:504 (1963).
68. Blore, W. E., and Royer, G. M., The Radar Cross Section of Bodies of Revolution, Defense Research Telecommunications Establishment, DRTE Report No. 1105 (March 1963), AD 406 112.
69. Smith, A. F., Radar Backscatter from Some Low Cross-Section Shapes, Massachusetts Institute of Technology, Group Report 1964-3 (January 1964), AF 19(628)-500.
70. Leontovich, G. A., Approximate Boundary Conditons for the Electromagnetic Field on the Surface of a Good Conductor. *In* Logan, N. A. (Ed.), *Diffraction, Refraction and Reflection of Radio Waves*, AFCRL-TN-57-102, AD 117 276, (1957), p. 383.
71. Pon, C. Y. and Angelakos, D. J., Backscattering from Dielectric Cones, University of California, Report No. 110 (July 1960), Nonr-222(74), AD 245 163.
72. Bradley, F. N., and Eastly, M. R., Radar Cross Section of Flat-Base Dielectric Cones, *Proc. IEEE* **53**:1123 (1965).
73. Garbacz, R. J., and Moffatt, D. L., An Experimental Study of Bistatic Scattering from Some Small, Absorber-Coated, Metal Shapes, The Ohio State University, Report No. 955-2 (May 1960), AF 30(602)-2042.
74. Weston, V. H., Scattering from a Circular Loop, University of Toronto, Research Report No. 12 (August 1957), AD 146 304.
75. Weston, V. H., Solutions of the Toroidal Wave Equation and Their Applications, Ph. D. thesis, University of Toronto (September 1956).
76. Weston, V. H., Solution of the Helmholtz Equation for a Class of Non-separable Cylincrical and Rotational Coordinate Systems, *Quart. Appl. Math.* **XV**:420 (1957).
77. Weston, V. H., Vector Wave Functions in Toroidal Coordinates, University of Toronto, Research Report No. 11 (August 1957), AD 146 303.
78. Weston, V. H., Toroidal Wave Functions, *Quart. Appl. Math.* **XVI**:237 (1958).
79. Weston, V. H., On Toroidal Wave Functions, *J. Math. Phys.* **XXXIX**:64 (1960).
80. King, R. W. P., and Wu, T. T. *The Scattering and Diffraction of Waves*, Harvard University Press, Cambridge, Massachusetts (1959).
81. Kouyoumjian, R. G., The Backscattering from a Circular Loop, the Ohio State University, Report No. 662-5 (April 1956), DA-36-039-sc-70174.
82. Kouyoumjian, R. G., The Backscattering from a Circular Loop (Antenna), The Ohio State University, Engineering Experimental Station Bulletin No. 162, **XXV** (1956).
83. Kouyoumjian, R. G., The Backscattering from a Circular Loop, *Appl. Sci. Res.* **6B**:165 (1956).
84. Stratton, J. A. and Chu, L. J., Steady-State Solutions of Electromagnetic Field Problems, *J. Appl. Phys.* **21**:230 (1941).
85. Philipson, L. L., An Analytical Study of Scattering by Thin Dielectric Rings, *IRE Trans. Antennas and Propagation* **AP-6**:3 (1958).
86. Chen, K.-M., Backscattering of a Loop and Minimization of Its Radar Cross Section, Michigan State University, Technical Report No 2 (May 1965) AF 33(615)–1656.
87. Baker, B. B., and Copson, E. T., *The Mathematical Theory of Huygen's Principle*, 2nd Ed. Oxford at the Clarendon Press, London (1950).
88. Oberhettinger, F., Diffraction of Waves by a Wedge, *Comm. Pure and Appl. Math.* **7**:551 (1954).
89. Oberhettinger, F., On the Diffraction and Reflection of Waves and Pulses by Wedges and Corners, *J. Res. Natl. Bur. Standards*, **61**:343 (1958).
90. Senior, T. B. A., Diffraction by a Semi-infinite Metallic Sheet, *Proc. Royal Soc.* **213**:436 (1952).
91. Williams, W. E., Diffraction by an Imperfectly Conducting Right-Angled Wedge, *Proc. Camb. Phil. Soc* **55**:195 (1959).
92. MacDonald, K. M., A Class of Diffraction Problems, *Proc. London Math. Soc.* **14**:410 (1915).

93. Carslaw, H. S., Diffraction of Waves by a Wedge of Any Angle, *Proc. London Math. Soc.* **18**:219 (1919).

94. Pauli, W., On Asymptotic Series for Functions in the Theory of Diffraction of Light, *Phys. Rev.* **54**:924 (1938).

95. Oberhettinger, F., On Asymptotic Series for Functions Occurring in the Theory of Diffraction of Waves by Wedges, *J. Math. Phys.* **XXXIV**:245 (1956).

96. Hansen, W. W., and Schiff, L. I., Theoretical Study of Electromagnetic Waves Scattered from Shaped Metal Surfaces, Stanford University, Quarterly Report No. 2 (February 1948), W28-099-ac-333, AD 133 921.

97. Harrington, R. F., *Time-Harmonic Electromagnetic Fields*, McGraw-Hill, New York (1961).

98. Jones, D. S., and Pidduck, F. B., Diffraction by Metal Wedges at Large Angles, *Quart. J. Mech. Appl. Math.* **1**:229 (1950).

99. Felsen, L. B., Diffraction by an Imperfectly Conducting Wedge, The McGill Symposium on Microwave Optics (1953) (April 1959), AFCRC-TR-59-118(II), AD 211 500.

100. Felsen, L. B., High-Freqyency Diffraction by a Wedge with a Linearly Varying Surface Impedance, *Appl. Sci. Res.* **9B**, 170 (1961).

101. Senior, T. B. A., Studies in Radar Cross Sections, XXV—Diffraction by an Imperfectly Conducting Wedge, The University of Michigan, Scientific Report No. 2 (October 1957), AF 19(604)-1949, AFCRL-TN-57-791, AD 133 746.

102. Senior, T. B. A., Studies in Radar Cross Sections, XXXI–Diffraction by an Imperfectly Conducting Half-Plane at Oblique Incidence, University of Michigan Radiation Laboratory (February 1959), AF 30(602)-1853, RADC-TN-59-105, AD 212 611.

103. Williams, W. E., Diffraction of an E-Polarized Plane Wave by an Imperfectly Conducting Wedge, *Proc. Royal Soc.* **252**A:376 (1959).

104. Malyuzhinets, G. D., Some Generalizations of the Method of Reflections in the Diffraction Theory of Sinusoidal Waves, Dissertation, Lebedev Physics Institute, Acad. Sci. USSR (1950). Not Generally Available.

105. Malyuzhinets, G. D., On One Weyl Formula Generalization for a Wave Field above an Absorption Plane, *Dokl. Akad. Nauk SSSR* **60**:367 (1948).

106. Malyuzhinets, G. D., Radiation of Sound by Vibrating Boundaries at an Arbitrary Wedge, *Sov. Phys. Acoust.* **1**:152 (1955).

107. Malyuzhinets, G. D., The Excitation, Reflection and Radiation of Surface Waves at a Wedge with Given Impedance Faces, *Soviet Phys. Dokl.* **3**:752 (1958).

108. Malyuzhinets, G. D., Inversion Formula for the Sommerfeld Integral, *Soviet Phys. Dokl.* **3**:52 (1958).

109. Malyuzhinets, G. D., Developments in Our Concepts of Diffraction Phenomena, *Soviet Phys. Uspekhi* **69**(2):749 (1959).

110. Malyuzhinets, G. D., The Sommerfeld Integral and the Solution of Diffraction Problems for Wedge-Shaped Regions, *Ann. Phys. (Germany)* **6**:107 (1960).

111. Bobrovnikov, M. S., and Kislitsyna, V. N., Diffraction of a Uniform Plane Electromagnetic Wave at an Impedance Step-Discontinuity, *Radio Eng. Electronic Phys.* **9**:1401 (1964).

112. Lebedev, N. N., and Skal'skaya, I. P., A New Method for Solving the Problems of Diffraction of Electromagnetic Waves by a Wedge with Finite Conductivity, *Soviet Phys. Tech. Phys.* **7**:867 (1963).

113. Radlow, J., Diffraction by a Right-Angled Dielectric Wedge, *Int. J. Eng. Sci.* **2**:275 (1964).

114. Khomazyuk, V. G., Fresnel Diffraction at the Edge of a Transparent Half-Plane of Finite Thickness, *Soviet Phys. Tech. Phys.* **8**:285 (1963).

115. Karal, F. C., Jr. and Karp, S. N., Diffraction of a Plane Wave by a Right-Angled Wedge Which Sustains Surface Waves on One Face, *Quart. Appl. Math.* **20**:97 (1962).

116. Felsen, L. B., and Marcuvitz, N., Diffraction by a "Perfectly Absorbing" Wedge (Scalar Problems), Polytechnic Institute of Brooklyn, Memorandum No. 14 (June 1956).

117. Felsen, L. B., and Marcuvitz, N., Diffraction by a Perfectly Reflecting Wedge (Scalar Problems), Polytechnic Institute of Brooklyn, Memorandum No. 15 (June 1956).

118. Marcinkowski, C. J., Diffraction Patterns for Reflecting and Absorbing Half-Planes, Polytechnic Institute of Brooklyn, Memorandum No. 17 (September 1956).
119. Marcinkowski, C. J., Some Diffraction Patterns of an Absorbing Half-Plane, *Appl. Sci. Res.* **9B**:189 (1961).
120. Row, R. V., Microwave Diffraction Measurements in a Parallel-Plate Region, Harvard University (May 1952), ATI 165 612.
121. Row, R. W., Microwave Diffraction Measurements in a Parallel-Plate Region, *J. Appl. Phys.* **24**:1448 (1953).
122. Sletten, C. J., Electromagnetic Scattering from Wedges and Cones, AF Cambridge Research Center–The Antenna Laboratory, E5090 (July 1952).
123. Hirt, P. T., Diffraction of Microwaves by Wedges and Wires, Master's thesis, University of Texas (August 1954), AD 37 572.
124. Watson, R. B., and Horton, C. W., On the Diffraction of a Radar Wave by a Conducting Wedge, *J. Appl. Phys.* **21**:802 (1950).
125. Horton, C. W., and Watson, R. B., On the Diffraction of Radar Waves by a Semi-infinite Conducting Screen, *J. Appl. Phys.* **21**:16 (1950).
126. Grannemann, W. W., and Watson, R. B., Diffraction of Electromagnetic Waves by a Metallic Wedge of Acute Dihedral Angle, *J. Appl. Phys.* **26**:392 (1955).
127. Fainberg, J., Bistatic Measurements of Scattering from Edges, *Proc. IEEE* **53**:1121 (1965).
128. Hutchins, D. L., Asymptotic Series Describing the Diffraction of a Plane Wave by a Two-Dimensional Wedge of Arbitrary Angle, Ph. D. Dissertation, The Ohio State University (1967).

www.ingramcontent.com/pod-product-compliance
Lightning Source LLC
Chambersburg PA
CBHW021024210326
41598CB00016B/900